CODED MODULATION SYSTEMS

Information Technology: Transmission, Processing, and Storage

Series Editor: **Jack Keil Wolf**
University of California at San Diego
La Jolla, California

Editorial Board: **Robert J. McEliece**
California Institute of Technology
Pasadena, California

John Proakis
Northeastern University
Boston, Massachusetts

William H. Tranter
Virginia Polytechnic Institute and State University
Blacksburg, Virginia

CODED MODULATION SYSTEMS
John B. Anderson and Arne Svensson

A FIRST COURSE IN INFORMATION THEORY
Raymond W. Yeung

MULTI-CARRIER DIGITAL COMMUNICATIONS: Theory and Applications of OFDM
Ahmad R. S. Bahai and Burton R. Saltzberg

NONUNIFORM SAMPLING: Theory and Practice
Edited by Farokh Marvasti

PRINCIPLES OF DIGITAL TRANSMISSION: With Wireless Applications
Sergio Benedetto and Ezio Biglieri

SIMULATION OF COMMUNICATION SYSTEMS, SECOND EDITION:
 Methodology, Modeling, and Techniques
Michel C. Jeruchim, Philip Balaban, and K. Sam Shanmugan

A Continuation Order Plan is available for this series. A continuation order will bring delivery of each new volume immediately upon publication. Volumes are billed only upon actual shipment. For further information please contact the publisher.

CODED MODULATION SYSTEMS

John B. Anderson
University of Lund
Lund, Sweden

and

Arne Svensson
Chalmers University of Technology
Göteborg, Sweden

Kluwer Academic / Plenum Publishers
NEW YORK, BOSTON, DORDRECHT, LONDON, MOSCOW

Library of Congress Cataloging-in-Publication Data

Anderson, John B., 1945–
   Coded modulation systems/John B. Anderson and Arne Svensson.
     p. cm. — (Information technology: transmission, processing, and storage)
   Includes bibliographical references and index.
   ISBN 0-306-47279-1
    1. Telecommunication systems.   2. Coding theory.   3. Modulation (Electronics)   I. Svensson, Arne, 1955– II. Title. III. Series.

TK5102.92 .A53 2002
621.382—dc21
                                                     2002066909

ISBN: 0-306-47279-1

© 2003 Kluwer Academic / Plenum Publishers, New York
233 Spring Street, New York, New York 10013

http://www.wkap.nl/

10 9 8 7 6 5 4 3 2 1

A C.I.P. record for this book is available from the Library of Congress

All rights reserved

No part of this book may be reproduced, stored in a retrieval system, or transmitted in any form or by any means, electronic, mechanical, photocopying, microfilming, recording, or otherwise, without written permission from the Publisher, with the exception of any material supplied specifically for the purpose of being entered and executed on a computer system, for exclusive use by the purchaser of the work.

Printed in the United Kingdom

To Janet, Kate and Alix

—jba

To my parents Nannie and Bertil; to Gun-Britt and Arvid

—as

Preface

Twenty-five years have passed since the first flowering of coded modulation, and sixteen since the book *Digital Phase Modulation* appeared. That book, the first of its kind and the antecedent of this one, focused mainly on phase coded modulation, although it did contain a few sections on what became known as TCM coding, and a whole chapter on Shannon theory topics. No one 25 years ago imagined how the field would grow. The driving force from the beginning can be said to be more efficient codes. At first, this meant codes that worked more directly with what the physical channel has to offer – phases, amplitudes, and the like. Rather quickly, it meant as well bandwidth-efficient coding, that is, codes that worked with little bandwidth or at least did not expand bandwidth.

Today we have much more complete ideas about how to code with physical channels. An array of techniques are available that are attuned to different physical realities and to varying availabilities of bandwidth and energy. The largest subfield is no longer phase coded modulation, but is codes for channels whose outputs can be directly seen as vectors in a Euclidean space. The ordinary example is the in-phase and quadrature carrier modulation channel; the Killer Application that arose is the telephone line modem. In addition, new ideas are entering coded modulation. A major one is that filtering and intersymbol interference are forms of channel coding, intentional in the first case and perhaps not so in the second. Other ideas, such as Euclidean-space lattice coding, predate coded modulation, but have now become successfully integrated. One such old idea is that of coding with real-number components in Euclidean space in the first place. Traditional parity-check coding was launched by Shannon's 1948 paper "A Mathematical Theory of Communication". Just as with parity-check coding, Shannon definitively launched the Euclidean concept, this time with his 1949 Gaussian channel paper "Communication in the Presence of Noise". As in 1948, Shannon's interest was in a probabilistic theory, and he specified no concrete codes. These arrived with the subject we call coded modulation.

This book surveys the main ideas of coded modulation as they have arisen in three large subfields, continuous-phase modulation (CPM) coding, set-partition and lattice coding (here unified under the title TCM), and filtering/intersymbol interference problems (under partial response signaling, or PRS). The core of this book comprises Chapters 4–6. Chapters 2 and 3 review modulation and traditional coding theory, respectively. They appear in order that the book be self-contained.

They are a complete review, but at the same time they focus on topics, such as quadrature amplitude modulation, discrete-time modeling of signals, trellis decoders, and Gaussian channel capacity, that lie at the heart of coded modulation. Many readers may thus choose to read them. The last two chapters of the book are devoted to properties, designs and performance on fading channels, areas that recently have become more important with the explosion of mobile radio communication.

The book is not a compendium of recent research results. It is intended to explain the basics, with exercises and a measured pace. It is our feeling that coded modulation is now a mature subject and no longer a collection of recent results, and it is time to think about how it can best be explained. By emphasizing pedagogy and underlying concepts, we have had to leave out much that is new and exciting. We feel some embarrassment at giving short shrift to such important topics as iterative decoding, concatenations with traditional coding, block coded modulation, multilevel coding, coding for optical channels, and new Shannon theory. One can name many more. Our long range plan is to prepare a second volume devoted to special topics, in which all these can play a role, and where the issues related to fading channels can be expanded and covered in more detail. Some recent advances in the PRS, CDMA, and ARQ fields were needed to give a complete picture of these fields and these do find inclusion.

In writing this book we have attempted to give an idea of the historical development of the subject. Many early contributors are now passing from the scene and there is a need to register this history. However, we have certainly not done a complete job as historians and we apologize to the many contributors who we have not referenced by name. The priority in the references cited in the text is first to establish the history and second to give the reader a good source of further information. Recent developments take third priority.

The book is designed for textbook use in a beginning graduate course of about 30 lecture hours, with somewhat more than this if significant time is spent on modulation and traditional coding. At Lund University, a quarter of the time is spent on each of introduction/review, TCM, CPM, and PRS coding. Full homework exercises are provided for the core Chapters 2–6. The prerequisites for such a course are simply good undergraduate courses in probability theory and communication engineering. Students without digital communication, coding and information theory will need to spend more time in Chapters 2 and 3 and perhaps study some of the reference books listed there. The book can be used as a text for a full course in coding by augmenting the coding coverage in Chapter 3.

It is a pleasure to acknowledge some special organizations and individuals. A critical role was played by L. M. Ericsson Company through its sponsorship of the Ericsson Chair in Digital Communication at Lund University. Without the time made available by this Chair to one of us (JBA), the book could not have been finished on time. Carl-Erik Sundberg, one of the pioneers of coded modulation, was to have been a co-author of the book, but had to withdraw because of other

Preface

commitments. We acknowledge years – in fact decades – of discussions with him. Rolf Johannesson and Kamil Zigangirov of Lund University were a daily source of advice on coding and Shannon theory, Göran Lindell of Lund University on digital modulation, and Erik Ström and Tony Ottosson of Chalmers University of Technology on channel coding, modulation, fading channels, spread spectrum, and CDMA. Colleagues of past and current years whose work plays an important role in these pages are Nambirajan Seshadri, Amir Said, Andrew Macdonald, Kumar and Krishna Balachandran, Ann-Louise Johansson, Pål Frenger, Pål Orten, and Sorour Falahati. We are indebted to many other former and current coworkers and students. The dedicated assistance of our respective departments, Information Technology in Lund and Signals and Systems at Chalmers, stretched over 7 years. We especially acknowledge the administrative assistance of Laila Lembke and Lena Månsson at home and our editors Ana Bozicevic, Tom Cohn, and Lucien Marchand at Plenum. The graduate students of Information Technology and the undergraduate students in Wireless Communications at Chalmers were the *försökskaniner* who first used the book in the classroom. All who read these pages benefit from their suggestions, corrections, and homework solutions.

JOHN B. ANDERSON
ARNE SVENSSON

Contents

1. **Introduction to Coded Modulation** 1
   1.1. Some Digital Communication Concepts 1
   1.2. A Brief History 8
   1.3. Classes of Coded Modulation 11
   1.4. The Plan of the Book 13
         Bibliography 15

2. **Modulation Theory** 17
   2.1. Introduction 17
   2.2. Baseband Pulses 19
         2.2.1. Nyquist Pulses 19
         2.2.2. Orthogonal Pulses 22
         2.2.3. Eye Patterns and Intersymbol Interference 24
   2.3. Signal Space Analysis 26
         2.3.1. The Maximum Likelihood Receiver and Signal Space ... 26
         2.3.2. AWGN Error Probability 29
   2.4. Basic Receivers 34
   2.5. Carrier Modulation 37
         2.5.1. Quadrature Modulation – PSK 38
         2.5.2. Quadrature Modulation – QAM 47
         2.5.3. Non-quadrature Modulation – FSK and CPM 49
   2.6. Synchronization 52
         2.6.1. Phase-lock Loops 53
         2.6.2. Synchronizer Circuits 56
   2.7. Spectra 58
         2.7.1. Linear Modulation Spectra 58
         2.7.2. The General Spectrum Problem 61
   2.8. Discrete-time Channel Models 65
         2.8.1. Models for Orthogonal Pulse Modulation 66
         2.8.2. Models for Non-orthogonal Pulse Signaling: ISI 68
   2.9. Problems 72
         Bibliography 73

3. Coding and Information Theory ... 75
- 3.1. Introduction ... 75
- 3.2. Parity-check Codes ... 76
  - 3.2.1. Parity-check Basics ... 76
  - 3.2.2. BCH and Reed–Solomon Codes ... 80
  - 3.2.3. Decoding Performance and Coding Gain ... 82
- 3.3. Trellis Codes ... 84
  - 3.3.1. Convolutional Codes ... 85
  - 3.3.2. Code Trellises ... 90
- 3.4. Decoding ... 95
  - 3.4.1. Trellis Decoders and the Viterbi Algorithm ... 95
  - 3.4.2. Iterative Decoding and the BCJR Algorithm ... 101
- 3.5. The Shannon Theory of Channels ... 106
- 3.6. Capacity, Cut-off Rate, and Error Exponent ... 110
  - 3.6.1. Channel Capacity ... 110
  - 3.6.2. Capacity for Channels with Defined Bandwidth ... 117
  - 3.6.3. Capacity of Gaussian Channels Incorporating a Linear Filter ... 121
  - 3.6.4. Cut-off Rate and Error Exponent ... 126
- 3.7. Problems ... 129
- Bibliography ... 130

4. Set-partition Coding ... 133
- 4.1. Introduction ... 133
- 4.2. Basics of Set Partitioning ... 136
  - 4.2.1. An Introductory Example ... 137
  - 4.2.2. Constellation and Subset Design ... 139
- 4.3. Set-partition Codes Based on Convolutional Codes ... 150
  - 4.3.1. Standard TCM Schemes ... 150
  - 4.3.2. Rotational Invariance ... 157
  - 4.3.3. Error Estimates, Viterbi Decoding, and the Free Distance Calculation ... 165
- 4.4. Lattice Codes ... 171
  - 4.4.1. Lattice Ideas ... 172
  - 4.4.2. Improved Lattices in Two or More Dimensions ... 175
  - 4.4.3. Set-partition Codes Based on Multidimensional Lattices ... 179
- 4.5. QAM-like Codes Without Set Partitioning ... 182
- 4.6. Problems ... 186
- Bibliography ... 188

5. Continuous-phase Modulation Coding ... 191
- 5.1. Introduction ... 191
- 5.2. CPM Distances ... 197
  - 5.2.1. Bounds on Minimum Euclidean Distance ... 197

Contents

    5.2.2. Calculation of Minimum Euclidean Distance................. 213
    5.2.3. Trellis Structure and Error Estimates......................... 220
  5.3 CPM Spectra 225
    5.3.1. A General Numerical Spectral Calculation 226
    5.3.2. Some Numerical Results.. 232
    5.3.3. Energy–Bandwidth Performance 240
  5.4 Receivers and Transmitters... 244
    5.4.1. Optimal Coherent Receivers................................... 246
    5.4.2. Partially Coherent and Noncoherent Receivers 251
    5.4.3. CPM Phase Synchronization 261
    5.4.4. Transmitters 266
  5.5. Simplified Receivers... 268
    5.5.1. Pulse Simplification at the Receiver........................... 269
    5.5.2. The Average-matched Filter Receiver......................... 272
    5.5.3. Reduced-search Receivers via the M-algorithm.............. 273
    5.5.4. MSK-type Receivers 275
  5.6. Problems 277
  Bibliography 279

6 PRS Coded Modulation 283
  6.1. Introduction 283
  6.2. Modeling and MLSE for ISI and Linear Coded Modulation 284
    6.2.1. A Modeling Framework for PRS Coding and ISI 285
    6.2.2. Maximum Likelihood Reception and Minimum Distance ... 289
  6.3. Distance and Spectrum in PRS Codes 293
    6.3.1. Basic PRS Transforms 294
    6.3.2. Autocorrelation and Euclidean Distance 298
    6.3.3. Bandwidth and Autocorrelation 303
  6.4. Optimal PRS Codes 309
  6.5. Coded Modulation by Outright Filtering 319
    6.5.1. Faster-than-Nyquist Signaling 320
    6.5.2. Euclidean Distance of Filtered CPM Signals 321
    6.5.3. Critical Difference Sequences at Narrow Bandwidth 326
    6.5.4. Simple Modulation Plus Severe Filtering 329
  6.6. PRS Receivers 333
    6.6.1. Review of Equalizers 333
    6.6.2. Reduced-search Trellis Decoders.............................. 338
    6.6.3. Breadth-first Decoding with Infinite Response Codes........ 345
  6.7. Problems 348
  Bibliography 350
  Appendix 6A Tables of Optimal PRS Codes 351
  Appendix 6B Said's Solution for Optimal Codes 358

Contents

7. Introduction to Fading Channels **363**
   7.1. Introduction 363
   7.2. Propagation Path Loss 364
       7.2.1. Free Space Path Loss 364
       7.2.2. Plane Earth Path Loss 365
       7.2.3. General Path Loss Model 366
   7.3 Fading Distributions 368
       7.3.1. Shadow Fading Distribution 369
       7.3.2. Multipath Fading Distribution 370
       7.3.3. Other Fading Distributions 372
   7.4 Frequency Selective Fading 375
       7.4.1. Doppler Frequency 375
       7.4.2. Delay Spread 376
       7.4.3. Coherence Bandwidth and Coherence Time 379
       7.4.4. Fading Spectrum 383
       7.4.5. Types of Multipath Fading 385
   7.5. Fading Simulators 386
       7.5.1. Flat Rayleigh Fading by the Filtering Method ... 387
       7.5.2. Other Methods for Generating a Rayleigh Fading Process .. 391
       7.5.3. Fading with Other Distributions 393
       7.5.4. Frequency Selective Fading 395
   7.6. Behavior of Modulation Under Fading 396
   7.7. Interleaving and Diversity 400
       7.7.1. Diversity Combining 400
       7.7.2. Ways to Obtain Diversity 408
       Bibliography 412

8. Trellis Coding on Fading Channels **415**
   8.1. Introduction 415
   8.2. Optimum Distance Spectrum Feed-forward
       Convolutional Codes... 416
   8.3. Rate Compatible Convolutional (RCC) Codes 419
       8.3.1. Rate Compatible Punctured Convolutional Codes............ 420
       8.3.2. Rate Compatible Repetition Convolutional Codes 422
       8.3.3. Rate Compatible Nested Convolutional Codes 423
   8.4. Rate Matching 424
   8.5. TCM on Fading Channels 426
       8.5.1. Performance of TCM on Fading Channels...... 427
       8.5.2. Design of TCM on Fading Channels 431
   8.6. DSSS and CDMA 435
       8.6.1. DSSS... 435
       8.6.2. Direct-Sequence CDMA 439
       8.6.3. Code Design 441

8.6.4. Multiuser Detection in CS-CDMA 449
8.6.5. Final Remark on SS and CDMA 454
8.7 Generalized Hybrid ARQ 454
    8.7.1. Simple ARQ 454
    8.7.2. Hybrid Type-I ARQ 458
    8.7.3. Hybrid Type-II ARQ 461
    8.7.4. Hybrid Type-II ARQ with Adaptive Modulation 468
    Bibliography 470

Index 475

1

Introduction to Coded Modulation

A coded modulation is a coded communication that is designed against the analog channel over which it is used. Since phase, amplitude, and continuous time play major roles in the analog world, a good coded modulation will tend to work with these. Coding that manipulates abstract symbols is not excluded; it is only that its effectiveness is measured in the analog world.

What are reasonable measures in an analog channel? A reasonable performance measure is the probability of error. To this we can add two measures of resources consumed, signal power, and bandwidth. Judging a system in the analog world means evaluating its power and bandwidth simultaneously. Traditionally, coded communication has been about reducing power for a given performance. A fundamental fact of communication – first shown by FM broadcasting – is that power and bandwidth may be traded against each other; that is, power may be reduced by augmenting bandwidth. Many channel coding schemes carry this out to some degree, but coding is actually more subtle than simply trading off. It can reduce power without increasing bandwidth, or for that matter, reduce bandwidth without increasing power. This is important in a bandwidth-hungry world. Coded modulation has brought power–bandwidth thinking to coded communication and focused attention on bandwidth efficiency. This book is about these themes: power and bandwidth in coding, schemes that perform well in the joint sense, narrowband coding and coding that is attuned to its channel.

In this introductory chapter we will discuss these notions in a general way and trace their history in digital communication. Part of the chronicle of the subject is the growth of coded modulation itself in three main branches. We will set out the main features of these. They form the organization of the main part of the book. The pace will assume some background in modulation and coding theory. Chapters 2 (modulation) and 3 (coding) are included for the reader who would like an independent review of these subjects.

1.1. Some Digital Communication Concepts

We first set out some major ideas of digital data transmission. Digital communication transmits information in discrete quanta. A discrete set of values is transmitted in discrete time, one of M values each T seconds. Associated with each value is an average symbol energy E_s, and the signal power is E_s/T. There are many good reasons to use this kind of format. Perhaps the major ones are that all data sources can be converted to a common bit format, that digital hardware is cheap, and the fact that error probability and reproduction quality can be relatively easily controlled throughout the communication system. Other motivations can exist as well: security is easier to maintain, switching and storage are easier, and many sources are symbolic to begin with.

Digital communication takes place over a variety of media. These can be roughly broken down as follows. Guided media include the *wire pair, glass fiber*, and *coaxial cable* channels; in these, the background noise is mainly Gaussian, and bandlimitation effects that grow with length cause signal portions that are nearby in time to interfere with each other, a process called intersymbol interference. The *space* channel only adds Gaussian noise to the signal, but it can happen that the channel responds only to signal phase. *Terrestrial microwave* channels are similar, but the signal is affected additionally by reflection, refraction, and diffraction. The *telephone line* channel is by definition a linear medium with a certain signal-to-noise ratio (SNR) (typically 30–40 dB) and a certain bandwidth (about 200–3300 Hz). It can be any physical medium with these properties, and its background noise is typically Gaussian. *Mobile channels* are subject to fading that stems from a rapidly changing signal path.

Except for the last, these channels can normally be well modeled by a stable signal with additive white Gaussian noise (AWGN). Chapters 1–6 in this book assume just that channel, sometimes with intersymbol interference. It will be called the *AWGN channel*. As a physical entity, it is characterized by the energy E_s applied to it, by the signal bandwidth W (positive frequencies), and by the power spectral density of the noise, N_0. The last chapters in the book will add the complication of fading.

In coded channel communication, the fundamental elements are the channel encoder, modulator, channel, demodulator, and decoder. Figure 1.1 shows this breakdown. The encoder produces an output stream $\{x_\ell\}$ in which each x_ℓ carries

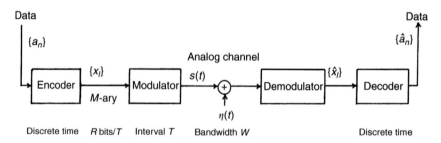

Figure 1.1 Traditional coded communication chain, showing discrete-time encoding, modulation, analog channel, demodulation, and discrete-time decoding.

Introduction to Coded Modulation

R data bits per modulator symbol interval T. The modulator is M-ary. In coded modulation, the first two boxes tend to appear as one integrated system, and so also do the last two. Here lies the key to more efficient transmission, and, unfortunately, the root of some confusion. As a start at resolving it, let us give some *traditional* definitions for coding and modulation.

1. *Channel encoding.* The introduction of redundant, usually binary, symbols to the data, so that future errors may be corrected.
2. *Modulation.* The conversion of symbols to an analog waveform, most often a sinusoid.
3. *Demodulation.* The conversion of the analog waveform back to symbols, usually one symbol at a time at the end of its interval.
4. *Channel decoding.* The use of redundant symbols to correct data errors.

A review of traditional binary channel coding is given in Chapter 3. The extra symbols in Hamming, convolutional, BCH, and other codes there are called "parity-check" symbols; these are related to the data symbols by algebraic constraint equations, and by solving those in the decoder, a certain number of codeword errors can be corrected. In traditional digital modulation, the symbols $\{x_\ell\}$ are converted one at a time to analog waveforms. The most common method, called linear modulation, simply forms a superposition of successive copies of a pulse $v(t)$ according to

$$s(t) = \sum_\ell x_\ell v(t - \ell T). \qquad (1.1\text{-}1)$$

Another method is phase- (PSK) or frequency-shift keying (FSK), in which a phase function $\phi(t)$ that depends on the $\{x_\ell\}$ modulates a carrier signal according to

$$s(t) = \cos[\omega_0 t + \phi(t)]. \qquad (1.1\text{-}2)$$

In traditional demodulation, one symbol at a time is extracted from $s(t)$, directly when the corresponding symbol interval finishes. Chapter 2 reviews this traditional view of modulation and demodulation, together with some important related topics, such as synchronization, detection theory, and modeling of signals in signal space.

Starting perhaps 30 years ago, the practice of digital communication began to diverge from this straightforward schemata. Coded modulation is one embodiment of that change. Increasingly, modulators and demodulators dealt with several symbols and their signaling intervals at a time, because of memory introduced in the modulation operation or in the channel. Combining modulation and coding introduces a third source of memory into the demodulation, that from the channel encoding. As well, methods of coding were introduced that did not work with binary symbol relationships. A final fact is that narrowband signaling makes it fundamentally difficult to force a clean separation of coding and modulation. In fact a new paradigm has emerged and we need to take a fresh view. We will organize a discussion about this around the three headings that follow.

Narrowband signaling. To start, it is worth recalling that a narrow-spectrum event is one that lasts a long time. As a modulated signal becomes more bandlimited, behavior in one signal interval comes to depend on neighboring ones. This dependence is theoretically unavoidable if the transmission method is to be both energy and bandwidth efficient at the same time. Correlation among signal intervals can be thought of as intentional, introduced, for example, through narrowband encoding or modulation, or unintentional, introduced perhaps by filtering in the channel. In either case, intersymbol interference appears. However correlation arises, a good receiver under these conditions must be a *sequence estimator*, a receiver that views a whole sequence of symbol intervals before deciding an individual symbol. Several examples of this receiver type are introduced in Section 3.4. An *equalizer* is a simple example consisting of a linear filter followed by a simple demodulator; a review of them appears in Section 6.6.1. When channel filtering cuts into the main part of the modulation spectrum, the result is quite a different signaling format, even if, like "filtered phase-shift keying," it still bears the name of the modulation. In this book we will think of it as a kind of coded modulation.

Extending coding beyond bit manipulation. The definition of simple binary coding seems to imply that coding increases transmission bandwidth through introduction of extra symbols. This is indeed true over binary-input channels, since there is no other way that the codeword set can differ from the data word set. In reality, coding need not widen bandwidth and can even reduce it, for a given signal energy. A better definition of coding avoids mention of extra symbols: coding is *the imposition of certain patterns onto the transmitted signal.* The decoder knows the set of patterns that are possible, and it chooses one close to the noisy received signal. The set is smaller than the set of all patterns that can be received. This set within a set construction is what is necessary in coded communication. Over a binary channel, we must create the larger set by adding redundant bits. Coded modulation envisions an analog channel; this is not binary and many other ways exist to create a codeword set without adding redundant bits. The new coding definition encourages the encoder–modulator and demodulator–decoder to be taken as single units.

An alternative word for pattern in the coding literature is *constraint*: Codewords can be bound by constraints on, for example, runs of 1s or 0s (compact disk coding) or spectrum (DC-free line coding). Reducing the bandwidth of $s(t)$ eventually constrains its values. Since the coded signals in Chapters 2–6 work in an AWGN channel with bandwidth and energy, it is interesting to read how Shannon framed the discussion of codes in this channel. In his first paper on this channel [2], in 1949, he suggests the modern idea that these signals may accurately be viewed as points in a Euclidean "signal space."[1] This notion is explained in Section 2.3,

---

[1] In this epic paper, Shannon presents the signal space idea (which had been advocated independently and two years earlier by Kotelnikov in Russia), gives what is arguably the first proof of the sampling theorem, and proves his famous Gaussian bandwidth–energy coding theorem. More on the last soon follows.

Introduction to Coded Modulation

and coded modulation analysis to this day employs signal space geometry when another view is not more convenient. To Shannon, a set of codewords is a collection of such points. The points are readily converted back and forth from continuous-time signals. In a later paper on the details of the Gaussian channel, Shannon writes as follows:

> A *codeword* of length n for such a channel is a sequence of n real numbers $(s_1, s_2, \ldots, s_n)$. This may be thought of geometrically as a point in n-dimensional Euclidean space...
>
> A *decoding system* for such a code is a partitioning of an n-dimensional space into M subsets... ([3], pp. 279–280).

Bandwidth vs energy vs complexity. The communication system designer works in a morass of tradeoffs that include government regulations, customer quirks, networking requirements, as well as hard facts of nature. Considering only the last, we can ask what are the basic engineering science tradeoffs that apply to a single link. We can define three major engineering commodities that must be "purchased" in order to achieve a given data bit rate and error performance: these are transmission bandwidth, transmission energy, and complexity of signal processing. One pays for complexity through parts cost, power consumption, and development costs. Transmission energy is paid for through DC power generation, satellite launch weight, and larger antenna size. Transmission bandwidth, probably the most expensive of the three, has a cost measured in lost message capacity and government regulatory approvals for wider bandwidth. Each of these three major factors has a different cost per unit consumed, and one seeks to minimize the total cost. In the present age, the complexity cost is dropping rapidly and bandwidth is hard to find. It seems clear that cost-effective systems will be narrowband, and they will achieve this by greatly augmented signal processing.[2] This is the economic picture.

Energy and bandwidth from an engineering science point of view can be said to have begun with Edwin Armstrong in the 1930s and his determined advocacy of the idea that power and bandwidth could be traded for each other. The particular system he had in mind was FM, which by expanding RF bandwidth achieved a much higher SNR after detection. Armstrong's doctrine was that distortion in received information could be reduced by augmenting transmission bandwidth, not that it could be reduced by processing complexity, or reduced to zero at a fixed power and bandwidth. That part of the picture came from Shannon [2] in the 1949 paper. He showed that communication systems could work error free

---

[2] The conclusion is modified in a multiuser system where many links are established over a given frequency band. Each link does not have to be narrowband as long as the whole frequency band is used efficiently. Now spectral efficiency, measured as the total bit rate of all users divided by the bandwidth, should be high and this can be obtained also with wideband carriers shared between many users as in CDMA. Augmented signal processing still plays an important role.

at rates up to the *channel capacity* C_W in data bits per second, and he gave this capacity as a function of the channel bandwidth W in Hz and the channel symbol energy-to-noise density ratio E_s/N_0. Sections 3.5 and 3.6 review Shannon's ideas, and we can can borrow from there his capacity formula

$$C_W = W \log[1 + 2E_s/N_0] \text{ (bits/s)} \qquad (1.1\text{-}3)$$

A modern way of presenting the capacity law is to express W and E_s on a per data bit basis as WT_b (Hz-s/bit) and E_b (joules/bit).[3] Then the law becomes a set of points in the energy–bandwidth plane. Figure 1.2 shows the energy–bandwidth performance of some coding systems in this book, and the law appears here as the heavy line.[4] Combinations of bit energy and bandwidth above and to the right of the line can be achieved by low error rate transmission systems while other combinations cannot be achieved. For a concrete system to approach the line at low error probability, more and more complex processing is needed. It must climb the contour lines shown in Fig. 1.2. The systems shown in Fig. 1.2 are practical ways to do so, which will be outlined in Section 1.3.

For another view of all these ideas, we can return to Fig. 1.1. We will assume linear modulation as in Eq. (1.1-1) and use the sampling theorem as a tool to analyze what options are available. We will find a close relationship between efficient narrow bandwidth signaling and a more general notion of coding and, in particular, coded modulation. To make this point, it will be useful to extend the notion of coding even further than the pattern definition, and let it mean any processing of significant size. Cost of processing, after all, is what matters in an implementation, not what the processing does.

---

[3] $T_b = T_s/R$ and $E_b = E_s/R$, where R is the transmission rate in data bits/symbol interval; the details of this conversion are given in Section 3.6.

[4] For future reference, we give additional details of the plot data. The interpretation of these requires knowledge of Chapters 2–6. Bandwidth is equivalent to RF bandwidth, positive frequencies only, normalized to data bit rate. E_b/N_0 is that needed to obtain bit error rate (BER) $\approx 10^{-5}$ as observed in concrete system tests. The continuous phase modulation (CPM) code region depicts the approximate performance region of 2-, 4-, and 8-ary codes with 1–3REC and 1–3RC phase pulses. Bandwidth is 99% power bandwidth; distance is full code free distance. The trellis-coded modulation (TCM) codes are displayed in three groups, corresponding to 16 quadrature amplitude modulation (QAM), 64QAM, and 256QAM master constellations (right, center, and left, respectively); within each group are shown best standard convolutional selector performances at 4, 16, and 128 state sizes. Bandwidth is full bandwidth, assuming 30% excess bandwidth root-RC pulses; distance is code free distance, degraded 1–2 dB to reflect BER curve offsets. The partial response signaling (PRS) codes are a selection of 2- and 4-ary best codes of memory ≤ 14. These lie along the trajectory shown, and certain codes are marked by symbols. Best modulation + filter codes from Section 6.5.4 lie along the 2PAM curve. Bandwidth is 99% power; distance is that part of the free distance which lies in this band, degraded 0.5–1 dB to reflect BER offset. Uncoded QAM has BER along the trajectory shown, with points marked at rectangular 4-, 16-, and 64QAM. Bandwidth is full bandwidth for 30% root-RC pulses; distance is full d_{min} degraded by 0.4 dB to reflect BER offset. Soft-decision parity-check coding is assumed to be 2.5 dB more energy efficient than hard-decision.

Introduction to Coded Modulation

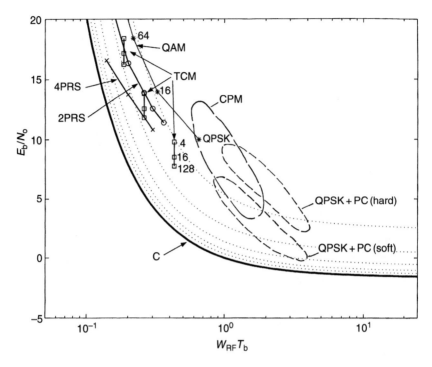

Figure 1.2 Energy–bandwidth performance of major classes of coded modulation, with comparison to Shannon AWGN capacity, ordinary QPSK/parity-check hard and soft decoding, and uncoded rectangular QAM. E_b/N_0 scale in dB.

Major classes portrayed: (i) CPM – general region shown; (ii) TCM – best 4-, 16-, 128-state standard codes with QAM constellation size $M_c = 16, 64, 256$ and rate $M_c - 1$ bits/QAM symbol [right, center, left groups denoted □, respectively]; (iii) PRS – selected best binary [o] and quaternary [x] codes.
Bandwidth is 99% RF-power bandwidth, positive frequencies only, normalized to the data bit rate. Energy is E_b/N_0 needed to achieve BER $\approx 10^{-5}$ in an AWGN channel. Dotted lines suggest areas equally distant from capacity and are approximate guidelines only.

Consider first such "coding" that works at rates $R < 2W \log_2 M$ bits/s, over a channel with bandwidth W Hz. At these R, sampling theory allows independent M-ary channel values x_ℓ to be recovered simply by sampling $s(t)$ each T seconds, subject to $T < 1/2W$. The sample stream has equivalent rate $\log_2 M/T$ bits/s, which allows the possibility of codes up to rate $2W \log_2 M$. The theoretical foundation for this is reviewed in Section 2.2. The simplest means is to let the pulses $v(t - \ell T)$ be orthogonal; an optimal detector for $\hat{x}_\ell$ is a simple filter matched to v followed by a sampler. We will pronounce the complexity of a circuit as "simple" and ignore it. There is every reason to let the encoder in this system work by symbolic manipulation and be binary, and if necessary several binary outputs can be combined to form an M-ary modulator input. The decoder box contains a Viterbi algorithm (VA) or other type of binary decoder. The system here is traditional coding.

Now consider coding in Fig. 1.1 at higher rates, such that $R \geq 2W \log_2 M$ bits/s. Now $\{\hat{x}_\ell\}$ cannot be recovered by a simple sampling. There is no T that allows it and still supports a code with rate as high as R. A significant computation is required in the analog part of the figure if $\{\hat{x}_\ell\}$ is to be recovered. It may resemble, for example, the Viterbi algorithm, and it may make sense to combine the demodulator and decoder boxes. Since decoding is in general much harder than encoding, there will be little loss overall if some analog elements are allowed in the encoding as well. The higher transmission rate here is what has made analog processing and joint coding–modulation the natural choices.

1.2. A Brief History

An appreciation of its history is a way to gain insight into a subject. Coded modulation is part of digital communication, a major event in intellectual and technological history that began in the 1940s. Digital communication arose out of the confluence of three major innovations: a new understanding of communication theory, whose largest single figure was Shannon, the advent of stored-program computing, whose initial figure was van Neumann, and the appearance of very low-cost digital hardware, with which to implement these ideas.

We pick up the communication part of the story in 1948. More detailed references appear in Chapters 2 and 3.

Advent of Digital Communication Theory

We have chosen 1948 because the largest single event in the theory occurred in 1948–1949, the publication by Shannon of two papers, "A mathematical theory of communication" [1], which introduced information theory and capacity, and "Coding in the presence of noise" [2], which introduced Gaussian channel capacity, the sampling theorem, and (to Western readers) a geometric signal space theory. These papers showed that bandwidth could not only be traded for energy, but that nearly error-free communication was possible for a given energy and bandwidth at data rates up to capacity. Further, these papers gave a conceptual framework to digital communication which it has retained to the present day. The geometric signal theory had also been proposed a few years before in the PhD thesis of Kotelnikov [4]. Further important events in the 1940s were the invention of the matched filter, originally as a radar receiver, the invention of pulse-code modulation, and the publication of the estimation and signal processing ideas of Wiener in his book [5]. In a separate stream from Shannon appeared the first error-correcting codes: Hamming codes (1950), Reed–Muller codes (1954), convolutional codes (1955), and BCH codes (1959).

Introduction to Coded Modulation

Phase-shift, Frequency-shift, and Linear Modulation (1955–1970)

In the 1960s basic circuits for binary and quaternary phase modulation and for simple frequency-shift modulation were worked out, including modulators, demodulators, and related circuits such as the phase-lock loop. A theory of pulse shaping described the interplay among pulse shape, intersymbol correlation, signal bandwidth, adjacent channel interference, and RF envelope variation. Effects of band and amplitude limitation were studied, and simple compensators invented. While strides were made at, for example, reducing adjacent channel interference, early phase-shift systems were wideband and limited to low-power wideband channels like the space channel. At the same time simple methods of intersymbol interference reduction were developed, centering on the zero-forcing equalizer of Lucky (1965). The decision-feedback equalizer, in which fed back decisions aided with interference cancelation, was devised (around 1970).

Maturation of Detection Theory (1960–1975)

The 1960s saw the growth and maturation of detection and estimation theory as it applies to digital communication. Analyses were given for optimal detection of symbols or waveforms in white or colored noise. Matched filter theory was applied to communication; many applications appeared in the paper and special issue edited by Turin [6]. Signal space analysis was popularized by the landmark 1965 text of Wozencraft and Jacobs [7]. In estimation theory, least squares, recursive, lattice, and gradient-following procedures were developed that could efficiently estimate signal and channel parameters. Adaptive receivers and equalizers were developed. The state of detection and estimation at the end of the 1960s was summarized in the influential three-volume treatise of van Trees [8].

Maturation of Coding Theory (1960–1975)

In channel coding theory, the 1960s saw the maturation of parity-check block coding and the invention of many decoders for it. Sequential decoding of convolutional codes was introduced. This method accepts channel outputs in a stream of short segments, searches only a small portion of the codebook, and decides earlier segments when they appear to be reliably known, all in an ongoing fashion. For the most part, these decoders viewed demodulator outputs as symbols and ignored the physical signals and channels that carried the symbols. Viterbi proposed the finite-state machine view of convolutional coding and the optimal decoder based on dynamic programming that bears his name (1967); soon after, Forney showed that the progression of states vs time could be drawn on a "trellis" diagram. This artifice proved useful at explaining a wide variety of coded communication systems; in particular, Forney (1971) gave a trellis interpretation of intersymbol interference, and suggested that such interference could be removed by the Viterbi algorithm.

Several researchers solved the problem of sequence estimation for general correlated interval and filtered signals. Coding theory was extended to broadcast and multiple-access channels.

The Advent of Coded Modulation (1975–1995)

Building on continuous-phase frequency-shift keying work in the early 1970s, methods were proposed after 1974 to encode increasingly complex phase patterns in carrier signals. These soon were viewed as trellis codes, with a standard distance and cut-off rate analysis [9] (1978); the field grew into the continuous-phase modulation (CPM) class with the thesis of Aulin [10], and become the first widely studied coded modulations. For the first time, practical codes were available that saved power without bandwidth expansion. Applications were to satellite and mobile communication. In parallel with this development, set-partition coding was proposed for the linear AWGN channel by Ungerboeck [11] (published after a delay, 1982); this ignited a huge study of "Trellis-Coded Modulation" (TCM) codes for Shannon's 1949 Euclidean-space channel with continuous letters and discrete time. Bandwidth efficient codes were achieved by encoding with large, non-binary alphabet sizes. Calderbank, Forney, and Sloane made a connection between TCM and Euclidean-space lattice codes (1987). In another, slower development, intersymbol interference and signal filtering in the linear AWGN channel came to be viewed as a form of linear ordinary-arithmetic coded modulation. Standard distance and trellis decoder analyses were performed for these "Partial Response Signaling" (PRS) codes; an analysis of very narrowband coding appeared and efficient reduced search decoders were discovered [12] (1986). Optimal linear coded modulations were derived [13] (1994). Coded modulations now became available at very narrow bandwidths.

Other Coded Communication Advances (1980–2000)

For completeness, we can summarize some other recent advances in coding that relate to coded modulation. "Reduced search" decoders, a modern form of sequential decoding with minimized search and no backtracking, have been applied to both ordinary convolutional codes and coded modulation. They are dramatically more efficient than the Viterbi algorithm for PRS and CPM coded modulations. Decoders using soft information, as opposed to hard symbol input, find increasing use. Concatenated coding, both in the serial form of the 1960s and in a new parallel form has been shown to perform close to the Shannon capacity limit (1993). The "Turbo Principle" – decoding in iterations with soft information feedback between two or more decoder parts – is finding wide application. All these innovations are being applied to coded modulation at the time of writing.

Introduction to Coded Modulation

1.3. Classes of Coded Modulation

We can now introduce in more detail the main classes of coded modulation that make up this book. Figure 1.3 gives a schematic diagram of each. In every case the transmitted data are denoted as $\{a_n\}$ and the output data as $\{\hat{a}_n\}$. These data can be binary, quaternary, or whatever, but if necessary they are converted before transmission. In what follows, the themes in the first part of the chapter are continued, but there are many new details, details that define how each class works. These can only be sketched now; they are explained in detail in Chapters 4–6.

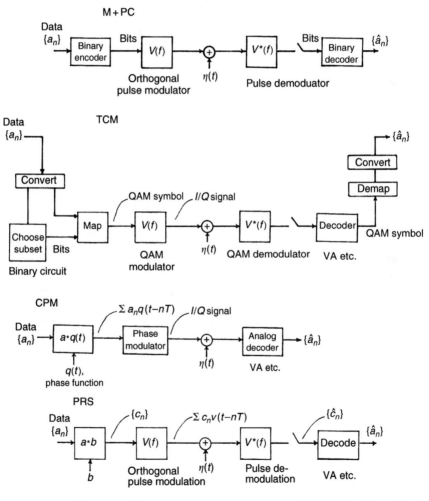

Figure 1.3 Block diagrams of the major coded modulation classes in this book, showing traditional modulator + parity check coding (M + PC), TCM, CPM, and PRS coding.

Figure 1.3 starts with a diagram of traditional modulation plus binary parity-check coding (denoted M + PC). The central part of that method is a basic orthogonal-pulse binary linear modulator. This simple scheme is reviewed in Section 2.2; it consists of a pulse forming filter $V(f)$, a matched filter $V^*(f)$, and a sampler/compare to zero, abbreviated here by just the sampler symbol. The last produces the estimate of the binary value ± 1, which is converted to standard symbols $\{0, 1\}$ by the conversion $-1 \rightarrow 1, +1 \rightarrow 0$. For short, we will call this scheme the M + PC "modem." Of course, many other means could have been used to transmit the codeword in M + PC transmission, but assuming that it was binary linear modulation will provide the most illuminating comparison to the remaining coded modulations.

The outer parts of the M+PC system are a binary encoder, that is, one that takes in K binary symbols and puts out N, with $K < N$, and a binary decoder, which does the opposite. The binary decoder is one of many types, the most common of which is perhaps the Viterbi algorithm. The M + PC method expands the linear modulator bandwidth by N/K; if the per-bit bandwidth of the modulator is WT Hz-s/bit, the per-databit bandwidth of the system is WTN/K. Despite the expansion, parity-check coding systems turn out to have an attractive energy–bandwidth performance. This is shown in Fig. 1.2, which actually gives two regions of good-code performance, one for the hard output binary-in/binary-out BSC and one for the binary-in/real-number-out AWGN channel (denoted "soft"). As will be discussed in Section 3.2, the second channel leads in theory to a 3 dB energy reduction. Within the global AWGN assumption in the book, it is fair to insist that M + PC coding should use its channel optimally, and therefore the soft region is the one we focus on. This region is the one of interest to those with little energy available and a lot of bandwidth. No other coding system seems to compete with it, given that the channel has that balance.

The TCM coded modulation class makes up Chapter 4. It is based on an in-phase and quadrature (I/Q) carrier modulator. The core modulator in this class is the M + PC modem, expanded to nonbinary quadrature amplitude modulation (QAM) form (the basics of QAM are explained in Section 2.5). TCM codes are based on a partition of the QAM constellation points into subsets. The encoder breaks the databit stream into two binary streams; the first selects a pattern of the subsets from interval to interval, and the bits in the second are carried by the subsets themselves. The decoder works by deciding which pattern of subsets and their individual points lies closest to the I/Q demodulator output values. The decoding problem is not different in essence from the M + PC one, although the inputs are QAM symbols and not binaries, and the decided symbol must be demapped and deconverted to recover the two databit streams. The Viterbi detector is almost exclusively used. In the hierarchy of coded modulation in Fig. 1.3, real-value processing enters the encoding and decoding for the first time. However, this work can take place in discrete time. Time-continuous signal processing can be kept within the central modem.

The TCM bandwidth per bit is the modulator WT in Hz-s/QAM symbol divided by the bits at the Map box. By using a large QAM symbol alphabet, good bandwidth efficiency can be achieved, and the patterning drives up the energy efficiency. The result is a coded modulation method that works in the relatively narrowband parts of the plane and is as much as 5 dB more energy efficient than the QAM on which the signaling is based.

The CPM class makes up Chapter 5. The encoding consists of a convolution of the data with a phase response function $q(t)$ to form a phase signal $\phi(t)$; this is followed by a standard phase modulator which forms $\sqrt{2E_s/T}\cos[\omega_0 t + \phi(t)]$. Like TCM this system fundamentally produces I/Q carrier signals, but they are now constant-envelope signals. The decoder again basically performs the Viterbi algorithm or a reduced version of it, but now with time-continuous signals. It is more sensible to think of CPM signaling as analog throughout. The analog domain has now completely taken over the coding system.

CPM is a nonlinear coded modulation, and consequently its energy and bandwidth properties are much more complex. The end result of the energy–bandwidth analysis in Chapter 5 is a region in Fig. 1.2. CPM occupies a center portion of the energy–bandwidth plane. It is somewhat further from capacity than the other classes, and one explanation for this is the constant envelope restriction on its signals. One way to factor this out of the CPM class performance is to subtract a handicap to account for the higher efficiency of class C (phase only) amplifiers: compared to the linear RF amplifiers that are needed for TCM and PRS coding, class C is 2–4 dB more efficient in its use of DC power. It can be argued that for space and battery-driven applications, then, the CPM class should be moved left by this 2–4 dB.

The last class in Fig. 1.3 is the PRS class, the subject of much of Chapter 6. As with M + PC and TCM coding, the core of this is the basic "modem," this time with pulse amplitudes c_n that take on fully continuous values. The encoder is a convolution of the data with a generator sequence b; these are thus straight convolutional codes, but with real arithmetic. The decoder is the Viterbi algorithm or a reduction. As with TCM, PRS encoding and decoding can work with sequences of real but discrete-time values, with the continuous-time signals kept within the modem. It can be more straightforward, however, to keep the processing in the continuous-time domain. The energy–bandwidth performance lies in a very narrowband region of the plane. The coding system here can be viewed as lowpass filtering with a maximum-likelihood estimation of the data at the receiver.

1.4. The Plan of the Book

The chapters of the book may be grouped into three parts:
  I. Review of coding and modulation.
  II. Methods of coded modulation.
  III. Fading channel problems.

The foundations are laid in Part I, which is Chapters 2 and 3. These introduce and review supporting ideas in first modulation, then coding and information theory. Chapter 2, Modulation Theory, focuses on linear pulse modulation, signal space, optimal receivers, phase modulation, and QAM, topics that are the building blocks of coded modulation. Section 2.7 treats spectrum calculation, with an emphasis on simple linear modulation spectra. This is sufficient for TCM and PRS coding, but CPM requires a special, more subtle calculation, which is delayed until Chapter 5. Section 2.8 introduces discrete-time modeling of continuous signals, a topic that supports PRS coding. Chapter 3, Coding and Information Theory, discusses first ordinary parity-check coding in Section 3.2. The notions of trellis and decoding based on trellises form Sections 3.3 and 3.4. Some basics of Shannon theory form Section 3.5 and Section 3.6 specializes to the Shannon capacity of coded modulation channels.

The centerpiece of the book is Part II, a survey of the main ideas of coded modulation as they have arisen in the subfields TCM, CPM, and PRS coding. The ideas in these have just been discussed. Part II comprises Chapters 4–6. In Sections 4.1–4.3 of Chapter 4, traditional TCM coding based on set partitions is described. Section 4.4 introduces the parallel subject of lattice coding. Section 4.5 extends TCM to codes without set partitioning. Chapter 5 is CPM coding. Because the signaling is nonlinear, distance and spectrum calculation is harder and requires special methods: Sections 5.2 and 5.3 are about distance and spectrum, respectively. The joint energy–bandwidth optimization for CPM codes is in Section 5.3. Many receivers have been developed for CPM, which correspond to varying degrees of knowledge about carrier phase; these are in Section 5.4. It is also possible to simplify the basic CPM Viterbi receiver in many ways, and these ideas appear in Section 5.5. Chapter 6 covers the general field of real-number convolutional coding, intersymbol interference, and heavily filtered modulation. Sections 6.1 and 6.2 return to the discrete-time modeling problem and distinguish these cases. Sections 6.3 and 6.4 calculate distance and bandwidth for real-number discrete-time convolutional coding and derive optimal codes in an energy–bandwidth sense. Section 6.5 turns to heavy filtering as a form of coded modulation. Simplified receivers are the key to PRS coding, and they are discussed in Section 6.6.

Part III extends the book in the direction of fading channels. We began this chapter by defining coded modulation to be coding that is evaluated and driven by channel conditions. Fading has a severe impact on what constitutes good coded communication. Chapter 7 is a review of fading channels. Sections 7.2–7.4 are about the properties of fading channels. Simulation of fading is treated in Section 7.5. The performance of uncoded modulation on fading channels is in Section 7.6, while Section 7.7 is devoted to methods for reducing the performance degradations due to fading. Chapter 8 reviews three different coding techniques for fading channels. After some improved convolutional codes are introduced in Sections 8.2 and 8.3, matching source data rates to channel rates is discussed in Section 8.4. Section 8.5 is devoted to design and performance of TCM on fading channels. Here it becomes clear that the design differs quite a lot from the AWGN case. Sections 8.6 and 8.7 focus on two coding techniques for fading channels, spread spectrum and repeat-request systems. In both cases, convolutional codes are taken as the heart of the system, and channel conditions and service requirements direct how they are used. This is coded modulation in a wider sense, in which the codes are traditional but the channel drives how they are used.

Bibliography

[1] C. E. Shannon, "A mathematical theory of communication," *Bell Syst. Tech. J.*, **27**, 379–429, 623–656, 1948; reprinted in *Claude Elwood Shannon: Collected Papers*, N. J. A. Sloane and A. D. Wyner, eds, IEEE Press, New York, 1993.

[2] C. E. Shannon, "Communication in the presence of noise," *Proc. IRE*, **37**, 10–21, 1949; in Sloane and Wyner, *ibid.*

[3] C. E. Shannon, "Probability of error for optimal codes in a Gaussian channel," *Bell Syst. Tech. J.*, **38**, 611–656, 1959; in Sloane and Wyner, *ibid.*

[4] V. A. Kotelnikov, "The theory of optimum noise immunity," PhD Thesis, Molotov Energy Institute, Moscow, Jan. 1947; available under the same title from Dover Books, New York, 1968 (R. A. Silverman, translator).

[5] N. Wiener, *The Extrapolation, Interpolation, and Smoothing of Stationary Time Series with Engineering Applications*. Wiley, New York, 1949.

[6] G. L. Turin, "An introduction to matched filters," Special Matched Filter Issue, *IRE Trans. Inf. Theory*, **IT-6**, 311–329, 1960.

[7] J. M. Wozencraft and I. M. Jacobs, *Principles of Communication Engineering*. Wiley, New York, 1965.

[8] H. L. van Trees, *Detection, Estimation, and Modulation Theory*, Part I. Wiley, New York, 1968.

[9] J. B. Anderson and D. P. Taylor, "A bandwidth-efficient class of signal space codes," *IEEE Trans. Inf. Theory*, **IT-24**, 703–712, Nov. 1978.

[10] T. Aulin, "CPM – A power and bandwidth efficient digital constant envelope modulation scheme," PhD Thesis, Telecommunication Theory Dept., Lund University, Lund, Sweden, Nov. 1979.

[11] G. Ungerboeck, "Channel coding with multilevel/phase signals," *IEEE Trans. Inf. Theory*, **IT-28**, 55–67, Jan. 1982.

[12] N. Seshadri, "Error performance of trellis modulation codes on channels with severe intersymbol interference," PhD Thesis, Elec., Computer and Systems Eng. Dept., Rensselaer Poly. Inst., Troy, NY, USA, Sept. 1986.

[13] A. Said, "Design of optimal signals for bandwidth-efficient linear coded modulation," PhD Thesis, Dept. Elec., Computer and Systems Eng., Rensselaer Poly. Inst., Troy, NY, USA, Feb. 1994.

2

Modulation Theory

2.1. Introduction

The purpose of this chapter is to review the main points of modulation and signal space theory, with an emphasis on those that bear on the coded modulation schemes that appear in later chapters. We need to discuss the basic signal types, and their error probability, synchronization and spectra. The chapter in no way provides a complete education in communication theory. For this, the reader is referred to the references mentioned in the text or to the starred references in the list at the end of the chapter.

We think of digital data as a sequence of symbols in time. A piece of transmission time, called the *symbol time* T_s, is devoted to each symbol. When no confusion will result, T_s will simply be written as T. The reciprocal $1/T_s$ is the rate of arrival of symbols in the channel and is called the transmission *symbol rate*, or *Baud rate*. Each symbol takes one of M values, where M is the size of the transmission symbol alphabet. Customer data may or may not arrive in the same alphabet that is used by the modulator. Generally, it arrives as bits, that is, as binary symbols, but even if it does not, it is convenient in comparing modulations and their costs to think of all incoming data streams as arriving at an equivalent data *bit* rate. The time devoted to each such bit is T_b. The modulator itself often works with quite a different symbol alphabet. T_s and T_b are related by $T_s = T_b \log_2 M$. Throughout this book, we will reserve the term *data symbol* for each customer data symbol coming in, whether binary or not. We will reserve *transmission symbol* for the means, binary or not, of carrying the information through the digital channel.

Since modulations and their symbols can differ greatly, and employ all sorts of coding, encryption, spreading, and so on, it is convenient to measure the transmission system in terms of resources consumed by the equivalent of one incoming data bit. We will measure bandwidth in this book in Hz-s/data bit and signal energy in joules/data bit. Similarly, system cost and complexity are measured per data bit. In the end the revenue produced by the system is measured this way, too.

Very often in modulation, transmission symbols are directly associated with *pulses* in some way. Suppose a sequence of transmission symbols

$a_0, a_1, a_2, \ldots$ scale a basic pulse $v(t)$ and superpose linearly to form the pulse train

$$s(t) = \sum_{n=0}^{N} a_n v(t - nT_\text{s}). \tag{2.1-1}$$

A modulation that works in this way is called a *linear modulation*. Many, but certainly not all, modulations are linear. The trellis coded modulation (TCM) and lattice coding schemes in Chapter 4 and the partial response schemes in Chapter 6 are constructions based on linear modulations. When they are linear, modulations have a relatively simple analysis, which devolves down to the properties of pulses. These properties are the subject of Section 2.2.

A *baseband* modulation is one for which $s(t)$ in Eq. (2.1-1) or in some other form is a signal with a lowpass spectrum. A nonlinear modulation will not have the superposition form but it can still be viewed as a baseband modulation. If the lowpass signal is translated in frequency to a new band, the spectrum becomes bandpass and the modulation is a *carrier* modulation. Most transmission systems must use carriers, because the new frequency band offers an important advantage, such as better propagation, or because the band has been pre-assigned to avoid interference. A linear carrier modulation is still compared to pulses, but now the $v(t)$ in (2.1-1) is a shaped burst of the carrier sinusoid. An important issue in carrier modulation is whether the modulated sinusoid has a constant envelope. With a few exceptions, schemes with a constant envelope are nonlinear. The CPM schemes in Chapter 5 are generally nonlinear modulation constructions and their signals have constant envelope. The basic carrier modulations will be reviewed in Section 2.5.

The fundamental measures of a modulation's virtue are its error probability, its bandwidth, and, of course, its implementation cost. The error probability of a set of signals is computed by means of signal space theory, which is reviewed in Section 2.3. The theory explains in geometric concepts the error properties of signals in additive Gaussian noise. A great many communication links are corrupted by Gaussian noise, but even when they are not, the Gaussian case provides an important worst-case benchmark evaluation of the link. Most of the evaluation in the first half of this book is in terms of Gaussian noise. Later, fading channels will become important, but we will hold off a review of these until Chapter 7.

An equally important property of a signal set is its bandwidth. Some methods to calculate bandwidth are summarized in Section 2.7. As a rule, when signals carrying data at a given rate become more narrow band, they also become more complex. We know from Fourier theory that the product of pulse bandwidth and time spread is approximately constant, and consequently, the bandwidth of a pulse train may be reduced only by dispersing the pulse over a longer time. Even with relatively wideband signals, it is necessary in practice that pulses overlap and interfere with each other. Another way to reduce the bandwidth in terms of Hz-s per data bit is to increase the symbol alphabet size. Whichever alternative is chosen, it becomes more difficult to build a good detector. Bandwidth, energy,

Modulation Theory

and cost, all of which we very much wish to reduce, in fact trade off against each other, and this fact of life drives much of what follows in the book.

2.2. Baseband Pulses

We want pulses that are narrowband but easily distinguished from one another. Nature dictates that ever more narrowband pulses in a train must overlap more and more in time. The theory of pulses studies of how to deal with this overlap.

The simplest pulses do not overlap at all, but these have such poor bandwidth properties that they are of no interest. We first investigate a class of pulses that overlap, but in such a way that the amplitudes of individual pulses in a train may be observed without errors from samples of the entire summation; these are called *Nyquist* pulses. It is not possible to base a good detector on just these samples, and so we next review the class of *orthogonal* pulses. These pulses overlap as well, but in such a way that all but one pulse at a time are invisible to a maximum likelihood (ML) detector. Nyquist, and especially orthogonal pulses thus act as if they do not overlap in the sense that matters, even though they do in other ways. As pulse bandwidth narrows, a point eventually is reached where Nyquist and orthogonal pulses can no longer exist; this bandwidth, called the *Nyquist bandwidth*, is $1/2T_s$ in Hz, where $1/T_s$ is the rate of appearance of pulses. A train made up of more narrow band pulses is said to be a faster-than-Nyquist transmission. These pulses play a role in partial response coding in Chapter 6.

2.2.1. Nyquist Pulses

Nyquist pulses obey a zero-crossing criterion. For convenience, let the pulse $v(t)$ be centered at time 0. Hereafter in this chapter, T denotes the transmission symbol time.

Definition 2.2-1. A pulse $v(t)$ satisfies the Nyquist Pulse Criterion if it crosses 0 at $t = nT, n = \pm 1, \pm 2, \ldots$, but not at $t = 0$.

Some examples of Nyquist pulses appear in Fig. 2.1, with a unit amplitude version of the pulse on the left and its Fourier transform on the right. The top pulse is $\text{sinc}(t/T)$,[1] which has the narrowest bandwidth of any Nyquist pulse (see Theorem 2.2-2). The second pulses have wider bandwidth and are members of a class called the raised-cosine (RC) pulses. These are defined in terms of

---

[1] $\text{sinc}(x)$ is defined to be $\sin(\pi x)/\pi x$.

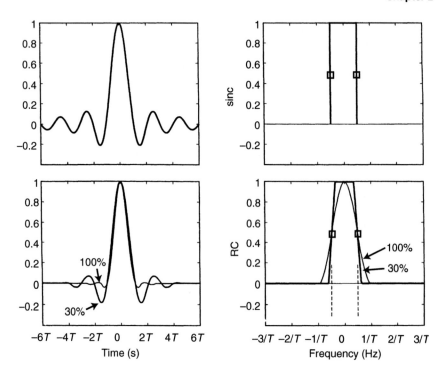

Figure 2.1 Nyquist pulses with their Fourier transforms at right. sinc(t/T) pulse (*top*) and RC pulse (*bottom*).

a frequency transform by

$$V(f) = \begin{cases} 1, & 0 \leq f \leq (1-\alpha)/2T \\ \cos^2\left[\dfrac{\pi T}{2\alpha}\left(f - \dfrac{1-\alpha}{2T}\right)\right], & (1-\alpha)/2T < f < (1+\alpha)/2T \\ 0, & \text{elsewhere} \end{cases}$$

(2.2-1)

and in the time domain by

$$v(t) = \frac{\sin(\pi t/T)}{\pi t/T} \frac{\cos(\alpha \pi t/T)}{1 - 4\alpha^2 t^2/T^2}.$$

(2.2-2)

(Note that both transform and pulse are scaled to unit amplitude.) The parameter $0 < \alpha \leq 1$, is called the "rolloff" or excess bandwidth factor. The bandwidth of the pulse is $(1+\alpha)/2T$, a fraction α greater than the narrowest possible Nyquist bandwidth, $1/2T$. Figure 2.1 shows the cases $\alpha = 0.3$ and 1. The extra RC bandwidth reduces the amplitude variation in the total pulse train and greatly reduces the temporal tails of the pulse.

Another Nyquist pulse is the simple square pulse defined by

$$v(t) = \begin{cases} 1, & -T/2 < t \leq T/2, \\ 0, & \text{otherwise.} \end{cases} \quad (2.2\text{-}3)$$

This pulse is called the NRZ pulse (for "non-return to zero") and it trivially satisfies the Nyquist Pulse Criterion, because its support lies in the interval $[-T/2, T/2]$. Such common pulses as the Manchester and RZ pulses lie in $[-T/2, T/2]$ as well; these are described in [1,2]. The penalty paid for the simple NRZ pulse is its spectrum, which is not only very wide but rolls off only as $1/f$. These simple pulses are useless in a bandwidth-efficient coding system, and we will not discuss them further.

A very simple kind of linear modulation can be constructed by using a Nyquist pulse in the standard linear form $\sum_n a_n v(t - nT)$. The detector can simply take samples at times nT. Such a detector is called a *sampling receiver*. If there is no noise in the received signal, the samples are precisely the transmission symbol stream $\{a_n\}$. Otherwise, the closest symbol value to the noisy sample can be taken as the detector output. As developed in Sections 2.3–2.5, good detectors for noisy signals need a filter before the sampler, and the sampling receiver error performance is in fact very poor with noisy signals. What is worse, a proper predetection filter will in general destroy the Nyquist sampling property of the pulse train. For these reasons, Nyquist Criterion pulses are generally not used over noisy channels.

Nyquist [3] in 1924 proposed the pulse criterion that bears his name[2] and gave a condition for the pulses in terms of their Fourier transform. He showed that a necessary condition for the zero-crossings was that $V(f)$ had to be symmetrical about the points $(1/2T, 1/2)$ and $(-1/2T, 1/2)$, assuming that $V(f)$ has peak value 1. This symmetry is illustrated by the two transforms in Fig. 2.1, with square blocks marking the symmetry points. Gibby and Smith [5] in 1965 stated the necessary and sufficient spectral condition as follows.

THEOREM 2.2-1 *(Nyquist Pulse Criterion).* $v(t)$ *satisfies the Nyquist Pulse Criterion (Definition 2.2-1) if and only if*

$$\sum_{n=-\infty}^{\infty} V(f - n/T) = K_0, \quad (2.2\text{-}4)$$

where $V(f)$ is the Fourier transform of $v(t)$ and K_0 is a real constant.

The theorem states that certain frequency shifts of $V(f)$ must sum to a constant. A proof of the theorem appears in [1,5].

In subsequent work [4], Nyquist suggested that there was a lower limit to the bandwidth of a Nyquist pulse, namely, $1/2T$ Hz. Formal proofs of this fact

---

[2] It subsequently became known as Nyquist's first criterion.

developed later, including particularly Shannon [9], and the result became known as the sampling theorem. We can state the version we need as follows. Proofs appear in any standard undergraduate text.

THEOREM 2.2-2. *The narrowest bandwidth of any Nyquist Criterion pulse is $1/2T$ Hz, and the pulse is $v(t) = A\,\mathrm{sinc}(t/T)$, where A is a real constant.*

2.2.2. Orthogonal Pulses

To summarize the above, there is no way to obtain good error performance under noise with bandwidth-efficient Nyquist pulses. The solution to the problem is to use orthogonal pulses, which are defined as follows.

Definition 2.2-2. A pulse $v(t)$ is orthogonal under T-shifts (or simply orthogonal, with the T understood) if

$$\int_{-\infty}^{\infty} v(t)v(t-nT)\,dt = 0, \quad n = \pm 1, \pm 2, \ldots,$$

where T is the symbol interval.

An orthogonal pulse is uncorrelated with a shift of itself by any multiple of T. Consequently, we can find any transmission symbol a_n in a pulse train $s(t)$ by performing the correlation integral

$$\int s(t)v(t-nT)\,dt = \int \left[\sum_m a_m v(t-mT)\right] v(t-nT)\,dt = a_n \int v^2(t-nT)\,dt.$$
(2.2-5)

If $v(t)$ has unit energy, the right-hand side is directly a_n. Some manipulations show that we can implement Eq. (2.2-5) by applying the train $s(t)$ to a filter with transfer function $V^*(f)$ and sampling the output at time nT. In fact, *all* of the $\{a_n\}$ are available from the same filtering, simply by sampling each T seconds. Pulse amplitude modulation, abbreviated PAM, is the generic name given to this kind of linear modulation signaling when it occurs at baseband.

The process just described is portrayed in Fig. 2.2. If there is no noise, the filter output sample is directly a_n. Otherwise, the sample is compared to the original symbol values in a threshold comparator, and the closest value is taken as the detector output. This simple detector is known as the *linear receiver*. By means of the signal space analysis in Section 2.3, it is possible to show that when M-ary orthogonal pulses are used, the error performance of the linear receiver is as good as that of any receiver with these pulses.

A necessary and sufficient condition on the transform of an orthogonal function is given by the next theorem.

Modulation Theory

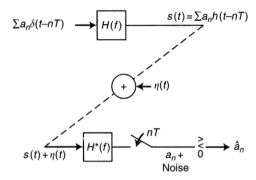

Figure 2.2 Linear modulation transmitter with linear receiver at bottom. $\eta(t)$ denotes additive channel noise.

THEOREM 2.2-3 *(Orthogonal Pulse Criterion)*. *$v(t)$ is orthogonal in the sense of Definition 2.2-2 if and only if*

$$\sum_{n=-\infty}^{\infty} |V(f - n/T)|^2 = K_0, \tag{2.2-6}$$

where $V(f)$ is the transform of $v(t)$ and K_0 is a real constant.

A proof appears in [1]. Note that Eq. (2.2-6) is the same as the Nyquist pulse condition of Theorem 2.2-1, except that the sum applies to the square magnitude of V rather than to V. In analogy to the case with Nyquist pulses, a sufficient condition for orthogonality is that $|V(f)|^2$ has the symmetry about the square blocks as shown in Fig. 2.1; this time, however, the symmetry applies to $|V(f)|^2$, not to $V(f)$. It is interesting to observe that if a modulation pulse $v(t)$ is orthogonal, then the waveform at the linear receiver filter output satisfies the Nyquist pulse criterion when there is no channel noise; that is, the filter outputs at successive times nT are directly the $\{a_n\}$ in the transmission (2.1-1).

The NRZ pulse is trivially orthogonal. The most commonly used orthogonal pulse in sophisticated modulation is the *root-RC* pulse, which takes its name from the fact that $V^2(f)$ is set equal to the RC formula in Eq. (2.2-1). $V(f)$ itself thus takes a root RC shape. The result is an orthogonal pulse that has the same excess bandwidth parameter α, $0 < \alpha \leq 1$. A sample pulse train appears in Fig. 2.3, made from the transmission symbols $\{+1, -1, +1, +1, -1, -1\}$; it can be seen that the train lacks the zero-crossing property. The time-domain formula

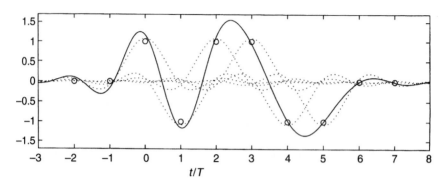

Figure 2.3 Total signal created from the pulses created by symbols $\{+1, -1, +1, +1, -1, -1\}$. Dotted curves show individual pulses. If the Nyquist property held, signal would pass through all the circles.

for the unit-energy $v(t)$ is

$$v(t) = \begin{cases} \frac{\sin[\pi(1-\alpha)t/T]+(4\alpha t/T)\cos[\pi(1+\alpha)t/T]}{\sqrt{T}(\pi t/T)[1-(4\alpha t/T)^2]}, & t \neq 0, \ t \neq \pm T/4\alpha, \\ (1/\sqrt{T})[1-\alpha+4\alpha/\pi], & t=0, \\ (\alpha/\sqrt{2T})[(1+2/\pi)\sin(\pi/4\alpha)+(1-2/\pi)\cos(\pi/4\alpha)], & t=\pm T/4\alpha. \end{cases}$$

(2.2-7)

2.2.3. Eye Patterns and Intersymbol Interference

The common impairments to a pulse waveform are easy to see from an eye pattern. To generate one, a plot of the pulse train waveform is triggered once each T by the receiver sampler timing and the results are superposed to form a single composite picture. Figure 2.4(a) shows what happens with a 30% excess bandwidth RC pulse train driven by 40 random data. The timing is arranged so that the times nT fall in the middle of the plot, and at these times all the superposed waveform sections pass through the transmission symbol values ± 1. It is clear that a sampling receiver that observes at these times will put out precisely the symbol values. On the other hand, if $s(t)$ is made up of orthogonal pulses and an eye plot taken at the output of the receive filter $V^*(f)$ in Fig. 2.2, a similar plot appears: The Nyquist pulse criterion applies at the filter output rather than directly to $s(t)$. Exactly Fig. 2.4(a) will appear if the pulses are 30% root RC.

The opening of an eye diagram is called the eye, and the one in Fig. 2.4(a) is said to be fully open. As long as the eye is always open at least slightly at the sampling time, the linear receiver (with orthogonal pulses) will detect correctly. The effect of most signal impairments is to close the eye some. If all the space

Modulation Theory

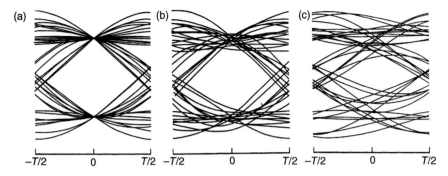

Figure 2.4 Eye patterns for 30% RC pulse train driven by 40 random data. (a) Undistorted pulse train; (b) Gaussian additive noise with bandwidth $1/T$ Hz and SNR $\approx$ 20 dB; (c) 6-pole Butterworth filtering with cut-off $0.4/T$.

is filled with signal transitions, then the comparator block can misread the symbol. One way the eye can close some is through jitter in the sampling time; some of the transitions shift left and right and the eye tends to close. Gaussian noise added to $s(t)$ can also close the eye. Figure 2.4(b) shows the effect of adding a noise with standard deviation 0.12 above and below the noise free values ± 1. The effect of either impairment is to reduce the eye opening, and effects from different sources tend to add.

Generally, as the pulse bandwidth declines, the open space in the eye plot reduces, although the waveform always passes through ± 1 if it satisfies the pulse criterion. The eye pattern for a non-binary transmission passes through the M symbol values, with open space otherwise. The eye pattern for an NRZ pulse train is an open rectangle of height ± 1 and width T.

The most common impairment to a signal other than noise is intersymbol interference, or ISI. Loosely defined, ISI is the effect of one pulse on the detection of pulses in neighboring symbol intervals. In the linear receiver with an orthogonal pulse train, sampling that is early or late pollutes the present symbol value with contributions from other transmission symbols, since the filter sample contributions from their pulses no longer necessarily pass through zero. Another major source of ISI is channel filtering, which destroys the precise orthogonality in the signal. Figure 2.4(c) shows the effect of a six-pole Butterworth lowpass filter on the 30% RC pulse train. The filter has 3 dB cut-off frequency $0.4/T$, while the pulse spectrum, shown in Fig. 2.1, runs out to $0.65/T$ Hz. Aside from ISI, filters also contribute delay to waveforms, but the delay is easily removed and in this case $1.7T$ has been subtracted out in Fig. 2.4(c). With the subtraction, the eye is still open at time 0 and the detection will always be correct in the absence of noise. But the eye is now narrower and more easily closed by another impairment. A cut-off of $0.35/T$ will completely close the eye even without noise.

It needs to be pointed out, especially in a coded modulation book, that ISI need not worsen the error performance of a properly designed receiver. The linear receiver will perform more poorly, but a receiver designed to be optimal in the presence of the ISI, based, for example, on the Viterbi algorithm (VA) of Chapter 3, may show no degradation at all. We will look at the relationship between ISI and such receivers in much more detail in Chapter 6.

2.3. Signal Space Analysis

The object of signal space analysis is to design an optimal receiver for a general set of signals and to calculate the receiver's error probability. The theory expresses signals and noise as components over a vector space and then calculates probabilities from these components. The modern theory stems from the 1947 thesis of Kotelnikov [7]; the theory was popularized by the classic 1965 text of Wozencraft and Jacobs [8]. When the channel disturbance is additive white Gaussian noise, henceforth abbreviated as AWGN, the vector space becomes the ordinary Euclidean one, and a great many results may be expressed in a simple geometric form, including particularly the calculation of error probability. One of the first investigators to espouse the geometric view was Shannon [9].

We begin by defining an optimal receiver.

2.3.1. The Maximum Likelihood Receiver and Signal Space

Suppose one of M messages, namely $m_1, m_2, \ldots, m_M$, is to be transmitted. These may be the M transmission symbols in the previous section, but they may also be a very large set of messages, that correspond to a block of many symbols. The transmitter converts the message m_i to $s_i(t)$, one of the set of signal waveforms $s_1(t), \ldots, s_M(t)$, and the channel adds noise $\eta(t)$ to form the received signal $r(t) = s_i(t) + \eta(t)$.

The receiver selects the most likely signal, from the information that it has available. This information is of two kinds, the *received signal* $r(t)$, which is an observation, and knowledge about the message source, which is the *a priori information*. The receiver then must calculate the largest probability in the set

$$P[s_i(t)|r(t)] = \frac{P[s_i(t) \text{ sent}, r(t) \text{ received}]}{P[r(t) \text{ received}]}, \quad i = 1, \ldots, M. \quad (2.3\text{-}1)$$

By means of Bayes Rule, Eq. (2.3-1) may be written as

$$P[s_i(t)|r(t)] = \frac{P[r(t) \text{ received} \mid s_i(t) \text{ sent}] P[s_i(t) \text{ sent}]}{P[r(t) \text{ received}]}. \quad (2.3\text{-}2)$$

Modulation Theory

The new form has several advantages. First, the probability $P[r(t)|s_i(t)]$ is simply the probability that the noise $\eta(t)$ equals $r(t) - s_i(t)$, since the channel noise is additive; the $r(t)$ is observed and $s_i(t)$ is hypothesized. Second, $P[s_i(t)]$ brings out explicitly the *a priori* information. Finally, $P[r(t)]$ does not depend on i and may be ignored while the receiver maximizes over i. What remains is a receiver that executes

$$\text{Find } i \text{ that achieves:} \quad \max_i P[\eta(t) = r(t) - s_i(t)] P[s_i(t)]. \qquad (2.3\text{-}3)$$

This detector is called the *maximum a posteriori*, or MAP, receiver. It takes into account both the observation and the *a priori* information.

When the *a priori* information is unknown, hard to define, or when the messages are all equally likely, the factors $P[s_i(t)]$ are all set to $1/M$ in Eq. (2.3-3). They thus do not figure in the receiver maximization and may be removed, leaving

$$\text{Find } i \text{ that achieves:} \quad \max_i P[\eta(t) = r(t) - s_i(t)]. \qquad (2.3\text{-}4)$$

This is called the maximum likelihood, or ML, receiver. It considers only the observed channel output. For equiprobable messages, it is also the MAP receiver.

The probabilities in Eqs (2.3-3) and (2.3-4) cannot be evaluated directly unless the noise takes discrete values. Otherwise, there is no consistent way to assign probability to the outcomes of the continuous random process $\eta(t)$. The way out of this difficulty is to construct an orthogonal basis for the outcomes and then work with the vector space components of the outcomes. This vector space is the *signal space*, which we now construct.

Assume that a set of orthonormal basis functions $\phi_1(t), \ldots, \phi_J(t)$ has been obtained. For white Gaussian noise, it can be shown that any basis set is acceptable for the noise if the basis is complete and orthonormal for the signal set alone. The AWGN basis set is often just a subset of the signals that happens to span the entire signal set. Otherwise, the Gram–Schmidt procedure is used to set up the basis, as explained in [1,6,8]. For colored noise, a Karhunen–Loeve expansion produces the basis, as shown in van Trees [10].

We proceed now as usual with a conventional inner produce space, with the basis just found. Express the ith transmitted signal $s_i(t)$ as the J-component vector

$$s_i = (s_{i1}, s_{i2}, \ldots, s_{iJ}), \quad i = 1, \ldots, M. \qquad (2.3\text{-}5)$$

Here the jth component is the inner product $\int_T s_i(t)\phi_j^*(t)\,dt$, where T is the interval over which the signals have their support and $1 \leq j \leq J$, with $J \leq M$. Each of the M signals satisfies

$$s_i(t) = \sum_{j=1}^{J} s_{ij}\phi_j(t), \quad i = 1, \ldots, M. \qquad (2.3\text{-}6)$$

In the same way, the noise waveform $\eta(t)$ is represented by the vector

$$\boldsymbol{\eta} = (\eta_1, \ldots, \eta_J, \eta_{J+1}, \ldots), \tag{2.3-7}$$

in which η_j is the inner product of $\eta(t)$ with $\phi_j(t)$. Extra dimensions beyond J are shown in Eq. (2.3-7) because the noise is not usually confined to the dimensions of the signals alone. But it will turn out that these dimensions play no role in the receiver decision. Similarly, the received waveform $r(t)$ is shown as

$$\boldsymbol{r} = (r_1, \ldots, r_J, r_{J+1}, \ldots), \tag{2.3-8}$$

although the components beyond the Jth will play no role.

In terms of vectors, the MAP and ML receivers of Eqs (2.3-3) and (2.3-4) are given as follows:

Find i that achieves: $\quad \max_i P[\boldsymbol{\eta} = \boldsymbol{r} - \boldsymbol{s}_i] P[\boldsymbol{s}_i] \quad$ MAP, $\tag{2.3-9}$

Find i that achieves: $\quad \max_i P[\boldsymbol{\eta} = \boldsymbol{r} - \boldsymbol{s}_i] \quad$ ML. $\tag{2.3-10}$

Sometimes, the noise components in $\boldsymbol{\eta}$ are discrete random variables, but with Gaussian noise, for example, they are real variables, and in this case the expression $P[\boldsymbol{\eta} = \boldsymbol{r} - \boldsymbol{s}_i]$ is to be interpreted as a probability density; the maximization then seeks the largest value for the density. To simplify the presentation, we will assume from here on that $\boldsymbol{\eta}$ is a real variable.

The key to finding the probabilities in (2.3-9)–(2.3-10) is the following theorem, a proof of which can be found in a stochastic processes text such as [11].

THEOREM 2.3-1. *If $\eta(t)$ is a white Gaussian random process with power spectral density (PSD) $N_0/2$ W/Hz, then the inner products of $\eta(t)$ with any set of orthonormal basis functions $\{\phi_j(t)\}$ are IID Gaussian variables that satisfy*

$$\mathcal{E}\{\eta_j\} = 0, \quad \text{all } j,$$

$$\operatorname{cov}(\eta_j, \eta_k) = \begin{cases} 0, & j \neq k, \\ N_0/2, & j = k. \end{cases} \tag{2.3-11}$$

Consequently, we can express the density $f[\boldsymbol{\eta} = \boldsymbol{r} - \boldsymbol{s}_i]$ for the AWGN case as the product of Gaussian density function factors

$$\prod_{j=1}^{\infty} f(r_j - s_{ij}) = \prod_{j=1}^{J} f(r_j - s_{ij}) \prod_{j>J} f(r_j), \tag{2.3-12}$$

where $f(\ )$ now denotes the common density function of the components. The second group of factors on the right in Eq. (2.3-12) forms a multiplicative constant that does not depend on i, and so this factor can be ignored during the maximization of Eqs (2.3-9) and (2.3-10) over i.

The final form for the ML receiver in AWGN follows after a few small steps. First, we have that (2.3-10) becomes

$$\prod_{j=1}^{J} f(r_j - s_{ij}) = (\pi N_0)^{-J/2} \exp\left[-\sum_{j=1}^{J}(r_j - s_{ij})^2/N_0\right]. \quad (2.3\text{-}13)$$

Second, the factor $(\pi N_0)^{-J/2}$ may be ignored. Third, taking the log of what remains in Eq. (2.3-13) will not change the maximizing i, since log is a monotone increasing function. Last, we can drop a factor of $-1/N_0$ in the exponent and take a min of the logs rather than a max. What remains for the ML receiver is

$$\text{Find } i \text{ that achieves:} \quad \min_{i} \sum_{j=1}^{J}(r_j - s_{ij})^2 = \min_{i} \|\mathbf{r} - \mathbf{s}_i\|^2. \quad (2.3\text{-}14)$$

A similar set of steps yields for the MAP receiver

$$\text{Find } i \text{ that achieves:} \quad \min_{i}\left[\sum_{j=1}^{J}(r_j - s_{ij})^2 - N_0 \ln(P[\mathbf{s}_i])\right]. \quad (2.3\text{-}15)$$

In both expressions, $\sum(r_j - s_{ij})^2$ is the ordinary Euclidean square distance between the vectors $\mathbf{r}$ and $\mathbf{s}_i$. In other words, the ML receiver finds the *closest signal to $\mathbf{r}$ in Euclidean space*; the MAP receiver does the same, but weights the signals by $-N_0 \ln(P[\mathbf{s}_i])$.

2.3.2. AWGN Error Probability

For any set of signals in signal space, the building block for computing the error probability is the two-signal error probability $p_2(k|i)$, the probability of deciding signal $s_k(t)$ when $s_i(t)$ was in fact sent. From the final form for the ML receiver in (2.3-14), this event will happen when the received vector $\mathbf{r}$ is closer to $\mathbf{s}_k$ than it is to $\mathbf{s}_i$. The situation in signal space is sketched in Fig. 2.5. Since by Theorem 2.3-1 the white noise projections are IID along any set of axes, let the first such axis be chosen to run along the line connecting $\mathbf{s}_i$ and $\mathbf{s}_k$; the other axes are orthogonal and the projections on them are independent. The perpendicular plane bisecting the line between the two signals is the locus of points equidistant between the signals. The probability that $\mathbf{r}$ lies on the $\mathbf{s}_k$ side of this plane is precisely the

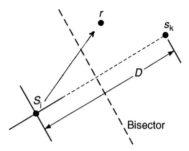

Figure 2.5 Calculation of the 2-signal error probability.

probability that the first component $r_1 - s_{i1}$ exceeds half the signal space distance $\Delta = \|s_i - s_k\|$. Thus our error probability is given by

$$p_2(k|i) = 1/\sqrt{\pi N_0} \int_{\Delta/2}^{\infty} \exp(-u^2/N_0)\, du. \tag{2.3-16}$$

Gaussian integrals like Eq. (2.3-16) occur so often that they are expressed in terms of the standard "Q-function," defined by

$$Q(x) = (1/\sqrt{2\pi}) \int_x^{\infty} \exp(-u^2/2)\, du, \tag{2.3-17}$$

which is the integral of the tail of the unit-variance zero-mean Gaussian density. A useful overbound to $Q(x)$ is

$$Q(x) < (1/2)e^{-x^2/2}, \quad x \geq 0. \tag{2.3-18}$$

We can go beyond this and write $Q(x) \sim (1/2)e^{-x^2/2}$. In this book, the symbol $\sim$, meaning asymptotically equal, states that in the limit $x \to \infty$ the ratio of the log of the left side to the log of the right side tends to 1. In terms of the standard erf(z) function, $Q(x)$ is $(1/2)[1 - \mathrm{erf}(x/\sqrt{2})]$.

Equation (2.3-16) becomes

$$p(k|i) = Q(\Delta/\sqrt{2N_0}), \tag{2.3-19}$$

which is the tail integral of the unit Gaussian density starting at $\Delta/\sqrt{2N_0}$. The quantity Δ is the Euclidean distance separating s_i and s_k. The convention in signal analysis is to normalize Δ in a certain way, namely by dividing it by twice the average signal energy per data bit. The average energy per signaling *symbol* for

Modulation Theory

the general case of M equiprobable symbols (i.e. signals) is[3]

$$E_s \triangleq \mathcal{E}_i\left[\int |s_i(t)|^2\, dt\right] = \frac{1}{M}\sum_{i=1}^{M}\|s_i\|^2, \qquad (2.3\text{-}20)$$

where the second expression uses Parseval's identity. The average energy per data bit is then $E_b = E_s/\log_2 M$. The normalized signal space square distance between s_i and s_k is thus

$$d^2 = \Delta^2/2E_b$$
$$= \log_2 M\, \Delta^2 / 2[(1/M)\sum_{i=1}^{M}\|s_i\|^2]. \qquad (2.3\text{-}21)$$

Throughout this book, the lower case d will mean such a normalized distance. In terms of d, the two-signal error probability becomes

$$p_2(k|i) = Q\left(\sqrt{d^2 E_b/N_0}\right). \qquad (2.3\text{-}22)$$

When $M = 2$ and there are really only two possible signals s_1 and s_2, we can imagine many arrangements in signal space such that s_1 and s_2 are distance Δ apart. From Eq. (2.3-19), however, all of these lead to an identical error probability. It is of interest which pair of points has the least energy, among all the possible translations and rotations of this signal set. For a general set of M points, the energy in Eq. (2.3-20) is minimized when the set $\{s_i\}$ is replaced by $\{s_i - z_0\}$, where z_0 is the centroid $z_0 = (1/M)\sum s_i$ of the signal set. Rotations of the centered set will not change the energy.

For a two-signal set, therefore, the origin of signal space should ideally lie half-way between the signals. We may as well rotate axis 1 to lie along $s_2 - s_1$, and so a minimum-energy set of two signals looks like Fig. 2.6(a). Such a balanced set

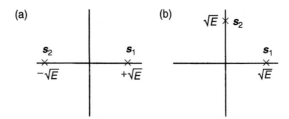

Figure 2.6 (a) Antipodal and (b) orthogonal binary signal constellations.

---

[3] Since E_s is an energy, it follows from Eq. (2.3-20) that the dimension of $s_i(t)$ is the square root of power.

is called an *antipodal* set. Since $d^2 = (2\sqrt{E})^2/2E = 2$ here, antipodal signaling always has error probability

$$p_e = p_2(2|1) = p_2(1|2) = Q\left(\sqrt{2E_b/N_0}\right). \quad (2.3\text{-}23)$$

Figure 2.6(b) shows the case of binary *orthogonal signaling*, where s_1 and s_2 are now two orthogonal signals. Now $d^2 = (\sqrt{2E})^2/2E = 1$ and

$$p_e = p_2(2|1) = p_2(1|2) = Q\left(\sqrt{E_b/N_0}\right). \quad (2.3\text{-}24)$$

This set will need twice the average bit energy in order to attain the same p_e. We see here the effect of not having minimized the energy in the set: s_1 and s_2 have the same energy but only half the normalized square distance as before. Plots like those in Fig. 2.6 are called *signal constellations*.

The Multi-signal Error Probability

When there are more than two signals, the general form for the error probability is

$$p_e = \sum_{i=1}^{M} P[\text{erroneous decision}|s_i \text{ sent}] P[s_i]. \quad (2.3\text{-}25)$$

We can think of signal space as divided up into M disjoint decision regions, $\mathcal{S}_1, \ldots, \mathcal{S}_M$, one for each s_i, that span the space. For the ML receiver case, $P[s_i] = 1/M$, all i, and Eq. (2.3-25) becomes in terms of decision regions

$$p_e = \frac{1}{M} \sum_{i=1}^{M} P[r \text{ not in } \mathcal{S}_i | s_i \text{ sent}]. \quad (2.3\text{-}26)$$

A very common multi-point constellation is the 4-point quaternary phase-shift keying (QPSK) one in Fig. 2.7; we will study QPSK in detail in Section 2.5.

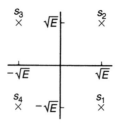

Figure 2.7 Four-point QPSK signal constellation.

Modulation Theory

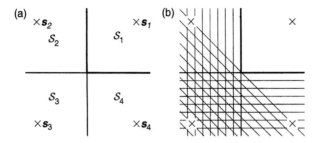

Figure 2.8 (a) Decision regions for the 4-point QPSK constellation. (b) Comparison of integration regions for the three terms in the union bound to p_e.

The QPSK constellation provides a good illustration of the details of a multisignal error calculation. The four decision regions are shown in Fig. 2.8(a). Integration of the Gaussian density over such regions can be complicated, and a better strategy is to replace each $P[r \text{ not in } S_i | s_i]$ in Eq. (2.3-26) with an overbound based on two-signal error probabilities. Since the S_i partition signal space,

$$P[r \text{ not in } S_i | s_i] = \sum_{k \neq i} P[r \text{ in } S_k | s_i].$$

Each S_k on the right here cannot be larger with M-ary signals than it would be if there were just two signals s_i and s_k in the set; consequently, each right hand term satisfies

$$P[r \text{ in } S_k | s_i] \leq p_2(k|i). \qquad (2.3\text{-}27)$$

Figure 2.8(b) illustrates how the three separate two-signal integrations overbound the true integration when s_1 is sent.

The calculation here is essentially the union bound of probability. A substitution of the overbound Eq. (2.3-27) into (2.3-26) gives

$$p_e \leq \frac{1}{M} \sum_{i=1}^{M} \sum_{k \neq i} p_2(k|i) = \frac{1}{M} \sum_{i=1}^{M} \sum_{k \neq i} Q\left[\frac{\|s_k - s_i\|}{\sqrt{2N_0}}\right]. \qquad (2.3\text{-}28)$$

By using the asymptotically tight estimate (2.3-18), we see that this is a sum of $M(M-1)$ exponentials that rapidly decay as the distances $\|s_k - s_i\|$ grow, and the term with the smallest distance will dominate the sum. This separation is called $D_{\min}$, the *minimum distance*. There can be several pairs at the smallest distance and the sum counts each distinct pair twice, and asymptotically in $D_{\min}/\sqrt{N_0}$ Eq. (2.3-28) has the form

$$p_e \sim (2K/M) \, Q\left(D_{\min}/\sqrt{2N_0}\right),$$

in which K is the number of distinct point pairs that lie $D_{\min}$ from each other. With the same normalization $d_{\min}^2 = D_{\min}^2/2E_b$ as in Eq. (2.3-21), we have as E_b/N_0 grows the asymptotic form

$$p_e \sim (2K/M)\, Q\left(d_{\min}\sqrt{E_b/N_0}\right). \qquad (2.3\text{-}29)$$

The multiplier $2K/M$ does not actually affect the asymptotic form here, but we retain it because it makes Eq. (2.3-29) more accurate at moderate E_b/N_0. In practical channels, Eq. (2.3-29) generally gives quite an accurate estimate of the E_b/N_0 needed to attain p_e.

We observe that the estimate depends on just a few variables, the average bit energy of the signals, E_b, the intensity of the noise, N_0, and M, K, and $d_{\min}$, which depend on the shape of the constellation. Of these last three, $d_{\min}$ is by far the most influential. In most of this book, E_b/N_0 will be taken as the *signal-to-noise ratio*. For an E_b/N_0 of at least moderate size, E_b/N_0 and the constellation $d_{\min}$ almost completely set the error probability.

2.4. Basic Receivers

We turn now to several basic receiver structures. We have already seen the linear receiver in Section 2.2, which is ML for orthogonal linear modulation. The receivers in this section are ML for arbitrary signals, whether baseband or carrier, in Gaussian noise, although the details of the carrier case are delayed to the next section. Our development here is admittedly theoretical and mathematical, but the receivers that come out of it are simple and optimal in many situations, and they work well enough in many others. It is often relatively easy to build a receiver that directly carries out the MAP receiver Eq. (2.3-15) or the ML receiver Eq. (2.3-14), and then, of course, one should do so.

The Correlator Receiver

From Eq. (2.3-14), the ML receiver finds the minimum over i of $\|s_i - r\|^2$. By Parseval's identity, this is directly

$$\int_T |r(t) - s_i(t)|^2\, dt = \int |r(t)|^2\, dt + \int |s_i(t)|^2\, dt - 2\int r(t)s_i(t)\, dt. \qquad (2.4\text{-}1)$$

The first term is constant during the minimization and can be dropped. The second term is the energy of $s_i(t)$, which we will denote ε_i for short. The third term is the correlation between $s_i(t)$ and the actual received signal $r(t)$. Finding the min of the left side of Eq. (2.4-1) is identical to finding the max of $\int r(t)s_i(t)\, dt - (1/2)\int |s_i(t)|^2\, dt$. This is the *correlator receiver*:

$$\text{Find } i \text{ that achieves:} \quad \max_i \int r(t)s_i(t)\, dt - \varepsilon_i/2. \qquad (2.4\text{-}2)$$

Modulation Theory

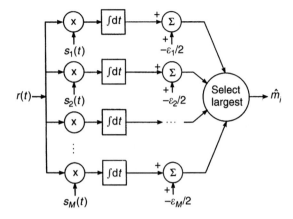

Figure 2.9 The correlator receiver structure. ε_i denotes the energy of signal i.

When all the signals have equal energy, $\varepsilon_i/2$ may be dropped from Eq. (2.4-2). A complete correlator receiver can be made up from a bank of signal correlators plus a Select-the-Largest block, as shown in Fig. 2.9. Some further derivation beginning from Eq. (2.3-15) shows that the MAP correlator receiver is:

$$\text{Find } i \text{ that achieves:} \quad \max_i \int r(t)s_i(t)\,dt - \varepsilon_i/2 + \frac{N_0}{2}\ln P[s_i]. \quad (2.4\text{-}3)$$

The Matched Filter Receiver

The correlation integral can be viewed as a convolution at time 0 of $r(\tau)$ with a certain function, namely $s_i(-\tau)$, since from the principles of convolution,

$$r(\tau) * s_i(-\tau)\big|_0 = \int r(\tau)s_i(\tau - t)\,d\tau\Big|_{t=0} = \int r(\tau)s_i(\tau)\,d\tau.$$

But this is the output at time 0 of the filter with impulse response $h(\tau) = s(-\tau)$, when $r(\tau)$ is the filter input. Equivalently, the filter has transform $H(f) = S_i^*(f)$. Actually, a few small repairs are needed if we are really to build a receiver by this means. First, the signal $s_i(\tau)$ is truncated at time NT; next the truncated signal $s_i^{tr}(\tau)$ is reversed to form $s_i^{tr}(-\tau)$; then this is shifted later by NT to produce finally a causal filter response $h(\tau) = s_i(NT - \tau)$. The desired correlation is now the filter output at time NT,

$$r(\tau) * s_i^{tr}(NT - \tau)\big|_{NT} = \int r(\tau)s_i^{tr}(NT + \tau - t)\big|_{t=NT} \approx \int r(\tau)s_i(\tau)\,d\tau. \quad (2.4\text{-}4)$$

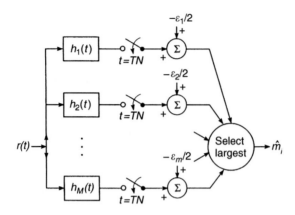

Figure 2.10 The matched filter receiver structure.

A receiver implemented by this filtering is called a *matched filter (MF) receiver*. Figure 2.10 shows such an ML receiver based on a bank of filters, one for each signal $s_i(t)$, and a Select-the-Largest block. Each filter response $h_i(t)$ is essentially the signal $s_i(t)$ time-reversed, and perhaps truncated in some harmless way.

A transmission of L M-ary symbols produces M^L signals, which could in principle lead to quite a large bank of matched filters. In many practical cases, however, the filters will be scaled or delayed versions of just a few basic responses. Even more important is the fact that the matched filter receiver often reduces to the linear receiver of Section 2.2.2, which has just a single matched filter.

It is a useful exercise to look at why binary antipodal signaling with equal-energy orthogonal pulses needs only a single matched filter. If the transmission were just one pulse, $a_1 v(t)$, the bank in Fig. 2.10 would be just the two filters with transforms $+V^*(f)$ and $-V^*(f)$ plus the Select-Largest block, but all of this can be replaced by one filter $V^*(f)$ and a block that compares the output sample to zero. If the complete pulse train $\sum a_n v(t - nT)$ appears at the filter instead of just one pulse, there will be no response at time NT from any pulse other than $a_1 v(t)$, as a consequence of the orthogonality property (2.2-5) applied to the matched filter formula (2.4-4). The matched filter ignores all but the present pulse, as long as the sample times are accurate multiples of T. The receiver is now precisely the linear receiver of Fig. 2.2. To implement a matched filter MAP receiver, the extra offsets $(N_0/2) \ln P[s_i]$ need to be subtracted from the filter sampler output.

A matched filter is fundamentally a lowpass filter, and a simple low pass with bandwidth similar to the s_i is often an admirable replacement for the true filter, provided that there are no other s_i in the same bandwidth.

An important property of a matched filter is that it optimizes the SNR at the sampler output. We can state this more formally as follows.

Property 2.4-1. Suppose $r(t) = s(t) + \eta(t)$ is received, where $\eta(t)$ is AWGN with density $N_0/2$. Pass $r(t)$ through the filter $H(f)$ and sample at t_0. The response to the signal alone is $g_s(t_0) = \int r(\tau)h(t_0 - \tau)d\tau$. The response to the noise alone is $g_n(t_0) = \int \eta(\tau)h(t_0 - \tau)d\tau$. The SNR, given by $|g_s(t_0)|^2/\mathcal{E}[|g_n(t_0)|^2]$, is maximized if and only if $H(f) = S^*(f)e^{-j2\pi f t_0}$; that is, $h(t)$ is $s(t)$ reversed and delayed by t_0.

When the matched filter response is orthogonal under T-shifts (i.e. $h(t)$ satisfies Definition (2.2-2)), another important property applies to the output noise samples: they are independent.

Property 2.4-2. Let a filter $H^*(f)$ be matched to $H(f)$, the impulse response of which satisfies Definition (2.2-2). Suppose AWGN of density $N_0/2$ is applied to $H^*(f)$. Then any set of filter output samples taken at T second intervals consists of IID Gaussian variates with zero mean. Their variance is $N_0/2$ if $H^*(f)$ has unit-energy response.

Proofs of these properties appear in [1,11,13]. Property 2.4-2 is important because it allows us to replace an orthogonal pulse transmission system in Gaussian noise with a simple discrete-time model. This is explained in Section 2.8.

Integrate-and-Dump Receiver

For the simple square NRZ transmission pulse, the matched filter response is also a (shifted) NRZ pulse. That is, the effect of the filter is to integrate the received signal over its length-T interval and then pass the result on to the comparator. This is called an integrate-and-dump receiver, since after the result is passed on, the integrator must be emptied to prepare for the next pulse. The receiver is ML for the NRZ pulse. For other pulses it is not so, but it performs a reasonable operation for a receiver – averaging the signal over the symbol time – and its error performance is often close to the ML one. Because of this and its low cost, it is often used. It is an example of so-called mismatched receiver; the performance of mismatched receivers in general can be derived by signal space analysis [1,14].

2.5. Carrier Modulation

A carrier modulation translates baseband signals such as those in Section 2.2 up to a new center frequency ω_0 called the carrier frequency. The signal has a bandpass rather than a lowpass spectrum. The baseband signals may or may not be pulse trains, but when they are, the carrier modulation is a train of carrier pulses. We will take up this kind of scheme first in Section 2.5.1.

A general form for a bandpass signal is

$$s(t) = A(t)\cos(\omega_0 t + \psi(t)), \tag{2.5-1}$$

in which $A(t) > 0$ is the signal amplitude and $\psi(t)$ is the phase. The bandwidth of $s(t)$ is about twice that of $A(t)$ and/or $\psi(t)$. In general it requires two lowpass signals to describe a bandpass signal, and when these are $A(t)$ and $\psi(t)$, they are said to form the *amplitude and phase description* of the signal. An equivalent description, which is more common in modulation theory, is the *in-phase and quadrature* baseband (or I/Q, or simply baseband) notation

$$s(t) = \sqrt{2E_s/T}\,[I(t)\cos\omega_0 t - Q(t)\sin\omega_0 t]. \qquad (2.5\text{-}2)$$

The conversion between the two forms is given by

$$I(t) = \frac{A(t)\cos\psi(t)}{\sqrt{2E_s/T}}$$
$$Q(t) = \frac{A(t)\sin\psi(t)}{\sqrt{2E_s/T}} \qquad (2.5\text{-}3)$$

or by

$$\psi(t) = \arctan[Q(t)/I(t)]$$
$$A(t) = \sqrt{2E_s/T}\sqrt{I^2(t) + Q^2(t)}. \qquad (2.5\text{-}4)$$

The quantity $\sqrt{I^2(t) + Q^2(t)}$ is the *envelope* of the modulation, and when it is constant in time the scheme is called a constant-envelope modulation. Coded modulations with a constant envelope comprise the continuous-phase modulation (CPM) class in Chapter 5. For all modulations, we will normally constrain the mean square of the envelope over a symbol interval to be unity, that is,

$$\mathcal{E}\left[(1/T)\int_T [I^2(t) + Q^2(t)]\,dt\right] = 1, \qquad (2.5\text{-}5)$$

because in this case E_s in Eq. (2.5-2) becomes the average energy per transmission symbol.

We will begin with *quadrature* digital modulations, which are ones for which $I(t)$ and $Q(t)$ are independent trains of orthogonal pulses of the form $\sum a_n v(t - nt)$. These are called linear carrier modulations, because the baseband pulse in the Section 2.2 linear modulations has effectively been replaced by a phased carrier pulse; the pulse superposition still applies.

2.5.1. Quadrature Modulation – PSK

In M-ary PSK, an M-ary transmission symbol drives the phase of the carrier signal. The carrier may simply hold one of M phases for the symbol time,

Modulation Theory

but more likely a shaped pulse of the carrier waveform is transmitted in one of M phase offsets. Binary PSK is denoted BPSK, and when $M = 4$, the PSK is called QPSK; for other symbol sizes, the PSK is denoted 8PSK, 16PSK, etc.

The general form of a binary PSK signal is

$$s(t) = \sqrt{2E_s/T} \sum_n a_n \sqrt{T} v(t - nT) \cos \omega_0 t, \qquad (2.5\text{-}6)$$

in which the transmission symbols $\{a_n\}$ take values ± 1 and $v(t)$ is a unit-energy pulse. $v(t)$ could be, for example, $(1/\sqrt{T})\text{sinc}(t/T)$ or the root RC pulse (2.2-7). When $v(t)$ is the NRZ pulse (2.2-3), $s(t)$ simply holds the phase 0 or π for the symbol interval, and has the I/Q form

$$\begin{aligned} s_1(t) &= \sqrt{2E/T} \cos \omega_0 t, \quad \text{for data } a_n = +1 \\ s_2(t) &= -\sqrt{2E/T} \cos \omega_0 t, \quad \text{for data } a_n = -1. \end{aligned} \qquad (2.5\text{-}7)$$

Form (2.5-6) takes the standard I/Q form (2.5-2) if we set $Q(t) = 0$ and take $I(t) = \sum_n a_n \sqrt{T} v(t-nT)$. We can view the signal (2.5-6) essentially as weighted time-shifts of a basic pulse $v(t) \cos \omega_0 t$, and whenever $v(t)$ is orthogonal, so is this pulse.[4]

The natural signal space basis for BPSK is the set of basis functions

$$\{\phi_n(t)\} = \{\sqrt{2} v(t - nT) \cos \omega_0 t\}, \quad -\infty < t < \infty,$$

which contains one function for each interval. These are orthonormal. The BPSK signal attributed to the nth interval is $\pm\sqrt{2E_s} v(t - nT) \cos \omega_0 t$, and the vector representation of a signal in that interval alone is $s_1 = (s_1(t), \phi_n(t)) = +\sqrt{E_s}$ and $s_2 = (s_2(t), \phi_n(t)) = -\sqrt{E_s}$. This is the antipodal signaling constellation in Fig. 2.6(a), whatever the details of $v(t)$ may be. Consequently, signal space analysis tells us that the error probability of the nth symbol is directly Eq. (2.3-23), $Q(\sqrt{2E_s/N_0})$, with $E_s = E_b$. From the discussion in Section 2.4, the simple linear receiver is ML for each orthogonal carrier pulse, during whose time it ignores all the others.

QPSK is a generalization of BPSK to two dimensions per symbol interval. The general I/Q form of QPSK for a unit energy pulse $v(t)$ is

$$s(t) = \sqrt{2E_s/T} \,[I(t) \cos \omega_0 t - Q(t) \sin \omega_0 t]$$

where

$$\begin{aligned} I(t) &= \sum_n a_n^I \sqrt{T} \, v(t - nT), \\ Q(t) &= \sum_n a_n^Q \sqrt{T} \, v(t - nT). \end{aligned} \qquad (2.5\text{-}8)$$

---

[4] Technically, the condition $\omega_0 \to \infty$ is needed for most of the claims of orthogonality in this section. But even ω_0 only modestly greater than $2\pi/T$ rad yields a virtually complete orthogonality.

Table 2.1 Symbol, angle and I/Q equivalences in QPSK

i	Angle	As Binaries	a_n^I	a_n^Q
0	45	00	$+1/\sqrt{2}$	$+1/\sqrt{2}$
1	135	10	$-1/\sqrt{2}$	$+1/\sqrt{2}$
2	225	11	$-1/\sqrt{2}$	$-1/\sqrt{2}$
3	315	01	$+1/\sqrt{2}$	$-1/\sqrt{2}$

Here E_s is the energy per quaternary symbol and $a_n^I = \pm 1/\sqrt{2}$ and $a_n^Q = \pm 1/\sqrt{2}$ are the independent I and Q transmission symbols. With the relations (2.5-4), we can write this in the amplitude and phase form

$$s(t) = \sqrt{2E_s} \sum_n v(t - nT) \cos(\omega_0 t + \pi/4 + i\pi/2), \quad i = 0, 1, 2, 3, \quad (2.5\text{-}9)$$

which shows the $s(t)$ as a sum of shaped cosine pulses whose phase offset takes one of four values. Table 2.1 shows the conventional correspondence between data bits, the phases in Eq. (2.5-9), and the transmission symbols in Eq. (2.5-8).

A Gram–Schmidt orthogonalization applied to the QPSK signals yields a two-dimensional orthonormal basis set for the signal space in each interval:

$$\{\phi_1(t), \phi_2(t)\} = \left\{ \sqrt{2} v(t - nT) \cos \omega_0 t, -\sqrt{2} v(t - nT) \sin \omega_0 t \right\}$$

Taking inner products $(s(t), \phi_1(t))$ and $(s(t), \phi_2(t))$ shows that the signal vectors for the nth interval are

$$s = \left(a_n^I \sqrt{E_s}, a_n^Q \sqrt{E_s} \right) = \left(\pm \sqrt{E_s/2}, \pm \sqrt{E_s/2} \right).$$

Each of these four has energy E_s, and the average bit energy is $E_b = E_s / \log_2 4 = E_s/2$. The constellation in each interval is that of Fig. 2.7, with E set to E_b.

We can use the more general multi-signal formula (2.3-29) with $K = 4$, $M = 4$ and $d_{\min}^2 = 4E_b/2E_b$, and get

$$p_e \sim (2 \cdot 4/4) Q((2\sqrt{E_b}/\sqrt{2E_b})\sqrt{E_b/N_0}) = 2Q(\sqrt{2E_b/N_0}). \quad (2.5\text{-}10)$$

A more careful derivation from first principles shows that the precise error probability is

$$p_e = 2Q(\sqrt{2E_b/N_0}) - Q^2(\sqrt{2E_b/N_0}), \quad (2.5\text{-}11)$$

which is very close to the estimate (2.5-10) for any reasonable p_e. Observe that the normalized square minimum distances of QPSK and BPSK are both 2, so that

Modulation Theory

their error probabilities are essentially the same, this despite the fact that QPSK carries two bits per interval while BPSK carries one. This fact largely eliminates BPSK in practical systems.

Figure 2.11 shows how a QPSK signal is constructed from two 30% root-RC pulse trains, with $I(t)$ driven by symbols $\{1, -1, 1, 1\}/(1/\sqrt{2})$ and $Q(t)$ driven by symbols $\{1, 1, -1, -1\}(1/\sqrt{2})$. Some properties are clear in the figure: the sinusoid occasionally drops to a very small amplitude and the pulses ring onward considerably outside the nominal interval $[-0.5, 3.5]$ that corresponds to the four symbols.

The standard circuits that generate and detect QPSK signals are shown in Figs. 2.12 and 2.13. These are the *quadrature transmitter* and the *quadrature receiver*. The names come from the fact that the signal processing is organized along two streams, or "rails," or "channels," one relating to the signal $\cos \omega_0 t$ and the other to $\sin \omega_0 t$. A BPSK transmitter/receiver consists of just the top rail in both cases.

The quadrature transmitter and receiver play a role in communication that goes far beyond QPSK. Many modulations carry data in independent I and Q signals, and all passband modulations can be expressed as possibly dependent I and Q. The transmitter serves to convert the baseband signals $I(t)$ and $Q(t)$ into a quadrature-modulated sinusoid, and the receiver converts the sinusoid back to I and Q and thereafter to the underlying transmission symbols. Because $\cos \omega_0 t$ and $\sin \omega_0 t$ are themselves orthogonal, the signals in the top and bottom rails behave independently. It can be shown that AWGN noise in the RF channel is converted by the receiver sin/cos multipliers into two independent zero-mean AWGN noises in the I and Q rails; furthermore, if a bit-normalized SNR of E_b/N_0 applies in the RF channel, then the SNR in the I-channel after the LPF, normalized to the bits carried in the I channel, is again E_b/N_0, and similarly for the Q channel (for a proof of this, see [1, Section 3.3]). In QPSK reception, for example, a signal $\sqrt{2E_s/T}[I(t)\cos \omega_0 t - Q(t)\sin \omega_0 t]$ that encounters an SNR of $E_b/N_0 = (E_s/N_0)/2$ in the RF channel will mix down to $\hat{I}(t)$ and $\hat{Q}(t)$, that both see an SNR of E_b/N_0 in their baseband channels. These facts make it easy to simulate a passband transmission scheme at baseband, without creating RF signals.

A plot of $Q(t)$ vs $I(t)$ as they evolve during a transmission is called an I/Q plot. The envelope of a passband signal at t is proportional to the radius of its I/Q plot. One of these plots is shown in Fig. 2.14 for 30% root-RC pulse QPSK. NRZ-pulse QPSK occupies the four corners of a square centered on the origin. Constant envelope schemes trace out a circle in the I/Q plane. An idea of the time spent at each power level is given by a histogram of the square envelope values, such as that in Fig. 2.15, which is for the 30% root RC signal in Fig. 2.14. Efficient RF amplifiers are nonlinear and consequently they cannot accurately amplify signals that have significant swings in RF power. Efficiency can be a critical issue in space and in battery-driven mobile communication. When such variations in power are significant, the transmission designer is forced to choose between an efficient

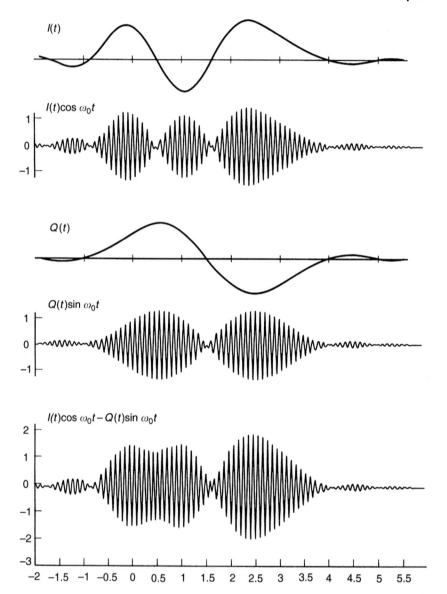

Figure 2.11 The formation of a QPSK signal with 30% root-RC pulse shaping from $I(t)$ and $Q(t)$. I-data is $+1, -1, +1, +1$ and the Q-data is $+1, +1, -1, -1$. Time axis in symbols; signal activity outside $[-2, 6]$ is truncated for clarity. (Reproduced from [1], chapter 3 with permission; copyright IEEE 1998.)

Modulation Theory

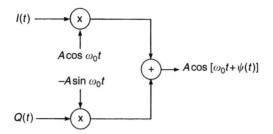

Figure 2.12 The quadrature transmitter.

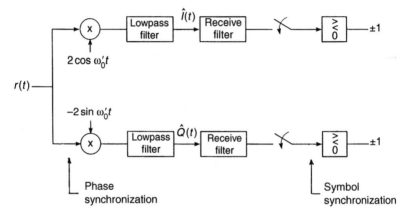

Figure 2.13 The quadrature receiver. The comparators are those for QPSK.

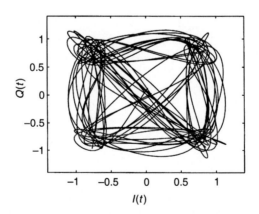

Figure 2.14 I/Q plot for QPSK with 30% root-RC pulses. 100 random quaternary symbols.

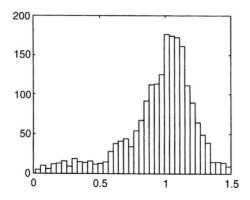

Figure 2.15 Histogram of the samples of the signal envelope in Fig. 2.14. 20 samples taken per second.

amplifier with high distortion and a less efficient but relatively linear amplifier. In practical linear modulation with the TCM and partial response signaling (PRS) of Chapters 4 and 6, the necessary loss is 2–4 dB. A way to escape the loss is constant-envelope signaling, and in the case of coded modulation, this implies CPM systems and Chapter 5. In a power-starved situation, CPM thus starts from a 2–4 dB advantage.

Envelope variation in a signal may be significantly reduced by delaying the Q signal $T/2$ s; this produces the *offset QPSK* signal

$$s(t) = \sqrt{2E_s/T}[I(t)\cos\omega_0 t - Q(t - T/2)\sin\omega_0 t]. \qquad (2.5\text{-}12)$$

Channel filtering is another common impairment that afflicts QPSK as well as other carrier modulations. Channel filtering leads to ISI in the baseband I and Q signals, as explained in Section 2.2.3, but it also worsens the envelope variation in the RF signal, or creates variation in a constant envelope signal where none existed before. In analyzing these effects, it is useful to convert from an RF passband filter to an equivalent baseband filter, and vice versa. The following two-step procedure does this. Consider the carrier modulated signal $s(t)$, which is filtered by $h(t)$ to produce $y = s*h$. $s(t)$ and $y(t)$ can be written in terms of baseband components as

$$s(t) = I(t)\cos\omega_0 t - Q(t)\sin\omega_0 t,$$
$$y(t) = y^I(t)\cos\omega_0 t - y^Q(t)\sin\omega_0 t.$$

Then:

(i) For a passband filter, express the impulse response $h(t)$ as

$$h(t) = h^I(t)\cos\omega_0 t - h^Q(t)\sin\omega_0 t, \qquad (2.5\text{-}13)$$

Modulation Theory

where $h^I(t)$ and $h^Q(t)$ are lowpass.

(ii) Find $y^I(t)$ and $y^Q(t)$ as

$$y^I(t) = \tfrac{1}{2}[I(t)h^I(t) - Q(t)h^Q(t)],$$
$$y^Q(t) = \tfrac{1}{2}[I(t)h^Q(t) + Q(t)h^I(t)].$$
(2.5-14)

The intersymbol interference caused by the passband RF filter may now be portrayed as the baseband signals y^I and y^Q. Given a pair of baseband filters $h^I(t)$ and $h^Q(t)$, reverse the steps to find the RF filter $h(t)$.

Nonlinearity generally widens the signal spectrum; if this widening is removed by subsequent filtering, the signal is further damaged. A third impairment is AM-to-PM conversion, a process in electronic devices whereby RF envelope variations are converted to phase variations. Constant-envelope signals are not affected by either nonlinearity or AM–PM conversion, and thus have a strong advantage in many practical channels. Further details about all these impairments appear in [1,6].

Two sophistications of simple PSK need to be mentioned. In *differentially encoded* PSK, the transmission symbol is sent in the difference of the phase from interval to interval, rather than in the absolute phase. A known initial phase must be sent to establish a reference. In *differentially decoded* PSK, the receiver extracts the phase difference between the present and past intervals. This is done as shown in Fig. 2.16, through multiplying the received signal by a T-delayed version of itself. *Differential PSK* is the name given to the binary scheme with both differential encoding and decoding, combined with a certain precoding of the data. The error performance is a little worse than BPSK and QPSK, because of noise propagation in the initial multiplier, but it is asymptotically as good. The precise formula is

$$p_e = (1/2)e^{-E_b/N_0}.$$
(2.5-15)

There are a great many differential encoding and decoding schemes and a complex theory exists about them. References [1,6], and particularly Lindsey et al. [15] provide details.

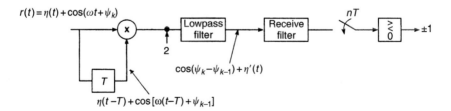

Figure 2.16 DPSK receiver operations in the kth signal interval.

A second extension of PSK is to more than four phases, and especially to 8 or 16. The latter two sometimes form the basis of TCM schemes. Now the data are mapped to the set of M two-tuples (a_n^I, a_n^Q) in such a way that $(a_n^I)^2 + (a_n^Q)^2$ is a constant and the signal constellation points are uniformly spaced around a circle. It can be shown [1,6] that $d_{\min}$ for a uniformly-spaced M-ary constellation is

$$d_{\min} = \sqrt{2 \log_2 M} \sin(\pi/M), \qquad (2.5\text{-}16)$$

and by using Eq. (2.3-28), a tight overbound to the symbol error probability is

$$p_e \leq 2Q\left[\sqrt{2(E_b/N_0) \log_2 M} \sin(\pi/M)\right]. \qquad (2.5\text{-}17)$$

The true symbol error probabilities – not the various bounds – are plotted in Fig. 2.17 for all these PSK schemes.

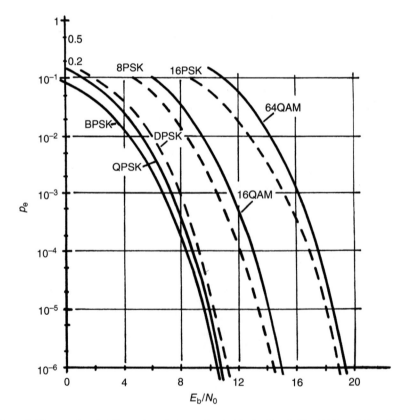

Figure 2.17 Actual probabilities of symbol error for BPSK, DPSK, QPSK, 8PSK, 16PSK, 16QAM and 64QAM. Data bit error rates will be higher. E_b/N_0 in dB.

Modulation Theory

It is important to realize that a circular constellation of points may or may not represent a constant envelope signal. The circular constellation only indicates that the different phased pulse manifestations have the same energy. The pulse shaping can still lead to a varying envelope; this is indeed the case with RC shaping.

2.5.2. Quadrature Modulation – QAM

Quadrature amplitude modulation, or QAM, extends the idea of PSK to modulating the pulse amplitudes as well as the phases. It forms the basis of TCM coded modulation, and so we will set down some analytical tools. As a generic communication term, QAM implies linear I and Q modulation and a carrier, in contrast to PAM, which implies single-channel linear baseband modulation. A general form for a QAM signal is once again Eq. (2.5-8), with the data mapped to M two-tuples (a_n^I, a_n^Q), but this time the resulting constellation is more than just the PSK circle. Some QAM signal space constellations are shown in Fig. 2.18. A common convention is to take the a_n^I and a_n^Q as whole integers, as in the figure, but this requires an extra step to normalize I and Q. To account for this, we can write the general form of QAM as $s(t) = \sqrt{2E_s/T}[I(t)\cos\omega_0 t - Q(t)\sin\omega_0 t]$, in which

$$I(t) = C_0 \sum_n a_n^I \sqrt{T} v(t - nT),$$

$$Q(t) = C_0 \sum_n a_n^Q \sqrt{T} v(t - nT). \qquad (2.5\text{-}18)$$

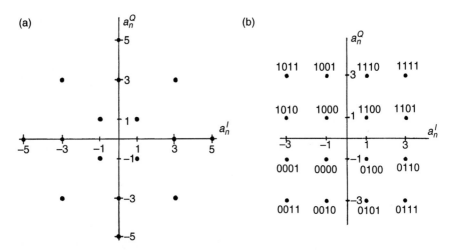

Figure 2.18 Sixteen-point rectangular QAM constellations. (a) The V.29 telephone line modem standard; (b) The V.32 ALT standard, which is based on rectangular 16QAM.

Here $v(t)$ has unit energy and C_0 is chosen so that the normalizing condition (2.5-5) is satisfied. This requires averaging over all the pairs (a_n^I, a_n^Q), weighting each by $1/M$; the integral in Eq. (2.5-5) is simply T, so

$$1/C_0 = \left[(1/M) \sum_{a^I, a^Q} [(a^I)^2 + (a^Q)^2] \right]^{1/2}. \qquad (2.5\text{-}19)$$

Each QAM symbol interval generates an independent two-dimensional constellation, provided that the basic pulse $v(t)$ is orthogonal. Both constellations in the examples of Fig. 2.18 have 16 points, but the (a_n^I, a_n^Q) in Fig. 2.18(a) are such that the I and Q signals are not independent of each other. When a_n^I and a_n^Q are independent and each takes values in the PAM set $\{\pm(\mu-1), \pm(\mu-3), \ldots, \pm 1\}$, a *rectangular QAM* constellation such as Fig. 2.18(b) results. In effect, $\log_2 \mu$ bits are sent in I and $\log_2 \mu$ in Q, and both are modulated onto the same carrier to form an MQAM with $M = \mu^2$. A labeling of data bits corresponding to the 16 constellation points is shown in Fig. 2.18(b), and the one here has the property[5] that a detection error to the nearest constellation point causes only a single bit error in all but two cases. We could label the 16 points in a such a way that independent I and Q bits were explicitly visible, but the bit error rate [BER] would then be higher.

The symbol-error probability of QAM with a ML receiver is dominated as usual by the distance between the closest pair of points. From Eq. (2.3-29), a good estimate for the rectangular 16QAM case is

$$p_e \sim (2K/M) Q\left(\sqrt{d_{\min}^2 E_b/N_0}\right) = 3 Q\left(\sqrt{d_{\min}^2 E_b/N_0}\right)$$

since there are 24 distinct point pairs lying at the minimum distance and M is 16. C_0 is $1/\sqrt{10}$ from Eq. (2.5-19), using the integer values in Fig. 2.18(b). Finding the unnormalized minimum distance $D_{\min}$ is made easier by two distance relations that hold as $\omega_0 \to \infty$:

$$\int [s_1(t) - s_2(t)]^2 \, dt = E_s/T \int [\Delta I(t)^2 + \Delta Q(t)^2] \, dt \qquad (2.5\text{-}20)$$

and

$$\frac{1}{T} \int [\Delta I(t)^2 + \Delta Q(t)^2] \, dt = C_0^2 \sum_n [(\Delta a_n^I)^2 + (\Delta a_n^Q)^2]. \qquad (2.5\text{-}21)$$

Δ indicates the difference between two outcomes of I, Q, a_n^I, or a_n^Q, as the case may be. In Eq. (2.5-20), $s_1(t)$ and $s_2(t)$ can be any two carrier signals, but

---

[5] A bit encoding that has this property in all cases and is called a Gray code.

Modulation Theory 49

Eq. (2.5-21) applies to QAM signals only, and the sum is over the succession of differences $\ldots, \Delta a_n, \Delta a_{n+1}, \ldots$ in the constellation points from interval to interval. With the help of these two relations, the $D_{\min}^2$ for a pair of closest signals is $4E_s C_0^2 = 4E_s/10$, since we take $(\Delta a_n^I)^2 + (\Delta a_n^Q)^2$ as 4 in one interval and 0 in all the others. Since $2E_b = 2E_s/\log_2 16 = E_s/2$, $d_{\min}^2$ is $D_{\min}^2/2E_b = 0.8$; the rectangular 16QAM estimate thus becomes

$$p_e \sim 3Q\bigl(\sqrt{0.8E_b/N_0}\bigr). \quad (2.5\text{-}22)$$

The true symbol probability for 16QAM and several other QAMs appears in Fig. 2.17. It shows that about 0.3 dB less energy is needed at moderately high E_b/N_0 than Eq. (2.5-20) estimates. It can be shown that the normalized minimum distance of QAM drops about 3 dB for each doubling of the constellation points. Thus much more energy per data bit is required for QAM, but in return the bandwidth per bit is reduced by a factor $1/\log_2 M$.

2.5.3. Non-quadrature Modulation – FSK and CPM

All of the earlier schemes in Section 2.5 can be thought of as linear modulations composed of pulses that are independent from interval to interval, and for the most part, $I(t)$ and $Q(t)$ act independently of each other. A major class of carrier modulations where neither holds is *frequency-shift keying* (FSK). This class generalizes to the CPM constructions in Chapter 5.

A general form for an FSK signal is

$$s(t) = \sqrt{2E_s/T}\cos(\omega_0 t + a_n h\pi(t-nT)/T + \psi_n), \quad nT < t < (n+1)T \quad (2.5\text{-}23)$$

for $s(t)$ in the nth interval. (In FSK analysis, intervals are most often centered at $(n+1/2)T$. E_s remains the symbol energy, a_n is an M-ary transmission symbol in the set $\{\pm(M-1), \pm(M-3), \ldots, \pm 1\}$. The second term in the cosine sets the frequency shift during the interval to $a_n h/2T$ Hz; h here is the *modulation index*, the constant of proportionality between the shift and the symbol a_n. With binary symbols and $h = 1/2$, for example, the shift from the carrier ω_0 will be $\pm 1/4T$ in Hz, or reckoned as a linear phase shift, it will be $\pm \pi/2$ rads over the interval.

FSK signals are categorized by the size of the shift. In *wideband* FSK, $h \gg 1$ and ψ_n is set to 0. The modulation is modeled by the orthogonal signaling constellation of Fig. 2.6, and the error probability is consequently $Q(\sqrt{E_b/N_0})$, which is 3 dB worse than QPSK, and the RF bandwidth is about Mh/T Hz, which is rather wide. In *discontinuous phase narrowband* FSK, ψ_n is again set to 0 and $0 < h < 1$. This FSK has discontinuities at each nT. An example signal is sketched in Fig. 2.19(a), which is a plot of $\psi(t)$, the excess phase in (2.5-23) over the carrier

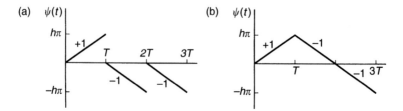

Figure 2.19 Excess phase $\psi(t)$ in FSK. (a) discontinuous phase FSK; (b) continuous phase.

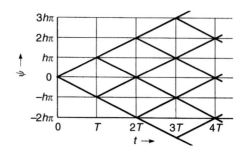

Figure 2.20 The phase tree of CPFSK, showing all phase trajectories starting at time 0.

phase $\omega_0 t$. Because the discontinuities lead to excessive bandwidth expansion, this FSK is seldom used. In *continuous phase* FSK, abbreviated CPFSK, $0 < h < 1$ and ψ_n is set to

$$\psi_n = \pi h \sum_{i<n} a_i, \qquad (2.5\text{-}24)$$

which guarantees phase continuity at every interval boundary. An example is shown in Fig. 2.19(b). CPFSK has attractive bandwidth and distance properties. These depend on h in a complicated way which is discussed in Chapter 5.

We can make a plot of all possible trajectories of the excess phase in CPFSK and get a picture like Fig. 2.20. This is called a phase tree, and is a useful analysis tool in later chapters.

All FSKs are constant-envelope modulations, and this can be a major advantage as was discussed in Section 2.5.1. For any two constant-envelope signals $s_1(t) = \sqrt{2E_s/T} \cos(\omega_0 t + \psi^{(1)}(t))$ and $s_2(t) = \sqrt{2E_s/T} \cos(\omega_0 t + \psi^{(2)}(t))$ with excess phases $\psi^{(1)}(t)$ and $\psi^{(2)}(t)$, it can be shown that the normalized distance between them is

$$d^2 = (\log_2 M)/T \int \left[1 - \cos[\psi^{(1)}(t) - \psi^{(2)}(t)]\right] dt \qquad (2.5\text{-}25)$$

Modulation Theory

as $\omega_0 \to \infty$. We can use Eq. (2.5-25) to estimate the error performance of CPFSK by finding the $d_{\min}$ for the closest two phase signals. Then the two-signal probability of deciding $s_2(t)$ given that $s_1(t)$ is sent is $Q(\sqrt{d_{\min}^2 E_b/N_0})$. The full error expression of the form (2.3-29) is covered in Chapter 5, but it can be said generally that p_e is overbounded by $2Q(\sqrt{d_{\min}^2 E_b/N_0})$. The $d_{\min}$ of CPFSK grows with M, whereas with QAM, it falls.

A special case of CPFSK is *minimum-shift keying* (MSK), which is binary CPFSK with $h = 1/2$. MSK can actually be generated and viewed as the offset QPSK in Eq. (2.5-12) running at half speed; that is, it can be viewed as a pulsed linear modulation. The unit-energy pulse shape is

$$v(t) = (1/\sqrt{T})\cos(\pi t/2T), \quad -T < t < T, \tag{2.5-26}$$

which is a cosine-shaped orthogonal pulse over an interval of width $2T$. As with all QPSKs, MSK has $d_{\min}^2 = 2$. The transmission symbols of the offset QPSK version of MSK are not the same as those in the CPFSK version; the relationship between the two is derived in [1,6,15] and elsewhere.

Continuous Phase Modulation Codes

CPM codes were the first widely studied class of coded modulations and they are the subject of Chapter 5. They can be thought of as an extension of CPFSK, first by smoothing further the linear phase transitions, and second by adding a higher order of memory. The CPM signal form is

$$s(t) = \sqrt{2E_s/T}\cos[\omega_0 t + 2\pi h \sum_n a_n q(t - nT)]. \tag{2.5-27}$$

Here the excess phase term is a convolution of the transmission symbols with a *phase response function*, $q(t)$. The function satisfies the conditions

$$q(t) = \begin{cases} 0, & t < 0; \\ \dfrac{1}{2} & t > LT, \end{cases} \tag{2.5-28}$$

the last of which insures that the signal phase is continuous whenever $q(t)$ is. As before, h is the modulation index and the $\{a_n\}$ are transmission symbols. By smoothing the transitions in $[0, LT]$ and spreading them over L intervals, $q(t)$ creates a signal whose phase depends on $L + 1$ symbols and whose energy and bandwidth properties are much improved. A CPM encoder amounts to a phase convolutional encoder.

Receivers for CPFSK and CPM can be directly the MF receiver of Fig. 2.10. Other simpler receivers are discussed in Chapter 5. A particularly simple receiver is the *discriminator* receiver, which tracks the instantaneous frequency of the signal. This is the time derivative of its phase. The frequencies in the nth interval are

$$\omega_0 + a_n h\pi/T, \quad \text{rads/s (FSK)},$$
$$\omega_0 + 2\pi h \sum_n a_n g(t - nT), \quad \text{rads/s (CPM)}, \quad (2.5\text{-}29)$$

in which $g(t) = dq(t)/dt$ is defined to be the frequency pulse of the CPM signal. Since $q(t)$ starts and ends with constant sections, $g(t)$ is a true pulse and Eq. (2.5-29) is a pulse train modulated by $\{a_n\}$ and offset by ω_0. FSK and CPM when used with discriminator detection are often called digital FM, because they can be viewed as the linear pulse modulations of Section 2.2 applied to the frequency of a signal.

The best known digital FM scheme other than CPFSK is GMSK, or Gaussian MSK. The frequency pulse has the Gaussian-integral shape

$$g(t) = \frac{1}{2T}[Q[(t - T/2)/\sigma] - Q[(t + T/2)/\sigma]], \quad (2.5\text{-}30)$$

where $\sigma = \sqrt{\ln 2}/2\pi B_b$, $0 < B_b < \infty$. B_b sets the signal bandwidth, and h in Eq. (2.5-29) is set to 1/2. The pulse is not orthogonal. It can be shown that $g(t)$ results when the ordinary MSK train (which is NRZ-pulse digital FM) passes through a filter whose impulse response is $\exp(-t^2/2\sigma^2)$. GMSK with $B_b = 0.3$ is the basis of the GSM digital mobile telephony standard. Discriminator detection destroys phase information and consequently exacts an SNR penalty that is typically 3 dB, but may be higher.

2.6. Synchronization

A critical aspect of carrier modulation detection is synchronization. A digital transmission network depends in reality on a long chain of synchronizers. The two that interest us now are synchronization to the *carrier phase*, which is $\omega_0 t$, and synchronization to the *symbol timing*, which is the set of instants nT. Without the first, $I(t)$ and $Q(t)$ cannot be obtained in the quadrature receiver and the excess phase cannot be known in the FSK/CPM receiver. Without the second, the proper sampling instant at the receiver filter output is unknown.

There are a great many sources of phase and timing disturbance, and it is safe to say that they can never be avoided entirely, since even the warming of a dielectric leads to some phase advance. One way or another, signal energy is spent to provide synchronization, either in pilot signals and training sequences, or in the extra energy that is needed to maintain the error probability when the phase must be extracted from the modulated signal itself. The signal energy required for good synchronization depends on the stability of the channel and the symbol rate. It is quite possible that a given stability/rate combination is simply unworkable.

We will review now the essentials of carrier phase synchronization, since it is the most challenging and the principles therein apply to all timing problems. References [1,17,19] contain full treatments.

Almost all synchronizers in practical applications depend on two subcircuits, one that produces a noisy sinusoidal signal, or "tone," from the received signal (the tone already exists if a pilot is present), and a phase-lock loop (PLL) that attempts to track the underlying tone. We will begin with the PLL.

2.6.1. Phase-lock Loops

We will discuss first the PLL. A standard PLL is shown in Fig. 2.21, that tracks an input sinusoid $A\cos(\omega_0 t + \psi_0)$, called the reference signal. It consists of a phase detector, which here is a multiplier, and a loop filter, and a controlled local oscillator, the VCO (for voltage controlled oscillator). The multiplier has as inputs the VCO output, $-2\sin(\omega_0 t + \theta_0)$, and the reference signal. By simple trigonometry, the multiplier output is $A\sin(\psi_0 - \theta_0)$, plus a sinusoid at frequency $2\omega_0$, which is removed by the lowpass filter. What remains is a slowly varying phase difference, or error, signal that is proportional to the reference phase offset ψ_0 minus the VCO offset θ_0. The variations in the error signal are smoothed by the low pass loop filter, whose output drives the VCO. If the error $\psi_0 - \theta_0$ is positive on the average, the VCO will be driven higher in phase by a positive control voltage; if $\psi_0 - \theta_0$ is negative, the VCO will be driven in a negative direction. Either way, the VCO will tend to lock to the reference signal phase.

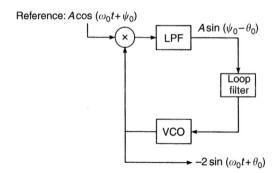

Figure 2.21 The basic PLL circuit.

The dynamics of a PLL can be analyzed by assuming the multiplier/LPF output is $\psi_0 - \theta_0$, rather than $\sin(\psi_0 - \theta_0)$. (In many PLLs it indeed is $\psi_0 - \theta_0$). Let the phase of the reference signal have Laplace transform $R(s)$ and the VCO phase have transform $V(s)$ and the error between them have $E(s)$; further, let the transfer function between $V(s)$ and $R(s)$ be

$$H(s) = V(s)/R(s) \quad (2.6\text{-}1)$$

and the error transfer function be

$$E(s)/R(s) = 1 - H(s). \quad (2.6\text{-}2)$$

If the loop filter is simply a gain block, and the gain around the loop is K, then $H(s)$ is $K/(s+K)$ and $1 - H(s) = s/(s+K)$. Consequently, the tracking error caused by a unit step transient in the reference signal is

$$e(t) = \mathcal{L}^{-1}[(1 - H(s))/s] = u(t)e^{-Kt}, \quad (2.6\text{-}3)$$

where $u(t)$ is the unit step function. This kind of PLL is called a first-order PLL. When the loop filter has transform $F(s) = 1/(1+as)$ and the overall loop gain is K, we have a second order PLL. The loop transfer function is then the second order response

$$H(s) = \frac{\omega_n^2}{s^2 + 2\zeta\omega_n s + \omega_n^2} \quad (2.6\text{-}4)$$

in which

$\omega_n = \sqrt{K/a}$, the undamped frequency.

$\zeta = \frac{1}{2}/\sqrt{Ka}$, the damping factor.

The error response to a step reference change is now

$$e(t) = \frac{e^{-\zeta\omega_n t}}{\sqrt{1-\zeta^2}} \sin\left(\omega_n\sqrt{1-\zeta^2}\,t + \cos^{-1}\zeta\right) \quad (2.6\text{-}5)$$

which is a decaying sinusoid.

Both step responses (2.6-3) and (2.6-5) decay to zero, but when the reference change is a ramp function – the case when there is a step change in the carrier frequency ω_0 – then neither loop's error response tends to zero. There is, in fact, a permanent phase error in the loop's tracking. In order for a second order loop to lock under a frequency error, it can be shown that a loop filter of the form $(1 + bs)/s$ is required. This effectively places an integrator in the loop.

The focus here on transient response has not yet made clear the real challenge to the PLL, and to the PLL designer, which is the fact that the reference signal is normally noisy. This can be background noise that arrives with the signal,

Modulation Theory

or it can be caused by the fact that the signal was not originally a perfect sinusoid, a kind of noise called *data noise*. The origin of the last will become clear shortly. A PLL can be made to ignore noise by designing a narrowband response $H(s)$; this is done by choosing an appropriate loop filter and loop gain K. However, the loop will now respond sluggishly to transients. The essence of synchronizer design is to find the best tradeoff between transient response and noise rejection. The critical parameter in the latter is the *noise equivalent bandwidth* of the PLL, or equivalently, of $H(s)$, a quantity equal to the bandwidth of an ideal lowpass filter with white noise input that passes the same power as $H(s)$ with this input. Mathematically,

$$B_N \triangleq \frac{1}{|H(0)|^2} \int |H(j2\pi f)|^2 \, df \quad \text{Hz.} \qquad (2.6\text{-}6)$$

It can be shown that noise at the PLL input produces a jitter in the VCO phase σ_θ^2, whose variance is approximately

$$\sigma_\theta^2 = 2B_N N_0 / A^2, \qquad (2.6\text{-}7)$$

where $N_0/2$ is the PSD of the noise and A is the reference signal amplitude. It can be shown further that $B_N = K/4$ for both the first order loop and the second order loop (2.6-4). The presence of data noise forces $B_N T \ll 1$, but transient response error, on the other hand, grows larger as B_N declines. Figure 2.22 shows some second-order error responses to a step (this is Eq. (2.6-5)) and to a ramp, for some choices of ζ and B_N.[6]

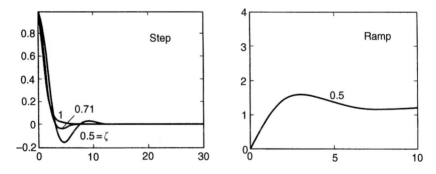

Figure 2.22 Step and ramp error responses of the second-order PLL. $B_N = 0.2$ Hz-s/symbol; ζ as shown. Horizontal axis in symbol intervals.

---

[6] A second-order loop is completely defined by its damping ζ and B_N.

2.6.2. Synchronizer Circuits

We now place the PLL in a complete phase synchronizer circuit. The most common circuit in use works by raising the modulated signal to a power. The effect if this is to produce a continuous sine wave tone together with artifacts of the operation, some of which are at far removed frequencies and can be filtered out, and some of which, the "data noise," cannot be removed and act as noise.

Figure 2.23 shows a fourth-power example of this circuit, together with the signals that would exist if the modulation were NRZ-pulse QPSK. At the input, $\psi(t)$ is the data-bearing phase $\pi/4 + i\pi/2$, i an integer, from Eq. (2.5-9), $\eta(t)$ is background noise, and ψ_0 is an unknown phase offset in the carrier. A fourth-power nonlinearity produces a number of signals, which can be revealed by some trigonometric exercises, and a bandpass filter allows through only components near $4\omega_0$. There is just one component it turns out, a cosine at four times the phase of the input cosine, namely

$$(A^4/8)\cos(4\omega_0 + 4\psi_0 + 4\pi/4 + 4i\pi/2),$$

which reduces to

$$-(A^4/8)\cos(4\omega_0 t + 4\psi_0). \qquad (2.6\text{-}8)$$

All trace of the modulation is removed, and what remains is a negative tone at $4\omega_0$, precisely four times the carrier frequency, with an offset of $4\psi_0$. The circuit is therefore called a frequency quadrupling loop.

The remainder of the circuit is a standard PLL that locks to the tone (2.6-8). Its job is made nontrivial by the noise $\eta(t)$, which propagates through the fourth-power nonlinearity; a rule of thumb for fourth-power loops is that the nonlinearity worsens the signal SNR by 6 dB [6,17]. When the pulses are not NRZ,

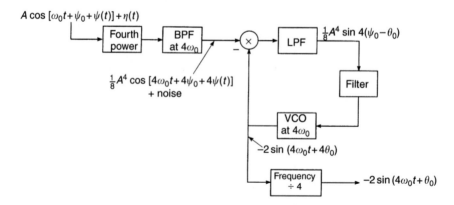

Figure 2.23 A fourth-power loop for synchronization of QPSK.

Modulation Theory

Eq. (2.6-8) again appears, but there will be a data noise disturbance as well that originates from the signal variations at the input. This noise can be severe and the need for a narrow loop bandwidth in order to smooth it out can in fact dictate the PLL design.

Fourth-power loops are standard with QPSK and work well enough with many other linear modulations such as QAM. Other powers can be used, for example, a second power for BPSK and an eighth power for 8PSK. As a consequence of the fourfold offset in Eq. (2.6-8), the PLL can lock to the true offset ψ_0 plus a multiple of $\pi/2$, a phenomenon called phase ambiguity. Ambiguity can be avoided by transmitting a pilot reference or by differentially encoding the data.

With nonlinear modulation or with coded modulation, phase synchronization is often not practical with power loops. Special circuits exist for CPFSK (see [14,15]). For CPM and for really complicated signals, it may be necessary to turn to the *remodulation loop*. An example is shown in Fig. 2.24. Now the transmission symbols are first detected using a guess at the carrier phase, a process that introduces some delay. A new, clean signal is created, or "remodulated," from the symbols. Finally, the new signal and the old with the same delay are multiplied to form a phase difference signal, which traverses around a PLL in the usual way. This synchronizer has a PLL, but the loop is broken, in effect, by a detector. The circuit simultaneously estimates the phase and detects the data. The remodulation loop suffers a start-up disturbance when its estimate of ω_0 and ψ_0 are not refined, and the detector delays its response to reference phase transients, but it can work with very complex signals.

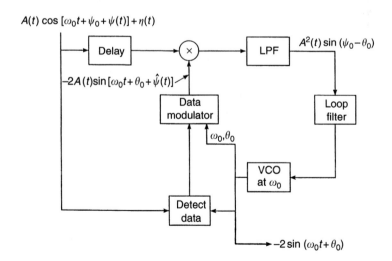

Figure 2.24 A remodulation loop for carrier synchronization to an arbitrary modulation.

Circuits that extract symbol timing consist of a phase-lock structure that tries to stay in lock with some characteristic of the I and Q signals. Most often, it is the zero-crossings of these signals that provide a series of marks. In PSK and QAM, these transitions occur $T/2$ s after the proper receive filter sampling time, plus or minus a small error (see, e.g. the eye plots in Fig. 2.4(a)). The timing marks so generated are thus noisy and many are simply missing, and the PLL tries to extract the average of the arrival pattern, which is the underlying symbol timing.

2.7. Spectra

The most important measures of a digital modulation are its error probability, its energy and its spectrum. It remains in this chapter to discuss the last, modulation spectra. Error probability, signal energy and bandwidth are intimately related, in theory as well as practice, and each trades off against the other two. It is easy to reduce, say, error probability at a fixed energy if bandwidth is allowed to grow, and bandwidth can easily be saved at the expense error of probability. Reducing both bandwidth and energy at a fixed probability is a heavy challenge. This is what is meant by bandwidth-efficient coding. The double reduction in fact was the original motivation for coded modulation.

Consider a modulated signal $s(t)$, either carrier or baseband. The general spectrum problem is to find the average of the energy spectrum of $s(t)$, that is, $(1/\tau_{tot})\mathcal{E}\{|S(f)|^2\}$, where the expectation is over the data symbols, $S(f)$ is the Fourier transform of $s(t)$, and τ_{tot} is the signal duration. This function of f, with the τ_{tot} division, is called the average PSD of the signal. The signal may be coded or not, a linear modulation or not, and the data themselves may have correlation. All of these may affect $S(f)$.

2.7.1. Linear Modulation Spectra

When a modulation is a linear superposition of pulses, as most simple modulations are, it is easy to find its PSD: it is simply the energy spectrum of an individual pulse. The same result holds most of the time for coded modulations built upon PSK and QAM. The easiest way to illustrate why this is so is to prove a formal theorem. First, we set out the assumptions, which directly hold for any proper linear modulation. Let

$$s(t) = \sum_{n=1}^{N} b_n w_n(t).$$

Now make two assumptions: (i), $|W_n(f)|$, the magnitude of a Fourier transform of $w_n(t)$, *is the same for all n*, and (ii), $\{b_n\}$ is any sequence of symbols that are IID

Modulation Theory

with $\mathcal{E}\{b_n]\} = 0$, all n. A pulse train $\sum b_n v(t - nT)$, whether carrier or baseband, satisfies (i), since the transforms of all the $w_n(t) = v(t - nT)$ differ only in phase. Condition (ii) holds for independent symbols in the standard alphabets (but excludes in general coded or otherwise correlated symbols).

THEOREM 2.7-1. *Suppose a signal satisfies (i) and (ii). Then its average energy spectrum satisfies*

$$\mathcal{E}\{|S(f)|^2\} = N\mathcal{E}\{|b_1|^2\}|W_1(f)|^2. \qquad (2.7\text{-}1)$$

The PSD is $1/NT$ times Eq. (2.7-1).

Proof. The transform of $s(t)$ is just

$$S(f) = \sum_{i=1}^{N} b_i W_i(f),$$

which has energy spectrum

$$|S(f)|^2 = \left[\sum_{i=1}^{N} b_i W_i(f)\right]\left[\sum_{k=1}^{N} b_k W_k(f)\right]^*$$

$$= \sum_{i=1}^{N}\sum_{k=1}^{N} b_i b_k^* W_i(f) W_k^*(f). \qquad (2.7\text{-}2)$$

Taking the expectation of this gives

$$\mathcal{E}\{|S(f)|^2\} = \sum_{i=1}^{N} \mathcal{E}\{b_i b_i^* W_i(f) W_i^*(f)\} = \sum_{i=1}^{N} \mathcal{E}\{|b_i|^2\}|W_i(f)|^2,$$

since condition (ii) implies that the off-diagonal terms are zero in the double sum (2.7-2). Condition (i) implies that this line is Eq. (2.7-1). □

As an example of the theorem, we can take the spectrum of NRZ-pulse QPSK. The modulated signal is Eq. (2.5-9), in which the transmission symbols are $\pm 1/\sqrt{2}$. This fits the assumptions, since the $\{b_n\}$ are $\pm\sqrt{E_b}$ and IID, with half of the $w_n(t)$ equal to $v(t - nT)\cos\omega_0 t$ and the other half $-v(t - nT)\sin\omega_0 t$. For all the $w_n(t)$, the transform absolute value is

$$W_n(f) = \tfrac{1}{2}[|V(f - f_0)| + |V(f + f_0)|], \quad f_0 = \omega_0/2\pi,$$

so long as f_0 is greater than the width of $V(f)$.[7] Thus, Theorem 2.7-1 says that the PSD is

$$\mathcal{E}\{|S(f)|^2\} = (E_s/2T)[|V(f - f_0)|^2 + |V(f + f_0)|^2]. \qquad (2.7\text{-}3)$$

[7] $V(f)$ for a practical pulse has unbounded support; in this case f_0 needs to be greater than the effective width of $V(f)$ and Eq. (2.7-3) becomes a close approximation.

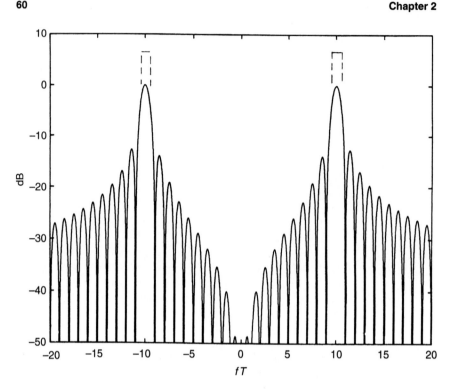

Figure 2.25 The actual PSD for PSK with NRZ pulses, when the carrier is 10 times the symbol rate. Sinc-pulse spectrum is indicated for comparison.

Equation (2.7-3) in fact even holds for a non-orthogonal pulse. For the NRZ pulse, the PSD becomes

$$\frac{E_s}{2T}[\text{sinc}^2(f - f_0)T + \text{sinc}^2(f + f_0)T]. \qquad (2.7\text{-}4)$$

Figure 2.25 plots the actual PSD of QPSK (or BPSK) with NRZ pulses for the case when the carrier frequency is 10 times the symbol rate $1/T$, which is a typical ratio in practice. Note that the sinc shapes in frequency are not quite symmetrical; this is because the spectrum $V(f)$ is not quite zero at $10/T$ and Eq. (2.7-3) does not quite hold. The asymmetry is ignored in practice.

Modulation spectra in general consist of a *main lobe*, in which most of the power lies, *nulls*, where the spectrum is zero, and a number of lesser peaks off to the side called *sidelobes*. As a rule, if the main lobe survives transmission intact, the receiver can detect the signal as if it were perfect; the significance of the sidelobes is that they interfere with neighboring transmissions. With the NRZ pulse, the main lobe has total width $2/T$ and nulls occur at $f = f_0 + n/T$, $n \neq 0$. The sidelobes fall off relatively slowly, with the first just 13 dB below the main

Modulation Theory

lobe peak. By comparison, a sinc time pulse has main lobe width just $1/T$ and no sidelobes at all.

The most used portrayal of bandwidth is the *power out-of-band*, or POB, plot. A POB plot differs in two ways from the true passband spectrum. First, it considers the baseband pulse, that is, the carrier frequency is set to zero. Such a spectrum is called a *baseband spectrum*. Second, it is a plot of the fraction of total energy that lies outside a frequency. Mathematically, it is

$$P_{\text{OB}}(f) = \frac{\int_{|u|>f} |S(u)|^2 \, du}{\int_{-\infty}^{\infty} |S(u)|^2 \, du}, \quad f \geq 0. \tag{2.7-5}$$

From the POB plot come some single-parameter measures of bandwidth. f is called the 99% power bandwidth when $P_{\text{OB}}(f) = 0.01$, that is, when 99% of the power is inside f. Typically, detection is unaffected when a signal passes through this bandwidth, and the signal is damaged at narrower bandwidths, so the 99% bandwidth is an often quoted measure. One can analogously define 99.9%, etc. bandwidths, and these are important as measures of interference to adjoining channels. The term RF bandwidth, applied to carrier signals, includes both the upper and lower sidebands of the passband spectrum; an RF power out-of-band is thus twice the width in Eq. (2.7-5).

Figure 2.26 shows the basic POB plot for NRZ and 30% root RC pulse modulation. The horizontal axis is fT, frequency normalized to the symbol time. When comparing QAMs and PSKs with different symbol alphabets, one normalizes to the per-data-bit time, T_b, where $T_b = T/\log_2 M$, and plots against fT_b, a measure with dimension Hz-s/bit. The POB plot for BPSK will again be Fig. 2.26, but the QPSK plot will be compacted twofold and 16QAM fourfold. With coded modulations based on QAM constellations, it is important to divide T by the true data bit rate per symbol interval and not by the log of the constellation points, which is a higher number.

The bandwidth of NRZ modulation is very much wider than the shaped pulse cases and its 99% RF bandwidth (twice the width at which the plot crosses -20 dB) exceeds $19/T$! It is clear why NRZ pulses are seldom used: their bandwidth is much too wide. Sinc and root-RC pulses in theory have no sidelobes, so that the POB curve drops to zero, but in practice there are always sidelobes because the pulse is generated by finite-order filtering or is subject to time truncation. Figure 2.26 illustrates this with a root-RC pulse that is truncated to total width $5T$. The truncation produces sidelobes which can pose severe difficulty in applications.

2.7.2. The General Spectrum Problem

When $s(t)$ carries correlated symbols, or is not a linear modulation, calculation of the spectrum is mathematically more difficult. In principle, an M-ary

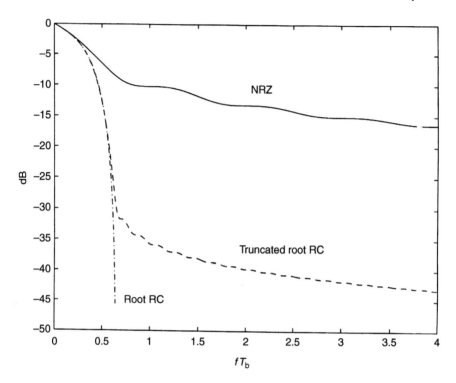

Figure 2.26 POB plots for linear modulation with NRZ and 30% root-RC pulses, and with 30% root-RC pulses time-truncated to $5T$ total support.

signal across L intervals takes on M^L possibilities, and weighting their transforms by their probabilities, one can compute $\mathcal{E}\{|S(f)|^2\}$. In times past, this was done. A modern approach to the problem is to find the PSD for a special version of $s(t)$ that has been *randomized in time*.[8] Suppose $s(t)$ is an ongoing signal. The new signal $\bar{s}(t) = s(t + \delta)$ is formed, where δ is an offset randomly distributed in $[0, T)$. The motivation for this is that $\bar{s}(t)$ is a stationary random process while $s(t)$ is not, since $s(t)$ has a structure of some sort based on the symbol times nT. The *time average autocorrelation* of $s(t)$ is defined as the autocorrelation of $\bar{s}(t)$,

$$r(\tau) = \mathcal{E}\{s(t + \delta + \tau)s^*(t + \delta)\}, \quad \text{for any } t, \tag{2.7-6}$$

where the expectation is now over the symbols and over δ. The PSD, as with any stationary process, is the Fourier transform of $r(\tau)$. We must hope that the PSDs of s and $\bar{s}$ are not different in an engineering sense, and this has proven to be the case.

---

[8] Another approach exists, based on cyclostationary random process theory.

Modulation Theory

We may as well take $t = 0$ in Eq. (2.7-6), since $\bar{s}(t)$ is stationary. By careful calculation of the autocorrelation, one can find the spectra for a great many cases of nonlinear modulation, correlated data and encoded input data. With correlated or encoded data, the modulator input can be modeled as a creation of a finite-state machine and $r(t)$ calculated by an averaging over the machine states. Often, properties of the modulation lead to considerable simplification of the calculation.

For the purposes of this book, the main spectra that we now lack are those of CPFSK and the CPM family, and we give those now in the form of a theorem. Let the FSK or CPM be generated by the standard phase response function $q(t)$, as defined in Section 2.5.3 and Eq. (2.5-27). The standard M-ary CPM transmission symbols are independent and equiprobable. Rather than use the time average autocorrelation $r(\tau)$, we define the *complex baseband autocorrelation*

$$R(\tau) = \mathcal{E}_{\delta,a}\{[I(t+\delta+\tau) + jQ(t+\delta+\tau)][I(t+\delta) - jQ(t+\delta)]\}, \quad \text{any } t, \tag{2.7-7}$$

in which I and Q are the baseband components of a carrier signal $s(t)$ and the expectation is over all the data symbols a and the random shift δ as usual. The actual carrier signal time average autocorrelation is given by

$$\sqrt{E_s/T}\ \Re\{R(\tau)e^{-j\omega_0\tau}\}, \tag{2.7-8}$$

where $\Re$ signifies real part. $R(\tau)$ is used in spectral calculations because it simplifies the algebra and because the transform $G(f) = \mathcal{F}\{R(\tau)\}$ is directly the desired PSD $\mathcal{E}\{|\bar{S}(f)|^2\}$, translated to frequency zero; that is,

$$\mathcal{E}\{|\bar{S}(f)|^2\} = G(f - f_0) + G(f + f_0),$$

where

$$G(f) = \int_{-\infty}^{\infty} R(\tau)e^{-j2\pi f\tau}\,d\tau. \tag{2.7-9}$$

Now for the CPM spectrum theorem.

THEOREM 2.7-2. *For the standard $q(t)$ in Eq. (2.5-27), M-ary IID equiprobable symbols, and h not equal an integer, the complex baseband autocorrelation is*

$$R(\tau) = \frac{1}{T}\int_0^T \prod_{i=1-L}^{\lceil \tau/T \rceil} \frac{\sin 2\pi h M[q(u+\tau-iT) - q(u-iT)]}{M\sin 2\pi h[q(u+\tau-iT) - q(u-iT)]}\,du, \tag{2.7-10}$$

for $0 \leq \tau \leq (L+1)T$ and the baseband PSD is

$$G(f) = 2 \int_0^{LT} R(\tau) \cos 2\pi f \tau \, d\tau$$
$$+ \frac{2 - 2C \cos 2\pi fT}{1 + C^2 - 2C \cos 2\pi fT} \int_{LT}^{(L+1)T} R(\tau) \cos 2\pi f \tau \, d\tau$$
$$- \frac{2C \sin 2\pi fT}{1 + C^2 - 2C \cos 2\pi fT} \int_{LT}^{(L+1)T} R(\tau) \sin 2\pi f \tau \, d\tau \qquad (2.7\text{-}11)$$

Here LT is the active length of the response $q(t)$, $\lceil \tau/T \rceil$ is the next integer larger than τ/T, and $C = (\sin M\pi h)/(M \sin \pi h)$.

The proof of this result is given in Section 5.3, and consists of working out the autocorrelation. The form of $q(t)$ leads to the reduction of $R(\tau)$ to the relatively simple form (2.7-10), and $G(f)$, which would otherwise be an integral over infinite time, reduces to the finite integral (2.7-11). Section 5.3 also gives more general formulas which cover a wider class of $q(t)$ and symbols with unequal probabilities.

The form of Eqs (2.7-10) and (2.7-11) is compatible with numerical integration, and accurate PSDs may be straightforwardly obtained even with complicated phase response functions. For some modulations, closed form expressions have been found. The baseband PSD of MSK is

$$G(f) = \frac{16}{\pi^2} \left[\frac{\cos 2\pi fT}{1 - 16 f^2 T^2} \right]^2. \qquad (2.7\text{-}12)$$

The spectrum of CPFSK is given by

$$G(f) = \frac{2}{M} \sum_{i=1}^{M} \left[\frac{(1/2) \sin^2 g_i}{g_i^2} + \frac{1}{M} \sum_{j=1}^{M} \frac{a_{ij} \sin g_i \sin g_j}{g_i g_j} \right] \qquad (2.7\text{-}13)$$

in which the following are defined:

$$g_i = [fT - (2i - M - 1)h/2]\pi, \quad i = 1, \ldots, M$$
$$a_{ij} = \frac{\cos(g_i + g_j) - C \cos(g_i + g_j - 2\pi fT)}{1 - 2C \cos(2\pi fT) + C^2}, \quad i, j = 1, \ldots, M$$
$$C = \frac{(1/M) \sin M\pi h}{\sin \pi h}.$$

Some spectra for binary CPFSK are shown in Fig. 2.27.

CPM spectral calculations can be complicated and numerical, but some simple rules apply nonetheless to their overall shape and form. The main lobe width, measured in terms of fT_b, depends mostly on $\log_2 M$, the transmission

Modulation Theory

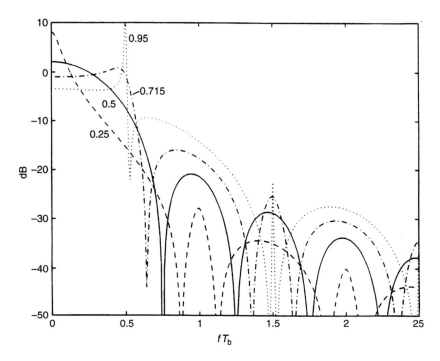

Figure 2.27 Baseband PSDs for binary CPFSK with modulation indexes $h = 0.25, 0.5, 0.715, 0.95$, normalized to the bit time. MSK is $h = 0.5$.

symbol rate in bits/symbol, and on the number of signal intervals affected by one symbol, which is the length of the phase convolution. The rolloff of the sidelobes depends, on the other hand, on the number of derivatives the phase possesses. For constant-envelope signaling, Baker [18] has shown that the PSD obeys

$$\mathcal{E}\{|S(f)|^2\} \approx |(f - f_0)T|^{-(2n+4)} \qquad (2.7\text{-}14)$$

in the limit as f tends away from the carrier, where n is the number of continuous derivatives in the excess phase. For discontinuous-phase PSK, n is -1 and the rolloff is only as $[|(f - f_0)|T]^{-2}$, or 6 dB per octave. For CPFSK, n is 0, which translates to 12 dB. These rolloffs are clear in Figs 2.25 and 2.27. For GMSK, Eq. (2.5-30) has all continuous derivatives, and so asymptotically in f its spectrum decays very fast.

2.8. Discrete-time Channel Models

It is convenient – but sometimes frustrating – to work with communication system models in discrete-time form. Real-world information bearing signals are ultimately analog, yet we would like to store and process signals as sequences of samples rather than as continuous functions of time, especially when the processor is a complicated software algorithm. This is surely the case in coded modulation. This section will set down the discrete-time models that are needed in the rest of the book.

With TCM, the modeling is particularly easy because the coding constructions are over the simple rate-T orthogonal modulation pulses of Section 2.2. Each I and Q pulse creates one new signal space dimension each T seconds. With partial response coding, there is the equivalent of intersymbol interference among the pulses; now we need to go carefully and think about how the modeling affects optimal detection of the original signal under these more difficult conditions. With CPM, it is generally easier to leave the signals in what is effectively continuous time. This means sampling the I and Q baseband signals at above their Nyquist rate[9] and working with these sample streams. A sampling at 2–5 times the modulation symbol rate $1/T$ almost always suffices. (Throughout this section, T is the modulation transmission symbol time.) It is possible in principle to construct signal space models for CPM signaling, or for any other digital signaling, but these obscure the issues in CPM enough that we will avoid such models in this book.

Because the subject of modeling can be so confusing, we will start off carefully, with the simplest of the modeling problems.

2.8.1. Models for Orthogonal Pulse Modulation

The key to modeling orthogonal-pulse modulations such as BPSK, QPSK, and QAM is Property 2.4-2, which states that the samples every T seconds of the output of a filter matched to an orthogonal pulse are IID Gaussians when the filter has a white Gaussian noise input. This is precisely the situation in the ML detection of the a_n in an orthogonal pulse train $\sum a_n v(t - nT)$ in AWGN. AWGN with PSD $N_0/2$ leads to IID Gaussian noise samples at times $T, 2T, \ldots$, whose variance is $N_0/2$; these noise variates $\eta_1, \eta_2, \ldots$, add to the values $a_1, a_2, \ldots$, which are the noise free matched-filter outputs.[10] Figure 2.28 shows the analog transmission system, together with the discrete-time model based directly on the sequences $\{a_n\}$ and $\{\eta_n\}$ alone.

The QAM transmission in Section 2.5.2 is precisely described by two copies of this discrete-time model, one that carries the I symbol sequence $\{a_n^I\}$ and one that carries the Q sequence $\{a_n^Q\}$. From the theory of bandpass Gaussian noise (see [11]), the noise variates in each copy are independent of each other, and it is assumed that a_n^I and a_n^Q are already normalized to unit energy. The analog signal

---

[9] That is, at twice their effective baseband bandwidth.

[10] We assume the signals are converted to baseband and the symbol synchronization is perfect; furthermore, the matched filter impulse response has unit energy.

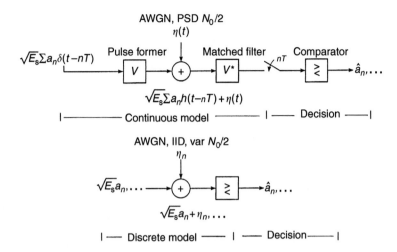

Figure 2.28 Orthogonal pulse linear modulation analog model for Gaussian noise (*top*), compared to equivalent discrete-time model (*bottom*).

is Eq. (2.5-18). Because of these facts we may ignore everything in that description except the symbols a_n^I and a_n^Q. These we can organize into the M possible two-dimensional symbols of the form (a_n^I, a_n^Q), one two-tuple for each symbol interval. K symbol intervals correspond to $2K$ signal space dimensions. TCM and lattice code constructions are built upon these M-ary symbols over a succession of intervals. While the contruction itself may be complicated, it is important to realize that the underlying modulation is fixed and simple and directly described by two copies of Fig. 2.28. In particular, the orthogonal pulse shape and its bandwidth are not part of a TCM code design.

The Section 2.3 signal space analysis for TCM and lattice signals stems from the fact that the $2K$-dimensional basis is created directly from the time shifts of the I and Q orthogonal pulses. The $2K$-dimensional vector r represents the K-symbol interval received signal $r(t)$. r is composed of I and Q pairs, which we can expand into components as $r = (r_1^I, r_1^Q, r_2^I, r_2^Q, \ldots, r_K^I, r_K^Q)$. The noise vector η is $(\eta_1^I, \eta_1^Q, \eta_2^I, \eta_2^Q, \ldots, \eta_K^I, \eta_K^Q)$, the components of which come directly from the two copies of Fig. 2.28. All these components are IID. The ML decision on the TCM word is the word s_i that minimizes $\|r - s_i\|$.

For completeness, we should point out that a signal space basis for any signal set, no matter how complex was the coding that produced it, may be created via the Gram–Schmidt process (see Section 2.3.1). In QAM modulation and the related TCM schemes, that process creates a pair of basis functions for each I/Q pulse pair, which is the pulse pair itself; for the entire transmission the process creates the Cartesian product space of all these pairs. This is just what we have described above. For other coded modulations, the process creates a basis that is mathematically valid but less convenient to use. In M-ary CPM for example, the process creates $2M$ basis functions for each symbol interval, in comparison to just 2 for TCM, and very little energy resides in some of these dimensions. In intersymbol interference and partial-response problems, a more direct modeling than the Gram–Schmidt process is preferred, and the next paragraphs turn to that subject. Nonetheless, a straightforward Gram–Schmidt modeling of CPM and ISI coding sometimes finds application.

2.8.2. Models for Non-orthogonal Pulse Signaling: ISI

When intersymbol interference affects QAM and the coded modulations built on it, then we should view the modulation as no longer based on orthogonal pulses. Furthermore, in Chapter 6, ISI will be exploited intentionally as a form of linear coded modulation. In both these cases the discrete-time channel modeling needs to be a step more subtle. We continue to assume an AWGN channel.

To begin, we need to set down a model for linear modulation with linear ISI. The result of both processes in tandem will be

$$s(t) = \sqrt{E_s} \sum a_n h(t - nT). \qquad (2.8\text{-}1)$$

Here the $\{a_n\}$ are transmission symbols as usual and $h = v * g$ is a non-orthogonal response which is the convolution of an orthogonal modulation pulse $v(t)$ and a function $g(t)$ that describes the ISI. The signal generation under AWGN is shown at the left in Fig. 2.29. The $g(t)$ can be unintentional ISI, or it can for example be an intentional PRS of the form

$$g(t) = \sum_{\ell=0}^{m} b_j \delta(t - \ell T), \qquad (2.8\text{-}2)$$

with $\delta(\ )$ the impulse function. PRS coding is the subject of Chapter 6. It creates the total response

$$s(t) = \sum_n c_n v(t - nT), \quad \boldsymbol{c} = \boldsymbol{a} * \boldsymbol{b}. \qquad (2.8\text{-}3)$$

Expressed in words, the last equation shows that PRS coding is built upon orthogonal pulses for each T, whose weights are the sequence $\boldsymbol{c} = c_1, c_2, \ldots$, which are the convolution of the data stream $\boldsymbol{a}$ and a generator sequence $\boldsymbol{b} = b_0, \ldots, b_m$. With certain restrictions, the creation of h from a general g may in fact be expressed in these terms, that is, a $\boldsymbol{b}$ and an orthogonal v in form (2.8-3), but we delay this discussion until Section 6.1.

Modulation Theory

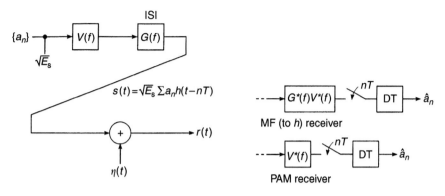

Figure 2.29 Generation and detection of linear modulation signals in linear ISI, showing the MF and PAM Receiver options. $\eta(t)$ is additive noise; DT refers to discrete-time processing of some kind.

To complete the modeling, we need to specify the receive end. In a communication link with ISI, much depends on whether the receiver has a description of the ISI. If it does, it may still be unwilling to change part of its architecture in response; for example, the receiver filter may continue to be matched to an undistorted orthogonal pulse. For the case when the receiver can configure itself as an ML receiver, we should recall Section 2.4, which showed that any set of N signal sequences may be ML-detected by a bank of N matched filters. This, however, is essentially analog processing; only the "Select Largest" decision in Fig. 2.10 is a discrete operation. But in this section, discrete-time modeling is taken to mean working with sequences of $1/T$-rate samples.

There are two natural ways that a $1/T$-rate receiver model might arise. These are shown at the right side in Fig. 2.29. In the first, which we will call a *PAM Receiver Model*, the receive filter is simply matched to the original PAM (or QAM) orthogonal pulse $v(t)$. ISI occurs but the modem carries on with the original matched filter. Its matched filter/sampler circuit produces a $1/T$-rate sample train, but even with a noise free channel, the samples are not the transmission symbols. The receiver may attempt to remove the ISI effect during later discrete-time processing; this is properly called *equalization* and is discussed in Chapter 6. Another application of the PAM Receiver is to the PRS coded signal (2.8-3); now the output is directly the sequence c, plus IID noise variates.

In the second receiver in Fig. 2.29, the receive filter is matched to the entire response $h = g * v$ in Eq. (2.8-1), not just to v; that is, its transform is $G^*(f)V^*(f)$. The filter produces as before a $1/T$-rate sample sequence. We will call this an *MF* model to distinguish it from the PAM model. There exists a processing of the MF model samples which performs ML detection of the $\{a_n\}$; this is not necessarily true with the PAM Model samples. The two receivers can create distinctly different discrete-time models and we now give some of the details.

The PAM Receiver Model $H_{\text{PAM}}(z)$

The impulse response of the entire analog transmit/receive system is $h(t) * v(-t)$, in which $v(t)$ is an orthogonal response and both v and h have unit energy. Let the value of this at time nT be denoted

$$\gamma_n = \int h(\tau) v(\tau - nT) \, d\tau, \quad \text{all } n. \tag{2.8-4}$$

The PAM sampler output in Fig. 2.29 in terms of the symbols $\{a_n\}$ is then

$$r_n = \sum_i a_n \gamma_{n-i} + \eta_n \tag{2.8-5}$$

in which the $\{\eta_n\}$ are IID zero-mean Gaussians with variance $N_0/2$. As a z-transform, the model is $Y_{\text{PAM}}(z) = H_{\text{PAM}}(z) A(z)$ plus noise, where $H_{\text{PAM}}(z) = \sum \gamma_k z^{-k}$ and $A(z) = \sum a_k z^{-k}$. Note that even with causal ISI and $g(t)$, γ_k can take values on negative k because the pulse $v(t)$ is in general noncausal.

The MF Receiver Model $H_{\text{MF}}(z)$

Now the entire analog chain has impulse response $h(t) * v(-t) * g(-t)$ and the sampler outputs are

$$r_n = \sum_i a_n \rho_{n-i} + \xi_n \tag{2.8-6}$$

in which $\{\xi_n\}$ is the response to noise and

$$\rho_k = \int h(\tau + kT) h^*(\tau) \, d\tau, \quad \text{all } k. \tag{2.8-7}$$

The sequence $\{\rho_k\}$ are thus samples of the autocorrelation of $h(\ )$. The noise components ξ_n of the sampler outputs under AWGN are no longer IID; it can be shown that the mean and variance of the ξ_n are 0 and $N_0/2$ but its autocorrelation is $\mathcal{E}[\xi_{n+k} \xi_n^*] = (N_0/2) \rho_k$ (a unit-energy h is assumed). The z-transform model is thus $Y_{\text{MF}}(z) = H_{\text{MF}}(z) A(z)$ plus noise, where $H_{\text{MF}}(z) = \sum \rho_k z^{-k}$.

Example 2.8-1 *(Models for Standard Butterworth Channel Filtering).* Detailed examples of these models, together with MATLAB calculation routines, are given in [1]. Here we briefly explore the case when the filter $G(f)$ is a 4-pole Butterworth filter with cut-off at $0.5/T$ Hz and $V(f)$ is a standard root RC pulse with 30% excess bandwidth. This represents a common PSK or QAM transmission with a relatively cheap bandlimitation filter and an RF bandwidth of about 1 Hz-s/transmission symbol. Figure 2.30 shows the basic pulse $v(t)$, plotted from

Modulation Theory

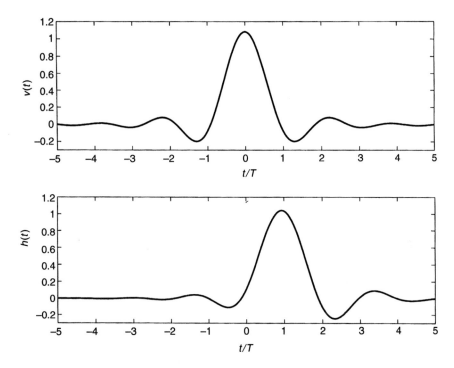

Figure 2.30 Four-pole Butterworth filter Example 2.8-1. 30% root-RC orthogonal pulse v and the total response $h = v * g$.

Eq. (2.2-7), and $h = v * g$.[11] In the calculation, $v(t)$ is truncated to the interval $[-5T, 5T]$, a limitation in line with practice; a narrower support generates unacceptable spectral sidelobes.

Figure 2.31(a) and (b) show the PAM and MF models $\{\gamma_n\}$ and $\{\rho_n\}$ as small circles. The models are truncated to $[-5T, 5T]$. The continuous curve is the full analog receive filter response to a unit symbol transmitted at time 0. The example here may be compared to the 6-pole Butterworth example in [1]. Several points need to be stressed. First, the models obviously differ. In particular, the MF model is symmetrical about time 0, and unlike the PAM model, the MF model has significant response before 0. Second, both models depend on the precise shape of the orthogonal pulse; were there no ISI, the models would be the same for all orthogonal pulses, namely, a single unit sample at time 0. The discussion of this model is continued in Example 6.2-2.

---

[11] $g(t)$ may be created by the MATLAB function $[b, a]$ = butter(4, 0.5 * 2/f_s), followed by g = filter(b, a, [1, zeros(1, 200)]). f_s is the number of samples of $g(t)$ per symbol time.

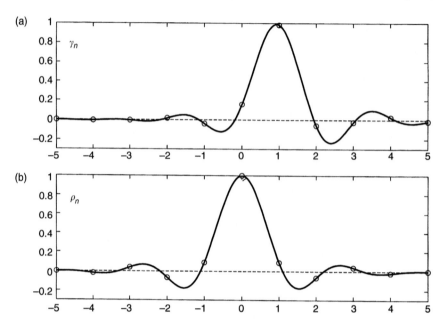

Figure 2.31 Example 2.8-1 continued. (a) (*top*), PAM receiver model $\{\gamma_n\}$, and (b) (*bottom*), MF receiver model $\{\rho_n\}$. "o" shows model; solid curve shows receive filter continuous output.

2.9. Problems

1. (*a*) Prove that $\sqrt{1/T}\ \text{sinc}(t/T)$ has unit energy. (*b*) Prove that $\sqrt{2/T}\ \text{sinc}(t/T)\cos(\omega_0 t)$ has unit energy, in the limit as $\omega_0 \to \infty$. (*c*) Show that BPSK basis function $\sqrt{2}v(t)\cos\omega_0 t$ has unit energy as $\omega_0 \to \infty$, if and only if $v(t)$ does.

2. Show that the RC pulse in Eq. (2.2-2) satisfies the Nyquist Pulse Criterion (Definition 2.2-1) without using Theorem 2.2-1.

3. Show that the ML decision boundaries for equal-energy signals always pass through or terminate at the origin of the signal constellation.

4. Suppose an integrate-and-dump receiver is used to detect the basic orthogonal pulse

$$v(t) = \begin{cases} e^{-t/T}, & -T/2 < t \leq T/2 \\ 0, & \text{otherwise} \end{cases} \quad (2.9\text{-}1)$$

Use signal space theory to derive the error probability in terms of E_b/N_0, for the case of binary transmission.

5. Suppose that orthogonal pulse 16PSK is transmitted and the basic linear receiver of Figs 2.2 and 2.13 is used for detection with the proper matched filter and sampler. Design the replacement for the threshold detectors, which will give ML detection of the 16PSK.

Modulation Theory

6. Verify by direct integration that E_s in Eq. (2.5-2) is the symbol energy as $\omega_0 \to \infty$.

7. Verify by direct integration that the $I(t)$ and $Q(t)$ given in Eq. (2.5-8) satisfy the normalizing condition (2.5-5).

8. Derive the probability expression (2.5-11) for QPSK. (*Hint*: Find the probabilities separately for the horizontal and vertical components in the four-point constellation.)

9. (*a*) By minimum distance calculations, show that compared to antipodal signaling, the asymptotic loss in E_b/N_0 for 8 PSK is 3.6 dB. (*b*) Find the loss of rectangular 64QAM.

10. Consider the rectangular 16QAM of Fig. 2.18(b). Derive the integral expressions that give the precise error probability of a transmitted point in AWGN. There are three cases: interval point, edge point, and corner point.

11. When the V.29 modem constellation in Fig. 2.18(a) encounters a poor channel, the 16-point constellation is reduced to either the innermost eight points or the innermost four points. By means of minimum distance calculations, give Q-function estimates of the error probability in all three cases. Estimate the error probability when $E_s/N_0 = 10$ dB.

12. Derive formula (2.5-25) for the minimum distance between two constant envelope signals, when $\omega_0 \to \infty$.

13. Consider a simple first-order PLL with gain K. Plot the error response $e(t)$ for the two excitations shown in Fig. 2.22, for $B_N = 0.2$ and $\zeta = 0.71$. Do this for several K and compare to the second-order case in the figure. Repeat for $B_N = 0.05$ and $\zeta = 0.71$.

14. Find the power spectral density for QPSK that uses the drooping pulse in Problem 4. Compare the results to the NRZ-pulse spectrum.

15. Use formula (2.7-13) to investigate the binary CPFSK spectrum when h is 0.1 and 0.01. What happens as $h \to 0$? What happens to the minimum distance?

16. Perform again Example 2.8-1, but with a 2-pole Butterworth filter; that is, find the MF and PAM models. How does the weaker spectral rolloff change the models?

17. Find the PAM model for the channel with total response

$$h(t) = \begin{cases} e^{-t/T}, & t > 0; \\ 0, & \text{otherwise.} \end{cases} \quad (2.9\text{-}2)$$

How does the answer differ from the PAM model for Problem 4? Why?

Bibliography

References marked with an asterix are recommended as supplementary reading.

[1] *J. B. Anderson, *Digital Transmission Engineering*. IEEE Press, New York, 1999.
[2] *J. G. Gibson, *Principles of Analog and Digital Communications*, 2nd Edn. Macmillan, New York, 1993.
[3] H. Nyquist, "Certain factors affecting telegraph speed," *Bell System Tech. J.*, 324–346, April 1924.
[4] H. Nyquist, "Certain topics on telegraph transmission theory," *Trans. AIEE*, **47**, 617–644, 1928.
[5] R. A. Gibby and J. W. Smith, "Some extensions of Nyquist's telegraph transmission theory," *Bell System Tech. J.*, **44**, 1487–1510, 1965.

[6] *J. G. Proakis, *Digital Communications*, 4th Edn. McGraw-Hill, New York, 2001.
[7] V. A. Kotelnikov, "The theory of optimum noise immunity," Ph.D. Thesis, Molotov Energy Institute, Moscow, January 1947; available under the same name from Dover Books, New York, 1968 (R.A. Silverman, translator).
[8] J. M. Wozencraft and I. M. Jacobs, *Principles of Communication Engineering*, Wiley, New York, 1965.
[9] C. E. Shannon, "Communication in the presence of noise," *Proc. IRE*, **37**, 10–21, 1949; reprinted in Claude Elwood Shannon: Collected Papers, N. J. A. Sloane and A. D. Wyner, eds., IEEE Press, New York, 1993.
[10] H. L. van Trees, *Detection, Estimation, and Modulation Theory*, Part I. Wiley, 1968.
[11] A. Papoulis, *Probability, Random Variables, and Stochastic Processes*, 2nd Edn. McGraw-Hill, New York, 1984.
[12] W.B. Davenport, Jr. and W.L. Root, *Random Signals and Noise*. McGraw-Hill, New York, 1958.
[13] *M. Schwartz, *Information, Transmission, Modulation and Noise*, 4th Edn. McGraw-Hill, New York, 1990.
[14] J. B. Anderson, T. Aulin and C.-E. Sundberg, *Digital Phase Modulation*. Plenum, New York, 1986.
[15] M. K. Simon, S. M. Hinedi and W. C. Lindsey, *Digital Communication Techniques*. Prentice-Hall, Englewood Cliffs, NJ, 1995.
[16] W. Webb and L. Hanso, *Quadrature Amplitude Modulation*. Pentech Press, London/IEEE Press, Piscataway, NJ, 1994.
[17] F. Gardner, *Phase-Lock Techniques*, 2nd Edn. Wiley, New York, 1979.
[18] T. J. Baker, "Asymptotic behavior of digital FM spectra," *IEEE Trans. Commun.*, **COM-22**, 1585–1594, 1974.
[19] R. E. Ziemer and R. L. Peterson, *Digital Communications and Spread Spectrum Systems*. Macmillan, New York, 1985.

3

Coding and Information Theory

3.1. Introduction

Just as Chapter 2 summarized modulation theory, we now review coding and information theory with an emphasis on what underlies the rest of the book. We will refine somewhat the ideas of data, channels and coding that were introduced in Chapter 1. A data source is a sequence of independent symbols in this chapter, and a channel converts these to another, possibly different, set of symbols. We will see in more concrete form how a channel code is a set of patterned sequences, each corresponding to a data sequence, and how the patterning allows errors to be recognized and corrected. Channel encoding and decoding are the study of schemes for creating these patterned sequences and for deciding the data therein after the channel has taken its toll. Coding involves signal distances and probabilities of error, as well as coding algorithms, their steps and complexity.

In a book about coded modulation it is worth repeating that with a binary channel the only way to impose a coded patterning on binary data is to add extra symbols to the transmission, symbols that we colloquially call parity checks. Section 3.2 introduces codes for this simple situation. Section 3.3 introduces an important way to view codes, the code trellis. The trellis idea also applies to most presently existing coded modulations, codes which are based only partly or not at all on parity checks. Since most of the codes in this book are most easily viewed as trellis codes, Section 3.4 introduces some basic trellis decoding algorithms. Other ideas in Section 3.4 are code concatenation, the combining of small codes into large ones, and iterative decoding. These concepts lie at the heart of many recent developments in coding.

The rest of Chapter 3 introduces some relevant parts of Shannon information theory, which is a subject quite different from coding. Shannon theory thinks of a data source not so much as symbols but as a probability distribution on the symbols; a channel likewise is a conditional distribution of the channel outputs given its inputs. Information is measured by a functional called entropy. The theory proves limit theorems about this measure before and after the channel and about what measure can flow through the channel. It is often difficult to infer hard conclusions from information theory about concrete encoders and decoders, but the theory's suggestions are provocative and roughly accurate, and this is the theory's appeal.

Information theory has three main divisions, a theory of channel transmission, of information in data sources and a theory of rate distortion, which is about the transmission of information when imperfect reproduction is allowed. Only the first division is of major interest in this book. We will focus expecially on the Gaussian channel models that underlie coded modulation.

3.2. Parity-check Codes

The chapter begins with codes that work by adding parity-check symbols to the data symbols. We will introduce only a few simple codes, but these nonetheless find direct use later in the book. The structure of more advanced parity-check codes is based on the algebraic theory of finite fields, a major topic that we cannot cover here. Fortunately, these details can usually be divorced from the rest of a coded modulation system. When not, we can recommend [2–4]. The section ends with the notion of a soft channel.

3.2.1. Parity-check Basics

We consider binary channel codes that encode a sequence $u^{(K)}$ of K data bits into a codeword $x^{(N)}$ of length N bits.[1] We will assume for now that the K data bits are gathered together and form the first K bits of the word; the last $N - K$ bits are the added parity-check bits. A code such as this, in which the data bits explicitly appear, is called a *systematic* code. The *rate* of this code, in data bits per channel use, is K/N.

When a codeword passes through a *binary symmetric channel* (BSC), the channel inverts some of the bits of x at random to form the length-N received vector y. In a standard BSC, these errors occur independently with probability p. We can write $y = x + e$, in which the addition is bit-wise mod-2 and e is another length-N sequence called the error sequence. An error in the jth position means that $e[j] = 1$; otherwise, $e[j] = 0$.

In analogy to the maximum likelihood (ML) receiver in Section 2.3, it is possible to derive an ML receiver for the BSC. As always, the receiver should find the maximum over i of the probability $P[y \mid x_i]$, where x_i denotes the ith of a set of words under consideration. Starting from Eq. (2.3-2) and adapting the steps thereafter, we get the analog to Eq. (2.3-4), that

$$P[y \mid x_i] = (1 - p)^{N - d_i} p^{d_i}, \quad p < 1/2, \quad (3.2\text{-}1)$$

---

[1] $x^{(N)}$ denotes a row vector of length N; x' denotes a column vector; x_i denotes the ith of several such vectors. $x[j]$ denotes the jth component of a sequence or vector x.

Coding and Information Theory

in which d_i is the number of bits in y and x_i that differ. This is simply the number of 1s in the sequence $y + x_i$, the *Hamming distance*, denoted $h_D(y, x_i)$. It is equivalent to maximize the log of Eq. (3.2-1) over i, which yields

$$\text{Find } i \text{ that achieves: } \max_i (N - d_i) \log(1 - p) + d_i \log p.$$

By eliminating $N \log(1 - p)$ and reversing the sign, this becomes

$$\text{Find } i \text{ that achieves: } \min_i d_i \log[(1 - p)/p]. \qquad (3.2\text{-}2)$$

For $p < 1/2$, this boils down to minimizing d_i, the Hamming distance $h_D(y, x_i)$ to y.

When the source probabilities for each x are available, it becomes possible to specify an MAP receiver. In analogy to Eq. (2.3-3), the receiver executes

$$\text{Find } i \text{ that achieves: } \max_i (1 - p)^{N-d_i} p^{d_i} P[x[i]],$$

which by the steps above becomes

$$\text{Find } i \text{ that achieves: } \min_i d_i \log[(1 - p)/p] - \log P[x[i]]. \qquad (3.2\text{-}3)$$

As $p \to 0$, the choice in Eq. (3.2-3) becomes the ML choice in Eq. (3.2-2).

In any real transmission system, a modulator somehow underlies the BSC, which is just saying that symbols have material existence when they are transmitted. A simple situation would be that antipodal signaling – or for carrier modulation, either BPSK or half of a QPSK – underlies the BSC. In these cases, p in a BSC is $Q(\sqrt{2E_b/N_0})$. It is common to plot error rate for a BSC decoder against the E_b/N_0 obtained from p through this equality. On the other hand, little is known sometimes about the underlying channel, and there may not even be a definable channel error probability. In such cases, one can still employ a minimum Hamming distance decoder to correct errors without, of course, defining a decoder error probability. Probabilities of any kind are not actually required in what follows.

The simplest parity-check code is the *repetition* code, that transmits one data bit per codeword by simply repeating the bit N times. Clearly, a simple majority decision will correct up to $\lfloor N/2 \rfloor$ transmission errors. The code rate is $1/N$. The opposite of a repetition code is a *single parity-check* code, which carries $N - 1$ data bits in each word, with the Nth a parity bit chosen so that the bits sum mod-2 to zero. This code cannot correct errors, but it can detect the presence of any odd number; in particular, it can detect a single error out of N, which with a short block length and a reasonable channel, can greatly reduce the undetected error rate. The rate is $(N - 1)/N$.

Effective codes most of the time need to avoid the extremes of these two codes. A parity-check code does this by defining the set of codewords to be the null space of a matrix, that is, the row vector x is a codeword if

$$xH = 0. \qquad (3.2\text{-}4)$$

Here H is an $N \times N - K$ binary matrix, $\mathbf{0}$ is the length $N - K$ all-zero row vector, and arithmetic is over the field of integers mod-2. In a single parity-check code, $H = (11\ldots 1)'$. For a systematic code with the first K bits in x being the data bits, H will have the form

$$H = \begin{bmatrix} P \\ I_{N-K} \end{bmatrix}, \qquad (3.2\text{-}5)$$

where I_{N-K} is the identity matrix of size $N - K$ and the $K \times N - K$ matrix P is what we are free to choose. Equation (3.2-4) expresses in matrix form a set of $N - K$ parity-check equations, all of which must be satisfied by x if x is a codeword.

The solutions to Eq. (3.2-4) form an algebraic group, since the bitwise mod-2 sum of any two solutions to Eq. (3.2-4) is itself a solution.[2] For this reason parity-check codes are called *group* codes, or more often, *linear* codes. The properties of the group so formed are the key to the decoder design for the different types of linear codes. The minimum distance of a linear code is in analogy to Section 2.3 the least Hamming distance between any pair of words. For word $\mathbf{0}$, the closest word is the one with the least weight (i.e. the fewest 1s). Call this one w and its weight d_w; since $x + w$ is a word lying this distance from word x, it is clear that any word has a neighbor at the same distance and no closer. d_w is, in fact, $d_{\min}$, the minimum distance between words that occurs in the whole set of codewords. Further arguments like this show that all words in a linear code have the identical constellation of neighbors.

From the triangle inequality, a linear code with a minimum distance decoder can correct all weight-t bit error patterns only if $t < d_{\min}/2$. When channel error sequence e occurs, the received y is $x + e$, and Eq. (3.2-4) becomes

$$yH = (x + e)H = \mathbf{0} + eH. \qquad (3.2\text{-}6)$$

eH is called the *syndrome* of x. If e consists of a single 1 in the jth place, the syndrome is the jth row of H; if e indicates errors in several places, eH is the sum of the respective rows of H. If all the rows of H are distinct, it is clear that all single errors in x can be distinguished by their syndromes and corrected. More generally, if every combination of t rows is distinct, then any combination of t errors can be corrected. A decoder that works by finding the syndrome yH, mapping to a set of error positions, and correcting these positions, is called a *syndrome* decoder. Such a simple table look-up procedure is practical for short codes. Much research has gone into finding less bulky procedures that are based on properties of the code group.

When $yH = \mathbf{0}$, y is already a codeword, and is usually the one sent. It will not be so if e itself is a codeword. When eH is not $\mathbf{0}$, an ML decoder,

---

[2] One needs to show as well that the codeword $\mathbf{0}$ is the identity element, that each element has an inverse (namely, itself) and that the addition is commutative.

in accordance with Eq. (3.2-2), will decide in favor of the codeword closest to y. In a syndrome decoder, it is this error pattern to which the syndrome is mapped. The syndrome decoder cannot guarantee correction of more than $\lfloor d_{\min}/2 \rfloor$ errors, because no decoder based on minimum distance can do so, but it remains to be shown that it really corrects up to t. This follows from the fact that $x_{\min} H = 0$ for the minimum weight codeword $x_{\min}$. Consequently, $d_{\min}$ rows of H sum to 0, but no fewer rows do, since then the weight of $x_{\min}$ would then be less. Thus, every sum of t or fewer rows must be unique, which implies that every e with t or fewer errors leads to its own syndrome.

The Hamming codes are a simple but effective class of codes that correct all single errors in the codeword. By definition, the rows of H for a Hamming code consist of all the length $N - K$ words except 0. There are $2^{N-K} - 1$ of these and so the length of a codeword is $N = 2^{N-K} - 1$. Each row of H is the syndrome for one of the N single-bit errors. Hamming codes have rates of the form $R = (2^\ell - 1 - \ell)/(2^\ell - 1)$, for $\ell = 2, 3, \ldots$

Example 3.2-1 *(The Length-7 Hamming Code).* The first non-trivial code is the one with $N - K = 3$, which creates the code and H shown in Fig. 3.1. The 16 words that satisfy $xH = 0$ are listed. The code is a systematic version of this Hamming code, with the first four bits equal to the data bits; H with a different arrangement of the rows will lead to a different version of the same code. The syndrome $(x+e)H = 110$ occurs when there is an error in the first place; although 110 also occurs when there are errors in places 2 and 4, the pattern $e = 1000000$ is the most likely one. All Hamming codes have minimum distance 3, which is clear in the figure for length 7.

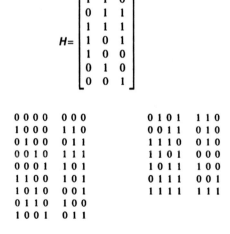

Figure 3.1 Parity-check matrix H and the codeword set for the systematic (7, 4) Hamming code. The first 4 bits in each word are the data.

A standard notation for block parity-check codes is the form (N, K), which means codeword length N and K data bits. The above example is a $(7, 4)$ code.

An alternate way to define a parity-check code is by its generator matrix, the $K \times N$ matrix that maps the data bit word $\boldsymbol{u}$ to a codeword $\boldsymbol{x}$ according to

$$\boldsymbol{x} = \boldsymbol{u}\boldsymbol{G}. \qquad (3.2\text{-}7)$$

For a systematic code, $\boldsymbol{G}$ has the form

$$\boldsymbol{G} = [\boldsymbol{I}_K \mid \boldsymbol{P}], \qquad (3.2\text{-}8)$$

where $\boldsymbol{P}$ is the $K \times (N - K)$ matrix in Eq. (3.2-5). From the form of Eq. (3.2-7), it must be that each row of $\boldsymbol{G}$ is a codeword, and the code must consist of $\boldsymbol{0}$ plus all $2^K - 1$ sums of the rows of $\boldsymbol{G}$. Since $\boldsymbol{x}\boldsymbol{H} = \boldsymbol{0}$, it must be that $\boldsymbol{u}\boldsymbol{G}\boldsymbol{H} = \boldsymbol{0}$ for all $\boldsymbol{u}$; consequently, it must be that $\boldsymbol{G}\boldsymbol{H} = \boldsymbol{0}$, with $\boldsymbol{0}$ here the $K \times (N - K)$ all-zero matrix. We could equally well write this as $(\boldsymbol{G}\boldsymbol{H})^{\text{tr}} = \boldsymbol{H}^{\text{tr}}\boldsymbol{G}^{\text{tr}} = \boldsymbol{0}$, which is an expression of the fact that $\boldsymbol{H}^{\text{tr}}$ can be a generator matrix and $\boldsymbol{G}^{\text{tr}}$ can be the parity-check matrix, for a length-N code with the new rate $(N - K)/N$. This code, which encodes $N - K$ data bits rather than K, is said to be the *dual code* to the one generated by $\boldsymbol{G}$.

3.2.2. BCH and Reed–Solomon Codes

These important codes are members of a large subclass of the linear codes called *cyclic* codes. These are codes whose words are all cyclic shifts of each other. The $(7, 4)$ Hamming code in Fig. 3.1 is one such code. Words number 2, 3, 7, 13, 8, 11 and 5 form a set of right shifts; there is one other seven-member set of shifts, and the remaining words 1111111 and 0000000 are shifts of themselves. It can be shown that for a cyclic code there exists a generator matrix whose rows are cyclic shifts as well. This new matrix may not set up the same correspondence between data words and codewords, but the overall set of codewords will be the same, and the code is therefore taken as equivalent.

The cyclic property, as well as many of the mechanics of cyclic codes, are easier to express with the delay polynomial notation. Consider the word $\boldsymbol{x} = x[N-1], x[N-2], \ldots, x[0]$, with bit number 0 taken as the rightmost one; its polynomial notation is given by

$$x(D) = \sum_{i=0}^{N-1} x[i]D^i, \qquad (3.2\text{-}9)$$

in which the polynomial variable D can be thought of also as signifying a delay of one symbol position. Yet another way to generate the words in a code if it is cyclic is by the polynomial multiplication $u(D)g(D)$, in which $u(D)$ represents the data word u, $g(D)$ is the first of the rows in the generator matrix, and operations on coefficients are mod-2. Some investigation shows that because the rows of G can be taken as cyclic shifts, the matrix operation $x = uG$ in Eq. (3.2-7) can indeed be replaced by $x(D) = u(D)g(D)$. $g(D)$ in the (7, 4) code, for example, is $D^3 + D^2 + 1$, which is the row vector 0001101; for the data word $u = 0010$, represented by the polynomial D, the polynomial product $u(D)g(D)$ is $D^4 + D^3 + D$, which is the codeword 0011010. This codeword corresponds to data word 0011 in the Fig. 3.1 list, which is not the u we have just taken, but the $u(D)g(D)$ process will eventually generate the same list if continued through all possible $u(D)$. The operation $u(D)g(D)$ has terms beyond the power $N - 1$, and we emulate the cyclic shift through the requirement that words of order greater than $N - 1$ be taken as

$$x(D) = u(D)g(D) \bmod(D^N - 1), \qquad (3.2\text{-}10)$$

that is, $x(D)$ is the remainder when $u(D)g(D)$ is divided by $D^N - 1$. It can be shown that $g(D)$ generates a cyclic code if and only if $g(D)$ is a factor of $D^N - 1$, and further that $g(D)$ is the unique nonzero codeword polynomial of least degree in the code. A rich coding theory grows out of these facts and leads to many clever decoders [2–4].

The BCH and Reed–Solomon codes are cyclic codes for which $g(D)$ breaks into factors taken from a certain kind of polynomials that stem from the theory of finite groups. These and the final $g(D)$ are listed in algebraic coding theory texts. In BCH codes the polynomials and the $g(D)$ take on binary coefficients, while RS code polynomial coefficients are non-binary. For any $m > 2$, there exists an (N, K) BCH code that corrects t errors, with $N = 2^m - 1$ and $K \geq 2^m - 1 - mt$. BCH codes are quasi-perfect, which means that for a given number 2^{N-K} of syndromes, they correct all patterns of t or fewer errors, plus a set of patterns of $t + 1$, and no others. For the BSC with $p < 1/2$, no code of length N and rate K/N has lower word error probability than such a code. The Hamming (7, 4) code is a BCH code with $m = 3$ and $t = 1$. A double error correcting (15, 7) BCH code (with $m = 4$ and $t = 2$) is generated by $g(D) = D^8 + D^7 + D^6 + D^4 + 1$; its 256 syndromes can be used to correct all 121 patterns of two or fewer errors, as well as 135 three-error patterns.

In RS codes, the coefficients of $x(D)$, $g(D)$ and the data $u(D)$ are taken as non-binary, and in fact are usually from alphabets of size 2^ℓ. In a way parallel to BCH codes, RS codes correct t or fewer 2^ℓ-ary symbol errors. As an example, there exists a (15, 9) RS code based on hexadecimal symbols, that corrects three symbol errors. The generator is

$$g(D) = D^6 + 7D^5 + 9D^4 + 3D^3 + cD^2 + aD + a,$$

where the hexadecimal coefficients are denoted in the usual way $0, 1, 2, \ldots,$ $a, b, \ldots, f$. Arithmetic rules between these symbols need also to be specified during the code construction. The code encodes nine hexadecimals, or 36 bits, into words of length 15 symbols (60 bits).

3.2.3. Decoding Performance and Coding Gain

Usually, but not always, decoder performance means probability of error. When probabilities cannot be defined, then the number of errors corrected is the performance measure. Several bounds on N, K and d can be constructed from the combinatorics and the algebra that govern parity-check codes. The most important of these is the Gilbert–Varsharmov bound, which is based on the fact that no combination of $d_{min} - 1$ or fewer rows of H may sum to zero. It can be shown [2–5] that this implies

$$\sum_{i=0}^{d_{min}-2} \binom{N-1}{i} < 2^{N-K}, \tag{3.2-11}$$

which is the precise form of the bound. It gives a limit on d_{min} in terms of N and K, or alternately, N and the rate R.

An interesting asymptotic form results when N and K become large at a fixed rate K/N. A result of combinatorics states that for $d_{min} < N/2$,

$$(1/N) \log \left[\sum_{i=0}^{d_{min}-2} \binom{N-1}{i} \right] \to h_B(d_{min}/N) \tag{3.2-12}$$

as $N \to \infty$, where $h_B(\ )$ is the binary entropy function (for a plot of this function, see Example 3.6-1). Thus, Eq. (3.2-12) becomes in the limit $h_B(d_{min}/N) < 1 - K/N = 1 - R$. Since $h_B(\ )$ is monotone in the range of interest, we have

$$d_{min}/N < h_B^{-1}(1 - R). \tag{3.2-13}$$

This is a bound on the achievable minimum distance as a fraction of block length.

When the channel is a BSC with a definable crossover p, the p and the capacity satisfy $p = h_B^{-1}(1 - C)$, as we will discuss in Section 3.6. Consider coding at a rate R close to C in this channel. On the average there will be about pN channel errors in each length-N codeword, and yet Eq. (3.2-13) states that d_{min} is less than $Nh_B^{-1}(1-R)$, which is about Np. Thus no parity-check code exists that *always* corrects even half the number of errors that we expect to occur! Fortunately, parity-check codes correct many error patterns at weights above $d_{min}/2$, even if they do not correct all of them. Both in theory and in practice, their error probability is set by how far beyond $d_{min}/2$ they continue to correct most errors.

Coding and Information Theory

When a full BSC with crossover p can be defined, one can compute an overbound to codeword error probability for a t-error-correcting code as

$$P_W \leq \sum_{n=t+1}^{N} \binom{N}{n} p^n (1-p)^{N-n} \qquad (3.2\text{-}14)$$

This estimate assumes that $t + 1$ or more channel errors always lead to a decoder error, which will be true for at least one such error pattern, but may not be true for too many others, particularly if the patterns have just $t + 1$ errors. BCH and Hamming codes correct all the t-error patterns, possibly some at $t + 1$, and no others; however, the convolutional codes in the next section correct many longer patterns. The only way to refine Eq. (3.2-14) perfectly is to enumerate the correctable patterns.

For a fixed N and t, as $p \to 0$ the log of the right side of Eq. (3.2-14) tends to $(t + 1) \log p$. The asymptotic form of P_W is thus[3]

$$p_W \sim p^{t+1}. \qquad (3.2\text{-}15)$$

When the decoder decides an incorrect word, data bit errors must necessarily occur, but there is no set rule about how the data bit error probability P_b compares to P_W. Obviously, $P_b \leq P_W$. Once again, an enumeration of the cases is needed and one such technique is the transfer function method (see [5,11,19]).

Since the output of a BSC is one of two symbols (namely, the two that can be sent), it is called a *hard decision channel*. A standard way to view the gain from a code over the BSC is to assume that an AWGN channel and a binary antipodal modulator underlie the BSC. Therefore, $p = Q(\sqrt{2E_s/N_0})$, in which E_s is the modulator's symbol energy. Since the code carries R data bits/channel use, our usual energy per data bit must be $E_b^c = E_s/R$, where the superscript c is a reminder that the transmission is coded. Substitution for E_s then gives $p = Q(\sqrt{2RE_b^c/N_0})$, and the tight approximation to $Q(\ )$ in Eq. (2.3-18) converts this asymptotically as $p \to 0$ to

$$p \sim (1/2) \exp(-RE_b^c/N_0). \qquad (3.2\text{-}16)$$

If we combine Eqs (3.2-15) and (3.2-16), and take $P_b = P_W$, then as $p \to 0$ the log of the data bit error probability P_b tends to $(t + 1)(-RE_b^c/N_0) \log e$.

Without coding, P_b is simply $Q(\sqrt{2E_b^{uc}/N_0})$, whose log tends to $-(E_b^{uc}/N_0) \log e$, where uc means uncoded. If we equate these two log probabilities, we obtain the asymptotic ratio E_b^c/E_b^{uc} that is required to maintain the same log probability; in dB it is

$$G_h = 10 \log_{10} R(t+1) \text{ dB}. \qquad (3.2\text{-}17)$$

---

[3] See Section 2.3.2 for the technical definition of $\sim$.

This is the asymptotic AWGN *hard decision coding gain*. It shows how much the uncoded system energy must be increased in a good channel in order to maintain the same error probability.

A *soft decision channel* takes in transmission symbols (binary here) but puts out symbols in a larger alphabet, for instance, over the real numbers. A standard way to view coding gain in this case is to once again assume binary antipodal signaling in AWGN, but now with real inputs to the decoder. One unit of Hamming distance between binary codewords translates to a Euclidean square distance $4E_s$; the minimum distance between codewords becomes $d_{min}4E_s$. When normalized to the data bit energy, the same distance becomes

$$4E_s d_{min}/(2E_s/R) = 2Rd_{min}. \tag{3.2-18}$$

The probability of deciding a word as its nearest neighbor is then from Eq. (2.3-22) the 2-signal probability $Q(\sqrt{2Rd_{min}E_b^c/N_0})$. Without coding, the antipodal signaling error probability is $Q(\sqrt{2E_b^{uc}/N_0})$ as usual. Following the same steps as for the hard channel case, we obtain the ratio

$$G_s = 10\log_{10} Rd_{min} \approx 10\log_{10} 2R(t+1/2) \text{ dB}. \tag{3.2-19}$$

This is the asymptotic *soft decision coding* gain for the AWGN channel. Comparing Eq. (3.2-17) with Eq. (3.2-19) shows that *soft decision leads in theory to a 3 dB coding gain*.

In Section 3.4, we will introduce the idea of a soft decision *decoder*, and it is easy to confuse this notion with a soft-output channel. The soft decoder puts out a probability that its data bit decisions are correct, together with the bits. To reiterate, the soft-output channel puts out a higher alphabet variable, which a decoder, itself either hard or soft, can use to make better decisions.

3.3. Trellis Codes

We now turn to codes whose words have a regularly repeating structure called a trellis. The trellis of a code closely relates to the Viterbi decoder for the code, but the notions of paths, states and branches in a code trellis also provide a language to describe other kinds of decoders. We will begin with convolutional codes, a class of parity-check codes that are also trellis codes. However, continuous-phase modulation (CPM) and partial response signaling (PRS) codes are not in any way parity check. It should be mentioned that all practical codes have some sort of trellis structure, and the study of these structures is an interesting subject, but in this book "trellis" codes will mean those with a certain highly regular structure.

3.3.1. Convolutional Codes

Viewed as a parity-check code, a convolutional code consists of words x defined through a code generator matrix by $x = uG$. Here the arithmetic operations are mod-2, u is the data bit word and G is the generator matrix as usual but with the special form:

$$G = \begin{bmatrix} G_0 & G_1 & G_2 & \cdots & G_m & & & \\ & G_0 & G_1 & G_2 & \cdots & G_m & & \\ & & G_0 & G_1 & G_2 & \cdots & G_m & \\ & & & \cdots & & & & \\ & & & & \cdots & & & \end{bmatrix} \qquad (3.3\text{-}1)$$

in which the successive rows consist of $G_0 G_1 \cdots G_m$ shifted right by a fixed step for each row and the pattern is truncated on the right. The elements not shown are zeros. Take a code of rate $R = (\log_2 \beta)/c$, where c and β are integers; in practical codes β is a power of 2 and so it will be convenient to denote $\log_2 \beta$ as b and take the rate as b/c. The building blocks G_j in Eq. (3.3-1) are $b \times c$ submatrices. It will be easier to work first with the case $b = 1$, which yields codes with rates of the form $1/2, 1/3, 1/4, \ldots$. Then each of the building blocks G_j consists of the c bits in the row vector $\{G_j[1], G_j[2], \ldots, G_j[c]\}$, and there are $m + 1$ such building blocks. m is called the *memory* of the convolutional code, since from the form of $x = uG$, a span of $m + 1$ data bits affects a given code symbol.[4] m sets the code's complexity and strength.

The concepts are much easier to visualize through an example such as the following, which is a standard example of a short systematic code of rate 1/3.

Example 3.3-1 *(Memory 2 Rate 1/3 Convolutional Code).* The submatrices will be $G_0 = 111$, $G_1 = 011$, $G_2 = 001$, each with length $m + 1 = 3$. From Eq. (3.3-1), the generator matrix is the top of Fig. 3.2. A length-N codeword is given by $x^{(N)} = u^{(K)} G$. N and K may take any convenient values, so long as K/N equals the rate. For example, with $N = 15$ and $K = 5$, G is truncated to the 5×15 matrix

$$\begin{bmatrix} 111 & 011 & 001 & & \\ & 111 & 011 & 001 & \\ & & 111 & 011 & 001 \\ & & & 111 & 011 \\ & & & & 111 \end{bmatrix}. \qquad (3.3\text{-}2)$$

The data 10011 will produce the codeword $\underline{1}$11 $\underline{0}$11 $\underline{0}$01 $\underline{1}$11 $\underline{1}$00. (The spaces here and in Eq. (3.3-2) are inserted only for legibility.) The underlined bits, it can be seen, are the data bits; in fact, the corresponding columns in Eq. (3.3-2) taken by themselves form an identity matrix, and so the code is evidently systematic.

---

[4] An older term for memory is *constraint length*, by which is meant the quantity $m + 1$.

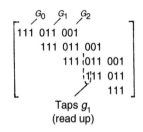

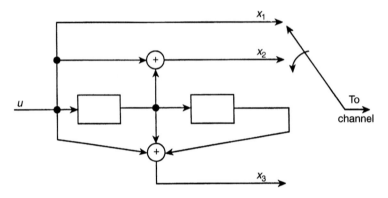

Figure 3.2 Rate 1/3 generator matrix G of width 15, for Example 3.3-1 (*top*); shift register encoder implementation (*bottom*).

The essence of the code generation in Eq. (3.3-2) is that codeword bits are produced in groups of c, groups of three in the example. The bottom of Fig. 3.2 is a circuit that produces the example code bits in a way that is considerably easier to follow than the generator method. The data bits enter a shift register on the left, and the code bits are generated by tap sets and leave on the right. The tap sets, or generators, as they are called, are themselves defined in vector notation as follows:

$$g_1 = \{G_0[1], G_1[1], \ldots, G_m[1]\}$$
$$g_2 = \{G_0[2], G_1[2], \ldots, G_m[2]\}$$
$$\ldots$$
$$g_c = \{G_0[c], G_1[c], \ldots, G_m[c]\}. \tag{3.3-3}$$

Here, $G_j[n]$ denotes the nth component of a $1 \times c$ submatrix G_j. The second generator, for example, consists of the second bits of each building block. The second tap set will consist of a tap at each '1' in this generator. In Example 3.3-1, the generators are $g_1 = 100$, $g_2 = 110$ and $g_3 = 111$. The top tap set implements g_1 and is just a single tap back to the data stream. It produces the first of each group of three codeword bits. Tap sets at the middle and bottom implement g_2 and g_3.

It will be convenient to denote the first tap set output sequence as x_1, the second as x_2, etc. up to x_c. The c sequences interleave to form the complete codeword.

Yet another way to generate convolutional codewords is by the simple formula

$$x_k[n] = \sum_{j=0}^{m} g_k[j]u[n-j], \qquad k = 1, \ldots, c. \tag{3.3-4}$$

In the notation here, $u[1]$ denotes the rightmost bit in the shift register, the oldest one, just as in the delay notation of Section 3.2.2. Each x_k equals the convolution $x_k = g_k * u$, a convolution of the kth generator with the data bits. In the succeeding chapters, we will exclusively use the convolutional form of Eq. (3.3-4). Generator vectors will be given in left-justified octal notation. For example, the rate 1/2 code with $g_1 = 1101$ and $g_2 = 1111$ has generators 64 and 74, and is called the (64,74) rate 1/2 code. The code in Example 3.3-1 is the (4, 6, 7) rate 1/3 code.

It remains to take the case of rates b/c, when $b > 2$. Now it is easiest to think of the data as arriving in blocks of b, which are de-interleaved into b streams, $u_1, \ldots, u_b$, as for instance in Fig. 3.3. Each stream feeds its own shift register, and taps on these registers feed a group of c summing junctions that produce the c output streams; any register can connect to any junction. Now there are as many as bc length $m+1$ generator sequences, which we can denote as $g_{k,\ell}$, for each $k = 1, \ldots, c$ and each $\ell = 1, \ldots, b$. The convolution form of Eq. (3.3-4) becomes

$$x_k[n] = \sum_{\ell=1}^{b} \sum_{j=0}^{m} g_{k,\ell}[j]u_\ell[n-j], \qquad k = 1, \ldots, c \tag{3.3-5}$$

with the x_k interleaved in the usual way to form the complete x. The G-matrix building blocks are now obtained from the generators by

$$G_j = \begin{bmatrix} g_{1,1}[j] & g_{2,1}[j] & \cdots & g_{c,1}[j] \\ g_{1,2}[j] & g_{2,2}[j] & \cdots & g_{c,2}[j] \\ & & \cdots & \\ & & \cdots & \\ g_{1,b}[j] & g_{2,b}[j] & \cdots & g_{c,b}[j] \end{bmatrix} \qquad j = 0, \ldots, m. \tag{3.3-6}$$

The notations here are easiest to learn by another standard example.

Example 3.3-2 *(Memory 2 Rate 2/3 Convolutional Code).* Take as generators the six sequences

$$g_{1,1} = 100, \qquad g_{2,1} = 010, \qquad g_{3,1} = 110,$$
$$g_{1,2} = 011, \qquad g_{2,2} = 000, \qquad g_{3,2} = 100.$$

The shift register circuit that produces the codewords appears at the bottom of Fig. 3.3. Commutators at the input and output de-interleave the input and interleave the output. Two bits enter for each three that leave. The second of each input

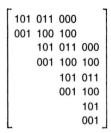

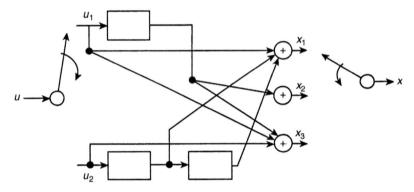

Figure 3.3 Rate 2/3 generator matrix G of width 12, for Example 3.3-2 (*top*); the shift register encoder implementation (*bottom*).

pair has no effect on the second of each output triple; this is signified by $g_{2,2} = 000$. Using Eq. (3.3-6), we can obtain the G-matrix shown at the top of the figure, when the block length is set to 12. Take the data input $u = 1001\,11\,00$. De-interleaved, this is $u_1 = 1010$ and $u_2 = 0110$. The output streams are $x_1 = 1000$, $x_2 = 0101$, $x_3 = 1001$; interleaving these gives $x = 101\,010\,000\,011$, which is the result of uG. (Here the first bit to enter or leave the encoder circuit is on the left.)

Immense literature exists about the properties of convolutional codes, a review of which can be found in [2–6], and [11] is a full length treatment. Our interest is mostly the concepts of minimum distance and error event, but these are easier to discuss in terms of a code trellis in Section 3.3.2. The references list codes with best minimum distance at each rate and memory m.

A systematic convolutional encoder produces the data bits explicitly as the first b bits of each output group of c. In the convolution formula (3.3-4), it is clear that g_1 must be $100\cdots 0$ in a rate $1/c$ systematic encoder. A generalization of this applies to Eq. (3.3-5). All of the convolutional encoders so far here have contained feedforward shift registers. It is also possible to generate feedback, or "recursive,"

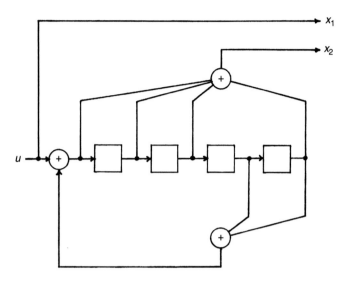

Figure 3.4 Shift register implementation of a rate 1/2 feedback systematic encoder (1 g_2/g_1), with $g_1 = 46$ and $g_2 = 72$.

convolutional codes by means of a feedback shift register like Fig. 3.4. The figure actually demonstrates a systematic recursive code with rate 1/2: The first bit in each output pair is directly the data bit and the shift register in the figure recursively generates the parity-check symbol.

The feedback register needs a little more explanation, and we will use as an aid the D-transform notation of Section 3.2.2. The register in Fig. 3.4 carries out the polynomial multiplication by $g_2(D)/g_1(D)$, in which the division is the polynomial kind, but the coefficients add and multiply by the mod-2 rule. The highest order term in $g_1(D)$ or $g_2(D)$ corresponds to the rightmost register tap. The division may carry on indefinitely, which is to say that the circuit has infinite unit response. The register has two tap sets, a feedforward one whose taps correspond to the 1s in g_2, and a feedback one whose taps are the 1s in g_1 ($g_1[0]$ must always be 1). With the corresponding tap sets, the circuit can carry out any $u(D)g_2(D)/g_1(D)$. A general rate $1/c$ feedback convolutional code would require c registers of the kind in Fig. 3.4.

A theory has developed that compares the different types of convolutional codes. We can summarize the important results briefly. For brevity, restrict the rate to $1/c$.

1. For a given rate R and memory m, nonsystematic feedback codes have in theory the largest distance. However, at short and medium m, few, if any, feedback codes are better than the best feedforward code at the same R and m. Thus, nonsystematic feedback codes are seldom used.

2. For any feedforward code, there exists a systematic feedforward encoder that generates the same set of words; however, its memory may be as large as N/c, the number of data symbols. If its memory is limited to that of the original code, the distance will be much worse.

3. For any feedforward code, there exists a systematic feedback code with *the same m*, that has the same codewords and therefore the same distance properties. The words in the two codes may correspond to different data sequences.

Since the feedforward code class includes essentially the best-distance codes, it follows that feedback is an important technique because it allows these best codes to be at the same time systematic. It can be shown that the feedback systematic code that corresponds to the rate $1/c$ feedforward code with generators $g_1(D), \ldots, g_c(D)$ has the generators

$$[1, g_2(D)/g_1(D), g_3(D)/g_1(D), \ldots, g_c(D)/g_1(D)] \qquad (3.3\text{-}7)$$

For example, the rate $1/2$ code $(46, 72)$, which has $g_1(D) = 1 + D^3 + D^4$ and $g_2(D) = 1 + D + D^2 + D^4$, has the equivalent systematic feedback code with generators 1 and $g_2(D)/g_1(D)$; the circuit for this is Fig. 3.4.

3.3.2. Code Trellises

Any realizable encoder is a finite-state machine (FSM) and a trellis is a graph of all possible state transversals against time. The trellis idea was suggested by G. D. Forney around 1970. The concept originally grew up as a way to visualize convolutional codes and we will start the discussion of trellises with these. The trellis concept leads directly to a basic decoder, the Viterbi algorithm (VA), which we take up in Section 3.4.

We begin with a FSM description of a convolutional encoder. The state of an encoder such as Fig. 3.2 can be defined at time n by the b-ary data symbols $u[n-1], u[n-2], \ldots, u[n-m]$. The output of the decoder (which is a c-tuple) depends on these symbols plus the present one, $u[n]$; that is, on the last $m+1$ data symbols. The trellis is created from the state transitions by plotting states vertically against time evolving to the right. Figure 3.5 shows the trellis of the machine in Fig. 3.2, assuming that the FSM begins in state 00. Nodes in the trellis represent states in the FSM. The trellis breaks down into *stages*. At each stage, two branches leave each state, one for each present data bit, and after two stages, two branches enter each state. After stage 2, the trellis is fully developed and repeats the same branch pattern indefinitely until the data stops. The branches each contain branch labels, which are the codeword c-tuples, and the data bits in parentheses drive the transition. The states are numbered in a binary coded

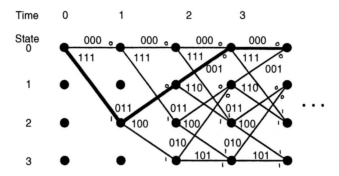

Figure 3.5 The code trellis for the rate 1/3 convolutional code generated in Fig. 3.2. Small figures are the data bits that cause each transition; large figures are branch labels.

decimal representation, with the least significant bit equal to the oldest bit in the shift register, $u[n-m]$.

Any path through the trellis spells out a codeword, and the whole trellis depicts all the possible words. As an example, the data bits $u[0], u[1], \ldots = 100\ldots$ produces the shaded path in Fig. 3.5. (Recall that bit $u[0]$ is the first to enter the encoder shift register.) The path rejoins the all-zero path, caused by $u = 00\ldots 0$, after three stages. The set of future paths is identical in front of these two three-branch codeword sections {000, 000, 000} and {111, 011, 001}; the two path sections are said to *merge*, in this case, to merge at state zero after stage 2.

The trellis makes it easy to visualize minimum distance. If convolutional codewords are of length N, which implies N/c trellis stages, the code minimum distance is the least nonzero weight of a path in the trellis. (We use here that the distance structure of the code is the same for all words, and assume that the all-zero codeword is sent.) The general problem of finding the least weight path through a trellis is solved by dynamic programming, which for a trellis is the VA. In finding the least weight path, we will consider only those that leave the all-zero path and later merge to it again; we will take care of the low weight paths at the far right in the trellis later. Some study of the example in Fig. 3.5 shows that among the paths that split and later merge, the least weight path has weight 6, no matter how long the trellis runs. The shaded path is one such path. A minimum distance has occurred rather soon in the trellis and no elongation will give a larger distance. This phenomenon occurs in general and the limit to distance is called the *free distance*, d_f. Not all trellis codes have the same distance structure around all their codewords, and so we give a more general definition as follows:

Definition 3.3-1. The free distance is the least distance that occurs between trellis codewords of unbounded length.

In a convolutional code trellis, the path pair that yields d_f lie separate at least $m + 1$ stages, but the separation is often somewhat longer. In general, the

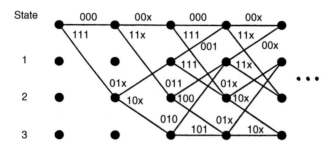

Figure 3.6 A punctured trellis code, using Fig. 3.5 as the mother code and a 2-stage puncture pattern that deletes every sixth bit. Bits denoted "X" are not transmitted.

dynamic program search for d_f must extend far beyond $m + 1$ in order to be sure of finding the minimum merged path pair.

With a convolutional feedforward encoder, the trellis state transitions are easily derived from the shift register contents and the present data symbol: if the present symbol is, for example binary 1, and the present state is 0110, then the next state is 1011, with the old present bit on the left. With recursive codes, the left bit is not necessarily the old present bit. The 2^m-state trellis can nonetheless be filled out by enumerating all the transitions, and the decoding, distance finding, etc., are then the same as always.

The trellis structure is convenient also for visualizing *punctured* codes. These are codes in which parity bits at certain times are simply deleted, or "punctured." Consider again the binary code in Fig. 3.5. If we simply ignore every sixth codeword bit, we have an encoder that puts out five bits for every two data bits, which is a rate 2/5 encoder. A section of this encoder trellis is shown in Fig. 3.6; the free distance is now reduced to five. The original code in Fig. 3.5 is called the mother, or parent, code and the every-sixth bit deletion scheme is called the puncture pattern. The pattern here extends over two trellis stages; other, two-stage patterns could puncture varying combinations of bits number 2, 3, 5 and 6, which would yield rates of 2/5, 2/4, 2/3, depending on how many were deleted. A set of codes like this that all stem from the same parent are called rate-compatible punctured codes. The set gives an easy way to shift the code rate in response to a change in channel quality, since the encoder and decoder for each code is the same except for which bits are to be ignored. Also, puncturing can provide high-rate convolutional codes, which are otherwise difficult to find.

Other Trellis Codes

The trellis structure notion is by no means limited to parity-check codes or to codes based on mod-2 arithmetic. We will close with examples of other kinds of trellis codes that appear in chapters to come.

Rather than mod-2 convolution, it is possible to construct trellis codewords by the real-number convolution

$$x[n] = \sum_{j=0}^{m} g[j]u[n-j]. \qquad (3.3\text{-}8)$$

This models intersymbol interference, ordinary linear filtering, and the PRS class of coded modulations, which are all discussed in Chapter 6. As a simple example, a trellis based on $g(D) = 1 + aD^2$ is shown[5] in Fig. 3.7; symbols in u take the values ± 1, and the branch labels are real numbers, and the encoder state $\{u[n-1]\,u[n-2]\}$ is one of the four two-tuples $\{-1-1, -1+1, +1-1, +1+1\}$, which are labeled respectively $\{0, 1, 2, 3\}$. The essential difference between this trellis and Fig. 3.5 is the arithmetic rule, and no more.

Another kind of trellis code is shown in Fig. 3.8. This is a CPM trellis, one that shows the codewords generated by Eq. (2.5-27) when the CPM coding is a kind called 3RC with modulation index $h = 1/2$. Now the data symbols in u are again ± 1, and the branch labels are excess phase transitions $\phi(t)$, where the entire signal is $s(t) = \sqrt{2E_s/T}\cos(\omega_0 t + \phi(t))$. A selection of these phase transitions is shown next to the trellis; respective transitions occur between the states that are shown. The states in this CPM trellis are enumerated by a phase that is a multiple of $2\pi/3$ radians and by the two previous data symbols. Many such CPM trellises are employed in Chapter 5.

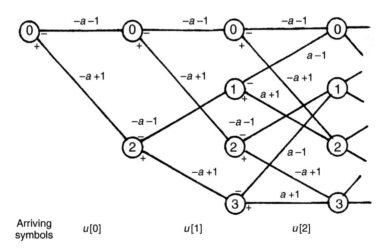

Figure 3.7 PRS code trellis generated by $1 + aD^2$. Branch labels are single real numbers. $\pm$ indicate data symbols that cause transitions. Symbols before $u[0]$ are -1.

---

[5] In Chapter 6, the delay notation will be exchanged for the z-transform notation $1 + az^{-2}$.

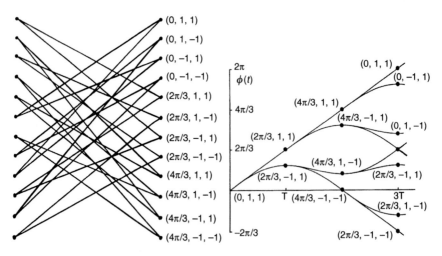

Figure 3.8 CPM trellis for the binary 3RC scheme with index $h = 2/3$ (memory 3); at right is a partial library of branch labels, which are T-second phase transitions. State descriptor consists of a phase and two data symbols. (Adapted from [8].)

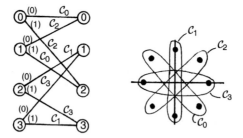

Figure 3.9 One stage of a TCM code trellis based on subsets of an 8PSK constellation. Branch labels are the four 2-point subsets C_0, C_1, C_2, C_3. Small figures are data bits.

Yet another trellis is the TCM trellis of Fig. 3.9. This one is binary and similar in appearance to Fig. 3.5, except that the branch labels are *sets*, in this case, subsets of points in an 8PSK constellation shown at the side. There are four two-point sets C_0, C_1, C_2, C_3, which are subsets of 8PSK as shown. In this code, one data bit drives a rate 1/2 encoder with generators $(g_1(D), g_2(D))$ equal to $(1 + D + D^2, 1 + D^2)$; this generates bit pairs $(x_1[n], x_2[n])$, which take values 00, 10, 11 and 01. This pair selects the subset C_0, C_1, C_2, C_3, respectively. The code carries a second data bit by selecting which of the two points in the subset is sent. The overall data rate is thus two data bits per 8PSK symbol interval. This kind of coding based on sets is introduced in Chapter 4.

3.4. Decoding

The maximum likelihood principle is an obvious design for a decoder: find the most likely transmitted message given the observed channel output. We developed an ML decoder for Gaussian noise in Chapter 2 and for the binary channel in Section 3.2. In both cases, as it turns out, the decoder seeks the closest codeword to the channel sequence, in a Euclidean distance sense or in a Hamming sense. We first present an efficient ML receiver for a trellis code, which is the VA.

There are, however, other design principles than the ML one. We might wish to use a non-ML receiver that has virtually ML error performance at, say, low error probabilities. Another design principle is bounded distance detection. Here the decoder is designed to correct all error patterns with a certain weight, or in the Euclidean distance case, all noises of a certain size. This principle is particularly useful when no definable channel error probabilities exist.

The section will focus on decoders for codes with a regular trellis structure. Most of this book is devoted to such codes. A large body of decoders exists for block parity-check codes, aside from the basic syndrome decoder in Section 3.2. For these we must defer to the references, especially [2–5].

3.4.1. Trellis Decoders and the Viterbi Algorithm

The aim of a trellis decoder is to find the closest path in the trellis to the received sequence. If it succeeds, it will have found the maximum likelihood codeword. One option is to search the entire trellis; this is the VA, to which we return momentarily. Another option is to search only a small region, where the nearest codeword is likely to be. An older term for these schemes is sequential decoding; a more modern one is *reduced-search* decoding. With some reduced-search decoders, it cannot be guaranteed that the ML path is found, although with most the error performance at moderate to high signal-to-noise ratio (SNR) is essentially that of an ML decoder.

There are many reduced-search schemes, but in their gross behavior they fall into two categories, *breadth-first* decoders, that search forward through the trellis in only one direction, and *backtracking* decoders, that can backtrack. Examples of the latter are the Fano and stack algorithms, and since these do not much figure in the rest of the book, we will not discuss them further. The best known breadth-first decoder, aside from the VA, is the M-algorithm.[6] This scheme works forward through the trellis, retaining just M paths; at each stage it makes the $2^b M$ extensions of these and keeps only the M paths with the least cumulative distance. Breadth-first decoders are inherently less efficient than backtracking ones, but

---

[6] A less common name for this scheme is list algorithm. However, the algorithm does not maintain a "list" in the computer science sense.

the complexity and variability of the latter have discouraged their use. Complete treatments of all these schemes appear in [5,11].

Whatever algorithm a decoder may follow, two basic measures of its complexity are the number of trellis paths it needs to view and store, and the length of these paths. The VA stores one path to each trellis state, or 2^m, altogether; the M-algorithm stores M. While it is true that the decoder must go through a number of steps to extend, view and sort each path, it is known that in theory these steps are in linear proportion to the number of paths. It is usually possible (see [5]) to relate the path number to the performance achieved, either a size of noise corrected or a probability of error. A reduced-search decoder needs to store much fewer paths than a VA, particularly in the case of the partial response codes in Chapter 6.

The length of the trellis paths stored by a decoder is its decoder decision depth, or alternately, its observation window, denoted N_{win}. A trellis decoder in the modern view moves its way down a trellis of indefinite depth, with new parts of the trellis entering the window at the right at stage n, and a final bit decision exiting on the left at stage $n - N_{\text{win}}$.[7] Hopefully, the remaining trellis paths in the decoder storage all share a single ancestor branch on the left. If not, one is chosen as the decoder output and all paths that stem from another branch are dropped from storage. The decoding delay is N_{win}. We will illustrate these concepts shortly with the VA.

The decision depth of a decoder should ideally be long enough so that the decoder error performance is the same as that of a decoder with unlimited depth. It is easiest to estimate a proper decision depth under the bounded distance design principle. The resulting depth is in fact a parameter of the code, not the decoder, and is called the code decision depth L_{dec}. Suppose it is desired to correct up to $d/2$ errors. Then when all paths are extended L_{dec} forward from some node, none should lie closer to the correct path than d; otherwise, the minimum distance as seen by the decoder will be less than d. Dynamic program algorithms exist that find the worst-case trellis path and starting state, and thus evaluate the $L_{\text{dec}}(d)$ to achieve a given d. Such decision depths are tabulated for many trellis codes (see [7] for convolutional codes and Section 5.2 for CPM codes). For good rate b/c convolutional codes and $d \leq d_{\text{f}}$, a simple law holds as d grows: The needed L_{dec} at d satisfies

$$L_{\text{dec}}(d) \approx \frac{d}{ch_{\text{B}}^{-1}(1-R)} \text{ stages.} \quad (3.4\text{-}1)$$

Here $h_{\text{B}}^{-1}()$ is the inverse of the binary entropy function. At $R = 1/2$, the law is $L_{\text{dec}}(d) \sim 4.5d$; at $R = 1/3$, it is $L_{\text{dec}}(d) \sim 1.9d$. It is important to note that when d is the full free distance, decision depth in Eq. (3.4-1) greatly exceeds encoder memory, and this is generally true for all coded modulation.

---

[7] A decoder with such a moving window is sometimes called a sliding block decoder. Originally, trellis decoders were considered to run to the end of the trellis and then decide the entire trellis path.

Coding and Information Theory

When the design principle is error probability, law (3.4-1) still roughly holds, but the probabilities of different events are such that in practice the decision depth can be a half or so of Eq. (3.4-1).

The Viterbi Algorithm

The VA searches the entire trellis for the least-distant path, in the most efficient manner possible. It finds wide use despite its exhaustive character for a number of reasons: it repeats a similar step at each node, its moves are synchronous and unidirectional, it adopts easily to soft channel outputs; finally, small trellises, where the exhaustiveness does not matter, often have attractive coding gains. The central idea in the algorithm, the optimality principle, was published by Viterbi in 1967 [9]. The full algorithm is an application of dynamic programming, an earlier procedure that finds the shortest route in a graph.[8]

It is easiest to describe the VA with respect to a fixed code, and we will use Fig. 3.5. The stage by stage progress of the algorithm is in Fig. 3.10. The encoder begins from state 0. In Fig. 3.10(a), the first bit triple received is 001 and the first two branches out of state 0 are drawn. The Hamming distances from these stage-0 branches (000 and 111 in Fig. 3.5) to the channel output are 1 and 2; the cumulative distances along the two paths in are shown in parentheses, and are just these same values. In Fig. 3.10(b), the channel output 110 arrives and two path extensions

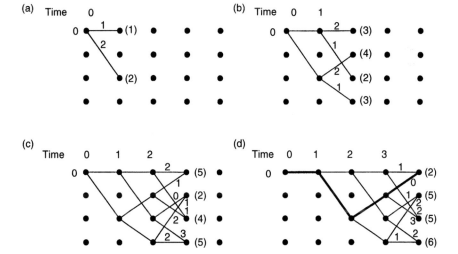

Figure 3.10 Viterbi decoding when the trellis is that of Fig. 3.5 and (a) 001, (b) 110, (c) 011, (d) 001 is received. Numbers on branches are distance increments; those in parentheses are distance accumulations.

---

[8] The dynamic programming observation was made by J. Omura around 1969.

from stage 1 have been made. The new branches create four paths in total, whose distance increments are shown on the new branches and whose cumulative distance again appears in parentheses. After the four stage-2 node extensions are made in Fig. 3.10(c), two paths merge at each node.

Now comes the critical idea in the VA. From the two paths entering each node at time 3, the one of those with the least total distance, called the *survivor*, is kept and the other path is deleted all the way back to the earlier node; if both have the same distance a survivor is selected at random. After all the path deletions, the distances of the best paths into each time-3 node are (5), (2), (4) and (5), and the surviving trellis looks like the first three stages of Fig. 3.10(d). The stage-4 path extension is shown in Fig. 3.10(d), along with the set of survivor distances (2), (5), (5) and (6). It appears that after four trellis stages, the least-distant path (heavier) starts from state 0 and runs to state 0 at time 4, and furthermore, it has total distance 2. This surviving path has path map $u[0]u[1]u[2]u[3] = 0100$, and corresponds to the partial codeword 000 111 011 001. There are two other words in the full trellis that enter and leave the same nodes, namely those with path maps 0000 and 1000. Their distances are both 6. One can trace the survivor paths to the other nodes at time 4 and find that no path has cumulative distance less than 5.

The VA procedure in Fig. 3.10 can be extended indefinitely. The reason that it finds the least-distant path to each state at each level is the so-called *optimality principle* of dynamic programming: when paths through a graph merge at a node, only a least-cost alternative need be retained, and a least-cost path through the entire graph exists that does not contain the dropped path sections.

In the decoding example here, all survivors after time 2 stem from the same initial branch, so that this branch, corresponding to data bit 0, could be released as output. In a software implementation the path maps into each node at time n and the cumulative distance of each node are stored in two arrays. In the extension to time $n + 1$, two new arrays are created as the path extensions are made, and survivors overwrite deletions. After the creation of stage $n + 1$, the new arrays contain the survivors and become the old arrays for the next stage. In a hardware implementation, an array of similar processor units corresponds to the state nodes at time n; these write to each other through a connection butterfly that mimics the trellis stage branches.

We have shown the VA search as beginning from a start state at time 0. Very often, a transmission has a defined end state as well, in which case the trellis collapses down to this single state. If so, the trellis codeword is said to be *terminated*. We can take the example of binary feedforward convolutional codes of rate $1/c$. In general, m extra stages are needed to terminate a codeword[9] and since this means adding m terminating 0s to the message, the overall rate is degraded. With length-K data frames, a rate b/c code degrades to $bK/c(K + m)$.

---

[9] In theory, L_{dec} terminating bits must be used, but a length equal to the shift register is the usual choice.

Coding and Information Theory

Another technique of termination is tailbiting, where the start state is defined to be the state at the end of the data symbols. Yet another strategy is to ignore the termination problem altogether. In this case the entire best path in the decoder storage is released when the front of the search reaches the last encoder depth. The path map bits near the front of the map will be less reliable in this case. Viewed as a block code, a trellis code without termination has a poor minimum distance, but the distance only affects the last few data symbols.

Although the discussion here has been carried out with a convolutional code example, the ideas of decision depth, paths, trellis searching, and termination apply to any code with a trellis, and in particular to those in Chapters 4–6.

Error Events in Trellis Decoding

We define an *error event* to begin when the decoder output trellis path splits from the correct trellis path and to end when it merges again. Such events are the basic mechanism of error in trellis coding, and other occurrences, such as data bit errors, derive from them. A bit error necessarily occurs when an event begins but may not occur when it ends.

Figure 3.11 portrays some different types of error events. When the decoder chooses a path at or near the minimum distance from the correct one, a "short" event occurs, whose length is close to the memory m of the code. These are the typical events in a good channel. In convolutional and PRS decoding, short events imply data symbol errors in the first few branches after the split, but none thereafter. The data symbol error rate tends in this case to be 1–2 times the event error rate. In CPM and TCM, bit errors occur at the beginning and end of a short event. In poorer channels, "long" error events occur, and typically these imply a 50% data error rate during the event. The overall BER is thus a multiple of the

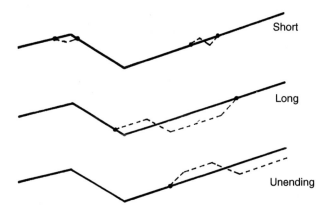

Figure 3.11 Types of error events: two short events (*top*); a long event (*middle*); an event that fails to terminate (*bottom*). Solid path is sent, dashed path is decoder output.

event rate that depends on the balance of short and long events. In reduced-search decoders with certain codes, indefinitely long events can happen. These occur because the decoder path storage has completely lost the correct path. In a properly designed decoder, these would occur only when a VA decoder would have failed as well, at least briefly.

We now turn to the calculation of probabilities for these decoder error events. As in the previous discussions, we will assume that the channel is AWGN or the BSC and that the decoder seeks a minimum distance path: that is to say, the decoder is ML or in the case of a reduced search, it at least attempts to be. Over the BSC, the signals are binary symbols and the distance can be Hamming; with AWGN, the signals are real functions of time and the distance is Euclidean. Since this is a coded modulation book, we will focus on the more general AWGN case.

Suppose that decoding is correct up to time n and that an alternative trellis path representing signal $s(t)$ exists and splits from the correct path $s_0(t)$ at this time. $s(t)$ lies distance d_i away from $s_0(t)$. Signal space theory in Section 2.3.2 then shows that in AWGN the probability is $Q(\sqrt{d_i^2 E_b/N_0})$ that the received $r(t)$ lies closer to $s(t)$ than to $s_0(t)$. Let P_{ev} be the probability that some event occurs at time n. If there are several alternatives, P_{ev} at this juncture is overbounded by $\sum A_i Q(\sqrt{d_i^2 E_b/N_0})$, where A_i is the number of alternatives at distance d_i. The sum is dominated as usual by the Q term for the closest alternative, and in the full unbounded trellis, its distance must be the free distance d_f. If there are many alternatives we can argue as in Section 2.3.2 that P_{ev} has the form $P_{ev} \sim A_0 Q(\sqrt{d_f^2 E_b/N_0})$, asymptotically in E_b/N_0.

The sort of argument here gives good estimates of P_{ev} for CPM and partial response codes at all useful E_b/N_0, if one notes carefully what alternatives cannot occur at nodes along the correct path. On the average over a long transmission it can be said that a log–log plot of P_{ev} vs E_b/N_0, a so-called "water fall curve" like those in Fig. 2.17, will tend at high E_b/N_0 toward the curve $AQ(\sqrt{d_f^2 E_b/N_0})$, with A the number of neighbors at d_f. The bit error rate (BER) plot will lie somewhat above, but will eventually converge to the same curve.

With convolutional codes, it is less easy to get tight estimates to P_{ev}. An approach called the transfer function bound has been devised which is partially successful. By the Mason gain technique, a transfer function

$$T(\delta) = \sum A_i \delta^i \qquad (3.4\text{-}2)$$

is obtained from the code generator state diagram, where A_i is the number of trellis paths of weight i. It can be shown that over a BSC with crossover p, P_{ev} has the estimate

$$P_{ev} \sim T(\delta)\big|_{\delta=\sqrt{4p(1-p)}} \qquad (3.4\text{-}3)$$

An analogous bound can be derived for BERs. Details may be found in [2]–[6].

3.4.2. Iterative Decoding and the BCJR Algorithm

The central theme in the preceding section was the decoding of a trellis path. The VA/dynamic program search of the code trellis yielded the maximum likelihood path. As for the data bits, these were taken to be the bits that drove the ML path. In some applications, it is necessary to know not the bits, but the *probability* of the bits. This section introduces some of these situations and the parallel to the VA called the BCJR algorithm, which produces these probabilities. Outputs such as these are an example of a *soft output*, an output that is not simply a data transmission symbol, but may even be a real number such as a probability. We have already seen soft-output channels; here we encounter soft-output decoders.

As an example of a transmission system that profits from soft-output decoding, we introduce concatenated coding systems. Two basic alternatives are shown in Fig. 3.12. The top system is *serial concatenation*, in which an outer encoder feeds an inner decoder, whose outputs pass through the physical channel. An inner decoder detects the inner encoding. This decoder could put out a hard estimate y_{out} of x_{out}, the inner encoder's input. Some thought, however, shows that the outer decoder might profit from some soft information about the symbols of y_{out}, such as the probability of each symbol. A way to visualize the situation is to imagine the inner encoder/channel/inner decoder as an "inner channel," the medium in fact seen by the outer encoder and decoder. The outer decoder will surely profit from outer channel soft outputs.

It is also conceivable that the inner decoder can profit from the deliberations of the outer decoder. The inner decoder output y_{out} may, for example,

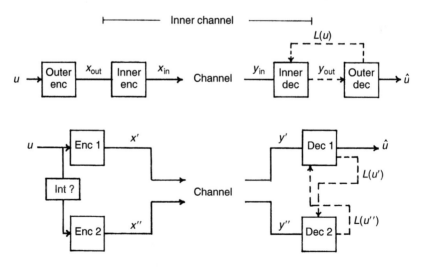

Figure 3.12 Comparison of basic serial (*top*) and parallel (*bottom*) code concatenation. Some possible soft information flows are shown.

somehow not be a legal outer codeword x_{out}. Information of this sort is shown as the feedback $L(u)$ in Fig. 3.12.

A second kind of concatenation is *parallel concatenation*. Here the data symbols are somehow encoded twice and sent at different times through the physical channel or through two separate channels. The two channel sequences are decoded, but it is clear that the two decoders can profit from an interchange, because both are guessing at the same data symbols. These flows are shown as $L(u')$ and $L(u'')$.

A common way to organize the feedback in these two systems is through *iterative decoding*: one decoder processes its inputs, then feeds information to the other, which does likewise, and so on back to the first, until both estimates stabilize. There are two important subtleties in this method. First, the decoding of a large code structure is broken down into small steps with small codes. Investigation has shown that large-code performance is obtained this way with relatively small decoding effort. Second, for the iterating to be effective, soft information must interchange.

The BCJR Algorithm

This scheme computes the data symbol probability at each trellis depth, or equivalently, the probability that each trellis branch was sent. The scheme accepts channel outputs as inputs but it can also accept a set of *a priori* probabilities about the symbols that originally drove the encoder, if these are available. By taking as output the most likely data symbol, we can obtain a hard symbol output from the BCJR algorithm, but if we take a sequence of most likely trellis branches as output, they will not necessarily form a valid trellis path. The algorithm was developed in the early 1970s by researchers who studied early iterative decoders, and it first appeared formally in 1974 in Bahl *et al.* [10], from which comes the algorithm's name. Since it puts out symbol by symbol probabilities, the algorithm is called a symbol by symbol MAP decoder. Provided that the start and stop states are known, the BCJR computes the true *a posteriori* probabilities of each symbol taken in isolation.[10]

The BCJR consists of two recursions on the trellis structure, instead of the one in the VA. These are pictured in Fig. 3.13. They calculate two working vectors called α and β. (The quantity λ will be defined later.) The components $\alpha[0], \alpha[1], \ldots, \alpha[\mu]$ and $\beta[0], \beta[1], \ldots, \beta[\mu]$ are special probabilities that apply to states $0, \ldots, \mu$ at some stage in a $\mu + 1$ state trellis. The row vector α is defined by

$$\alpha_k[j] \triangleq P[\text{Observe } y(1:k), \text{ Encoder in state } j \text{ at time } k]. \quad (3.4\text{-}4)$$

---

[10] Otherwise, BCJR gives a slightly biased calculation (see [11,22]). It is also possible to design a sequence MAP algorithm, whose output is the *a posteriori* most likely codeword and its probability.

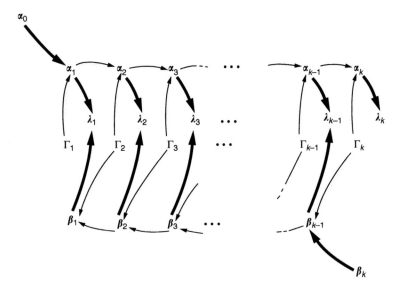

Figure 3.13 A picture of the BCJR algorithm. $\Gamma_1, \ldots, \Gamma_K$ are first generated from the observed outputs $y[1], \ldots, y[K]$. Forward α recursion runs across the top; backwards β recursion across the bottom. The scheme is initialized by $\boldsymbol{\alpha}_0$ and $\boldsymbol{\beta}_k$.

Here $\boldsymbol{\alpha}_k$ is the α-vector at time k and the special notation $y(1:k)$ means the sequence of stage by stage channel observations $y[1], \ldots, y[k]$. Each $y[k]$ is the entire, possibly composite, observation for one stage; in the case of a rate $1/c$ convolutional code, this would comprise c channel outputs. The column vector at stage k is defined by

$$\beta_k[i] \triangleq P[\text{Observe } y(k+1:K) \mid \text{Encoder in state } i \text{ at time } k]. \quad (3.4\text{-}5)$$

We also need a matrix Γ at stage k, whose i, j element is

$\Gamma_k[i, j]$

$$\triangleq P[\text{Observe } y(k), \text{Encoder in state } j \text{ at time } k \mid \text{Encoder in state } i \text{ at } k-1]. \quad (3.4\text{-}6)$$

Note that $\alpha_k[j]$ is the probability of the channel observations at times $1, \ldots, k$ and at the same time state j at k; $\beta_k[j]$ is the probability of the observations at $k+1, \ldots, K$ given state j at k; Γ_k is a modification of the usual state transition matrix to include the kth observation. Technically, the algorithm seeks at every step the probabilities of states *given* the entire observations $\boldsymbol{y}$, but since it is true for any event $\mathcal{E}$ that $P[\mathcal{E} \mid \boldsymbol{y}] = P[\mathcal{E}, \boldsymbol{y}]/P[\boldsymbol{y}]$, and since $P[\boldsymbol{y}]$ is fixed throughout, the algorithm needs only to find the probabilities of the states *and* $\boldsymbol{y}$. In some applications one needs to divide by $P[\boldsymbol{y}]$ before using the BCJR result.

Now we can define the two recursions. We assume that the encoding starts and ends in state 0. The *forward recursion* of the BCJR algorithm is

$$\alpha_k = \alpha_{k-1}\Gamma_k, \quad k = 1, \ldots, K \tag{3.4-7}$$

with α_0 set to $(1, 0, \ldots, 0)$. The *backward recursion* is

$$\beta_k = \Gamma_{k+1}\beta_{k+1}, \quad k = K-1, \ldots, 1 \tag{3.4-8}$$

with β_K set to $(1, 0, \ldots, 0)'$. The algorithm can account for different initial and final encoder conditions by a different α_0 and β_K. The object of the BCJR is to find the probability of the transition from state i to state j ending at time k, while observing the entire channel output y. We abbreviate the last as

$$\sigma_k[i \to j] = P\left[\text{Observe } y, \text{ Encoder in state } j \text{ at } k, \text{ Encoder at state } i \text{ at } k-1\right]. \tag{3.4-9}$$

It will be easiest to organize the calculation of $\sigma_k(i \to j)$ as a theorem and we will justify there also the recursions (3.4-7) and (3.4-8).

THEOREM 3.4-1. *Suppose codeword symbols are generated via a Markov chain and sent over a memoryless channel. Then recursions (3.4-7) and (3.4-8) hold. Furthermore,*

$$\sigma_k[i \to j] = \alpha_{k-1}[i]\Gamma_k[i, j]\beta_k[j]. \tag{3.4-10}$$

Proof. We first show Eq. (3.4-10). For brevity, replace the statement "Encoder in state j at time k" with the event $\{S_k = j\}$. The proof works by breaking the event (3.4-9) into five simultaneous events as follows:

$$P[S_{k-1} = i, S_k = j, y]$$
$$= P[S_{k-1} = i, S_k = j, y(1:k-1), y(k), y(k+1:K)].$$

It may be helpful to denote these five events in the abstract as, respectively, $\mathcal{A}, \mathcal{B}, \mathcal{C}, \mathcal{D}, \mathcal{E}$. For any five events it is true that $P[\mathcal{ABCDE}] = P[\mathcal{AC}]P[\mathcal{BD} \mid \mathcal{AC}]P[\mathcal{E} \mid \mathcal{ABCD}]$. For our specific events, we may drop event $\mathcal{C}$ in the conditioning $\mathcal{AC}$ in the probability $P[\mathcal{BD} \mid \mathcal{AC}]$. The reason is fundamental and rather subtle: with a Markov chain, future outcomes, conditioned on the present outcome, are independent of the past. The same applies when another sequence of independent variables, the channel noise, are added to the Markov outcomes. In this probability, the future event $\mathcal{BD} = \{S_k = j, y(k)\}$, when conditioned on $\mathcal{A} = \{S_{k-1} = i\}$, is independent of the past event $\mathcal{C} = \{y(1:k-1)\}$. By the same logic, the probability $P[\mathcal{E} \mid \mathcal{ABCD}]$ is equal to $P[\mathcal{E} \mid \mathcal{B}]$. What remains is

$P[\mathcal{AC}]P[\mathcal{BD} \mid \mathcal{A}]P[\mathcal{E} \mid \mathcal{B}]$

$= P[S_{k-1} = i, y(1:k-1)]P[S_k = j, y(k) \mid S_{k-1}=i] P[y(k+1:K) \mid S_k=j]$,

which is precisely Eq. (3.4-10).

Coding and Information Theory

To demonstrate the recursion (3.4-7), we start by writing

$$\alpha_k[j] = P[S_k = j, y(1:k)] = \sum_i P[S_{k-1} = i, S_k = j, y(1:k)].$$

We can break the event $\{y(1:k)\}$ into $\{y(1:k-1), y(k)\}$ and write the right-hand sum as

$$\sum_i P[S_{k-1} = i, S_k = j, y(1:k-1), y(k)].$$

As before we can denote these four events respectively as $\mathcal{A}, \mathcal{B}, \mathcal{C}, \mathcal{D}$ and note that in general $P[\mathcal{ABCD}] = P[\mathcal{AC}]P[\mathcal{BD} \mid \mathcal{AC}]$. By the Markov argument, $\mathcal{C}$ may be dropped in the second probability. What remains is

$$\sum_i P[S_{k-1} = i, y(1:k-1)] \, P[S_k = j, y(k) \mid S_{k-1} = i],$$

which is $\sum_i \alpha_{k-1}[i]\Gamma_k[i,j]$. This sum is just the jth component of the row vector $\boldsymbol{\alpha}_{k-1}\boldsymbol{\Gamma}_k$, from which Eq. (3.4-7) follows. A similar reasoning shows Eq. (3.4-8). □

Since the basic BCJR finds just the trellis transition (i.e. branch) probabilities, some further calculation may be required in a particular application. The probability that y occurs and the encoder was in state j at time k (i.e. of a *node*) is, using Eq. (3.4-7),

$$\sum_i \sigma_k[i \to j] = \sum_i \alpha_{k-1}[i]\Gamma_k[i,j]\beta_k[j] = \alpha_k[j]\beta_k[j], \tag{3.4-11}$$

that is, the product of the jth components of vectors $\boldsymbol{\alpha}_k$ and $\boldsymbol{\beta}_k$. The componentwise product of $\boldsymbol{\alpha}_k$ and $\boldsymbol{\beta}'_k$ is given the special name $\boldsymbol{\lambda}_k$; this vector is a list of all the node probabilities. In order to find the probability that a given data symbol $u(k)$ was sent, we would sum the probabilities of all transitions that imply this symbol:

$$P[u(k) = \ell] = \sum_{i,j \text{ in } \mathcal{L}} \sigma_k[i \to j]. \tag{3.4-12}$$

Here $\mathcal{L}$ is the set of trellis transitions that correspond to data value ℓ.

Finally, the BCJR can be easily extended to the case where *a priori* probabilities are available for the data symbols. In an iterative decoder, for example, these could be outputs from a previous BCJR algorithm. A close look at the foregoing will show that only the transition probabilities $\Gamma_k[i,j]$ are affected, and so the construction of this series of matrices needs to be modified. This kind of BCJR algorithm is variously called the Apri-BCJR, the Apri-MAP or the soft-in soft-out BCJR algorithm.

There is admittedly more calculation and storage in the BCJR algorithm than in the VA and this is its chief drawback. At each trellis depth, a matrix Γ_k needs to be constructed and stored; then the forward and backward recursions must be run. Fortunately, some applications use short codes and then there is no reason not to simply run the algorithm. The development of simpler near-optimal substitutes for BCJR is an active research area.

3.5. The Shannon Theory of Channels

As remarked at the beginning of the chapter, only one part of information theory is of interest to us in this book, namely the probabilistic theory of channels. This theory, like the rest of Shannon theory, is about limit theorems for random variables. It needs to be distinguished carefully from the coding theory in Sections 3.2–3.4, which concerns itself with codeword symbols, noise, distance and error correction. Shannon theory makes many provocative *suggestions* about encoding and decoding, but it is a free-standing theory based on its own premises. A full treatment of information theory is available in the classic text of Gallager [12]; a newer text with a strongly information-theoretic point of view is Cover and Thomas [13].

In information theory, an *information source* X is a random variable defined by a probability distribution $P[x]$. The source can consist of a sequence of variables $X[1], \ldots, X[K]$, which we can take as the one object $X^{(K)}$ with distribution $P[x^{(K)}]$. The $X[k]$ may also be correlated, although in this book we will assume the source outcomes are independent so that $P[X^{(K)}] = \prod_k P[x[k]]$; then the distribution of just the kth variable defines the measure $P[\ ]$ for the entire sequence.

The measure of information in a source X is its *entropy*, defined for discrete variables as[11]

$$H(X) = -\sum_i P[x_i] \log_2 P[x_i] \text{ (bits/outcome)}, \qquad (3.5\text{-}1)$$

where the subscript i now enumerates the different outcomes of x. If X is a continuous variable, an integration replaces the sum sign in Eq. (3.5-1). If the variable is $X^{(K)}$ in the IID sequence $X[1], \ldots, X[K]$, then it is easy to show that $H(X^{(K)}) = KH(X)$, where X is one variable in the sequence. For discrete X with I outcomes, $H(X) \leq \log_2 I$, with equality if and only if the outcomes all have probability $1/I$. A variable with outcomes in $\{-3, -1, +1, +3\}$, for example, has entropy 2 bits/outcome or less. It needs to be stressed that "bits" here is a real-number measure of information, just as meters are a measure of length; this is not to be confused with the earlier meaning of bits as symbols.

---

[11] Here and throughout, capital X denotes a random variable and lower case x denotes an actual outcome.

When there are two information sources X and Y, they may be correlated. For example, when X passes through a distorting medium to become Y, we hope that X and Y are quite closely correlated. For a particular outcome y, the *conditional entropy* of X is defined to be

$$H(X \mid y) \triangleq -\sum_i P[x_i \mid y] \log_2 P[x_i \mid y] \text{ (bits/outcome)}, \qquad (3.5\text{-}2)$$

in which $P[x_i \mid y]$ is the probability of outcome x_i given y. The expectation of $H(X \mid y)$ over y,

$$H(X \mid Y) = \sum_i H(X \mid y_j) P[y_j] \text{ (bits/outcome)}, \qquad (3.5\text{-}3)$$

is called the *equivocation* between X and Y. If X and Y are independent, the equivocation $H(X \mid Y)$ is simply $H(X)$.

A transmission or storage medium in Shannon theory is called a channel. The transmission of a source outcome comprises one use of the channel. When successive uses of a channel are independent, the channel is said to be memoryless. As with an information source, Shannon theory defines a channel in terms of probabilities. This time, the definition is by means of the conditional probability distribution $P[y \mid x]$, the probability the channel output is y given that its input is x. As with any conditional distribution, $\sum_j P[y_j \mid x] = 1$, for each input x. A memoryless channel for which both X and Y are discrete is called a discrete memoryless channel (DMC). The channel output is another information source, Y, whose defining distribution is $P[y_j] = \sum_i P[y_j \mid x_i] P[x_i]$. If the channel and the input source are both memoryless, then so is source Y. When Y is a continuous variable, the channel is input-discrete only, and both the foregoing summations become integrals.

In this book we make use of only a few classic channels. The simplest is the BSC, already introduced in Section 3.2.1, for which inputs and outputs are taken from $\{0, 1\}$ and for which $P[1 \mid 1] = P[0 \mid 0] = 1 - p$ and $P[1 \mid 0] = P[0 \mid 1] = p$. A convenient way to diagram the BSC probabilities is shown in Fig. 3.14. In other channels, symbols are either received perfectly or erased, that is, there is no hint of the input at the output. In information theory this is the binary erasure channel (BEC). Its model is shown in Fig. 3.14; with probability p the inputs are converted to a special erasure output, denoted e.[12] A combination of the BEC and the BSC is the binary symmetric erasure channel (BSEC), in Fig. 3.14. As the figure shows, the binary input can be received as either 1 or 0 or e, with the last signifying in effect that the reception was unreliable. The BSEC is a one-step-softened BSC: the more outcomes that Y can take, the "softer" is the channel output.

---

[12] It is a powerful advantage to know that any 1 or 0 received is necessarily correct. Decoders that fill in erasures are called erasure decoders. In general, a decoder can correct twice as many erasures as it can errors.

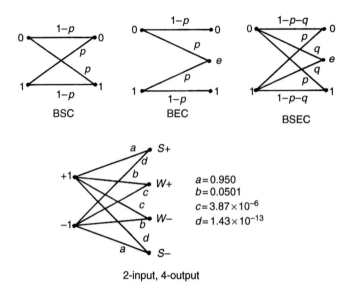

Figure 3.14 Some basic channels: the binary symmetric, binary erasure, binary symmetric erasure, and a 2-input 4-output channel derived from an underlying AWGN.

The most important case of a soft-output channel is the one for which the underlying physical channel is the discrete-time AWGN channel. This model originates in the discussion of Section 2.4. In the binary case, the source X takes values in the binary set $\{-1, +1\}$. Since the symbol energy is E_s and the noise is zero-mean independent Gaussian with variance $N_0/2$, the channel is defined by the distribution,

$$P[y \mid x] = \left(1/\sqrt{2\pi\sigma^2}\right) e^{-(y-x)^2/2\sigma^2}, \quad \sigma^2 = N_0/2E_s. \qquad (3.5\text{-}4)$$

We can call the output of this channel totally soft, since it is in fact a continuous variable that is exactly the physical output.

A less-soft-output channel would be the one at the bottom of Fig. 3.14, which has inputs ± 1 and four outputs, that we might call "strong -1," "weak -1," "weak $+1$," and "strong $+1$," or simply the set $\{S^-, W^-, W^+, S^+\}$ for short. The physical channel in Fig. 3.14 has $\sigma^2 = 0.1/2$, which is an $E_s/N_0 = 10$ dB. In order to create a two-input four-output DMC, the model must also assume a definite quantization of y into four decision regions. This transpires when we assume an orthogonal transmission pulse and the linear receiver of Fig. 2.2, but with a quantization to four outputs instead of the threshold comparison to zero. Because of symmetry, the quantization regions must be $\{(-\infty, -A), [-A, 0), [0, A), [A, \infty)\}$, for which a single parameter A needs to be specified. If A is set to $2/\sqrt{10}$, the four probabilities, given that $x = +1$, are

$$P[S^+ \mid +1] = 1 - Q[(1 - 2/\sqrt{10})\sqrt{20}] = 0.9499,$$
$$P[W^+ \mid +1] = Q[(1 - 2/\sqrt{10})\sqrt{20}] - Q[(1)\sqrt{20}] = 0.0501,$$
$$P[W^- \mid +1] = Q[(1)\sqrt{20}] - Q[(1 + 2/\sqrt{10})\sqrt{20}] = 3.87 \times 10^{-6},$$
$$P[S^- \mid +1] = Q[(1 + 2/\sqrt{10})\sqrt{20}] = 1.43 \times 10^{-13}.$$

The $x = -1$ probabilities are the same in reverse order. Thus, the DMC is obtained in Fig. 3.14.

The best quantization of y, that is, the choice of A, is unfortunately not easy to specify. It is known in general that differing J-region quantizations of the AWGN channel optimize the resulting DMC channel capacity, the channel cut-off rate (see Section 3.6) and the sequential decoding effort!

The channels just discussed account for most of those in this book. The main exception is the frequency-selective fading channel, which is a channel with memory. The information theory of channels with memory is considerably less developed; the cut-off rate in Section 3.6 can be calculated but little else is known.

Channels with Bandwidth

Until now, we have made no mention of bandwidth. Channels in information theory are simply "used," and symbols pass in discrete time once each use. A physical reality lies behind an abstract channel model, of course, a reality where bandwidth and spectrum play a critical role. How can we make the bridge from bandwidth and continuous time to an abstract Shannon model?

The most important physical channel is the white Gaussian one, with a bandwidth of W Hz (counting positive frequencies), used for τ s, with a noise density of $N_0/2$ W/Hz. Fortunately, the signal space theory in Chapter 2 provides the bridge we seek, because it is possible to specify how many orthogonal signal space dimensions can be packed into W Hz and τ s. From Section 2.3, each dimension in the space is affected by an independent additive noise variate, that is, each is separate AWGN Shannon channel, available for one "use". In order to count the signal space dimensions, let the physical time duration τ grow large. Then we can take as the signal space basis the set of orthogonal functions

$$\text{sinc}[(t - jT)/T], \quad j = 0, \ldots, \tau/T, \quad T = 1/2W, \qquad (3.5\text{-}5)$$

which occupies total time $\approx \tau$. This is a sequence of pulses appearing at rate $1/T$ and the individual Shannon AWGN channel inputs $x[1], x[2], \ldots$ are carried by the pulse train $\sum_k x[k]\text{sinc}[(t - kT)/T]$. From Theorem 2.2-2, we know that no faster sequence of orthogonal pulses exists. Furthermore, the linear receiver (Section 2.2.2) detects each $x[k]$ in isolation, subject to independent additive

Gaussian noise (Property 2.4-2). After approximately τ, then, a total

$$\tau/T = 2W\tau \quad \text{dimensions} \tag{3.5-6}$$

will have occurred, and each is a separate Shannon AWGN channel.

Equation (3.5-6) will be important in the coming section when we calculate the capacity of time T and bandwidth W to carry data bits. Actually, the formula is only asymptotically correct, because the sinc pulses in Eq. (3.5-5) cannot be both time- and frequency-limited. The formal proof of Eq. (3.5-6) – and thus the extension of the Shannon theory to a Gaussian channel with bandwidth and continuous time – originated in the early 1960s with Landau, Slepian and Pollak [see 14] and Reza [20].

3.6. Capacity, Cut-off Rate, and Error Exponent

The central accomplishment of channel information theory is the computation of capacity. This is the largest information measure that a channel may carry with vanishing error probability, in bits/channel use. That such a limit even existed was revolutionary in 1948, when Shannon's channel coding paper [15] appeared. We will calculate now the capacities for the main channels in this book and look also into two related channel quantities, the cut-off rate and the error exponent.

3.6.1. Channel Capacity

Channel capacity is based on the idea of the mutual information between two information sources X and Y, which is defined to be the alternate forms

$$\begin{aligned} I(X;Y) &\triangleq H(X) - H(X\mid Y), \\ I(X;Y) &\triangleq H(Y) - H(Y\mid X). \end{aligned} \tag{3.6-1}$$

Yet another expression is

$$I(X;Y) = \sum_x \sum_y P[x,y] \log \frac{P[x,y]}{P[x]P[y]}. \tag{3.6-2}$$

When X and Y are the input and output of a channel, respectively, these quantities take on special meanings as diagrammed in Fig. 3.15. The equivocation $H(X\mid Y)$ can be viewed as the information measure in $H(X)$ which is lost in the channel. The irrelevance $H(Y\mid X)$ is the extraneous information that the channel contributes to the measure in Y, which is $H(Y)$. The useful information that survives the passage is $I(X;Y)$.

Coding and Information Theory

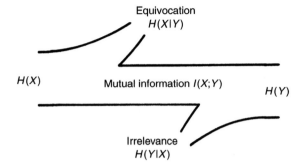

Figure 3.15 An illustration of equivocation, irrelevance, and mutual information in a channel.

The formal definition of channel capacity is

$$C = \max_{P[X]} I(X; Y) \text{ (bits/channel use)}, \tag{3.6-3}$$

the largest information that can pass through the X-to-Y relationship, when one is free to vary the distribution of X. To truly optimize transmission, we must be free to change the form of the information source; this is a fact not only in information theory, but for practical channels as well.

The simplest calculation is for the BSC.

Example 3.6-1 *(BSC Channel Capacity).* The mutual information is

$$\begin{aligned} I(X;Y) &= H(Y) - H(Y\mid X), \\ &= H(Y) + \sum_x P[x] \sum_y P[y\mid x]\log P[y\mid x], \\ &= H(Y) + \sum_x P[x]\left[p\log p + (1-p)\log(1-p)\right], \\ &= H(Y) - h_B(p), \end{aligned} \tag{3.6-4}$$

where $h_B(\ )$ is the binary entropy function. It remains to optimize Eq. (3.6-4) over $P[X]$. $H(Y)$ is largest in Eq. (3.6-4) when $P[y=1] = P[y=0] = 1/2$. This can only occur when $P[x=1] = P[x=0] = 1/2$. So the optimizing X-distribution is the equiprobable one, and the capacity of a BSC with crossover p is

$$C(p) = 1 - h_B(p). \tag{3.6-5}$$

$C(p)$ is plotted in Fig. 3.16; for example, C is 1/2 at $p = 0.110$, that is, at an 11% error rate.

Another attitude toward the capacity calculation is that $P[X]$ is fixed and is part of the channel definition. For instance, we might define a "16QAM Channel"

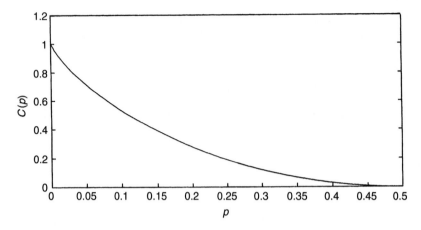

Figure 3.16 The capacity $1 - h_B(p)$ for the BSC with crossover probability p.

to be the rectangular point array in Fig. 2.18(b) with points selected equiprobably, since these are the conditions of practical 16QAM. The "capacity" would then be $I(X; Y)$ under these conditions, and the max in Eq. (3.6-3) will be removed. In Shannon theory, this type of capacity is more properly called an *information rate*, with the term capacity reserved for the information flow after the fullest possible optimization. We will often have use for the information rate concept.

The significance of the channel capacity stems from Shannon's channel coding theorem, which states that in a certain probability sense codes exist at all rates less than C, that can carry information through a channel virtually error free. One formal statement for the DMC is as follows.

THEOREM 3.6-1. *Consider transmitting μ equiprobable messages through a DMC, where $\mu = 2^N$ and N is the codeword length. Then a code exists with μ words and rate $R < C$ bits/channel use, such that the probability of error tends to zero as N grows.*

Theorems of this type exist for any DMC and for the continuous-output AWGN channel; they exist for restricted classes of codes, such as trellis, convolutional or parity-check codes with the BSC (technically, these capacities should be called information rates); they are known as well for many types of channels with memory. The proofs of these theorems are quite technical and except for two remarks, we will defer to the references, especially [12,13]. The first remark is that the theorem at its heart is about probability distributions and their limiting form, and not about concrete codes. Second, most proofs work by imagining that the symbols in the codewords are selected at random, a method Shannon called random coding. A very large number of words are so selected, and there is no guarantee that all of these words can be decoded properly, only that this happens with probability tending to one. It is interesting to observe that classes of concrete

Coding and Information Theory

codes with much structure, however cryptic, all fail to perform near capacity as their word length grows; randomness is thus not just a tool in the coding theorem proof, but also essential to near-optimal performance.

As complements to the channel coding theorem, there are two types of converse theorems. The first, the Weak Converse, states that at rates above capacity, the decoding error probability is bounded away from zero. Often a second type, the Strong Converse, can be proven, which states that the probability tends actually to one.

As further illustrations of capacity, we turn to two examples that explore the theme of soft-output channels. These capacities underlie in various ways the coded modulations in Chapters 4–6.

Example 3.6-2 (*Soft-output BSC Capacities, Derived from the AWGN Channel*). Very often, a BSC is in fact derived from an underlying AWGN channel. A hard-output BSC occurs when antipodal modulation is applied over the AWGN channel with a matched filter receiver and a hard threshold detector, to produce a BSC with crossover p, where $p = Q(\sqrt{2E_s/N_0})$. We recall from Section 3.2.3 that soft-output AWGN models can yield up to 3 dB coding gain over the BSC model. A softer output channel has been given in Fig. 3.14, in the form of a two-input four-output channel derived from the underlying AWGN. We can generalize it one step further and allow for an arbitrary SNR E_s/N_0: The inputs are $x = \pm 1$, E_s is frozen at 1, and the noise variance is $\sigma^2 = 2/(E_s/N_0) = 2N_0$; the threshold parameter A is still taken as $2/\sqrt{10}$. Since the channel is symmetric, we can specify it as in Section 3.5 by the $P[y \mid x = +1]$, which are:

$$\begin{aligned}
p_1 &= P[S^+ \mid +1] = 1 - Q[(1-A)\sqrt{2E_s/N_0}], \\
p_2 &= P[W^+ \mid +1] = Q[(1-A)\sqrt{2E_s/N_0}] - Q[\sqrt{2E_s/N_0}], \\
p_3 &= P[W^- \mid +1] = Q[\sqrt{2E_s/N_0}] - Q[(1+A)\sqrt{2E_s/N_0}], \\
p_4 &= P[S^- \mid +1] = Q[(1+A)\sqrt{2E_s/N_0}].
\end{aligned} \quad (3.6\text{-}6)$$

To find the capacity of channel (3.6-6), we start with Eq. (3.6-1). Since the channel is symmetric and has only two inputs, the $P[X]$ that optimizes $I(X; Y)$ must be $P[x = +1] = P[x = -1] = 1/2$. Furthermore, $P[y] = \tfrac{1}{2} P[y \mid +1] + \tfrac{1}{2} P[y \mid -1]$. Thus,

$$\begin{aligned}
\max I(X; Y) &= H(Y) + \sum_x P[x] \sum_y P[y \mid x] \log_2 P[y \mid x] \Big|_{P[x]=1/2} \\
&= 2\left[-\frac{p_1 + p_4}{2} \log\left[\frac{p_1 + p_4}{2}\right] - \frac{p_2 + p_3}{2} \log\left[\frac{p_2 + p_3}{2}\right] \right] + \sum_{i=1}^{4} p_i \log p_i
\end{aligned}$$

$$(3.6\text{-}7)$$

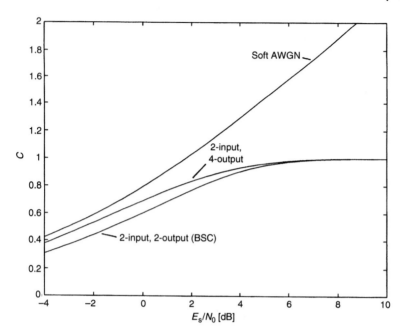

Figure 3.17 Capacities in the AWGN channel as a function of E_s/N_0 for the BSC(p) with $p = Q(\sqrt{2E_s/N_0})$, the 2-input 4-output channel in Example 3.6-2, and the continuous-in continuous-out AWGN channel.

Figure 3.17 shows a plot of capacity (3.6-7) against E_s/N_0. For comparison, the hard-output BSC capacity $C(p)$ in Eq. (3.6-5) is shown, in which p is set to $Q(\sqrt{2E_s/N_0})$. It is clear that the provision for soft-output improves capacity.[13] No matter how soft the channel output quantization is, C can never exceed one because the channel input rate itself is limited to one.

Example 3.6-3 *(Continuous-input Continuous-output Memoryless AWGN Channel).* When the channel input is not a discrete variable, the capacity is restricted only by the signaling energy. Now the capacity is computed from a continuous-time version of Eq. (3.6-3), namely,

$$C = \max_{P[x]} \iint P[x]P[y\,|\,x]\log_2 \frac{P[y\,|\,x]}{\int P[x']P[y\,|\,x']\,dx'}\,dy\,dx,$$

$$= \max_{P[x]} \int P[x] \int P[y\,|\,x]\log_2 P[y\,|\,x]\,dy\,dx - P[y]\log_2 P[y]\,dy. \quad (3.6\text{-}8)$$

---

[13] Although the change in C is relatively small, the switch to a soft-output model strongly improves the error probability of practical codes. Here is an example of Shannon theory insight that is somewhat misleading.

Coding and Information Theory

Assume the noise variate is zero-mean with variance $N_0/2$ and that x has average energy E_s. Then, in this expression,

$$\int P[y|x] \log_2 P[y|x] \, dy$$
$$= \int \left[(1/\sqrt{2\pi})e^{-(y-x)^2/N_0}\right] \log_2 \left[\left(1/\sqrt{2\pi}\right) e^{-(y-x)^2/N_0}\right] dy$$
$$= (-1/2) \log_2(2\pi e N_0/2),$$

independently of x. Consequently, the first term in Eq. (3.6-8) integrates to $(-1/2) \log_2(2\pi e N_0/2)$ no matter what $P[x]$ is. The second term is an entropy; it can be shown [12] that the entropy of a continuous variable with fixed energy is maximum when the variable is Gaussian. Thus a Gaussian y must maximize Eq. (3.6-8). Since $x = y - \eta$ and the noise η is Gaussian, x must also be Gaussian if y is; that is, a Gaussian x maximizes $I(X; Y)$. The Gaussian x with energy E_s has

$$P[x] = (1/\sqrt{2\pi E_s})e^{-x^2/2E_s}.$$

Since x and η add independently, it must be that

$$P[y] = \left[1/\sqrt{2\pi(E_s + N_0/2)}\right] e^{-y^2/2(E_s + N_0/2)}$$

It can be shown that the second integral in Eq. (3.6-8) is then $-\int P[y] \log_2 P[y] \, dy = (1/2) \log[2\pi e(E_s + N_0/2)]$. Thus the AWGN capacity is the sum of the two terms, or

$$C = (1/2) \log[1 + 2E_s/N_0] \text{ (bits/channel use)}. \tag{3.6-9}$$

This capacity is also shown in Fig. 3.17.

In the final example we explore an information rate for a channel model that is typical of the TCM codes in Chapter 4. Suppose that a concrete TCM code is based on an M-point rectangular QAM constellation in which each point is used equally often on the average. The noise is Gaussian. Effectively, there are two Shannon channel models that might be considered for this I/Q transmission: (i) two uses of a real-in real-out AWGN model, whose capacity is twice Eq. (3.6-9), and, (ii) one use of the actual QAM constellation points with two-dimensional AWGN. The second model is closer to the TCM reality and to obtain more instructive conclusions, it is better to use it and furthermore to require that all points be equiprobable. C will then be for codes whose symbols have the same constraints, but, of course, have no particular resemblance to any concrete TCM code. C is properly called an information rate here because there is no optimization over the distribution of the M inputs to the model.

Example 3.6-4 *(Rectangular QAM Information Rates in AWGN).* The M outcomes of X are now points in Euclidean two-space, in particular, the I/Q space in Section 2.5.2. X is thus discrete. The two-dimensional AWGN noise η has zero mean and variance $N_0/2$ in each component. We take the two-dimensional channel output $y = x + \eta$ to be real-valued, that is, the channel output is totally soft. Take E_s as the average energy in x. Since the channel inputs are constrained to be equally likely, we take Eq. (3.6-3) with the max deleted; it becomes

$$C_{\text{QAM}} = \sum_{i=1}^{M} \iint P[x_i]P[y \mid x_i] \log_2 \frac{P[y \mid x_i]P[x_i]}{P[x_i]P[y]} \, dy,$$

in which $P[y] = \sum_i P[y \mid x_i]P[x_i]$ and $P[x_i] = 1/M$, all i. Some particular E_s/N_0 holds. The calculation is tedious but straightforward. Kressel [17] performs it, together with calculations for many other channels models, and Fig. 3.18 is adapted from his data. Shown for comparison is the 2-dimensional capacity, twice $(1/2) \log_2[1 + 2(E_s/2)/N_0]$, from the unrestricted AWGN formula (3.6-9). It can

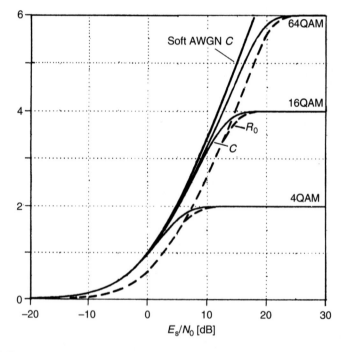

Figure 3.18 Capacity and cut-off rate (dashed) for rectangular QAM channels with alphabet size 4, 16, 64, as a function of E_s/N_0, where E_s is the total energy in the two QAM channel dimensions. AWGN capacity (3.6-9) shown for comparison. C/R_0 tends to 1 with E_s/N_0 for each alphabet size.

Coding and Information Theory

be seen that the equiprobable QAM channel restriction hardly affects the channel capacity, as long as the QAM alphabet size is large enough for the selected E_s/N_0.

3.6.2. Capacity for Channels with Defined Bandwidth

As in Section 3.5, the foregoing discussion does not explicitly consider bandwidth and continuous time. We can add these elements by means of formula (3.5-6), which states that $2W\tau$ AWGN channel uses can be extracted from W Hz (W measures positive frequencies) and τ s. Starting from Eq. (3.6-9), we can express the capacity C_W in bits/s for a bandwidth W channel as

$$C_W = (1/2)\log[1 + 2E_s/N_0] \text{ (data bits/channel use)} \cdot 2W \text{ (channel uses/s)}$$
$$= W\log[1 + 2E_s/N_0] \text{ (bits/s)}. \tag{3.6-10}$$

The energy here is measured per channel use.

Formula (3.6-10) may be recast in terms of signal power, a form in which it is sometimes more useful. In Shannon analysis, the encoded signal is a random process, in this case a Gaussian one $x(t)$, and it has a power spectral density $S_{xx}(f)$ with dimension watts/Hz. Here we take $S_{xx}(f)$ to be a constant function. The total signal power in watts must count positive and negative frequencies and is $P = 2W \times$ (PSD) watts. Since there are $2W\tau$ channel uses in W Hz and τ s, the energy per channel use in terms of P is

$$E_s = \frac{P\tau}{2W\tau} = \frac{P}{2W} \text{ (joules)}.$$

Substitution of this into Eq. (3.6-10) gives the alternate form

$$C_W = W\log_2(1 + P/WN_0) \text{ (bits/s)} \tag{3.6-11}$$

for the capacity at power P uniformly distributed over W Hz, counting positive frequencies.

It is also useful to have the capacity in terms of energy and bandwidth per data bit, just as the modulation quantities were expressed in Chapter 2. To begin this conversion, divide E_s/N_0 by both sides of Eq. (3.6-9), to obtain

$$E_b/N_0 = E_s/CN_0 = \frac{E_s/N_0}{(1/2)\log(1 + 2E_s/N_0)} \quad (E_b \text{ in joules/data bit}) \tag{3.6-12}$$

for a code working at rate C bits/channel use. Dividing W by both sides of Eq. (3.6-10) gives

$$W_b = W/C_W = \frac{1}{\log(1 + 2E_s/N_0)} \text{ (Hz-s/data bit)}. \tag{3.6-13}$$

These equations in effect give the per-bit quantities E_b/N_0 and W_b in terms of a common parameter, which is E_s/N_0, the channel symbol SNR. By stepping through the values of E_s/N_0, one generates the solid curve of Fig. 3.19. Combinations of W_b and E_b/N_0 above and to the right of the curve can transmit information at an error probability that tends to zero with codeword length.

Equations (3.6-12) and (3.6-13) can be combined into the form,

$$E_b/N_0 = (2^{1/W_b} - 1)W_b, \qquad (3.6\text{-}14)$$

a plot of which is Fig. 3.19. Since the per data bit bandwidth consumption W_b in Hz-s/data bit is $(W \text{ Hz})/(C \text{ data bits/channel use})(2W \text{ available channel uses/s}) = 1/2C$, Eq. (3.6-14) yields another form,

$$E_b/N_0 = \frac{2^{2C} - 1}{2C}, \qquad (3.6\text{-}15)$$

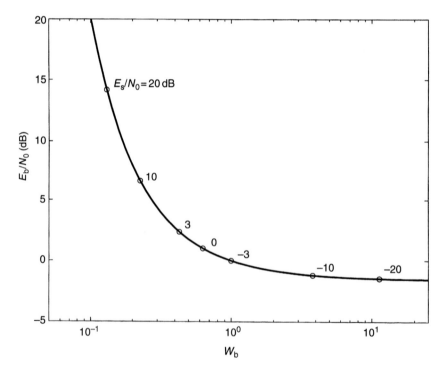

Figure 3.19 Locus of capacity-achieving combinations of W_b and E_b/N_0, parameterized in E_s/N_0, where E_s is the energy per channel use. Combinations above and right of the curve can achieve arbitrarily low error probability with long codes.

Coding and Information Theory

in which C is directly the code rate in bits/channel use. From either of these, it can be shown that E_b/N_0 tends to -1.59 dB in the limit, as either W_b becomes large or C tends to 0. At other W_b and C, the formulas give a least E_b/N_0 that can in principle lead to low error probability. At rate 0.5 bits/channel use, for example, E_b/N_0 must, from Eq. (3.6-15), exceed 0 dB, and the bandwidth per data bit, from Eq. (3.6-13), must exceed 1 Hz-s/bit. We can thus expect that very long good rate 1/2 convolutional codes might come close to $E_b/N_0 = 0$ dB at low error rates, and they might require 1 Hz-s/data bit bandwidth in a Gaussian channel.[14]

Extension to Channels with a Defined Error Rate

The foregoing assumes an decoder error rate that is essentially zero. When the data is binary and a nonzero BER is acceptable, the following argument gives an alternate Shannon capacity in form (3.6-14)–(3.6-15) which holds when the BER is some $P_{ber} > 0$.

Suppose an error free Shannon capacity C_0 bits/s is available. According to the Shannon theory of rate distortion, source codes exist in principle that encode up to

$$C_{ber} = C_0/[1 - h_B(P_{ber})] \text{ (bits/s)} \quad (3.6\text{-}16)$$

into a stream at C_0 bits/s, subject to the condition that the C_{ber} stream, when reproduced, is allowed to have long-term error rate P_{ber}. The demonstration of Eq. (3.6-16) can be found in [5,12,19]. Repeating now the argument that led to Eq. (3.6-12)–(3.6-14), we obtain the energy and bandwidth per data bit

$$E_b/N_0\big|_{P_{ber}} = \frac{[1 - h_B(P_{ber})]E_s/N_0}{(1/2)\log(1 + 2E_s/N_0)} \quad (E_b \text{ in joules/data bit}), \quad (3.6\text{-}17)$$

$$W_b\big|_{P_{ber}} = \frac{[1 - h_B(P_{ber})]}{\log(1 + 2E_s/N_0)} \text{ (Hz-s/data bit)} \quad (3.6\text{-}18)$$

in terms of the parameter E_s/N_0; furthermore, the new E_b/N_0 and W_b are related by

$$E_b/N_0 = (2^{[1-h_B(P_{ber})]/W_b} - 1)W_b \quad (3.6\text{-}19)$$

instead of Eq. (3.6-14). The same argument after Eq. (3.6-14) then gives

$$E_b/N_0\big|_{P_{ber}} = \frac{2^{2C[1-h_B(P_{ber})]} - 1}{2C} \quad (3.6\text{-}20)$$

as the ultimate minimum energy needed by a rate C data bits/channel use code that attains P_{ber}.

The change in E_b/N_0 and W_b from introducing the nonzero BER constraint is slight at reasonable BER. Setting $P_{ber} = 10^{-5}$, for example, reduces

---

[14] To achieve this bandwidth, the underlying modulation would have to employ sinc( ) pulses.

E_b/N_0 and W_b by the factor $1 - h_B(10^{-5}) = 0.99982$, which is invisible on Fig. 3.19. Setting $P_{ber} = 0.1$ reduces both by the factor 0.53, which translates both axes by this factor.

Equation (3.6-20), with R substituted for C, provides a way to estimate the ultimate minimum BER that codes of rate R data bits/channel use can produce at per-data-bit SNR E_b/N_0. This P_{ber} vs E_b/N_0 curve is often called the *Shannon limit* for rate R transmission methods. Figure 3.20 shows such curves at a number of rates. These may be compared directly to standard waterfall curves like those in Fig. 2.17, and the BPSK curve is shown on the figure as one example. BPSK without coding is a rate 1 system, and its E_b/N_0 at error rate 10^{-5} is 9.6 dB, compared with a rate 1 Shannon limit figure of 1.7 dB, a 7.9 dB disparity.[15]

Note that a transmission method that appears to be near the Shannon limit in Fig. 3.20 may not be near the capacity curve in Fig. 3.19 if it does not make efficient use of bandwidth. For methods that utilize pulse modulation, this means

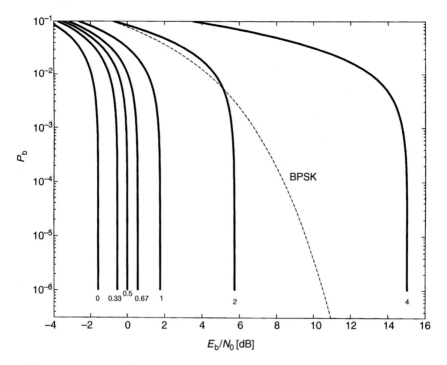

Figure 3.20 The Shannon limit on the bit error rate P_b as a function of E_b/N_0 for codes of rates ≈ 0, 0.33, 0.5, 0.67, 1, 2, 4. Ordinary rate 1 BPSK is shown for comparison.

---

[15] One must take care to choose the right Shannon limit curve. For example, 16QAM modulation consumes two "channel uses" per constellation, and so its rate per channel use is 2, not 4.

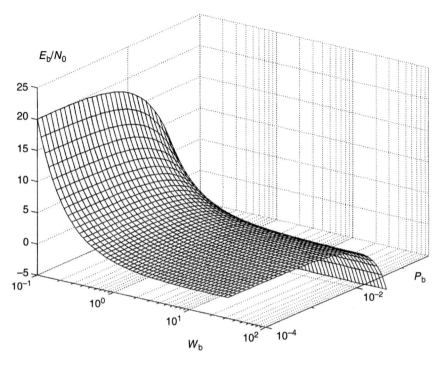

Figure 3.21 The Shannon limit surface in the space formed by data bit SNR E_b/N_0, bandwidth W_b Hz-s/data bit, and data bit error rate P_b. E_b/N_0 in dB.

that sinc( ) pulses need to be used. The combined energy–bandwidth Shannon limit for systems that achieve P_{ber} is a surface in 3-space, as shown in Fig. 3.21. The front plane in the figure, at $P_{ber} = 10^{-4}$, cuts the surface in a curve that resembles Fig. 3.19.

3.6.3. Capacity of Gaussian Channels Incorporating a Linear Filter

The foregoing section derived the Shannon capacity of a block of time and bandwidth. It often happens that bandlimitation occurs through linear filtering and we review now some facts about the information that can flow through this sort of arrangement. The capacities so calculated are related to the PRS coding in Chapter 6.

In actuality, there are several ways to incorporate a filter into a Gaussian channel. Each has a distinct capacity and relates to a separate practical situation. Two classic models are shown in Fig. 3.22. We will call them the Filter-in-the-Medium Model (contracted to Filter/Medium Model) and the Filter-in-the-Transmitter Model (Filter/Transmitter Model), after the position of the filter

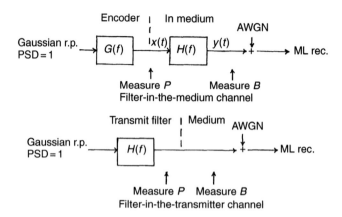

Figure 3.22 Basic models for a Gaussian channel containing a filter.

in the transmission chain. In the first, a known and fixed filter $H(f)$ is part of the channel proper; $H(f)$ filters the continuous coded signal $x(t)$, after which is added AWGN. In this Shannon communications model, the codewords $x(t)$ are created by means of filter $G(f)$ acting on a white Gaussian process.[16] It is important to be precise about where the power constraint is applied, and here it is *before* $H(f)$. The everyday example of this channel is the telephone "local loop" channel, consisting of a long pair of wires, or other linear bandlimiting, whose effect is $H(f)$. The local loop can withstand a certain maximum power, measured at its input. This is P. The physical reality of the wire pair local loop is discussed in [1,21].

In the Filter/Transmitter Model, codewords are a white process filtered by $H(f)$, but now the measurement of power occurs *after* the known filter $H(f)$. The standard practical exemplar is a radio channel in which $H(f)$ is the transmitter bandlimitation filter. Power is measured after, in fact at the receiving antenna, and the AWGN occurs at the receiver front end.[17] It makes more sense to talk about an information rate through this channel, rather than a capacity, since there is no $G(f)$ to optimize over.

The models here are sufficient for a meaningful Shannon analysis, but they are incomplete and easily misleading in practice because they do not include certain subtleties, notably whether both ends have knowledge of $H(f)$ and where and how the bandwidth is reckoned. The wire pair and radio channels will provide an example of these puzzles. In the wire pair local loop, we can consider that we "own" the Filter/Medium model: there is a physical wire pair that we may more or less use as we please so far as bandwidth is concerned. $H(f)$ rolls off with frequency

---

[16] It needs to be shown that such Gaussian signals achieve capacity in the limit as the signaling time grows. We omit these important details here. See [12,19].

[17] The calculation of E_s/N_0 here via a link budget is described in [1], Chapter 5.

Coding and Information Theory

and constrains the bandwidth, but we might precode by something like $1/H(f)$ and gain some of this bandwidth back. On the other hand, one certainly does not own the medium in the radio channel; the bandwidth is now *shared*. A $1/H(f)$ prewidening strategy is not allowed. Whatever encoding is done is filtered by $H(f)$ and the result probably has to lie within an allocated bandwidth.

Information rates sustained by a transmission model depend also on the knowledge of $H(f)$ at each end. If $H(f)$ stems from a known filter or from a wire pair that is dedicated and measured, then both ends can know it. If the medium is "owned," prefiltering is allowed. In other cases $H(f)$ varies and is measured at the receiver, which may or may not tell the transmitter. These rules are discussed further under the heading of equalization in Section 6.6.

What we write here is itself a simplification, and every application has its own peculiarities. Our aim now is to set out a few clear models and let their Shannon analysis yield some insights. Accordingly, we will assume as follows:

1. In the Filter/Medium Model, power is measured before the filter H; the channel is owned. There is no set limit on the positive-frequency transmission bandwidth B, aside from that which may be imposed by regions where $H(f) = 0$, although we are nonetheless interested in how capacity depends on B.
2. In the Filter/Transmitter Model, power is measured after the filter H; the channel is shared. The bandwidth is limited intentionally by H, although H may roll off more or less slowly and there may not exist a B outside which transmission is zero. Still, we think of B not as a free parameter but as dictated by H.

There has been relatively little coding theory developed around the Filter/Medium Model, other than some methods of precoding for, say, telephone modems (see Section 6.6). As for the Filter/Transmitter Model, we can take this as the model underlying the coded modulations of Chapter 6.

Capacity for the Filter/Medium Model

We seek the capacity in bits/s, as a function of bandwidth B and power P. In the standard Shannon argument, codewords $x(t)$ are produced through filtering a white Gaussian input process by $G(f)$, which then passes through the channel $H(f)$ to become the process $y(t)$. The input process has unit PSD, so that $x(t)$ has PSD equal $|G(f)|^2$, with $P = \int |G(f)|^2 df$. The receiver observes $y(t)$ plus noise; the noise has PSD $N_0/2$ and $y(t)$ has PSD $|G(f)|^2 |H(f)|^2$.

The capacity is calculated by dividing the bandpass into many small pieces of width Δf and then applying the so-called waterfilling argument. Figure 3.23(a) shows one of the pieces. The argument works as follows.

Within each bandwidth $(f_0, f_0 + \Delta f)$ taken in isolation, the signal power is approximately $|G(f_0)|^2 |H(f_0)|^2$, and from Eq. (3.6-11) the capacity of this bandwidth piece is about

$$\Delta f \log_2 \left(1 + \frac{\Delta f \, |G(f_0)|^2 |H(f_0)|^2}{\Delta f \, N_0/2}\right) \text{ (bits/s)}.$$

Now let the transmission time $\tau \to \infty$ and pass to the integral to get the capacity of the total positive-frequency passband $\mathcal{B}$:

$$C = \int_{\mathcal{B}} \log_2 \left(1 + \frac{|G(f)|^2 |H(f)|^2}{N_0/2}\right) df \text{ (bits/s)}. \qquad (3.6\text{-}21)$$

Which $G(f)$ maximizes C subject to $\int_{-\infty}^{\infty} |G(f)|^2 df = P$? Clearly, power is best spent where $|H(f)|^2$ is highest. Yet capacity grows nearly linearly with bandwidth but only logarithmically with power, and so a point must be reached where it is best to spread the power over a wider bandwidth. Variational calculus shows that the solution is parameterized by a constant $K > 0$ as follows:

Choose $K > \min_f 2/N_0 |H(f)|^2$; form the set of frequencies $\mathcal{B}_K = \{f : 2/N_0 |H(f)|^2 < K\}$.

Then the optimal encoder $G(f)$ satisfies

$$|G(f)|^2 = \begin{cases} K - 2/N_0 |H(f)|^2, & f \in \mathcal{B}_K, \\ 0, & \text{otherwise}. \end{cases}$$

for the power

$$P = \int_{-\infty}^{\infty} |G(f)|^2 df. \qquad (3.6\text{-}22)$$

The result is conveniently sketched as Fig. 3.23(b). It visualizes water as filling the valleys of the function $2/N_0 |H(f)|^2$ (only positive frequencies are shown in the figure, but all frequencies are taken). The depth of the water is $|G(f)|^2$; its total quantity is P. The signaling bandpass $\mathcal{B}_K$ comprises the flooded

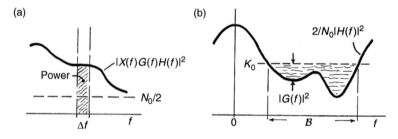

Figure 3.23 Illustration of the waterfilling argument. (a) calculation of the capacity in a small bandwidth Δf; (b) water filling a bandpass $\mathcal{B}$.

Coding and Information Theory

regions. When P is small, only a small bandwidth is occupied by the encoded signal. Both C and the occupied bandwidth grow with P and the bandwidth may comprise disjoint regions if $|H(f)|^2$ has multiple peaks. Since we consider the Filter/Medium channel to be owned, the growing bandwidth is not a worry.

Calculation of Capacity for the Filter/Transmitter Model

One can take the point of view that a channel which is created by a filter with output power P and passband width approximately B consumes a channel of capacity (3.6-11), namely, $B \log_2(1 + P/BN_0)$ bits. For transmit filters with an approximately rectangular passband shape, this attitude makes sense in some situations. However, as some of the free distance optimizing filters in Chapter 6 show, transmit filters can have complicated shapes. Another attitude is to compute the information rate that can flow through a transmit filter followed by AWGN; this is not technically a capacity, since there is no optimization over the form of the channel input.

By the same argument that leads to Eq. (3.6-21), we can conclude that subject to $P = \int_{-\infty}^{\infty} |H(f)|^2 df$ watts,

$$R = \int_0^{\infty} \log_2\left(1 + \frac{|H(f)|^2}{N_0/2}\right) df \text{ (bits/s)} \quad (3.6\text{-}23)$$

can flow through a filter $H(f)$ followed by AWGN. Suppose on the other hand that we may vary $H(f)$ within the constraints power P and bandwidth B, then variational calculus will show that R is maximized by an ideal brickwall filter and that the rate is therefore $B \log_2(1 + P/BN_0)$. Either view is a useful insight into the codes of Chapter 6.

A simple example will illustrate these ideas.

Example 3.6-5 *(Capacities and Information Rates for a Two-Band Filter).* Let $|H(f)|^2$ have the two-level shape in Fig. 3.24(a). Set the power P equal 3 watts and let $N_0/2 = 1$. The waterfilling solution to the Filter/Medium channel capacity is shown in Fig. 3.24(b). The "reservoir" $N_0/2|H(f)|^2$ is filled with 3 watts of

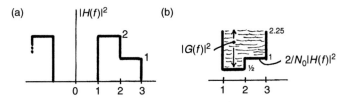

Figure 3.24 Illustration of Example 3.6-5. (a) Plot of the channel spectrum $|H(f)|^2$. (b) Waterfilling calculation at 3 watts power.

power, an amount that fills it to level 2.25. The encoder $G(f)$ therefore satisfies

$$|G(f)|^2 = \begin{cases} 1.75, & 1 \le |f| < 2, \\ 1.25, & 2 \le |f| < 3, \\ 0, & \text{otherwise} \end{cases}$$

and total bandwidth B is consumed (positive frequencies). The capacity is

$$R = \int_1^3 \log_2\left(1 + \frac{|G(f)|^2 |H(f)|^2}{N_0/2}\right) df = 3.34 \text{ bits/s.} \qquad (3.6\text{-}24)$$

By contrast, if $P \le 1/2$ watt is available, then all transmission should be limited to the 1 Hz band $1 \le |f| < 2$: transmission should take place exclusively in the part of the band with highest the SNR. The Filter/Transmitter channel information rate, by comparison, is

$$R = \int_1^3 \log_2\left(1 + \frac{|H(f)|^2}{N_0/2}\right) df = 2.58 \text{ bits/s};$$

the highest this can be raised in bandwidth 2 Hz and power 3 watts is the brickwall H result,

$$R = B \log_2\left(1 + \frac{P}{BN_0/2}\right) df = 2.64 \text{ bits/s.}$$

These are both less than the Filter/Medium channel result in Eq. (3.6-24). However, the combination $|G(f)|^2 |H(f)|^2$ there applies power 4.75 watts at the point where the AWGN occurs. If the Filter/Transmitter power is raised to 4.75, its information rate grows to 3.43. Thus, more information can flow through this particular H in 2 Hz in the Filter/Transmitter configuration, if powers are equated at the noise insertion.

3.6.4. Cut-off Rate and Error Exponent

An estimate of decoding error probability and at the same time an underbound to capacity may be obtained by the following simple argument. Suppose that codeword symbols are selected at random by $P[X]$, to form 2^{RN} words $x^{(N)}$. The proof mechanism in Shannon's channel coding theorem shows that this method, when applied to produce a code at any rate R below capacity, yields a code whose probability of error tends to zero with the word length N. Now let one of these words, x_1, be sent through the channel. The other words are all selected independently from x_1, and the probability of deciding each of these, given x_1 is sent,

Coding and Information Theory

is the two-signal error probability $p_{2,i} = P[\text{Decide } x_i \mid x_1 \text{ sent}]$. Since all the $p_{2,i}$ are the same, we replace them by the expression

$$2^{-NR_0} \triangleq p_{2,i}, \quad \text{all } i \neq 1. \tag{3.6-25}$$

By the union bound of probability, the probability P_e of deciding any of the other $2^{NR} - 1$ codewords is overbounded as

$$P_e \leq \sum_{i=1}^{2^{NR}-1} p_{2,i} = [2^{NR} - 1] 2^{-NR_0} = 2^{-N(R_0-R)}. \tag{3.6-26}$$

The parameter R_0, assigned the dimensions bits/channel use, is called the *cut-off rate* of the channel. Through the codeword selection process, R_0 depends only on the channel. The bound Eq. (3.6-26) shows that P_e tends exponentially to zero at least with exponent $-N(R_0 - R)$, whenever $R < R_0$. R_0 is thus an underbound to capacity.

The R_0 parameter pops up in many places in the Shannon theory of channels, and its ultimate meaning is still the subject of controversy. For almost all channels, it can be shown that R_0 is also the rate above which sequential decoding procedures fail in one sense to decode a symbol in finite expected work.[18] In the present day, it is becoming accepted that sequential and "reduced search" decoders, as well as for example turbo decoders, can indeed work in a practical sense at rates above R_0. Aside from its role in decoder effort, R_0 appears in a number of analytical bounds on the error probability of various classes of random codes and channels (see especially Viterbi and Omura [19]). Perhaps its significance can be summed up in two ways. First, decoding does get rapidly harder as R grows beyond R_0, although it may still be economic. Second, R_0 is much easier to compute than capacity, and for many channels with memory, R_0 is our best estimate at present for the value of C.

A cutoff rate of particular interest is the one for the BSC(p),

$$R_0(p) = 1 - \log_2[1 + 2\sqrt{p(1-p)}]. \tag{3.6-27}$$

The cut-off rates for several QAM channels are compared to the capacities in Fig. 3.18.

Much research has gone into the technical problem of finding a tighter bound than Eq. (3.6-26) to the asymptotic probability of decoding error. The general form for such a bound is

$$P_e \sim 2^{-NE(R)}. \tag{3.6-28}$$

---

[18] The sense is that the decoder error probability must tend to zero as $N \to \infty$; if the error instead may tend to an $\epsilon > 0$ no matter how small, the expected decoding work is usually finite for all $R < C$. See [5].

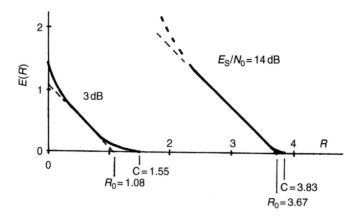

Figure 3.25 Error exponent $E(R)$ for the 16QAM channel at $E_s/N_0 = 3$ and 14 dB, compared to the exponent $R_0 - R$. E_s is total energy in two QAM dimensions. Data from Kressel [17].

Here N is the codeword length in channel uses and $E(R)$, sometimes called the Gallager exponent, is a function of the code rate R and the E_s/N_0 for an underlying physical channel, although the latter dependence is not explicitly shown. For most memoryless channels of importance, the exponent $E(R)$ is accurately known for $0 < R < C$. For rates in the important middle part of the range $[0, C]$, Eq. (3.6-28) is in fact Eq. (3.6-26), which underscores the significance of R_0. At low and high rates, $E(R)$ lies above $R_0 - R$.

Figure 3.25 shows both $E(R)$ and $R_0 - R$ at E_s/N_0 equal 3 and 14 dB for the 16QAM channel coding case that appears (with others) in Fig. 3.18; E_s is the energy in a 16QAM symbol. The random codes here are made up of words that look like 16QAM transmissions, and the R_0, C and $E(R)$ can be thought of as those that underlie TCM codes based on 16QAM in Chapter 4. The lower E_s/N_0 (3 dB) leads to a very poor error rate with 16QAM itself (see Fig. 2.17; note $E_b = E_s/4$), while the 14 dB value leads to error rate $\approx 10^{-5}$. The figure shows a number of properties that apply widely to other cases and gives us good insights for the next chapters:

1. $E(R)$ and $R_0 - R$ are the same in a certain middle range.
2. R_0 becomes close to C as E_s/N_0 grows. In fact, for almost all channels, R_0 tends to C as $E_s/N_0 \to \infty$ and to $C/2$ as $E_s/N_0 \to 0$.
3. R_0 is thus a good indication of C in a good channel.
4. Furthermore, if one accepts that decoding is "easy" at rates below R_0, then decoding close to capacity is "easy" in channels with large E_s/N_0. These are the channels to the right in Fig. 3.18. In particular, Fig. 3.18 shows that it is not hard to approach C with 16QAM codes.

Coding and Information Theory

The mathematical details of $E(R)$ for memoryless channels are treated in [11,12,19]. Actual values of R_0, C and $E(R)$ are listed for all the important memoryless channels in Kressel [17]; included particularly there are the channels underlying parity-check and TCM codes. Channel models with memory underlie CPM and PRS codes, and cut-off rates for CPM codes, together with the general calculation method for channels with Markovian memory, appear in [8].

3.7. Problems

1. Consider the rate 1/2 convolutional code with generators (40,64). (*a*) Draw a shift-register encoder circuit. (*b*) Give the generator matrix G when the code is used in block form with $N = 10$ bits per codeword. (*c*) From G, find H, the parity-check matrix. (*d*) Find the minimum distance of this code.

2. Show that $g(D) = D^4 + D^3 + 1$ generates the (15,11) binary cyclic Hamming code. Find the parity-check polynomial $h(D)$. Express the code in terms of parity check and generator matrices, both in systematic form.

3. Consider the extended binary (N, K) parity-check code. In such a code a single bit is added to the end of each original codeword, to form a $(N + 1, K)$ code; this bit is the mod-2 sum of the N original bits. (*a*) Form the (8,4) extension of the (7, 4) Hamming code and list its words. (*b*) Find the minimum distance. (*c*) Show that the minimum distance of an extended code is, in general, one more than the original code.

4. The optimal memory-2 rate-1/2 convolutional feed-forward code is the code with generators (7,5) and free distance 5. (*a*) Draw the shift register generator circuit for this code. (*b*) Draw the shift register circuit for the equivalent systematic feedback encoder. (*c*) Draw the trellises for both codes. (*d*) Give an asymptotically correct expression for the probability that the first data symbol is decoded in error in AWGN.

5. Consider the memory-2 rate 1/3 code with generators (7,7,5) and free distance 8. (*a*) Draw the first six sections of the code trellis. (*b*) Use the VA to decode the least distant path when the received sequence is 000 111 000 000 000. (The first bit triple corresponds to the leftmost trellis stage.)

6. Find the hard and soft decision AWGN coding gains G_h and G_s for the memory-2 rate 1/2 and 1/3 codes in Problems 4 and 5. Compare these. Rate 1/3 represents a 50% bandwidth expansion over rate 1/2. Is it worthwhile here?

7. A fair coin is tossed until a head appears. If ℓ is the length of the run of tails, $\ell = 0, 1, \ldots$, find the entropy of the random variable L.

8. Prove that the entropy of a single Gaussian variate is $(1/2) \log[2\pi e \sigma^2]$, where σ^2 is its variance.

9. Prove that all three definitions given in Eqs (3.6-1) and (3.6-2) for the mutual information are identical.

10. Find the capacity of the BEC in terms of p and q.

11. A binary channel has transition matrix

$$p(y\,|\,x) = \begin{bmatrix} 0.9 & 0.1 \\ 0.2 & 0.8 \end{bmatrix}$$

   (a) Find the mutual information $I(X;Y)$, the equivocation $H(X\,|\,Y)$, and the irrelevance $H(Y\,|\,X)$, if the inputs X are equally likely. (b) What is the capacity of this channel? (c) Compare this answer to the capacity of a BSC with crossover 0.2.

12. An exceptionally good BSC is one with crossover probability 0.0001. (a) Find its capacity, C. (b) Find its cut-off rate and compare to C. (c) Using the estimate found in Problem 4(d), give the approximate probability of error for an ML decoder and the (7,5) rate 1/2 convolutional code. (d) For what crossover probability is the capacity of the BSC equal to the rate of this code?

13. Suppose a BSC is implemented by BPSK and E_b/N_0 is 5 dB. (a) Find the probability of bit error in this channel. (b) Find the BSC channel capacity and cut-off rate. (Note that with long enough codewords, information may be carried essentially error free at these rates.) (c) In the spirit of Eq. (3.6-16), find the Shannon capacity of this channel subject to the constraint that it may have average BER equal to 0.01. Compare to the capacity in (b).

14. Find the information rate in bits/s for the 4-pole Butterworth filter in Example 2.8-1, when the filter is employed in a Filter/Transmitter channel model system. Let the filter cutoff frequency be $0.5/T$ Hz, with $T = 1$, and set the power parameter P to 3 watts. Express the information rate as a function of the allowed bandwidth W. (This problem will require Matlab or a similar tool.)

15. In Example 3.6-5, the Filter/Medium channel capacity turned out to be 3.34 bits/s. By trying out several other allocations of the power in $G(f)$ to the two spectral bands, verify that this rate is indeed the maximum that can occur at power 3 watts.

16. Write a program for your toolbox that finds the Shannon limit in Fig. 3.20 (P_{ber} vs E_b/N_0) at a specified coding rate R. Compare the curve for $R = 3$ bits/channel use to the approximate BER for 64QAM. What is the energy loss at BER 10^{-5} of 64QAM compared to the Shannon limit?

Bibliography

References marked with an asterix are recommended as supplementary reading.

[1] *J. B. Anderson, *Digital Transmission Engineering*. IEEE Press, New York, 1999.
[2] A. M. Michelson and A. H. Levesque, *Error-Control Techniques for Digital Communication*. Wiley, New York, 1985.
[3] S. Lin and D. J. Costello, Jr., *Error Control Coding*. Prentice-Hall, Englewood Cliffs, NJ, 1983.
[4] R. E. Blahut, *Theory and Practice of Error Control Codes*. Addison-Wesley, Reading, MA, 1983.
[5] J. B. Anderson and S. Mohan, *Source and Channel Coding*. Kluwer, Boston, MA, 1991.
[6] *G. C. Clark, Jr. and J. Bibb Cain, *Error-Correction Coding for Digital Communications*. Plenum, New York, 1981.
[7] J. B. Anderson and K. Balachandran, "Decision depths of convolutional codes," *IEEE Trans. Inf. Theory*, **IT-35**, 455–459, March 1989.

[8] J. B. Anderson, T. Aulin and C.-E. Sundberg, *Digital Phase Modulation*. Plenum, New York, 1986.
[9] A. J. Viterbi, "Error bounds for convolutional codes and an asymptotically optimal decoding algorithm," *IEEE Trans. Inf. Theory*, **IT-13**, 260–269, April 1967.
[10] L. R. Bahl, J. Cocke, F. Jelinek and J. Raviv, "Optimal decoding of linear codes for minimizing symbol error rate," *IEEE Trans. Inf. Theory*, **IT-20**, 284–287, March 1974.
[11] R. Johannesson and K. Zigangirov, *Fundamentals of Convolutional Coding*. IEEE Press, New York, 1999.
[12] *R. G. Gallager, *Information Theory and Reliable Communication*. McGraw-Hill, New York, 1968.
[13] T. M. Cover and J. A. Thomas, *Elements of Information Theory*. Wiley, New York, 1991.
[14] H. J. Landau and H. O. Pollak, "Prolate spheroidal wave functions, Fourier analysis and uncertainty, III: the dimension of the space of essentially time- and bandlimited signals," *Bell Syst. Tech. J.*, **41**, 1295–1336, July 1962.
[15] *C. E. Shannon, "A mathematical theory of communication," *Bell Syst. Tech. J.*, **27**, 379–429 (Part I) and 623–656 (Part II), July 1948; reprinted in *Claude Elwood Shannon: Collected Papers*, Sloane and Wyner, (eds), IEEE Press, NY, 1993.
[16] C. E. Shannon, "Communication in the presence of noise," *Proc. IRE*, **37**, 10–21, 1949; reprinted as cited in [15].
[17] U. Kressel, "Informationstheoretische Beurteilung digitaler Uebertragungsverfahren mit Hilfe des Fehlerexponenten," ("Information theoretic evaluation of digital transmission methods with the help of error exponents"), Fortschritt-Berichte VDI, Reihe 10, Nr. 121, VDI-Verlag, Duesseldorf, 1989 (in German).
[18] B. Friedrichs, *Kanalcodierung*. Springer-Verlag, Berlin, 1995 (in German).
[19] A. Viterbi and J. Omura, *Principles of Digital Communication and Coding*. McGraw-Hill, New York, 1979.
[20] F. M. Reza, *An Introduction to Information Theory*. McGraw-Hill, New York, 1961.
[21] E. A. Lee and D. G. Messerschmitt, *Digital Communication*. Kluwer, Boston, 1988 and later editions.
[22] J. B. Anderson and K. Tepe, "Properties of the tailbiting BCJR decoder," in *Codes, Systems and Graphical Models*, Marcus and Rosenthal (eds), IMA Volumes in Mathematics and its Applications, vol. 123, Springer-Verlag, New York, 2001.

4

Set-partition Coding

4.1. Introduction

The first class of coded modulations in this book was the second to be extensively studied but the one that finds the greatest use at present. It is based on coding the amplitudes of an orthogonal-pulse linear modulation. These values can be viewed as the components of the signal vector in signal space. As mentioned in Chapter 1, Shannon introduced the concept of coding these components in 1949. The properties of pulse trains were discussed in Section 2.2. Set-partition coding normally envisions a pair of such rate $1/T$ pulse trains in an orthogonal modulation I/Q combination, although more than two can be employed and there is no requirement to take them in pairs. Still, it is convenient to speak of constellation points in two dimensions, points that in polar notation have amplitude and phase. The values of the trains are assumed to form a quadrature amplitude modulation (QAM)-like collection of points (see Section 2.5.2) called the *master constellation*. Codewords are patterns of these values: from symbol time to symbol time only certain sequences of QAM values are allowed. Set-partition coding works by organizing the QAM values into certain *subsets* at certain times. Codeword letters are constrained to take values in these subsets, and the choice of the subset is what is sequenced in a coded way. The choice is set by a finite state machine and in most practical schemes it is a simple tapped shift register already familiar to us, a convolutional encoder. The process is explained in detail in Sections 4.2 and 4.3.

The idea of sequencing through QAM subsets under the direction of a convolutional shift register was suggested by G. Ungerboeck in the 1970s, although he did not publish it in journal form until 1982 [4].[1] An older idea called lattice coding works by defining codewords to be points of a mathematical lattice in N-dimensional Euclidean space. Successive dimensions of the space can be set up as above, by pulse amplitudes in successive symbol intervals. The code consists of all points in some volume of space, and the most reasonable such volume is an N-sphere of radius $\sqrt{NE_s}$. Lattice codes were originally studied by theoreticians who were looking for alternate proofs of the Gaussian channel coding theorem, but in time the ideas of partitioning the set of points into subsets came to be applied to lattices as well. For a subset the natural object was a lattice coset. Cosets, whether

---

[1] However, Ungerboeck's work was described earlier in the journal literature by others [6].

of lattices or of group codes, are another older means of code design. Lattice coding ideas form Section 4.4.

By the 1990s the ideas of lattice, set partition, and coset coding were more or less all contained under the banner of set-partition coding. The field has taken on the perhaps misleading[2] name trellis coded modulation (TCM) coding. We will continue the usage and take it to mean coding based on a QAM-like constellation in Euclidean N-space, usually but not always coupled with a subset idea. Codewords $x = x[1]x[2]\ldots$ are a sequence of points x from such a constellation. The axes of N-space are created through N uses of a memoryless, probably additive white Gaussian noise (AWGN) channel. When there are subsets, and the term is appropriate, the term set-partition coding will be used. Coding in Euclidean N-space, without set partitioning, is another older idea that goes back to Shannon. Even though Chapter 4 is mostly set-partition coding, the chapter seems a reasonable place for Euclidean coding without set partitioning, and that is the subject of Section 4.5. N-space Euclidean coding, with and without set partitioning, has seen a remarkable growth of interest since the early 1980s. In part, this stems from a very important application, the telephone modem. Among researchers, an attraction has been that the coding is a straightforward yet interesting extension of traditional BSC coding, now with a new symbol arithmetic and a new distance measure. By the end of the 1990s, there were several book or chapter treatments of TCM [1–3].

TCM makes use of a channel that preserves both phase and amplitude variation. As such, it is quite different from CPM coding in the next chapter, which exploits a constant-envelope channel that preserves only phase. The PRS and related coding systems in Chapter 6 require a phase and amplitude channel, and are therefore more allied to TCM coding. If need be, both can be modeled in terms of successive uses of a memoryless, real-number channel with Euclidean metric. The differences are that TCM usually exploits a large data symbol alphabet and the set-partition idea, and TCM's pattern-setting structures are binary-arithmetic processors, whereas the systems in Chapter 6 employ a small alphabet and code directly over the reals.

Energy, Bandwidth, and Capacity

Energy and bandwidth are particularly simple in a TCM code. Since its constellation stems from a train $\sqrt{E_s} \sum a_n v(t-nT)$ of T-orthonormal pulses, the energy in an N-dimensional constellation is simply $NE_s\mathcal{E}\{|a|^2\}$. Distance between signals can be computed either from the analog signals or from the respective modulation symbols $\{a_n\}$, in the manner discussed in Section 2.5.2. It is expressed in this chapter as it is in the rest of the book, as a value normalized by twice E_b,

---

[2] TCM originally stood for Trellis Coded Modulation, yet any practical coding scheme has a trellis, all coded modulations are coded, and all surely use a modulator. The phrase is thus meaningless in the context of this book if taken literally.

Set-partition Coding 135

the energy expended per data bit. Occasionally, it is natural to express distance as a gain over ordinary QAM signaling at the same rate. The bandwidth of signals is the bandwidth of $v(t)$ (see Section 2.7). Aside from changing the constellation symbol alphabet or the subset configuration, there is *no design of bandwidth* in TCM coding. With a 30% root RC pulse, for example, baseband transmission bandwidth is $(1 + 0.3)1/2T = 0.65/T$ Hz in an absolute sense, or $0.575/T$ Hz in a 99% sense. The search for good codes of a certain configuration is an exercise in Euclidean distance optimization. This is certainly not the case in CPM and PRS coding, where the optimization is simultaneously over distance and bandwidth.

The rate R of a TCM code will ordinarily be given in the most natural way for applications, which is in data bits per appearance of a 2-dimensional QAM-like constellation, shortened to bits/2-dim. In QAM, one constellation appears each T-second symbol interval, and so it is natural to express the same rate as bits/QAM interval. These rates are typically 2 or 3 or more, in comparison to $0 < R < 1$ for parity-check codes; the point of TCM coding in transmission engineering is that it allows such rates. But a QAM interval comprises two signal space dimensions, and a lattice may exist over any number of dimensions, and so in some contexts rates in bits per signal space dimension (abbreviated bits/dim) are appropriate. An equivalent phrase in Chapters 4 and 6 is bits/channel use. These rates are half those just discussed.

An archaic usage, which we avoid, is to call the rate of a TCM code the rate of the convolutional shift register or other mechanism that sets up the subset sequencing. This rate is not the code rate, and the extra bits that appear in the processing are not properly called parity checks. TCM codes are not parity-check codes.

Rates that can be compared across very different coding schemes and a proper measure of bandwidth are a particular need in this book. In comparing TCM to other methods where an independent AWGN "channel use" can be defined, the rate per single dimension or single channel use is usually the important one. In comparing TCM to CPM and PRS coding, it is not natural to speak of independent AWGN channel uses. Here we will use measures of consumption like the bandwidth and energy per data bit W_b and E_b that were defined in Section 3.6.2. It is clear that W_b is proportional to the reciprocal of R, but a more precise definition of bandwidth needs special care. There are several ways to measure a spectrum, and some applications are baseband and some are bandpass. A pair of dimensions in QAM, for example, can just as well come from two unrelated 1-dimensional channels as from a 2-dimensional bandpass one. The *normalized bandwidth* (NBW) of a scheme is given by

$$\text{NBW} = \frac{TWN_{\text{dim}}}{R} \text{ (Hz-s/data bit)}, \qquad (4.1\text{-}1)$$

where T is the modulation symbol time, R is the rate, W is a measure of bandwidth, and N_{dim} is the number of independent dimensions the system uses. For baseband transmission, $N_{\text{dim}} = 1$; for a passband system, $N_{\text{dim}} = 2$. Two independent dimensions may be transmitted in a carrier system and the scheme thus consumes twice the NBW. The quantity W can measure the equivalent baseband spectrum, the equivalent RF spectrum, the 99% power bandwidth, etc.; the NBW then becomes the NBW in this same sense.[3] For TCM coding, $N_{\text{dim}} = 2$ and W will ordinarily be taken as the baseband positive-frequency bandwidth of the modulation pulse. It may be useful to compare the beginning of Section 6.4, where these issues are discussed as they apply to PRS coding, which is most often a $N_{\text{dim}} = 1$ scheme.

The subject of Shannon capacity for channel models employed with TCM is treated in Sections 3.6.1 and 3.6.2, and will not be discussed at length in this chapter. One approach, given in Section 3.6.2, is to compute the Shannon capacity of the AWGN channel with the same bandwidth and E_b/N_0 as the TCM code. A more focused approach, in Section 3.6.1, is to compute a Shannon information rate for equi-probable data and codes constrained to the points in the master constellation (see Example 3.6.4). Some cut-off and information rates of this type for representative constellations are shown in Fig. 3.18. Some basic properties of these are: (i) At practical E_b/N_0, cut-off rate is nearly the same as information rate, (ii) both are nearly $\log_2 M_c$ bits/2-dim, and (iii) they lie far below the capacity of the AWGN channel without the master constellation constraint. These computations take no notice of the subset sequencing that is characteristic of TCM codes, and indeed, they should not. We will find that adherence to the wrong size and number of subsets – or to the subset idea at all – can unnecessarily degrade the code rate and error performance. From a coding theory point of view, the subset idea has mainly *implementation* significance: it provides a relatively simple and cost-effective method of high-rate coding.

The capacity of an AWGN channel with bandwidth W_b, as a function of E_b and W_b, is another way to compute capacity. Section 3.6.2 computes this. In TCM coding W_b depends on the subset size and number. Once that configuration is set, W_b is still proportional to the bandwidth of the underlying orthogonal pulses. A practical scheme is always bounded away from capacity by at least the excess bandwidth in the pulses, no matter how clever is the coding in the N-dimensional space that the pulses create. But we leave Shannon theory now, and turn to more concrete coding schemes.

4.2. Basics of Set Partitioning

In order to create a TCM code, three jobs need to be done:

1. define a master QAM-like constellation (it may have more than two dimensions);

---

[3] W_b in Section 3.6.2 can be considered to measure positive baseband frequencies over which the signal has nonzero power spectral density.

Set-partition Coding

2. partition the constellation into subsets;
3. design a machine that creates a stream of subsets in response to some of the data.

The master constellation has M_c points. Each subset has M_s points, their union is the master constellation and there are M_c/M_s subsets. A TCM code carries data bits in two ways. First, b bits per QAM interval drive the subset selection machine; these determine *which* subset applies to an interval. Second, $\log_2 M_s$ bits are carried *in* the subset, once selected. The code rate overall is thus:

$$R = b + \log_2 M_s \text{ (bits/2-dim)}. \tag{4.2-1}$$

This is $R = (b + \log_2 M_s)/2$ bits per independent AWGN channel use, or equivalently, per dimension of signal space.

4.2.1. An Introductory Example

We turn immediately to a simple example, which first appeared in Section 3.3, Fig. 3.9. Those who do not need this introduction can skip directly to Section 4.2.2.

The master constellation is 8PSK and the subsets, labeled $C_0, C_1, C_2,$ and C_3, are the four antipodal point pairs in that constellation (see Fig. 4.1). Since it has two points, a subset can carry one data bit. For the subset selection machine we will borrow the (7, 5) feedforward rate 1/2 convolutional encoder. This simple device takes in one bit, a data bit, and puts out two, which will select the subset according to a table (again see Fig. 4.1). It is important to realize that even though the device can be used to generate error-correcting codewords, it is not so used here; rather, it generates a pseudorandom sequence of subsets. The overall rate of the coding system is 2 data bits/2-dim (or 1 bit per signal space dimension).

How are energy and distance reckoned in this code? 8PSK has the general expression (2.5-8), with 8 modulation symbol pairs $\left(a_n^I, a_n^Q\right)$ spaced uniformly around a circle of radius E_s, as in the figure; the symbol energy is E_s per two dimensions. The TCM transmission is a sequence of 8PSK constellation points. To find the minimum distance between two TCM transmissions, we must consider neighboring point pairs of two types, those within a subset and those in two different subsets. Denote by d_{ss} the *same subset minimum distance*, the minimum distance of points within a subset. In every case in the present example, a subset point is square-distant $4E_s$ from one other subset point. The energy per data bit in the TCM system is $E_b = E_s/R = E_s/2$ joules/data bit, and the standard division by $2E_b$ gives a normalized square distance of $4E_s/2E_b = 4E_s/E_s = 4$. In AWGN, an event of this type occurs with probability $\sim Q(\sqrt{4E_b/N_0})$.

The second type of neighboring pair is more subtle to think about, and as with many other coding methods, we organize our thoughts with a trellis structure.

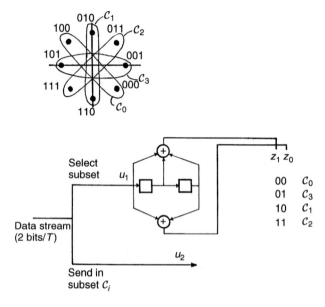

Figure 4.1 Introductory TCM scheme with rate 2 bits/2-dim: formation of four subsets from an 8PSK constellation, division of data stream into subset selection bit u_1 and subset-transported bit u_2, and subset selector machine.

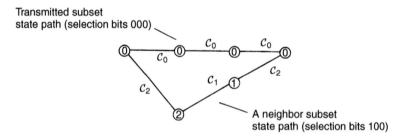

Figure 4.2 Two neighboring subset paths in the subset selection trellis for the code in Fig. 4.1.

But now the trellis will show the time evolutions not of codeword symbols, but of *subsets*. It is these evolutions that are driven by the subset selection machine, and the trellis shows states vs time of that machine. A section of the trellis was given in Fig. 3.9. The selection machine here is a (7, 5) convolutional encoder, and not surprisingly, the subset trellis is the same as the convolutional one, but it now has subset branch labels. The points in the second type of pair lie along different paths in the trellis.

One such pair is shown in Fig. 4.2: one transmission produces the subset selection bits $u_1[1]u_1[2]u_1[3] = 000$, the other produces 100, and the subset

Set-partition Coding

sequences are $C_0C_0C_0$ and $C_2C_1C_2$, respectively. Between the state path split and merge 3 symbol times later, there are two intervals in which the transmitted and neighbor path subsets are C_0 and C_2, and one for which it is C_0 and C_1. In the intervals of the C_0/C_2 kind, some study of the subsets in Fig. 4.1 shows that no transmitted point in subset in C_0 can come closer to a neighbor point in C_2 than square distance $2E_s$ (two points away in the master constellation). This is an example of an *inter-subset* distance: given two subsets C_i and C_j, it is the minimum distance between two constellation points, given that one is in C_i and one is in C_j. In the C_0/C_1 case, no points can lie closer than $0.586E_s$. Note that this is also the square minimum distance of 8PSK. With the worst occurrence of the subset data bits, then, the total square distance between these trellis paths is $(2+2+0.586)E_s$, or $4.586E_s$. Repeating the calculation for other path pairs in the selector state trellis shows that no worse case occurs than this one.

What, then, is the TCM code minimum distance? For transmissions along the same state trellis path, neighbors can lie $D_{ss}^2 = 4E_s$ apart and otherwise they can lie $4.586E_s$; the square minimum distance is thus $4E_s$, or after normalizing, just 4. An algorithm to find the minimum distance would repeat the example here systematically throughout the trellis structure, in the same way that minimum distance algorithms work with other trellis codes, but employing intra-subset distances. Some details about algorithms are given in Section 4.3.3. If the trellis is finite state, the minimum distance so obtained will not increase after a certain trellis depth, and it is proper to call the distance a free distance.

The search for minimum distance does differ in one way from ordinary trellis coding: the outcome can never exceed the same-subset minimum distance. However clever the subset selection is, the free distance in the example cannot exceed 4. Still, the scheme has a 3 dB asymptotic energy advantage over QPSK, a scheme with the same per data bit bandwidth[4] but whose d_{min}^2 is only 2. The TCM scheme has a 6.6 dB distance advantage over 8PSK, but it consumes also 50% more bandwidth, so the comparison is unfair in a combined bandwidth–energy sense.

The example here is simple and effective as a coding scheme, but it is *ad hoc* and we need to look more carefully at how to design the constellation, subsets, and selector circuit.

4.2.2. Constellation and Subset Design

We look first at how to design a master constellation and subsets that will work together to give a good code performance. Many combinations are possible. The same subset minimum distance d_{ss} will play the major role in evaluating them. The selector design problem will be delayed until Section 4.3.

---

[4] The 99% baseband NBW of both schemes is $(2/2)(1+\alpha/2)/2T$ Hz-s/bit, where α is the pulse excess bandwidth factor. For root RC pulses and $\alpha = 0.3$, this is 0.575 Hz-s/bit.

The first task is to choose an agreeable master constellation. Very often, this is simply rectangular M-ary QAM, since modulations, synchronizers, and analog systems exist for this, leaving only the software and logic design needed for the TCM coding. A 16-point rectangular QAM is shown in Fig. 2.18(b). Two important parameters of a QAM constellation are its minimum distance $D_{\min}$ and its average energy $\bar{E}_s$. Henceforth, we will assume that the master constellation is unit spaced, that is, $D_{\min} = 1$. The average energy of an M-ary constellation is thus,

$$\bar{E}_s \triangleq (1/M) \sum \|x_i\|^2, \qquad (4.2\text{-}2)$$

where $\|x_i\|^2$ is the square length of point x_i. For the same physical minimum distance $D_{\min}$, however, some constellations have lower $\bar{E}_s$, and therefore the normalized square minimum distance, $d_{\min}^2 = RD_{\min}^2/2\bar{E}_s$, grows larger. Constellations with roughly uniform points inside a circular perimeter tend to have the least average energy. Because of their "corners", standard QAM constellations are thus not quite ideal. The 8PSK constellation in Section 4.2.1 is perfectly circular, but it falls short because the space inside is not occupied.

Some practical compromises are shown in Fig. 4.3. In addition to standard QAM constellations, denoted MQAM, some "cross" correlations appear there;

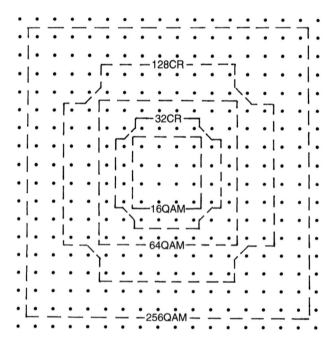

Figure 4.3 A rectangular array of constellation points and the standard 16QAM, 32CR, 64QAM, 128CR, and 256QAM master constellations. All constellations have minimum distance 1.

Set-partition Coding

Table 4.1 Average energy, energy per data bit, and minimum distance for standard TCM master constellations, plus QPSK

Constellation	$\log_2 M_c$	Average energy, E_s	Energy/bit E_b	$d_{\min}^2$	[dB below 2]
4QAM [QPSK]	2	0.5	0.25	2	[0 dB]
8PSK	3	1	0.333	0.879	[−3.6]
16PSK	4	1	0.25	0.304	[−8.2]
16QAM	4	2.5	0.625	0.8	[−4.0]
32CR	5	5	1	0.5	[−6.0]
64QAM	6	10.5	1.75	2/7	[−8.5]
128CR	7	20.5	41/14	7/41	[−10.7]
256QAM	8	42.5	85/16	8/85	[−13.3]

these are QAM-like constellations with the corners removed and will be denoted MCR. The standard cross correlations with $M \leq 256$ used in TCM coding are 32CR and 128CR, and some non-square form is required for these, but it is also true that the cross shape lowers the average energy. All constellations in Fig. 4.3 have the same $D_{\min}$ and their normalized minimum distance therefore depends solely on $\bar{E}_s/R$.

When $D_{\min}$ is set to 1, Table 4.1 shows $\bar{E}_s$ and the $d_{\min}^2$ that ultimately results, for all these constellations. $d_{\min}^2$ falls off rapidly with the constellation size M_c. This is simply a tradeoff of bandwidth for energy. With a rectangular grid, QAM satisfies

$$\bar{E}_s = (M_c - 1)/6 \tag{4.2-3}$$

and $d_{\min}^2$ is

$$d_{\min}^2 = \frac{3 \log_2 M_c}{M_c - 1} \approx \frac{3\mu}{2^\mu}, \tag{4.2-4}$$

where $\mu = \log_2 M_c$ is the number of bits in the constellation. Each doubling of the constellation size approximately doubles the average energy and halves the normalized square distance. The same happens with the cross family, although the law that replaces Eq. (4.2-4) is

$$\bar{E}_s \approx (31/32) M_c/6 \tag{4.2-5}$$

and the normalized minimum distance satisfies

$$d_{\min}^2 \approx (32/31) \frac{3 \log_2 M_c}{M_c - 1} \approx (32/31) \frac{3\mu}{2^\mu}, \tag{4.2-6}$$

which represents a gain of 32/31 (0.14 dB) over Eq. (4.2-4) [1]. It can be verified from the table that approximations (4.2-4)–(4.2-6) are very tight. Some further perturbation of the point locations can raise this gain slightly higher, to about 0.20 dB (details of this will appear in Section 4.4.2).

Subset Configurations: The d_{ss} Bound

The second task is setting up the subsets and here is where the art of TCM design really centers. The most important aspect is the *number* of subsets, and therefore their size. Exactly which points go to which subset is ordinarily done via a simple repeating scheme due to Ungerboeck, and we will focus on that. The subset distance d_{ss} for those subsets creates a series of simple, tight bounds, which control the distance behavior of the TCM method.

A generic TCM encoder is defined in Fig. 4.4. It takes in R bits per QAM/PSK interval, $R - b$ that will be carried by a subset and b that help to select which subset. There are 2^c subsets and it takes c bits to select one. Some thought will show that $c > b$; otherwise, d_f will turn out to be just d_{min} for the master constellation. A total of $\log_2 M_c$ bits pass to a mapper that selects the QAM point. Figure 4.5 shows schematically the actual configurations that are possible with 16 and 32 point master constellations.[5] The first $R - b$ bits are sometimes called the uncoded bits; correspondingly, the selection bits are called the coded bits. We will continue this traditional terminology, but strictly speaking, all bits are jointly coded and none consistently appear as themselves in the coded modulation signal. The block produces $\log_2 M_c$ bits which taken together specify a constellation point. A mapper box converts block outputs to respective constellation points.

Consider now a scheme of rate R, R an integer. Clearly R satisfies $0 < R \leq \log_2 M_c$. But $R = 1$ would mean no uncoded bits and $R = \log_2 M_c$ is uncoded QAM. Thus, R will satisfy

$$2 \leq R \leq \log_2 M_c - 1. \quad (4.2\text{-}7)$$

Figure 4.4 A generic TCM encoder, with b bits that take part in the subset selection and $R - b$ that are carried uncoded in the subset. A total $\log_2 M_c$ bits go to the mapper.

---

[5] Throughout, we consider only integer rates, $\log_2 M_c = 2^k$, and 2^c subsets, k and c integers; other numbers are possible, of course, but seldom used.

Set-partition Coding

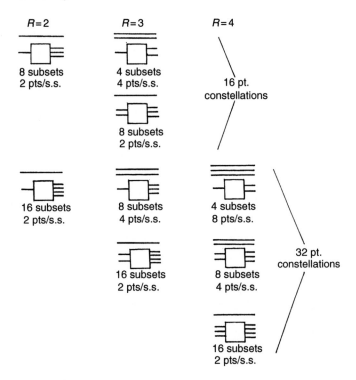

Figure 4.5 Possible TCM subset configurations for 16 and 32 point master constellations and rates 2, 3, 4. Box denotes subset selection machine.

Figure 4.5 shows all the allowed configurations for these rates. At each R, the coded bit number b can be $1, \ldots, R-1$, giving $M_c/2^{R-b}$ subsets of 2^{R-b} points. Most practical TCM schemes work at the highest allowed rate, $R = \log_2 M_c - 1$, the rightmost configurations in the figure. The most common breakdown is $b = 1$; that is,

$$\log_2 M_c = R + 1;$$
$$b = 1 \text{ (coded bits)}, \qquad (4.2\text{-}8)$$
$$R - b = R - 1 \text{ (uncoded bits)},$$

or a master constellation of 2^{R+1} points with 4 subsets of 2^{R-1} points. These configurations are the upper right corner of each M_c-group. This configuration is chosen because its implementation is simplest; we will see, however, that it is not usually the configuration with the best performance potential. This is reasonable, because the schemes lower in a group (with higher b) in a sense include those higher, since they can set up the higher subset definitions as trivial special cases.

The most powerful codes are likely to be those that fully encode all R bits, and are not of the set-partition type at all. But we will delay taking up these until Section 4.5.

Once the configuration is set, it remains to define the $2^{\log_2 M_c - R + b}$ actual subsets. Within each subset, points should spread out uniformly so that they have the largest possible minimum distance: the d_f of the final TCM code cannot exceed this distance. For creating the separate subsets, it is difficult to improve on the following set of *Ungerboeck Rules*, suggested by him in his original paper [4]. The procedure is illustrated for 16QAM in Fig. 4.6. It consists of successively splitting the master constellation in half, according to these rules.

1. Points in any subset are distributed as uniformly as possible.
2. At each level in the partition, split all sets into two subsets containing "alternate" points.
3. Associate binary symbols z_0 with the first partition level, z_1 with the second, etc., as shown in Fig. 4.6. These symbols select the subset.

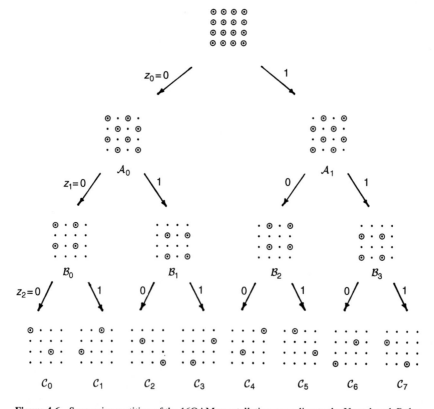

Figure 4.6 Successive partition of the 16QAM constellation according to the Ungerboeck Rules.

Set-partition Coding

The process may be continued $\log_2 M_c - 1$ times, producing successively $2, 4, 8, \ldots, M_c/2$ subsets. The subset square minimum distance doubles after each split. The split outcomes that produce 2^c subsets can be identified by c bits, $z_0 z_1 \ldots z_{c-1}$. Furthermore, every constellation point is used once, and if the subsets are used with equal probability, the constellation points all appear with equal probability. At a given level, we can number these subsets left to right as $z_0 z_1 \ldots z_{c-1}$, from $00 \ldots 0$ to $11 \ldots 1$ (z_{c-1}, the least significant bit, is on the right). This is called a *natural mapping*, since it is a straightforward count in binaries.

The meaning of alternate points in rule 2 is that a set is split such that the points in each new set are the companion set's nearest neighbors. Starting from a standard rectangular array, successive splits produce a checkerboard pattern, a square pattern, a checkerboard, and so on, as is clear in the figure. The concept of alternate points will be defined more precisely in the lattice discussion in Section 4.4.

One more rule defines the assignment of subsets to selector trellis branches:

4. The inter-subset distances among the 2^b subsets leaving a trellis node should be as large as possible, and similarly for the subsets entering a node.

If the rules are followed, the 2^b subsets in rule 4 will be those that have a common father ($b = 1$), a common grandfather ($b = 2$), and so on for higher b. Rule 4 is borrowed from convolutional coding; the idea is that a code, or a subset selection procedure, that starts and ends well is likely to be a good procedure.

To illustrate the Ungerboeck Rules, we will work through subset design examples based on Fig. 4.6 and on the 32CR constellation. Several important properties will emerge and these will be generalized at the end.

Example 4.2-1 *(Subset Configurations for 16QAM)*. The master constellation is the one at the top of Fig. 4.6. Its square distance on an absolute scale is taken as $D_{\min} = 1$, which yields average energy $\bar{E}_s = 2.5$. After successive splits, the Ungerboeck Rules yield 2, 4 and 8 subsets, with the same subset square minimum distances $D_{ss}^2 = 2$, 4 and 8, respectively. At a code rate of 2 bits/QAM interval (1 bit/dim), a look at Fig. 4.5 shows that only one configuration is usable, the one using the bottom row of 8 subsets $C_0, \ldots, C_7$ in Fig. 4.6. The subset sequence machine is a 1-for-3 machine; a convolutional encoder machine will have rate 1/3. The TCM code square free distance cannot exceed the same subset minimum distance D_{ss}^2, namely 8; in normalized terms this limit is $d_{ss}^2 = 8/2(\bar{E}_s/R) = 8/2(2.5/2) = 3.2$. This is a gain of 2.0 dB over QPSK, an uncoded scheme with the same NBW. With a complex enough subset sequence machine, this gain can presumably be attained.

Next we explore rate 3. Figure 4.5 shows two configurations, one with 4 subsets ($b = 1$) and one with 8 ($b = 2$). The Ungerboeck Rules after two splits yield 4 subsets with normalized $d_{ss}^2 = 4/2(\bar{E}_s/R) = 4/2(2.5/3) = 2.4$; after 3 splits they yield 8 subsets with $d_{ss}^2 = 8/2(\bar{E}_s/R) = 4.8$. We see that the second configuration, 1 uncoded bit/2 subset selection bits, has potentially 3 dB higher d_f^2, and is clearly the better one. However, in Section 4.3 it will develop that the 2-for-3 sequence machine is much more complex. There is a distance–complexity tradeoff here.

Example 4.2-2 *(Subset Configurations for 32 Cross).* The master constellation is the 32-point one in Fig. 4.3. Partitions into 4, 8, 16 subsets are usable, if the partition is carried out in the style of Fig. 4.6. Even though the constellation is not square QAM, the points still form a rectangular grid and the same Ungerboeck Rules may be applied. The non-normalized square subset minimum distances that result are respectively $D_{ss}^2 = 4, 8, 16$. The master constellation average energy is $\bar{E}_s = 5$. At the highest allowed code rate, which is 4 bits/QAM interval, the normalized subset distances are

$$d_{ss}^2 \triangleq \frac{D_{ss}^2}{2\bar{E}_s/R} = 1.6, 3.2, 6.4. \qquad (4.2\text{-}9)$$

In comparison, ordinary 16QAM, a rate 4 scheme working in the same bandwidth, has $d_{min}^2 = 0.8$. We can draw two interesting conclusions: the simplest TCM code structure with 4 subsets of 8 points and 1-for-2 subset selector, has d_f^2 potentially as large as 1.6, a 3 dB improvement over 16QAM; and further subset splits to 16 subsets of 2 points can create codes with d_f^2 as big as 6.4, a 9 dB gain. This is a huge gain if the subset selectors can be found. The configurations here are those in the rightmost column of Fig. 4.5.

With rate 3, the same calculation as in Eq. (4.2-9) yields $d_{ss}^2 = 2.4, 4.8$ for 8- and 16-subset configurations (middle column in Fig. 4.5). These are sizable gains over various 8-point PSK and QAM modulations, but they are just the same as the distances for rate 3 TCM based on 16QAM in Example 4.2-1: the more complex constellation has gained nothing. But a close look shows that there is a gain in another sense. By skipping rate 3 and going to rate 4, higher distances are possible in theory, even though the rate is higher. It will become clear later that the payment for this is a much higher code complexity.

Now we will collect the trends in these examples together and look for some rules for TCM configurations. We will list the trends first and then give some further explanations.

1. Each subset split contributes 3 dB coding gain. But we can be more explicit. The QAM and CR constellation average energies closely follow the law $\bar{E}_s \approx 2^\mu/6$, with $\mu = \log_2 M_c$ (see Eqs. (4.2-3) and (4.2-5)). Under the Ungerboeck Rules, D_{ss}^2 doubles each split. Let S denote the number of splits and R the code rate in bits/2-dim. Then,

$$d_{ss}^2 \triangleq \frac{D_{ss}^2}{2\bar{E}_s/R} \approx \frac{3 \cdot 2^S R}{2^\mu}, \qquad S \leq \mu - 1. \qquad (4.2\text{-}10)$$

Ordinary QAM obeys Eq. (4.2-4).

Set-partition Coding

2. A complicated enough subset selector will have a long-constraint state trellis, so that d_f reaches d_{ss}. The details of these machines appear in Section 4.3; for now, we can say such selectors seem always to exist and that more splits lead to more complex selection.

3. *The most effective all-around coding* seems to occur at rate

$$R = \log_2 M_c - 1 = \mu - 1. \qquad (4.2\text{-}11)$$

4. The largest number of splits allowed is $\mu - 1$. Therefore, Eq. (4.2-10) implies that the *largest* d_{ss}^2, hence the largest upper limit to d_f^2, available from TCM with a 2^μ-point 2-dimensional constellation is

$$\max d_{ss}^2 \approx 3R/2. \qquad (4.2\text{-}12)$$

This grows linearly with R; since R is upperbounded by μ, there is good reason to allow the *largest possible* μ. These codes are complex, both because the selector will turn out to be complex and because 2^μ is large.

5. The least number of splits in a bona fide TCM code is 2. The simplest subset selector has one input ($b = 1$). The selector is therefore 1-for-c, with $c = \mu - R + 1$; with the condition $R = \mu - 1$, c is 2, and the selector is 1-for-2. From Eq. (4.2-10), this simplest scheme has

$$\text{Simplest code: } d_{ss}^2 \approx 3R/2^{R-1}, \qquad (4.2\text{-}13)$$

and from Eq. (4.2-4), its gain over rate R QAM or CR is about 2 (3 dB).

To make these rules concrete and for reference in the following, we can calculate the actual d_{ss}^2 available for practical TCM configurations. Most of these are listed in Table 4.2. The table is organized into subtables by code rate, in contrast to Fig. 4.5, which is organized by constellation type. The subtables display d_{ss}^2 on a subset-carriage bits vs constellation grid. The constellations are the standard ones of Table 4.1, with 2^μ points, $\mu = 3\text{–}8$. Each entry is a triple

$$\begin{bmatrix} S \\ b/c \\ d_{ss}^2 \end{bmatrix}$$

which denotes that S splits form the subsets (therefore $2^{\mu-S}$ points per subset) and that the subset selection is b bits in, c bits out (therefore $R - b$ bits are carried by a subset). A few entries, denoted unc, give the uncoded constellation performance for comparison; otherwise, $2^R < 2^\mu$ always. The notation $*/*$ means that R is such that not enough bits remain for subset selection; now it is assumed that rate R is sent by cycling in turn through the $2^{\mu-R}$ subsets of size 2^R that

Table 4.2 Same subset minimum distances of standard TCM configurations at rates 2–6 bits/2-dim. Distances for each rate are presented on a constellation vs $R - b$ grid, where $R - b$ is the number of bits carried by a subset. Each entry consists of the number of partition splits S, the subset selection rate b/c, and the square distance d_{ss}^2. unc denotes uncoded transmission

Subset bits, $(R - b)$ …

	Rate 2		Rate 3			Rate 4				Rate 5					Rate 6					
	2	1	3	2	1	4	3	2	1	5	4	3	2	1	6	5	4	3	2	1
8PSK	1	2	0																	
	/	1/2	unc																	
	2.0	4.0	0.88																	
16QAM	2	3	1	2	3	0														
	/	1/3	*/*	1/2	2/3	unc														
	1.6	3.2	1.2	2.4	4.8	0.8														
32CR			2	3	4	1	2	3	4	0										
			/	1/3	2/4	*/*	1/2	2/3	3/4	unc										
			1.2	2.4	4.8	0.8	1.6	3.2	6.4	0.5										
64QAM			3	4	5	2	3	4	5	1	2	3	4	5	0					
			/	1/4	2/5	*/*	1/3	2/4	3/5	*/*	1/2	2/3	3/4	4/5	unc					
			1.14	2.29	4.57	0.76	1.52	3.1	6.10	0.48	0.95	1.90	3.81	7.62	0.28					
128CR			4			3	4	5	6	2	3	4	5	6	1	2	3	4	5	6
			/			*/*	1/4	2/5	3/6	*/*	1/3	2/4	3/5	4/6	*/*	1/2	2/3	3/4	4/5	5/6
			1.17			0.78	1.56	3.1	6.24	0.49	0.98	1.95	3.90	7.80	0.29	0.59	1.17	2.34	4.68	9.37
256QAM						4	5	6	7	3	4	5	6	7	2	3	4	5	6	7
						/	1/5	2/6	3/7	*/*	1/4	2/5	3/6	4/7	*/*	1/3	2/4	3/5	4/6	5/7
						0.75	1.51	3.0	6.02	0.47	0.94	1.88	3.76	7.53	0.28	0.56	1.13	2.26	4.52	9.04

Set-partition Coding

exist in the constellation.[6] In a given row of the table, then, the horizontal axis $R - b$ passes through $R, R - 1, R - 2, \ldots$; the subset splits proceed through $(\log M_c - R), (\log M_c - R) + 1, \ldots$ and the subset selector rate entry begins with $*/*$ and is thereafter $b/c = \ell/(\mu - R + \ell), \ell = 1, \ldots, R - 1$.

The subset distance in Table 4.2 behaves in a clear and simple manner. The entries down each column are almost identical; this is the embodiment of Eq. (4.2-10), when $\mu - S$ is constant. The lead configuration in each column – the entry for the simplest constellation – is usually best; this is the $R = \log_2 M_c - 1$ rule of thumb. (The mild exception in the rate 5 table occurs because the CR constellations are slightly more efficient than rectangular QAM ones.) Finally, there should be as many subset splits as possible, since d_{ss}^2 doubles with each one. This insight has not figured so much in practical implementations, because complexity also grows with the number of splits.

The d_{ss}^2 available also grows with rate. This is startling in any case, and the more so when it is compared with some competing systems in this book. At rate 6 potentially 15 dB coding gain over 64QAM exists. How can such large coding gains exist? First, it should be said that the gains may apply only at very large E_b/N_0. Second, almost no high-rate many-split configurations have actually been built, for reasons of complexity. But a fundamental reason that such gains can exist is that they do not in fact bring the signaling closer to capacity. Equation (3.6-9) gives the capacity C in bits/dim that accrues from energy E_s. TCM occupies two such channels, with each getting $\bar{E}_s/2$; from Eq. (4.4-11), $\bar{E}_s$ grows as $2^\mu/2\pi$, assuming a circular boundary. Take $R = \mu - 1$. C is thus

$$C = \log_2[1 + 2(\bar{E}_s/2)/N_0] \to R/\log_2(2\pi N_0), \qquad (4.2\text{-}14)$$

where the convergence is in ratio as μ grows. In fact, both d_{ss}^2 and capacity grow linearly with R, and are not in conflict. Equation (4.2-14) is precisely the "Soft AWGN C" curve in Fig. 3.18.

The NBW corresponding to a position in Table 4.2 is the NBW of the modulation pulse, divided by $R/2$. To continue the 99% bandwidth/30% root RC example from earlier pages, this NBW at rate 4 is $(2/R)[(2 \cdot T/2T)(1 + \alpha/2)] = 0.29$ Hz-s/data bit. Table 4.2 shows that the simplest 32CR configuration at rate 4 can lead to square distance 1.6, a 3 dB gain over uncoded 16QAM. Asymptotic gains as high as 9 dB can exist in principle.

This then is the shape of what may be expected from traditional TCM configurations. A small distance gain can be obtained from perturbations of the basic rectangular QAM boundary. Some improvement in decoder bit error rate (BER) comes from modifying the natural bit labeling of the subsets or the Ungerboeck Rules used in their definition, as we will see in later sections. Some further gains are possible by forming higher dimensional constellations, which we will take up

---

[6] For most constellations, symmetries are such that all the subsets ordinarily have the same average energy; in this case it does not actually matter which is used.

in Section 4.4. Before that, a major remaining topic is how to implement the subset selector.

4.3. Set-partition Codes Based on Convolutional Codes

After the discussion of constellations and subsets in the previous section, what remains is to design the subset selector mechanism. From b data bits, a means is needed to select one of $M_c/2^{R-b}$ subsets during each QAM interval. In most practical TCM codes today, the means is an ordinary convolutional encoder with the needed rate, $R_{\text{conv}} = b/c = b/(\log_2 M_c - R + b)$. The encoder does not directly provide error correction here; rather, it generates a pseudorandom sequence of subsets via a simple finite-state machine. Its combination of simplicity and randomness is what makes a convolutional encoder effective. Practical schemes employ encoders with R_{conv} equal 1/2 or sometimes 2/3. These are the simplest ones possible at the most effective overall code rate, $\log_2 M_c - 1$ bits/QAM interval. The d_{ss}^2 Table 4.2 shows that typically 3 dB gain over an uncoded constellation is available at rates 4, 5, 6 when $R_{\text{conv}} = 1/2$, and 6 dB when $R_{\text{conv}} = 2/3$ (at rate 3 the gains are closer to 4 and 7 dB). We want now to achieve these potentials with as simple a convolutional encoder as possible.

After a study of convolutional subset selectors, we will look at a related topic, rotational invariance. A TCM code with this property can have its I/Q coordinate system rotated through for example, 90 degrees, without damage to the decoding. The section ends with another related topic, algorithms that find TCM code free distance.

4.3.1. Standard TCM Schemes

Convolutional encoders are introduced in Section 3.3. Codewords x there, are generated by a tapped shift register, either feedforward, as in Figs 3.2 and 3.3, or feedback as in Fig. 3.4. The tradition in TCM coding has been to depict the encoder as feedback systematic (Fig. 3.4 is a rate 1/2 example) with one additional characteristic: each tapped register is shown in observer canonical form. The difference between this circuit form and the one in Section 3.3, which is called the controller canonical form, is shown in Fig. 4.7. Either form implements the operation $z(D) = u(D)f(D)/q(D)$, taking the generator sequences in the standard delay notation with $f(D) = f_0 + f_1 D + f_2 D^2 + \cdots + f_m D^m$, and similarly for $q(D)$. The figure generates one output stream from one input stream. The taps are the same in either form, except that they come in reverse order. q_0 is always 1 in either form, and is therefore drawn as a fixed line. Further information on these and other forms appears in [14], Chapter 2. To keep with tradition, we will show encoders in observer form, with h denoting such a tap set. But to avoid confusion, tap sets for the controller form, denoted by g, will be given as well in the tables.

Set-partition Coding

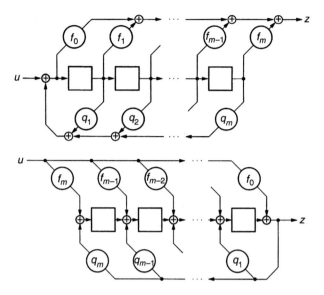

Figure 4.7 Controller (*top*) and observer (*bottom*) canonical forms of the rational transfer function $f(D)/q(D)$. (Adapted from Johannesson and Zigangirov [14], copyright 1999 IEEE; used with permission.)

As mentioned earlier, the encoders in TCM have been mostly of a single type, the feedback systematic encoder. A standard rate 2/3 TCM subset selector in observer form is shown at the top of Fig. 4.8. The two input lines u_2 and u_1 carry directly over to the two subset selection lines z_2 and z_1; they also tap into the observer form summing junctions via tap sets $\boldsymbol{h}_2 = h_2[m], \ldots, h_2[0]$ and $\boldsymbol{h}_1 = h_1[m], \ldots, h_1[0]$, respectively, reading left to right. The register output becomes z_0, which is also fed back according to tap set $\boldsymbol{h}_0 = h_0[m], \ldots, h_0[0]$. $h_0[m]$ and $h_0[0]$ are taken always as 1 and are therefore shown as hard wired. By removing the u_2 line and tap set $\boldsymbol{h}_2$, one obtains a rate 1/2 feedback systematic encoder. In either case, the single extra line z_0 creates a rate $\log_2 M_c - 1$ coding system. Rates of this form are nearly universal in practical TCM. The same selector circuit can be used with different constellations, which have different M_c.

When the subset selection bits $z_0, z_1, \ldots$ in Fig. 4.8 are directly those of the first, second, ... subset splits via the Ungerboeck Rules, as illustrated in Fig. 4.6, we can call the encoder standard TCM. This mapping is often but by no means always followed in applications. Sometimes, as in Section 4.3.2, another mapping is required for some reason. There is no guarantee that a certain mapping scheme has the optimal distance or BER, but at the same time, searches for better mappings have not yielded much gain (see [1], Chapter 3, and [2]).

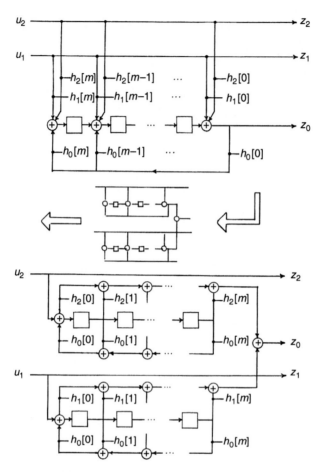

Figure 4.8 *(Top)* A typical subset selection circuit for rate $\log_2 M_c - 1$ TCM, realized as a feedback systematic convolutional encoder in observer form. A convolutional encoder with rate $k/k+1$ is obtained by implementing k systematic bit lines together with tap sets $\boldsymbol{h}_0, \boldsymbol{h}_1, \ldots$ ($k = 2$ is shown). $\boldsymbol{h}_\ell$ in the text denotes taps in descending order, as shown. The circuit may be converted as shown to controller form *(bottom)*. Tap sets $\boldsymbol{g}_0, \boldsymbol{g}_1, \ldots$ in the text are the respective $\boldsymbol{h}_\ell$ taken in forward order, as shown. $h_0[m]$ and $h_0[0]$ are taken as 1, but are shown for clarity.

Encoder forms other than feedback systematic encoders exist. In ordinary convolutional error correction, a (nonsystematic) feedforward convolutional always exists that creates the same codeword set as a feedback systematic encoder, with of course the same free distance. In TCM these words map to subset sequences, but both encoder types create the same codeword set, and hence the same subset sequence set, with the same inter-subset minimum distance. Only the BER in the overall TCM scheme can vary to a small degree. There has been some experimentation with feedforward encoder forms.

Set-partition Coding 153

Table 4.3 is a selection of good subset selectors, together with the performance of the overall TCM encoders. The content of the table will be explained in the following paragraphs.

The table is organized into QAM and PSK families. The selector convolutional code rates are 1/2 and 2/3 only, the simplest possible structures. The master constellations are those in Table 4.1. All entries in the table are for codes of rate $R = \log_2 M_c - 1$ bits/QAM interval. For the rectangular-grid QAM family, the square free distances of the codes based on 16QAM, ..., 256QAM (which have rates $\log_2 M_c - 1 = 3, \ldots, 7$) are all shown against the same subset selector, under the heading d_f^2. This is because of the following generally observed fact: since all these constellations are subsets of the same rectangular point pattern (as in Fig. 4.3), they all have the same worst-case local neighbor structure after one or two splits, and the same subset selector is therefore the best one for all of them.

The free distance for a code is normalized in the same manner as in Eq. (4.2-9), but now the factor $2\bar{E}_s/R$ divides the unnormalized worst-case square distance instead of dividing D_{ss}. The inter-subset distance may be found through a systematic comparison of all pairs of paths in the subset selector trellis, as will be discussed in detail in Section 4.3.3. Under our assumption that the QAM master constellations have unit spacing, the square minimum is an integer; that is, the worst-case path pair represents two sequences of constellation points that must be an integer total square distance apart. The integer appears in the QAM family table against each subset selector in the column labeled D_{is}. d_f^2 itself must be a multiple of $R/2\bar{E}_s$, with an upper limit of d_{ss}^2; that is, it has the form

$$d_f^2 = \frac{kR}{2\bar{E}_s}, \quad k \text{ integer}, \quad k \leq 2^S, \qquad (4.3\text{-}1)$$

with S the number of partition splits and $\bar{E}_s$ from Table 4.1. Reading down a QAM family distance column clearly shows the form (4.3-1). The PSK family distances follow rules, too, but they are more complex and we will leave them as exercises.

The tap sets h_0, h_1, h_2 in the table are defined as in Fig. 4.8, with the taps in each h listed in reverse order as right-justified octals. A memory 3 tap set with octal notation 11 is 1001 in binary, meaning taps in the first and last positions. The equivalent controller form circuit may be obtained as shown in Fig. 4.8, by breaking the standard observer-form circuit into two registers, converting each to controller form by Fig. 4.7, then combining the two new forms into one. The sets g_0, g_1, g_2 are made up of h values as shown at the bottom of Fig. 4.8, with the taps now listed in forward order as left-justified octals. Now octal 11 is the memory-5 tap set 001001. Some other examples are given in the table. As the state space size grows in the table, there is a first size that achieves each inter-subset distance for the first time, and eventually a first size that achieves the same subset bound d_{ss}^2 (marked by the symbol #). The latter size for the rectangular family is 4 states for 2 splits and 128 for 3 splits; for 8PSK only the 2 split case is allowed and the size

Table 4.3 Best convolutional feedback systematic subset selectors for standard TCM based on PSK and rectangular master constellations. Code rate is $\log_2 M_c - 1$ bits/QAM interval in all cases. Observer form taps h shown as right-justified octals; controller form taps g shown as left-justified octals (note conversion example). m is convolutional selector memory; d_f is standard normalized free distance; A_f is average multiplicity of nearest neighbors. # denotes same subset bound is achieved. [ ] denotes gain in dB over next smallest uncoded constellation in PSK or QAM heirarchy. Data on master constellations appears in Table 4.1

States	m	h_0	h_1	h_2	g_0	g_1	g_2	D_{is}	N_d	A_f	d_f^2
						PSK Family					
8PSK constellation											
4	2	5	2	—	5	2	—			1	4# [3.0]
16PSK constellation											
4	2	5	2	—	5	2	—			4	1.99 [3.5]
8	3	13	04	—	64	10	—			4	2.21 [4.0]
16	4	23	04	—	62	10	—			8	2.44 [4.4]
32	5	45	10	—	51	04	—			8	2.87 [5.1]
64	6	103	024	—	604	120	—			2	3# [5.3]
128	7	024	203	—	120	602	—			2	3# [5.3]
256	8	427	176	374	721	374	176			8	3.13 [5.5]

Set-partition Coding

Rectangular Family

Tap	Ex:	5	2	—	5	2	—	4#	4	d_f^2 for constellation				
										16Q	32CR	64Q	128CR	256Q
4	2							4#	4	2.4 [4.4]	1.6 [3.0]	0.95 [2.8]	0.59 [3.1]	0.33 [2.9]
8	3	11	02	04	44	20	10	5	16	3.0 [5.3]	2.0 [4.0]	1.19 [3.8]	0.73 [4.1]	0.41 [3.8]
16	4	1001 / 23	0010 / 04	0100 / 16	1001 / 62	0100 / 10	0010 / 34	6	56	3.6 [6.1]	2.4 [4.8]	1.43 [4.6]	0.88 [4.9]	0.49 [4.6]
32	5	41	06	10	41	30	04	6	16			Same as above		
64	6	101	016	064	404	340	130	7	56	4.2 [6.8]	2.8 [5.4]	1.67 [5.2]	1.02 [5.5]	0.58 [5.3]
128	7	203	014	042	606	140	210	8#	344	4.8 [7.4]	3.2 [6.0]	1.90 [5.8]	1.17 [6.1]	0.68 [5.9]
256	8	401	056	304	401	350	106	8#	44			Same as above		
512	9	1001	0346	0510	4004	3160	0450	8#	4			Same as above		

is just 4 states. Even though a certain state size can achieve d_{ss}, larger sizes can often achieve it with fewer nearest neighbors. The number of nearest neighbors is shown in the column headed A_f. Since these codes often have lower BER as well, they are worth listing in the table. The subjects of neighbor numbers and BER are discussed later in Section 4.3.3.

The table shows in brackets the coding gain in dB that each d_f^2 represents, compared to uncoded transmission with the next smaller constellation in the respective hierarchies

$$4PSK/8PSK/16PSK$$

or

$$8PSK/16QAM/32CR/64QAM/128CR/256QAM.$$

The comparison is fair in the bandwidth–energy sense because both schemes have the same spectrum. It is clear that significant gains are available with relatively modest selector trellises. Unfortunately, they tend to appear only at high E_b/N_0, so that coding gains at practical BER can lag significantly. BER gains need to be measured carefully by simulation. It is also clear that the QAM family has much greater distance than the PSK family, for the same rate and number of trellis states. This is due to the lower energy efficiency of the PSK master constellation.

Roughly speaking, the more narrowband TCM schemes achieve a bandwidth–energy performance similar to the partial response signaling (PRS) codes in Chapter 6. The overview of Fig. 1.2 gives some indication of this. Figure 4.9 plots distance vs NBW bandwidth in Hz-s/data bit for good QAM-family TCM codes, with comparison to some good binary and quaternary PRS codes.[7] TCM codes achieve their gains via a large alphabet and orthogonal pulses, while PRS codes have binary or quaternary alphabets and much more strongly bandlimited pulses.

Extensions

Set partitions can be performed on other types of 1- and 2-dimensional master constellations, but for one reason or another, they are not as effective as the QAM partitions in Table 4.3. One-dimensional equi-spaced constellations with 4, 8, ... points have been studied; lists of good subset selectors appear in [2], Chapter 5 and in [5]. For the same state size, these codes have worse free distance by a fraction of a dB. An explanation of this might be that coding takes place over half as many signal space dimensions. In any case, fewer subset splits are available from, say, an 8-point 1-dimensional array than from a 64-point 2-dimensional one, which limits the same subset bound possibilities.

---

[7] PRS codes with larger alphabets have better performance, but precisely how much is unknown at the time of writing.

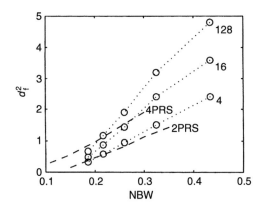

Figure 4.9 Free distance vs NBW for TCM codes in the QAM family; natural mapping and best 4-, 16-, and 128-state subset selectors. Master constellations indicated by circles: right to left, 16QAM, 32CR, 64QAM, 128CR, 256QAM. Good binary and quaternary PRS codes from Section 6.4 shown (*dashed*) for comparison. TCM assumes 30% excess bandwidth pulses. PRS assumes 99% bandwidth and d_f inside this.

Circular PSK constellations with more than 8 points[8] have also been studied [2,5]. Eight-phase codes work well, but higher master constellations suffer because of the ill-distribution of the points (Cf. Table 4.3).

"Asymmetric" constellations, which are non-uniform circular or rectangular arrangements, have been studied [2]. So also have non-standard mappings from subset selector bits to sets. But almost none of this work applies to true set-partition codes, that is, ones that work with subsets, and so we delay discussion until Section 4.5.

Subset selector circuits can also be nonlinear. This kind of circuit is most often associated with rotational invariance, which is discussed next.

4.3.2. Rotational Invariance

It often happens that a receiver can obtain the I/Q plane reference phase only to a multiple of 2π. For example, the fourth power synchronizer loop, the one most often employed in QAM, has a phase ambiguity of $k\pi/2$ radians, $k = 0, 1, 2, 3$. Once the loop is locked, k may be identified via a training sequence or made moot by differential encoding. In any case, it is convenient if a TCM scheme has *rotational invariance* to some useful angle φ; that is, if a sequence of QAM

---

[8] It is worth stressing again that equi-energy points do not imply constant envelope signaling. In general this happens in PSK only when NRZ pulses are employed, which is usually impractical. Cost-effective constant-envelope modulation requires CPM. See Chapter 2.

points $x[1], x[2], \ldots$ is a codeword, then so also is the word $x[1]^{\angle\varphi}, x[2]^{\angle\varphi}, \ldots,$ in which each point is rotated by φ. Upon decoding, the rotated codeword is at least a legitimate one, even if it maps to the wrong data symbols. A combination of training sequence, differential encoding, and mapping can then be used to correct the data. Several important modem standards are based on these ideas.

The CCITT V.32 and V.33 standards for 9600 and 14,400 bit/s telephone line transmission are prime examples. As they illustrate both rotational invariance and non-convolutional subset selection – and because it is time for another detailed example – we will investigate them in detail.

The V.32 standard employs a 1800 Hz carrier and orthogonal pulses to transmit a 32-point constellation at a 2400 QAM interval/s rate. It has rate 4 bits/QAM interval and is invariant to rotations of $\pi/2$ radians. The partitioning is shown in Fig. 4.10; the master constellation is the 32CR one in Fig. 4.3 rotated 45°. After three standard partition splits, 8 subsets are formed with 4 points each, but the mapping to subset selector bits does not follow the Ungerboeck Rules. Rather, the subsets are labeled $C_0, \ldots, C_7$ as shown, with selector bits $z_2 z_1 z_0$ mapping to the subscript of C according to the natural binary representation in which 000

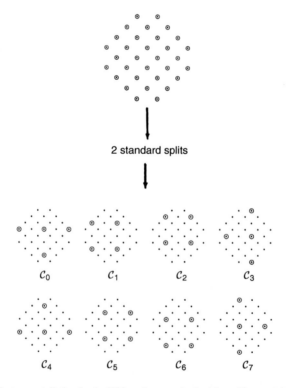

Figure 4.10 Master constellation for the V.32 modem standard, with partition to eight 4-point subsets.

Set-partition Coding

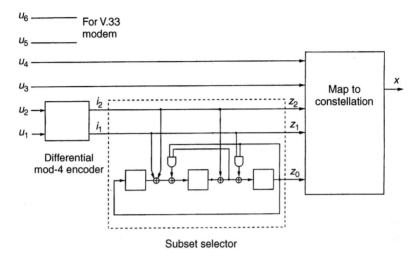

Figure 4.11 The V.32 modem standard: overall encoding circuit, showing differential encoder, 2-for-3 subset selector and mapping to 32CR point x. Bits $u_4 u_3$ are carried in the subset. Rate 4 bits/QAM interval. Mapping is defined in Fig. 4.12. The rate 6 V.33 standard adds bits $u_6 u_5$ to the subset bits.

selects C_0, 001 selects C_1, etc. Figure 4.11 shows the overall encoding circuit. It consists of a differential encoder, a 2-for-3 subset selector, and a mapping to a 32CR point. Data appears at the left in 4-tuples $u_4 \cdots u_1$. u_4 and u_3 are carried directly in a 4-point subset; u_2 and u_1 pass through the differential encoder, whose output, labeled $i_2 i_1$ forms the subset selector's input; the selector outputs $z_2 z_1 z_0$ choose the subset. The mapping from $u_4 u_3 z_2 z_1 z_0$ to 32CR points is defined by Fig. 4.12. The first two symbols in a point label are the bits carried by the subset, while the last three, shown for clarity in brackets, are the binary-coded integer subset label. The TCM encoder here was proposed by Wei along with other rotationally invariant codes in [7], the paper that introduced the first such practical codes. When combined with standard Viterbi decoding to produce $u_4 u_3 i_2 i_1$, and differential decoding to produce $u_2 u_1$ from $i_2 i_1$, the overall scheme exhibits rotational invariance to any multiple of 90°.

We will now explain the parts of Fig. 4.11 in detail and show that the encoder has rotational invariance.

The differential encoder works by computing a mod-4 running sum in binaries of the inputs $u_2 u_1$, as shown in Fig. 4.13. As a decimal at time n, this number is $\sigma[n] = 2i_2[n] + i_1[n] = 2i_2[n-1] + i_1[n-1] + 2u_2[n] + u_1[n] \bmod 4$, taking the binaries here as ordinary integers 0 and 1. $\sigma[n]$ is converted back to binaries.[9] (The differential decoder in Fig. 4.13 will be discussed later.) The subset

---

[9] For example, if the running sum $i_2[n-1]i_1[n-1]$ is binary 10 at time $n-1$ and $u_2[n]u_1[n]$ is 11, $i_2[n]i_1[n]$ is 5 mod 4 = 1 as a decimal integer, or 01 in binary.

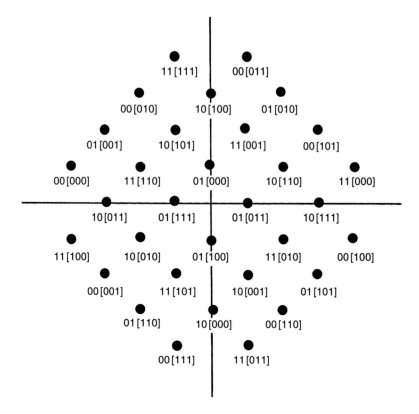

Figure 4.12 V.32 modem standard: mapping from bits $u_4 u_3 z_2 z_1 z_0$ to 32CR constellation points. Bits in brackets are the subset labels as a binary coded integer with MSB at left.

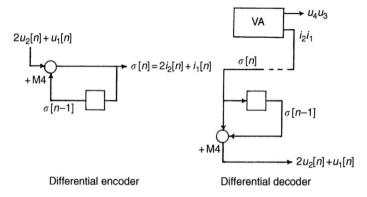

Figure 4.13 V.32 and V.33 modem standards: Differential encoder and decoder circuits. Variables u and i are defined in Fig. 4.11.

Set-partition Coding

	Input i_2i_1			
State	00	01	10	11
000	000 [0]	010 [2]	011 [4]	001 [6]
001	100 [1]	111 [3]	101 [5]	110 [7]
010	001 [0]	011 [2]	010 [4]	000 [6]
011	111 [1]	100 [3]	110 [5]	101 [7]
100	010 [0]	000 [2]	001 [4]	011 [6]
101	110 [1]	101 [3]	111 [5]	100 [7]
110	011 [0]	001 [2]	000 [4]	010 [6]
111	101 [1]	110 [3]	100 [5]	111 [7]

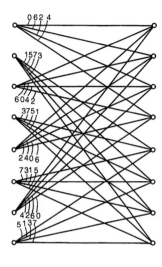

Figure 4.14 V.32 and V.33 modem standards: state transition table for the subset selector in Fig. 4.11. Bracket values are the subset label output $z_2z_1z_0$, expressed as a decimal (MSB z_2). Equivalent trellis section at right.

selector circuit contains AND gates and is not any kind of linear convolutional encoder. Nor does it need to be. It is perhaps easiest to define it by the equivalent table of state transitions, which is given in Fig. 4.14. The table assumes that the selector state is $s = s_1s_2s_3$, the contents of the three binary memory blocks in Fig. 4.11, taken left to right. Such a table lookup may well be the physical implementation of choice. Across the top of the table are the binary input pairs $i_2[n]i_1[n]$ to the circuit, and the table entries are the next state (as a bit triple) and the subset label that results, expressed as $z_2z_1z_0$ in decimal form in [ ].[10] This decimal is the $\mathcal{C}$ subscript in Fig. 4.10. Next to the table appears a standard trellis section that depicts the same thing. Wei's problem in the abstract was to find a state transition table, which when combined possibly with differential coding, yielded rotational invariance. From Table 4.2, such a code's d_f^2 is limited to 3.2; Wei's code achieves 2.0, but this is still a 4.0 dB improvement over uncoded rate 4 QAM.

To demonstrate the invariance, we need to show first that subsets rotate into same shaped subsets with the same bits carried in the subset, and second that

---

[10] The decimal integer is $4z_2[n] + 2z_1[n] + z_0$, which is also $4i_2[n] + 2i_1[n] + s_3$, where s_3 is the present state variable.

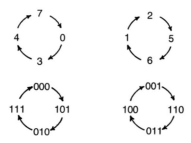

Figure 4.15 V.32 and V.33 modem standards: shifts of subset labels under 90° CW rotations *(top)* and shifts of states along paths *(bottom)*.

coded sequences of subsets map into new sequences with the same distance and merge properties. Last, we need to find out how the b subset selector bits change. To take up the first question, consider the subsets in Fig. 4.10. Take for example subset C_7. Rotate the subset 90° clockwise (CW) with the axes fixed, or equivalently, rotate the coordinate system 90° counterclockwise (CCW). Under the rotation, C_7 assumes the position previously occupied by C_0; rotations by 180° and 270° place it in the positions of C_3 and C_4. Rotations of C_0, C_3, and C_4 also have outcome in the group C_0, C_3, C_4, C_7. The group of subsets C_1, C_2, C_5, C_6 shows a similar behavior. Some study shows that the bits carried by the subset – the left two in each label in Fig. 4.12 – are always invariant under these rotations. These symmetries can be summarized in the two top diagrams of Fig. 4.15; it shows how rotation by 90° steps changes the label of any given subset. For example, subset C_7 under 90° CW rotation becomes C_0; subset C_2 under 270° CCW rotation becomes C_5.

Next, we explore the effect of rotations on the trellis transitions. As an example of a path, take the pair

$$000 \xrightarrow[6]{11} 001 \xrightarrow[7]{11} 110 \xrightarrow[4]{10} 000 \xrightarrow[0]{00} 000,$$
$$000 \xrightarrow[0]{00} 000 \xrightarrow[0]{00} 000 \xrightarrow[0]{00} 000. \qquad (4.3\text{-}2)$$

These two split at the first transition and merge after the third. The data bits driving the transition and the subset label number are shown as two bits above the transition and an integer below, respectively. Now rotate the constellation points in these paths 90° CW (equivalently, the axes 90° CCW). The subset progressions along the two paths are now (using Fig. 4.15)

$$1, 0, 7, 3, \qquad 3, 3, 3.$$

How do the states progress? Is there another trellis that keeps track of these states, and is the evolution of subsets in it consistent?

Set-partition Coding

Wei disclosed a 1-to-1 state mapping that guarantees consistency in every such case for rotations of 90°, 180°, and 270°; that is, he found a map $f_{\Delta\varphi}(s) \longrightarrow s_{\Delta\varphi}$, where $s_{\Delta\varphi}$ denotes the state in a rotated path, which moves a path to the equivalent place in a new trellis. We can summarize all three maps graphically in the lower half of Fig. 4.15. Under a 90° CW rotation, a state in the original trellis should be replaced by the next state CW from it in the figure; under a 180° CW rotation, the replacement is the state two steps CW, and so on; under CCW rotations, the replacements are CCW steps in the figure. For example, trellis state 000 before a 90° CW rotation becomes state 101 in the after-rotation trellis; 100 before a 270° CCW rotation becomes 001 after.

By applying the state map in Fig. 4.15 once to the entire transition table in Fig. 4.14, we obtain a new table for use after a 90° rotation. The result is shown in Table 4.4. If the same subset selector machine is to be used as before, the subset outcomes and the input bit pairs must be filled in as the ones shown in [ ] and ( ), respectively. The rows of Table 4.4 are out of natural binary order, but this does not matter. Once they are ordered, the table differs from Fig. 4.14 in only one way: the

Table 4.4 V.32 and V.33 modem standards: application of Wei's state map to the state transition table of Fig. 4.14. Input pairs ( ) and subset labels [ ] are those required by the selector circuit in Fig. 4.11

	Input $i_2 i_1$			
State	(01)	(10)	(11)	(00)
101	101	111	100	110
	[3]	[5]	[7]	[1]
110	001	000	010	011
	[2]	[4]	[6]	[0]
111	110	100	111	101
	[3]	[5]	[7]	[1]
100	000	001	011	010
	[2]	[4]	[6]	[0]
001	111	101	110	100
	[3]	[5]	[7]	[1]
010	011	010	000	001
	[2]	[4]	[6]	[0]
011	100	110	101	111
	[3]	[5]	[7]	[1]
000	010	011	001	000
	[2]	[4]	[6]	[0]

rows are cyclically shifted one place right, in the same way that the parenthesized input pairs are. In fact, Figs 4.14 and 4.15 are the same.

We can observe the same end result by applying the Fig. 4.15 shifts and mappings to the path pair given before in Eq. (4.3-2). It becomes

$$101 \xrightarrow[1]{00} 110 \xrightarrow[0]{00} 011 \xrightarrow[7]{11} 101 \xrightarrow[3]{01} 101,$$
$$101 \xrightarrow[3]{01} 101 \xrightarrow[3]{01} 101 \xrightarrow[3]{01} 101. \qquad (4.3\text{-}3)$$

The input pairs $i_2 i_1$ here are not those in Eq. (4.3-2), but they must be as shown if the same subset selector machine is be retained. This can be verified by checking against Fig. 4.14 or Table 4.4. There is in fact a cyclic mod-4 rotation of the bits according to the rule $00 \to 01 \to 10 \to 11 \to 00$. A second rotation by 90° CW means another rotation of subsets and states as in Fig. 4.15, and the result is

$$010 \xrightarrow[2]{01} 011 \xrightarrow[3]{01} 100 \xrightarrow[0]{00} 010 \xrightarrow[4]{10} 010,$$
$$010 \xrightarrow[4]{10} 010 \xrightarrow[4]{10} 010 \xrightarrow[4]{10} 010. \qquad (4.3\text{-}4)$$

The transitions here are again consistent with Fig. 4.14, if the input bit pairs are as shown. There is another cyclic mod-4 rotation of the input pairs, compared to Eq. (4.3-3).

All path pairs show a similar behavior. We can conclude that the coding system can be made immune to these rotations by differentially encoding the data. This is done by the circuit in Fig. 4.13. Before encoding the nth data pair, form a mod-4 sum $\sigma(n)$ of the present and previous inputs taken as integers, and convert $\sigma(n)$ to the natural binary form $i_2[n]i_1[n]$. Since the data is encoded in the difference of $i_2[n]i_1[n]$ from $n-1$ to n, the cyclic shift caused by a rotation will not lead to error if the decoding of the rotated transmission is correct.

A standard decoder, Viterbi or otherwise, estimates the bits $i_2[n]i_1[n]$. A differential decoder, implemented as in Fig. 4.13 by a mod-4 subtractor, recreates $u_2[n]u_1[n]$, assuming that $i_2[n-1]i_1[n-1]$ and $i_2[n]i_1[n]$ are correct. Note that error events in the sequences $u_4 u_3 i_2 i_1$ and their distances are preserved under rotations, but once an error event terminates, there is ordinarily one extra erroneous $u_2 u_1$, because of the differential encoding.

This ends our example of rotationally invariant TCM coding. We have seen how differential encoding and more general subset selector machines play important roles. By introducing these features, practical schemes with good gains become available. An extension of the V.32 coding system with two more bits $u_6 u_5$ carried in the subsets gives the V.33 standard (see Fig. 4.11). A 128CR master constellation is partitioned into 8 subsets of 16 points, to give a rate 6 bits/QAM

Set-partition Coding

165

interval system with $d_f^2 = (5/8)1.17$ and overall transmission rate 14.7 kb/s.[11] The subset selector is unchanged. Several other TCM modem standards exist. The V.34 family has constellations of 960 or 1664 points, baud rates up to 3429 QAM interval/s, and subset selectors to 64 states. The family supports data bit rates as high as 33.6 kb/s.

Full invariance to all 90° multiples is not possible with a 2-dimensional constellation and ordinary linear convolutional subset selection (this was first asserted by Wei [7]). Such linear selectors do exist with 180° invariance. With higher dimension master constellations, like those in Section 4.4, linear convolutional selectors exist with 90° invariance [9,11]; these papers also discuss PSK-constellation codes with 2^n points and $k2\pi/n$ invariance, $k = 0, \ldots, n-1$. A treatment of rotational invariance with some recent innovations can be found in [12]. An extensive chapter appears in [2], Chapter 8.

4.3.3. Error Estimates, Viterbi Decoding, and the Free Distance Calculation

Consider a TCM code with words $s_1, s_2, \ldots, s_\mathcal{M}$ in AWGN and received signal r. The notation is that of Section 2.3: we consider s and r to be vectors of components in signal space and each component of r is a component of s_0, the transmitted signal, plus a zero-mean Gaussian variate with variance $N_0/2$. With TCM coding, it is both convenient and natural that each use of the underlying QAM system corresponds to a pair of these components; they are the I and Q components of a constellation point. We do not have to deal with continuous-time signals.

The basic estimate of error probability is Eq. (2.3-28). If there are $\mathcal{K}$ distinct pairs of words lying at the minimum distance $d_{\min}$, then asymptotically in E_b/N_0,

$$p_e \sim \frac{2\mathcal{K}}{\mathcal{M}} Q\left(\sqrt{d_{\min}^2 E_b/N_0}\right). \quad (4.3\text{-}5)$$

For coded modulation in general, the factor in front of $Q(\ )$ can be anything from a fraction of one to quite a large number. At moderately small p_e, it can lower the practical coding gain several dB from the value of $Q(\ )$ alone; this almost always happens in TCM coding (since $\mathcal{K}/\mathcal{M} \gg 1$).[12] In addition, there may be word pairs whose distance only slightly exceeds $d_{\min}$, which further raise the true error probability.

With the structural symmetries that exist in trellis codes, Eq. (4.3-5) is an unnecessarily complex way to look at the error calculation. Rather, we can interpret Eq. (4.3-5) in tune with the trellis ideas of Section 3.4.1, as the average *first event*

---
[11] A picture of the master constellation and subset assignment appears in Couch [13].
[12] The factor is often less than 1 in, for example, CPM coding.

probability. Suppose that decoding is correct up to a trellis node at time n, and that among the signals running out of the node into the infinite future, the closest ones lie at total distance d_f from the correct signal, some lie at a next-to-nearest distance d_1, some at d_2, and so on. The structure of such distances forward of each node is not necessarily the same, and so we average uniformly over all the nodes at time n. This leads to the first event expression

$$p_{ev} \sim \sum A_i Q\left(\sqrt{d_i^2 E_b/N_0}\right), \qquad (4.3\text{-}6)$$

in which $d_f, d_1, d_2, \ldots$ are the possible (merged) distances and $A_f, A_1, A_2, \ldots$ are called the *average multiplicities* of the distances. The collection is called the *spectrum*. Equation (4.3-6) expresses the union bound of probability.

Asymptotically in E_b/N_0, Eq. (4.3-6) reduces to $A_f Q(\sqrt{d_f^2 E_b/N_0})$. At more moderate E_b/N_0, higher terms have an effect and adding them can lead to a much more accurate estimate. As E_b/N_0 drops further, there is generally a point where Eq. (4.3-6) fails to converge; this happens for fundamental reasons, because the multiplicities grow faster with d_i than the terms $Q(\sqrt{d_i^2 E_b/N_0})$ drop. Multiplicities need not be integer, and can be less than one; A_f would be 3/2, for example, if three transmitted/neighbor pairs exist that achieve d_f forward of half the state nodes at n, but none exist in front of the other half. A TCM code is said to be *regular* if the structure of distances around each codeword is the same. Codes based on PSK master constellations tend to be regular if they have symmetric subsets and subset selection.

The average multiplicity A_f is shown in the code tables. It can be seen that the addition of a unit of memory to the subset selection can lead to a large drop in A_f. This can create significant coding gain even if d_f remains unchanged. By counting the data bit errors that correspond to different error events, one can develop an expression $\sum B_i Q(\sqrt{d_i^2 E_b/N_0})$ similar in spirit to Eq. (4.3-6) but for the BER. A typical TCM spectrum for A_i and B_i, taken from Schlegel [1], is shown in Fig. 4.16 (the code here is discussed in Section 4.5). The plot is not different in principle from that of an ordinary convolutional code: many neighbors of the correct path lie a little further out than d_f and their numbers grow exponentially with distance. More sophisticated error estimates, such as transfer function bounds, can be derived; these are beyond our scope and we defer to [1, Section 5.2] and [2, Sections 4.1 and 4.2].

Viterbi Decoding

Some typical Viterbi decoder error rates measured for rate 2 TCM codes are shown in Figs 4.17 and 4.18. In TCM decoding, the Viterbi algorithm (VA) searches a trellis with subsets on each branch. Each trellis stage aligns with a QAM symbol. The I and Q components of r that apply to a branch are compared to each

Set-partition Coding

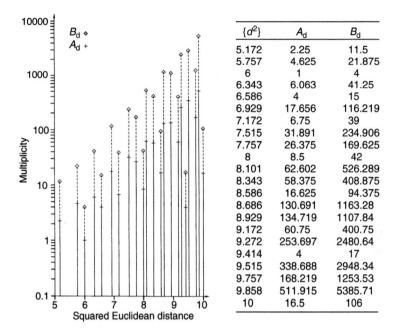

$\{d^2\}$	A_d	B_d
5.172	2.25	11.5
5.757	4.625	21.875
6	1	4
6.343	6.063	41.25
6.586	4	15
6.929	17.656	116.219
7.172	6.75	39
7.515	31.891	234.906
7.757	26.375	169.625
8	8.5	42
8.101	62.602	526.289
8.343	58.375	408.875
8.586	16.625	94.375
8.686	130.691	1163.28
8.929	134.719	1107.84
9.172	60.75	400.75
9.272	253.697	2480.64
9.414	4	17
9.515	338.688	2948.34
9.757	168.219	1253.53
9.858	511.915	5385.71
10	16.5	106

Figure 4.16 The distance spectrum for the 8PSK coded modulation with $(h_0, h_2, h_3) = (23, 04, 16)$ proposed by Ungerboeck in [4,5]. Notation as in Table 4.8. (From Schlegel [1], copyright 1997 IEEE; used with permission.)

point in the branch subset. The distance increment for this branch is the distance from the r-components to the *closest point in the set*. That point and its corresponding $R - b$ "uncoded" TCM bits are associated with the branch. The trellis search continues and one path of branches presents itself as the trellis decoder output. Associated with the path are the sequence of "coded" bits that drove the subset selection, and a sequence of QAM points with their corresponding uncoded bits. These two streams are interleaved appropriately to produce the decoded data bit stream.

The TCM Viterbi decoder search thus works in a way similar to ordinary convolutional decoding with soft AWGN channel outputs, except for the closest-point distance calculation and the path interpretation at the end. In ordinary TCM decoding, the continuous-time nature of the modulated signals never manifests itself. Signal space components are a sufficient signal description. In contrast, continuous time is the natural mode for a Viterbi CPM decoder, and the method in Section 5.4.1 may be compared to that just described here.

Figure 4.17 shows decoder BER for the original 4-, 8- and 128-state codes due to Ungerboeck, based on the 8PSK constellation. The 8- and 128-state codes

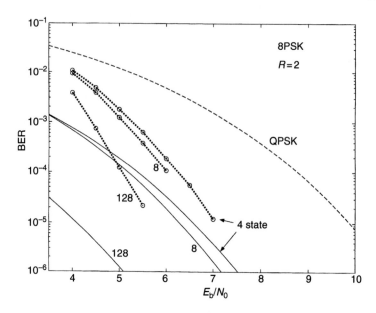

Figure 4.17 Measured BER for rate 2 TCM codes over the 8PSK master constellation. Subset selectors: 4-state, $(h_0, h_1) = (5, 2)$; 8-state, $(h_0, h_1, h_2) = (11, 02, 04)$; 128-state, $(h_0, h_1, h_2) = (277, 054, 122)$. Dashed curve is BER for same-bandwidth QPSK; solid curves are estimates of form $A_f Q(\sqrt{d_f^2 E_b/N_0})$. Data adapted from Zhang [30].

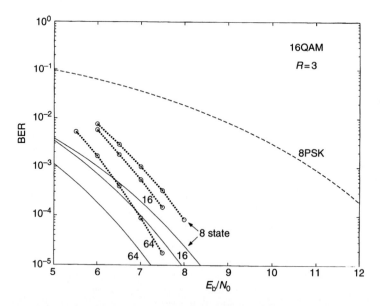

Figure 4.18 Measured BER for rate 3 TCM codes over the 16QAM master constellation. Subset selectors: 8-state, $(h_0, h_1, h_2) = (11, 02, 04)$; 16-state, $(h_0, h_1, h_2) = (23, 04, 16)$; 64-state, $(h_0, h_1, h_2) = (101, 016, 064)$. Dashed curve is BER for same-bandwidth 8PSK; solid curves are estimates $A_f Q(\sqrt{d_f^2 E_b/N_0})$. Data adapted from Zhang [30].

Set-partition Coding 169

are not actually set-partition codes, and details of these will be given in Section 4.5. (The generators of all codes are in the figure caption; distance parameters are the same as shown for the same-size codes in Tables 4.3 and 4.8.) In all cases, the estimate of form $A_f Q(\sqrt{d_f^2 E_b/N_0})$ is shown as a solid curve, using the A_f in Schlegel [1]. The dashed curve shows the BER of QPSK, a rate 2 bits/QAM interval scheme with the same NBW. The 4-state code, which has the same d_f and A_f as the introductory code in Section 4.2, has nearly the theoretical 3 dB coding gain at BER 10^{-5}; the 8-state code also has nearly the full asymptotic coding gain (ACG). The 128-state code falls short of its ACG by about 1 dB. The measured BERs lie about 0.5 dB above the estimates, except for the 128-state code.

Figure 4.18 repeats Fig. 4.17, but for rate 3 TCM over the 16QAM constellation. Again, the codes shown are original ones of Ungerboeck, but the parameters continue to be those in Table 4.3. The coding gain comparison is now to the rate 3 scheme 8PSK. At BER 10^{-4}, the 8-, 16- and 64-state codes again lie about 0.5 dB above their estimates. The coding gains fall 1–1.5 dB short of the ACG. The shortfall grows at higher BER.

Algorithms to Find Free Distance

With TCM and most other coded modulations, the distances between a codeword and its neighbors can depend on the codeword itself. This is distinct from ordinary convolutional codes, whose words have a linearity property that makes the distance structure the same for every codeword. Since the allzero word is a word in those codebooks, the free distance problem reduces to finding the heaviest weight word in an S-state trellis, which splits from and later merges to the all-zeros state path. This search has complexity equal to S times the search depth. Such a simple approach is not possible in general with TCM, but a complexity S^2 procedure exists that is practical and not harder, at least in principle. The scheme has been in use since the mid-1970s, when it was called double dynamic programming (DDP) in [16,17], and it has also been called the pair-state algorithm or superstate algorithm since then.

DDP is one of the two main approaches to finding coded modulation free distances, the other being a branch and bound technique that is discussed in Section 6.3.2. TCM distances have chiefly been found by the DDP algorithm, and so it will be described now. CPM distances (in Chapter 5) have been found by both procedures and PRS code distances (Chapter 6) chiefly by the branch and bound method.

The DDP procedure to find d_f is as follows. Construct a trellis in which the nodes are *pairs* of states (σ_T, σ_N), one called the transmitted state σ_T (T-state) and one the neighbor state σ_N (N-state). Associated with each pair state is the distance of the worst-case combination of transmitted and neighbor paths, which once split, pass through states σ_T and σ_N at the present time. The trellis keeps

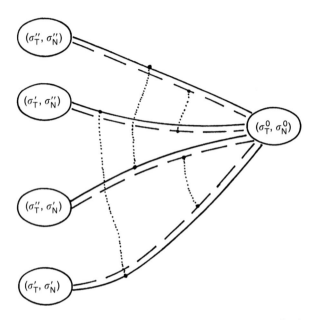

Figure 4.19 Transitions in the pair state trellis that terminate at pair state (σ_T^0, σ_N^0). Within a pair transition, solid lines trace σ_T, dashed lines trace σ_N; dotted lines tie together transitions with the same σ_T or σ_N.

track of these distances. The procedure executes a dynamic program on the trellis, consisting of the following major steps.

Extension to a new stage. From each pair state node at time n, make extensions to $n + 1$. With a b-branching code trellis, the combination (σ_T, σ_N) has b^2 pair state extensions. When $b = 2$, the branches extending to pair state (σ_T^0, σ_N^0) at time $n + 1$ are shown in Fig. 4.19. T-path states σ_T' and σ_T'' can reach σ_T^0, and N-path states σ_N' and σ_N'' can reach σ_N^0; altogether, four path pairs reach (σ_T^0, σ_N^0). Like T- and N-states are linked by dotted lines in the figure. Associated with each pair state at N is a worst-case square distance for path pairs up to that state, and each type of transition adds its own increment. The pair-survivor into (σ_T^0, σ_N^0) is the pair with the least total.

Initial extension. Active pair states at time 0 are all those with $\sigma_T = \sigma_N$; that is, those for which the T- and N-history are the same. Effectively, the dynamic program starts its search for the worst case simultaneously and in parallel from all start positions. Placing a large initial distance (>100) on each pair state with $\sigma_T \neq \sigma_N$ will cause the program to ignore these.

Termination. Path pairs merge when pair states satisfy $\sigma_T = \sigma_N$ after time 0. The distance of a pair survivor into any such node is a candidate for d_f. The program is canceled forward of these nodes (by assignment of a large distance).

Set-partition Coding 171

The least such distance found is d_f. The program terminates when the distance associated with every node at some time is more than the present candidate for d_f. If desired, the identity of survivor pair candidates can be stored along with their distances; those leading to d_f are the worst case signal pairs in the coded modulation. The trellis depth at termination is the decision depth L_{dec} required if the decoder is to achieve d_f (see Section 3.4.1).

With TCM codes, a distance increment along a pair state trellis branch is the least distance between a point in the subset of the T-transition and a point in the subset of the N-transition. For coded modulations without subsets, the increment is simply the distance between the pieces of signal corresponding to the branches. Some computation can be saved by stopping the program forward of any node whose associated distance exceeds the present estimate of d_f. For regular codes, the program's complexity drops from S^2 to S. However, regularity can be tricky to prove, software time savers can be tedious to prove out, and modern computers are fast, so that it is often best to let one general double dynamic program take care of all cases.

Book discussions of the DDP method are available in [1,2,15]. A readable review appeared in the journal literature in [18].

4.4. Lattice Codes

A lattice is a regular arrangement of points in space. In this section we view TCM codes as codes over lattices. Lattice coding is an older idea than TCM, which began to merge with set-partition coding in the 1980s. It provides a succinct and elegant view of set-partition coding and makes clear some interesting generalizations. In particular, it shows ways to extend TCM to constellations over spaces with more than two dimensions.

If the noise is Gaussian so that the space is Euclidean signal space, then the coding problem has a direct connection to a lattice of points: we wish to pack N-space as efficiently as possible with balls of radius ρ; such packings are defined by a lattice; codewords placed at the centers of the balls will then correct noises of length ρ. The optimal coding problem may be stated thus: given equiprobable words and average energy $\bar{E}_s$ per dimension, how many ρ-balls can be packed? The $\log_2$ of this number divided by the number of dimensions is the code rate in bits/dimension. Stated this way, the lattice coding problem is the classic sphere packing problem and the codeword points form a sphere packing lattice.

In a relatively less known paper, Shannon [20] in 1959 proved that in the limit $N \to \infty$ there exists a code with points distributed in signal N-space and average energy $\bar{E}_s$ which has rate as high as the Gaussian capacity (3.6-9). He did this by projecting the points onto the surface of the N-hypersphere with radius $\sqrt{N\bar{E}_s}$ and then giving a proof for the projected points. He gave a number of properties, including a reliability exponent (Cf. Section 3.6.4), but the code structure was purely random. By using the classical Minkowski–Hlawka theorem from the geometry of numbers, deBuda and Kassem (see in [21,22]) showed in 1975–1989 that essentially the same result applied to a code whose words were not random but were points on a lattice. These papers deal with coding theorems. Framing the ideas of set-partition coding in a lattice context came in the late 1980s with the papers of Forney, Sloane, Calderbank and Wei [24–26]. The role of what we have called the master constellation was taken by a lattice; subsets became cosets of the lattice. It becomes easier to see how to improve the master constellation and how to make it multidimensional.

We will first review some lattice ideas.

4.4.1. Lattice Ideas

Consider the N-dimensional Euclidean space $\mathbb{R}^N$. A lattice Λ is the set of points $\boldsymbol{w} \in \mathbb{R}^N$

$$\Lambda = \left\{ \boldsymbol{w} : \boldsymbol{w} = \sum_{j=1}^{N} i_j \lambda_j \right\}, \tag{4.4-1}$$

where coefficients $i_1, \ldots, i_N$ take all the integer values and vectors $\lambda_1, \ldots, \lambda_N$ are a linearly independent set of *lattice basis vectors*.[13] The integer nature of the coefficients $i_1, \ldots, i_N$ creates a regular, repeating structure of points. A familiar example by now is the rectangular QAM family lattice of the previous sections; here $\boldsymbol{w} \in \mathbb{R}^2$, λ_1 and λ_2 are the vectors $(1, 0)$ and $(0, 1)$. Some 300 points of this lattice offset by $(1/2, 1/2)$ – that is, $\Lambda + (1/2, 1/2)$ – are shown in Fig. 4.3.

A minimum distance d_Λ can be defined for a lattice; this is simply the minimum Euclidean distance between any two of its points. The number of nearest neighbors to a point is κ, the *kissing number* of the lattice. If N-spheres of radius $d_\Lambda/2$ are placed at each lattice point, each will barely touch, or "kiss," κ others. As mentioned before, the coding problem may be viewed as looking for dense packing of spheres. A lattice has a *fundamental volume* $V(\Lambda)$, which is the volume in N-space per lattice point; this is also the volume of the space closest to each point, a repeating polytope called the *Voronoi region*.

The lattice in Fig. 4.3 has $d_\Lambda = 1$, kissing number $\kappa = 4$, square Voronoi regions, and fundamental volume 1. It will provide a benchmark for the more subtle lattices to come. The extensions of the rectangular lattice to N-dimensions are denoted $\mathbf{Z}^N$ and called cubic lattices; the one here is $\mathbf{Z}^2$. The points in $\mathbf{Z}^N$ can be denoted by all the integer N-tuples. $\mathbf{Z}^N$ has $d_\Lambda = 1$, all N, and $\kappa = 2N$.

Like a linear code, a lattice has a generator matrix G_Λ whose ith row is the vector λ_i. Lattice points satisfy $\boldsymbol{w} = i G_\Lambda$, with i the integer row vector $(i_1, \ldots, i_N)$. $G_\mathbf{Z}$ for $\mathbf{Z}^2$, for example, is the 2×2 identity matrix $\boldsymbol{I}$. Any full

---

[13] A more general definition is possible, but we restrict the definition to present needs.

Set-partition Coding

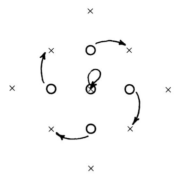

Figure 4.20 The mapping from Z^2 (○ points) to D_2 (× points) caused by the rotation operation R. Innermost 5 points of Z^2 are shown.

rank $N \times N$ matrix generates a lattice. Operations on a lattice such as rotation and expansion can be conveniently described as matrix operations on G_Λ. An important transformation in what follows is

$$R \triangleq \begin{bmatrix} 1 & -1 \\ 1 & 1 \end{bmatrix}. \tag{4.4-2}$$

A lattice with generator RG is a rotation CW by 45° of the lattice with G, combined with an expansion by $\sqrt{2}$. R is illustrated in Fig. 4.20; the innermost 5 points of Z^2 (shown by ○) rotate to the innermost 5 points of the lattice with generator $RI = R$ (shown by ×). Some further points of the new lattice are shown. It is called the *checkerboard* lattice D_2 in two dimensions. Symbolically, we can write $D_2 = RZ^2$, meaning that D_2 is the rotation by R of lattice Z^2. The checkerboard lattice D_N in N dimensions can be defined as those points in Z^N for which $\sum_{j=1}^{N} i_j$ is even. For D_N, $d_\Lambda = \sqrt{2}$ and $\kappa = 2N(N-1)$.

To express set-partition codes in lattice terms, we need one more concept, that of a partition of a lattice. This act starts by taking a sublattice of the original one, that is, a regular structure Λ', itself a lattice, which is a subset of the original Λ. Next, cosets of the sublattice are created. A *coset* of a lattice Λ' is a translate $\Lambda' + c$ of all the points by c, with the condition that c itself is not a lattice point. There will be certain c such that $c + \Lambda' \in \Lambda$, that is, the points of $c + \Lambda'$ are points in Λ. With enough of these, namely those from $c^{(1)}, c^{(2)}, \ldots$, it will be possible to recreate the original lattice from a union of disjoint cosets of the sublattice:

$$\Lambda = \Lambda' \cup \{c^{(1)} + \Lambda'\} \cup \{c^{(2)} + \Lambda'\} \cup \cdots. \tag{4.4-3}$$

If there are $p - 1$ such cosets plus Λ', we say that a partition of Λ of order p has been performed. The notation Λ/Λ' denotes this p-way partition, the one that

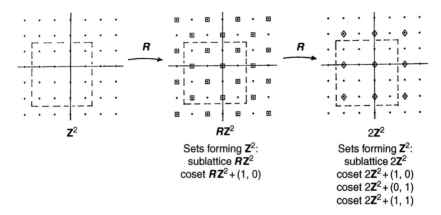

Figure 4.21 The partition chain $\mathbf{Z}^2 / R\mathbf{Z}^2 / 2\mathbf{Z}^2$, showing two lattice rotations by R. Points in dashed box will become TCM code points, after translation by $(1/2, 1/2)$. Compare to Fig. 4.6 ($d_\Lambda = 1$).

comprises

$$\Lambda', c^{(1)} + \Lambda', \ldots, c^{(p-1)} + \Lambda'. \tag{4.4-4}$$

We can also define a partition *chain* $\Lambda/\Lambda'/\Lambda''$, which means that the partition process with respect to sublattice Λ'' is performed on Λ' and each coset in Eq. (4.4-4); that is, for each of these, disjoint cosets of Λ'' are formed whose union is the lattice or coset. If these are each q-way partitions, we could as well think of one pq-way partition Λ/Λ''. The notion can be extended to chains of any depth.

We can take as an example set partitioning by the Ungerboeck Rules for the standard QAM family of codes in Section 4.2. We will use the concepts of lattice rotation, sublattice, partition, and partition chain. In so doing, we will express the QAM family of TCM codes in lattice language. Figure 4.6 showed the Ungerboeck partitioning process for the 16QAM master constellation. That constellation consists of the inner 16 points of the translated lattice $\mathbf{Z}^2 + (1/2, 1/2)$. The first two partitions in Fig. 4.6 are the same as the partition chain $\mathbf{Z}^2/R\mathbf{Z}^2/2\mathbf{Z}^2$ shown in Fig 4.21, after translations by $(1/2, 1/2)$. The figure shows the progression $\mathbf{Z}^2, R\mathbf{Z}^2, RR\mathbf{Z}^2 = 2\mathbf{Z}^2$, with dashed lines enclosing the part that will become the 16QAM partition. Each partition is 2-way. Underneath $R\mathbf{Z}^2$ is the coset and sublattice which make up $\mathbf{Z}^2$ in the sense of Eq. (4.4-3). Underneath $2\mathbf{Z}^2$ are the 3 cosets and sublattice Λ'' whose union is $\mathbf{Z}^2$. The chain $\Lambda^2/R\Lambda^2/2\Lambda^2$ expresses the fact that the partition consists of two 2-way partitions, but it could as well be written as one 4-way partition $\Lambda^2/2\Lambda^2$, if the aim is four subsets. Every sublattice or coset is translated by $(1/2, 1/2)$ before use as a TCM subset.

A basic lattice theory reference with a coding point of view is Conway and Sloane [23].

Set-partition Coding

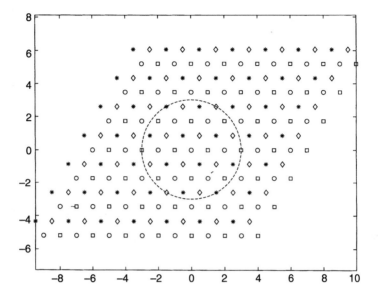

Figure 4.22 Part of the sphere packing lattice $\mathbf{A}_2$, with the 4-way partition $\mathbf{A}_2/2\mathbf{A}_2$ indicated by symbols $\circ, \diamond, *, \square$. The circle encloses the 32 points with least energy.

4.4.2. Improved Lattices in Two or More Dimensions

Can the multidimensional lattice view improve the TCM codes in earlier sections? As a warmup for the multidimensional case, we start with lattices that are over the standard two dimensions, but are not rectangular.

An interesting example is the well-known sphere packing lattice $\mathbf{A}_2$ in two dimensions, with its hexagonal pattern. It is shown in Fig. 4.22 in a version with $d_\Lambda = 1$. Its generator matrix is

$$G_A = \begin{bmatrix} 1 & 0 \\ 1/2 & \sqrt{3}/2 \end{bmatrix}, \tag{4.4-5}$$

with $\kappa = 6$ and $V(\mathbf{A}_2) = 0.866$. There is no interesting 2-way partition of $\mathbf{A}_2$, but the 4-way partition $\mathbf{A}_2/2\mathbf{A}_2$ quadruples square distance, just as did the standard rectangular-lattice partition $\mathbf{Z}_2/2\mathbf{Z}_2$; so also do further 4-way partitions by $2^k \mathbf{A}_2, k = 2, 3, \ldots$. This allows, for example, TCM codes with $c = 2, 4, 6, \ldots$ subset selector output bits.

Figure 4.22 shows $\mathbf{A}_2$ with the 4-way partition $\mathbf{A}_2/2\mathbf{A}_2$, centered on $(0, 0)$. The dashed circle is a boundary that contains 32 points. With the help of G_A,

Table 4.5 $\bar{E}_s$ and $d^2_{\min}$ for circular-boundary lattice codes based on $\mathbf{A}_2$ and $\mathbf{Z}^2 + (1/2, 1/2)$, compared to QAM family numbers from Table 4.1. Brackets show gains over standard QAM with rectangular or CR boundaries

		E_s			$d^2_{\min}$	
M_c	Std. QAM	Circular boundary, rect. QAM	Circular boundary, lattice $\mathbf{A}_2$	Std. QAM	Circular boundary, rect. QAM	Circular boundary, lattice $\mathbf{A}_2$
16	2.5	2.5	2.25	0.8	0.8 [0 dB]	0.889 [0.46 dB]
32	5(CR)	5	4.41	0.5	0.5 [0]	0.567 [0.55]
64	10.5	10.25	8.86	0.286	0.293 [0.11]	0.339 [0.74]
128	20.5 (CR)	20.44	17.68	0.1707	0.1713 [0.02]	0.198 [0.64]
256	42.5	40.69	35.26	0.0941	0.0983 [0.19]	0.113 [0.81]

it is not difficult to find the average energy $\bar{E}_s$ for such a constellation.[14] Table 4.5 lists $\bar{E}_s$ for circular-boundary constellations with sizes 16–256 and compares it to $\bar{E}_s$ for circular sets of points from $\mathbf{Z}^2 + (1/2, 1/2)$ as well as the standard QAM-family $\bar{E}_s$ numbers from Table 4.1. The square minimum distances of all these signal sets are given by the standard formula $d^2_{\min} = D^2_{\min} \log_2 M_c / 2\bar{E}_s$ with $D^2_{\min} = 1$. The distances are shown together with dB gains (in [ ]) over the standard distances in Table 4.1. The gains here stem from two sources. First, the lattice is more compact in area by a factor 0.886, which reduces average energy by approximately this factor, which in turn increases $d^2_{\min}$ by $10 \log_{10} 0.866$, or 0.62 dB. Second, the constellation boundary is circular, which reduces the energy of the M_c points a little; for large M_c this effect is worth 0.20 dB. This second effect is less noticeable when compared to CR constellations.

The same subset minimum distance bound for TCM codes based on these new constellations, as computed from Eq. (4.2-10), improves by these same two amounts. So also does the distance of any concrete scheme. Both are given special names: the first is the *lattice coding gain*, the second is the *shape gain*. We will now extend these notions to N-dimensional lattice codes with $N > 2$. It is proper to call lattice structures codes in themselves, because there is a non-trivial structure in an energy-efficient lattice, which becomes more subtle and attractive as N grows. However, we will also apply set partitioning and subset selection, which will make the lattice idea even more attractive.

---

[14] Proceed as follows: create a suitably large lattice via $z = iG$; compute energies of all z; sort in order of energy and take the lowest M_c points.

Set-partition Coding 177

The multidimensional lattice will extend over N signal space dimensions, where N is ordinarily even, since we assume the dimensions are provided by $N/2$ uses of a QAM channel. It is convenient to retain the rate R in bits/QAM interval, which means there are $(N/2)R$ bits carried per N-dimensional lattice codeword. Similarly, M_c will denote the constellation size per QAM interval; $M_{cN} \triangleq (M_c)^{N/2}$ is the true multidimensional constellation size, and M_c may not be an integer. In the spirit of the earlier part of the chapter, the standard 2-dimensional QAM lattice with rectangular boundary and $d_{\min} = 1$ will be a benchmark. $N/2$ independent uses of this benchmark can be viewed as a rectangularly bounded subset of the cubic lattice $\mathbf{Z}^N$ in N-space. From Eqs (4.2-3) and (4.2-4), $N/2$ independent uses of the benchmark create a transmission with average energy $\bar{E}_{sN} = (N/2)(M_c - 1)/6$, with M_c the square of an integer; the composite constellation size is $(M_c)^{N/2}$ and the total rate is $(N/2) \log_2 M_c$ bits/composite constellation. The square minimum distance is

$$d_{\text{std}}^2 = \frac{1}{2\bar{E}_{sN}/(N/2)R} = \frac{(N/2)R}{(N/6)(M_c - 1)}, \qquad (4.4\text{-}6)$$
$$= \frac{3R}{M_c - 1}, \quad \text{with } R = \log_2 M_c \text{ bits/QAM interval.}$$

This is the same as Eq. (4.2-4) for the 2-dimensional standard lattice alone. d_{std}^2 will be the standard of comparison in what follows.

Now consider a general N-dimensional lattice code. It is the set $\mathcal{W}$ containing the M_{cN} points of Λ that lie within a boundary $\mathcal{V}$. This set of points probably has a roughly spherical boundary in order to minimize energy, but we will not yet make that assumption. Assume as usual that the lattice minimum distance d_Λ is 1. Then, the normalized square minimum distance of this structure is $d_{\min}^2 = (RN/2)/2\bar{E}_{sN}$. We are interested in the ratio $d_{\min,\Lambda}^2/d_{\text{std}}^2$, since this will be the improvement of the new code compared to the QAM benchmark. The ratio is[15]

$$\frac{[M_c - 1](N/2)}{6\bar{E}_{sN}} = \frac{[(M_{cN})^{2/N} - 1](N/2)}{6\bar{E}_{sN}}. \qquad (4.4\text{-}7)$$

Before proceeding further, we would like to separate out the effect of lattice packing from this. The new lattice Λ may pack points more efficiently in N-space than $\mathbf{Z}^N$; that is, its fundamental volume $V(\Lambda)$ may be less than $V(\mathbf{Z})$, which is 1. A consequence of this would be that $V(\Lambda)$ may be expanded by a factor $1/V(\Lambda)$ compared to $\mathbf{Z}$, without significant change to the overall energy or volume of the

---

[15] M_c may no longer be the square of an integer; nonetheless, we continue to use Eq. (4.4-6) as a standard, in order to obtain a well-defined number for γ_s in Eq. (4.4-9). The continuous approximation of Eq. (4.4-12) is unaffected.

code. A volume expansion by $1/V(\Lambda)$ is equivalent to a square distance expansion by $1/V(\Lambda)^{2/N}$. We will define the lattice coding gain γ_{cg} to be this factor

$$\gamma_{cg} \triangleq 1/V(\Lambda)^{2/N}, \quad \text{for } d_\Lambda = 1 \quad (4.4\text{-}8)$$

and take it out of Eq. (4.4-7). What remains is defined to be the shape gain

$$\gamma_s \triangleq \frac{(N/2)\left[(M_{cN})^{2/N} - 1\right] V(\Lambda)^{2/N}}{6\bar{E}_{sN}}. \quad (4.4\text{-}9)$$

This is because it is set by the shape of the bounding volume and to a first approximation not by the lattice definition. What we now have is the minimum distance of the lattice code in the form

$$d_{\min}^2 = \gamma_{cg}\gamma_s d_{std}^2 = \gamma_{cg}\gamma_s \frac{3R}{M_c - 1} \quad (R \text{ in bits/2-dim}), \quad (4.4\text{-}10)$$

where there are M_{cN} total points and $M_c = (M_{cN})^{2/N}$.

Closed-form approximation to γ_s for spherical boundary. Equation (4.4-9) gives the precise value for the shape gain but a tight approximation to γ_s may be obtained in simple form. It depends on the fact that

$$\bar{E}_{sN} \triangleq \frac{\sum_{w_j \in \mathcal{W}} \|w_j\|^2}{M_{cN}} \approx \frac{\int_\mathcal{V} \|v\|^2 dV}{V(\Lambda) M_{cN}}, \quad (4.4\text{-}11)$$

$$M_{cN} \approx \frac{\int_\mathcal{V} dV}{V(\Lambda)}. \quad (4.4\text{-}12)$$

These results from lattice theory are called the "continuous approximations" to $\sum \|w\|^2$ and M_{cN}.[16] $\mathcal{V}$ need not be a sphere, but if it is, energy is minimized. The integral in Eq. (4.4-12) is then the volume of an N-sphere, namely

$$\int_\mathcal{V} dV = \frac{\pi^{N/2}\rho^N}{(N/2)!}, \quad \rho, \text{ the radius of } \mathcal{V}, \quad (4.4\text{-}13)$$

with $dV = N\pi^{N/2}r^{N-1}/(N/2)!dr$. Furthermore, the (4.4-11) integral may be evaluated as

$$\int_\mathcal{V} \|v\|^2 dV = \int_0^\rho \frac{r^2 N\pi^{N/2} r^{N-1}}{(N/2)!} dr = \frac{\pi^{N/2} N\rho^{N+2}}{(N+2)(N/2)!}. \quad (4.4\text{-}14)$$

We can use these in Eq. (4.4-9) and form the estimate

$$\gamma_s \approx \frac{N/2 \left[\int_\mathcal{V} dV\right]^{2/N}}{6\int_\mathcal{V} \|v\|^2 dV / \int_\mathcal{V} dV} = \frac{\pi(1 + N/2)}{6[(N/2)!]^{2/N}}. \quad (4.4\text{-}15)$$

---

[16] The meaning of $\approx$ is that the two sides in Eqs (4.4-11) and (4.4-12) have ratio tending to unity as the volume of $\mathcal{V}$ grows.

Set-partition Coding

Table 4.6 Important coding lattices and their parameters

Lattice	Common name	Kissing number	Lattice coding gain
$\mathbf{Z}^N$	Cubic	$2N$	1
$\mathbf{D}_N$	Checkerboard	$2N(N-1)$	$2^{1-2/N}$
$\mathbf{A}_2$	2D Sphere packing	6	1.155
$\mathbf{E}_8$	Gosset	240	2
$\mathbf{\Lambda}_{16}$	Barnes–Wall	4320	2.829
$\mathbf{\Lambda}_{24}$	Leech	196560	4
$\mathbf{D}_{32}$	Barnes–Wall	208320	4

For $N = 2$ dimensions, Eq. (4.4-15) is $\pi/3$, or 0.20 dB, the amount cited before in Section 4.2.2. It is visible in the circular boundary gain for 256QAM in Table 4.5. At $N = 4$, Eq. (4.4-15) rises to 0.35 dB, and at $N = 8$ to 0.76 dB. Using Stirling's approximation, it can be shown that $\gamma_s \to \pi e/6$ (1.53 dB) as $N \to \infty$.

Shape gains can thus be significant and they are "free," so long as the peculiar spherical set of code points can be managed. It can be shown, however, that the peak-to-average power ratio in each constituent 2-dimensional constellation is not ideal [25].

A larger source of gain is γ_{cg}, the gain from more efficient N-space packing of the lattice points. Table 4.6 identifies some interesting higher dimensional lattices and their parameters. All of these are sublattices of $\mathbf{Z}^N$ and can thus be based on a processor designed for ordinary rectangular N-dimensional QAM. Lattice coding gains as large as 6 dB are available.

4.4.3. Set-partition Codes Based on Multidimensional Lattices

Practical TCM codes have been created by partitioning a region of points of one of the higher dimension lattices in the previous subsection. Both the shape and the lattice coding gains improve with dimension, and in addition, rotation invariance is easier to obtain. Processing complexity is much higher, but is not a great problem at, for example, telephone modem speeds.

As with 2-dimensional lattices, one must design a partition, and with higher N it is convenient to use the lattice and partition chain notation of Section 4.4.1. One hopes to obtain a chain that doubles square distance at each 2-way partition. These sometimes, but not always, exist. One such chain, given by Schlegel [1] that begins with master constellation $\mathbf{Z}^4$ is

$$\mathbf{Z}^4/\mathbf{D}_4/R\mathbf{Z}^4/R\mathbf{D}_4/2\mathbf{Z}^4, \tag{4.4-16}$$

where R is a 4-dimensional rotation operation. But an attractive distance growth often requires partitioning into many more subsets than would be needed in 2-dimensional coding. Hopefully, the master constellation and partition chain can be designed for a rapid distance increase and not too many nearest neighbors.

Table 4.7 shows a number of good multidimensional TCM codes and we will organize the discussion around it. The table is broken into lists of 4-, 8- and 16-dimensional codes. The codes are taken from a variety of sources as indicated.

Table 4.7 Parameters of good multidimensional set-partition codes over 4, 8, and 16 dimensions. Lattice and subset parameters are grouped at left, ending with the modified same subset bound $d_{ss}^2/\gamma_s d_{std}^2$. Distance parameters are grouped at right, ending with ACG; the ACG is limited to the modified same subset bound. Key to sources: U = Ungerboeck [5], W = Wei [8], CS1 = [27], CS2 = [26]

	Lattice and subset parameters					Selector and distance gain					
Λ	Subsets	γ_{cg}	$(d'_\Lambda/d_\Lambda)^2$	$d_{ss}^2/d_{std}^2\gamma_s$		States	b	ζ	A_f	ACG (dB)	Ref.
4 Dimensions – shape gain: add 0.35 dB											
Z^4/RD_4	8	1	4	2.83 (4.52 dB)		8	2	1	44	4.52	W
"	8	1	4	"		16	2	1	12	"	W
$Z^4/2Z^4$	16	1	4	"		32	3	1	4	"	W
$Z^4/2D_4$	32	1	8	5.66		64	4 5/8		72	5.48	W
"	32	1	8	"		128	4 6/8		728	6.28	U
$D_4/2D_4$	16	1.414	4	4 (6.02 dB)		16	3 3/4		152	4.77	CS2
"	16	"	4	"		64	3	1	828	6.02	CS2
8 Dimensions – shape gain: add 0.76 dB											
Z^8/E_8	—	1	4	3.36 (5.27 dB)		16	—	1	316	5.27	W
"	16	1	4	"		32	3	1	124	"	W
"	16	1	4	"		64	3	1	60	"	W
Z^8/RD_8	16	1	4	"		128	3	1	28	"	U
RD_8/RE_8	32	1.682	2.83	4 (6.02 dB)		32	4	1	>500	6.02	W
"	32	"	"	"		64	4	1	316	"	W
"	32	"	"	"		128	4	1	124	"	W
E_8/RE_8	16	2	2	3.36 (5.27 dB)		8	3	1	764	5.27	CS1
"	16	2	2	"		16	3	1	316	"	CS1
"	16	2	2	"		32	3	1	124	"	CS1
"	16	2	2	"		64	3	1	60	"	CS1
16 Dimensions – shape gain: add 0.98 dB											
Z^{16}/H_{16}	32	1	4	3.67 (5.64 dB)		32	4	1	>900	5.64	W
"	32	1	4	"		64	4	1	796	"	W
"	32	1	4	"		128	4	1	412	"	W

Set-partition Coding

All codes have rate $(N/2)R = \log_2 M_{cN} - 1$ bits/N-dim. If there are b subset selection bits, then the code carries $(N/2)R - b = \log_2 M_{cN} - 1 - b$ data bits in a subset, per N dimensions. The number of subsets in the scheme is 2^{b+1} and the rate of the subset selector is $b/(b+1)$. For practical rates, these constraints can lead to quite a complex coding scheme. At 8 dimensions and rate 4 bits/2-dim and partition to 32 subsets, for example, the subset selector rate is 4/5 and the 8-dimensional master constellation has 2^{17} points.

Since the subset generation is often rather special with, for example, circuitry that ensures rotation invariance, the reader must turn to the respective reference for details. The set partition used is given in terms of one overall partition Λ/Λ', which may be 2-, 4- or more ways; it is often achieved by a long chain of component partitions, for which, once again, the reader must consult the references. In any case, the lattice coding gain γ_{cg} is given for Λ, and the final sublattice Λ' and its translates produce the listed number of TCM subsets. The square distance gain in one of these subsets is also listed as $(d_{\Lambda'}/d_\Lambda)^2$, the multiple of the original lattice square minimum distance.

Distance gains in these TCM constructions and how to compute them need some careful discussion. We first consider the same subset minimum distance d_{ss} that applies in the N-dimensional case. The calculation is the one in Section 4.4.2 that produced the relation $d_{\min}^2 = \gamma_s \gamma_{cg} d_{std}^2$ for straight lattices, except that we now apply it to the sublattice Λ' and we must account for the fact that the rate $(N/2)R$ is 1 less than $\log_2 M_{cN}$. The last means that the master constellation has twice the points that it would have under lattice coding without set partitioning. The M_{cN} points thus enclose twice the volume in N-space. A sphere has volume Eq. (4.4-13); doubling it requires that the radius ρ grow by a factor $2^{1/N}$; this increases the estimate of $\bar{E}_{sN}$ computed in Eqs (4.4-11)–(4.4-14) by $2^{2/N}$; the estimate of d_{ss}^2 is therefore reduced by the same number.

The factor d_{std}^2 is defined as $3R/(M_c - 1)$ in Eq. (4.4-6), with $M_c = (M_{cN})^{2/N}$; when M_c is the square of an integer, d_{std} is the distance of standard QAM with M_c points. As R grows, $d_{std}^2 \approx 3R/2^R$. It remains to account for Λ' and its wider spacing. For this we need the expansion factor in the square minimum distance of Λ', compared to Λ: this is $(d_{\Lambda'}/d_\Lambda)^2$. Including all these factors, we get the same subset distance expression

$$d_{ss}^2 = \frac{\gamma_s \gamma_{cg} d_{std}^2}{2^{2/N}} (d_{\Lambda'}/d_\Lambda)^2 \approx \frac{3\gamma_s \gamma_{cg} R}{2^{R+2/N}} (d_{\Lambda'}/d_\Lambda)^2. \qquad (4.4\text{-}17)$$

By computing the precise $\bar{E}_{sN}$, a precise γ_s may be found, and formula Eq. (4.4-17) may be applied for a specific rate and master constellation. But this is hardly necessary, since Eq. (4.4-17) with the spherical γ_s from Eq. (4.4-15) is a good estimate at $R \geq 4$ bits/2-dim and any reasonable constellation. Table 4.7 lists d_{ss}^2 as a multiple of d_{std}^2 without the shape gain; that is, it lists $\gamma_{cg}(d_{\Lambda'}/d_\Lambda)^2/2^{2/N}$. The shape gain itself depends to a first approximation only on N and is listed separately for each section of the table. Note that these shape gains assume a spherical boundary to Λ.

It remains to list the free distance of the actual TCM scheme. As always, the subset selector is designed to maximize square inter-subset distance, with d_{ss} as the upper limit. The degree to which it succeeds is listed in the table as a fraction ζ of the partition expansion $(d_{\Lambda'}/d_\Lambda)^2$ that is available. A_f is the nearest neighbor number of the TCM code, divided by $N/2$, as given in [1]. Rather than list the actual free distance, the ACG in dB relative to standard QAM is given. The reason for this is that to a first (and close) approximation, *the ACG is independent of the per 2-dimension code rate R.* Thus, one approximate ACG can be given for a given lattice, subset configuration, and subset selector, which applies for any reasonable rate. If an absolute square free distance is needed, it may be obtained from

$$d_f^2 = \left[10^{\text{ACG}/10} \right] d_{\text{std}}^2. \qquad (4.4\text{-}18)$$

A formal definition of the ACG in terms of the basic lattice coding parameters is

$$\text{ACG} \triangleq 10 \log_{10} \left[\zeta \frac{d_{ss}^2}{d_{\text{std}}^2} \frac{1}{\gamma_s} \right] = \zeta \frac{\gamma_{cg}}{2^{2/N}} (d_{\Lambda'}/d_\Lambda)^2. \qquad (4.4\text{-}19)$$

Note that the shape gain γ_s is once again factored out. For a spherically bounded lattice, the γ_s values in the table sections should be added to the ACG; for other boundaries, a corresponding value can be added.

In much of the table, the subset selector achieves $\zeta = 1$, and the ACG is $\gamma_{cg}(d_{\Lambda'}/d_\Lambda)^2/2^{2/N}$. By comparing with earlier tables, we see that the passage to higher dimensions gains perhaps 1 dB including shape gain, for the same selector state complexity. It is interesting that a large part of this is the shape gain available at higher dimensions. Perhaps the 1 dB will increase in the future if better partition chains are discovered. A final caveat is that many of these codes have a very complex neighbor structure, and their decoder error event rate and especially their BER can be poorly predicted by distance-based expressions such as $Q(\sqrt{d_f^2 E_b/N_0})$. Little is actually known about the BER performance of higher dimension codes.

4.5. QAM-like Codes Without Set Partitioning

The limiting case of the set-partition process in the foregoing sections is subsets containing a single point. This brings us full circle, back to the fundamental problem of imposing a coded pattern on the points in a succession of QAM-like constellations. Equivalently, this is the problem coding component values in signal space.

N-tuple codewords, whose letters are points in Euclidean N-space, date back as a concept to Shannon's introductory paper [19] in 1949. A typical early

Set-partition Coding

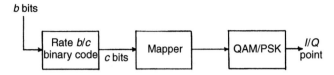

Figure 4.23 Rate b/c binary encoder–c-ary modulator combination for coded modulation over I/Q signal sets.

effort to develop concrete codes over PSK constellations was that of Zetterberg [32] in the 1960s. Another stream was the permutation coding work of Slepian [33]. We have already discussed early lattice code developments in Section 4.4. A parallel stream to the lattice work was multidimensional coding by relatively *ad hoc* methods; some representative schemes from the 1970s appear in [34,35]. By the late 1970s, however, the focus was on schemes of the form in Fig. 4.23, consisting of an ordinary rate b/c binary code, a binary-to-constellation-point mapper, and a QAM or PSK modulator. If the modulator is BPSK or QPSK, then Euclidean distance in the overall system is equivalent to Hamming distance in the binary code alone, and so we leave this case to a traditional coding text. Otherwise, any optimizing of the binary code and mapper needs to be performed with regard to Euclidean distance.

PSK constellations have equal-energy points, but in applications the signals are not constant envelope.[17] Almost always, PSK consists of phased pulses which for bandwidth reasons need to have significant envelope variation (see Section 2.5.1). In reality, QAM and PSK are alike in that they are built upon pulses that have both amplitude and phase variation. Such modulation combined with binary coding is our subject here.

Binary coding may also be combined with true constant-envelope modulation. Due mostly to Lindell and Sundberg, this subject was already well developed in the early 1980s, and it culminated in [15,28]. To be even moderately bandwidth-efficient, this kind of coded modulation requires a partial response CPM or at least a non-binary CPFSK modulator. CPM is a coded modulation in itself, and so, unlike the binary coding–QAM constellation combination, the binary coding–CPM combination is a form of concatenating coding.

8PSK Systems

We begin with some good convolutional code–8PSK combinations that have been discovered at rate 2 bits/2-dim. They are summarized in Table 4.8. The column d_f^2 gives the normalized square Euclidean free distance, computed in the

---

[17] For a picture of QPSK envelope variation, see Fig. 2.14.

Table 4.8 Good systematic feedback convolutional encoders for the encoder–8PSK and encoder–16QAM coded modulation combinations. d_f^2 shown is standard normalized free distance; gain (dB) for 8PSK combinations is over QPSK. Tap sets are defined in Fig. 4.8. Mapper designations: N = natural, Z = Table 4.9. *Sources*: Z = Zhang [30], P = Porath [31]. (*) indicates outcome of exhaustive search

States	m	h_0	h_1	h_2	h_3	g_0	g_1	g_2	g_3	d_f^2	Gain(dB)	A_f	Map	Ref.
8	3	17	02	06		74	20	30		4.59*	3.6	2	N	Z
Tap	Ex.	1111	0010	0110		1111	0100	0110						
						8PSK constellation; rate 2								
16	4	27	04	12		72	10	24		5.17*	4.1	2.25	N	Z
32	5	43	04	24		61	10	12		5.76*	4.6	2.38	N	Z
64	6	147	012	066		714	240	330		6.34*	5.0	3.25	N	Z
128	7	277	054	176		772	150	374		6.59*	5.2	0.5	N	Z
256	8	435	072	142		561	270	214		7.52*	5.8	1.5	N	Z
512	9	1377	0304	0350		7764	1060	0560		7.52*	5.8	0.031	N	Z
1024	10	2077	0630	1132		7702	0630	2644		8.10*	6.1	0.28	N	Z
2048	11	4041	1212	0330		4101	2424	0660		8.34	6.2	3.88	N	P
4096	12	15201	06306	04112		40254	30630	24410		8.68	6.4	1.41	N	P
8192	13	20201	12746	00304		40202	31724	10600		8.68	6.4	0.62	N	P
						Standard 16QAM constellation; rate 3								
8	3	13	04	02	06	64	10	20	30	3.0		3.66	N	Z
16	4	25	12	06	14	52	24	30	14	3.6		9.16	N	Z
32	5	47	22	16	34	71	22	34	16	3.6		2	N	Z
64	6	117	026	074	052	744	320	170	250	4.2		5.08	N	Z
128	7	313	176	154	022	646	374	154	220	4.8		20.3	N	Z
256	8	417	266	040	226	741	332	010	322	4.8		3.27	N	Z

Set-partition Coding

standard way[18] (see Section 2.3.2). Then the distance gain in dB over uncoded QPSK, which has the same NBW, is shown. The convolutional rate 2/3 encoders are the systematic feedback ones in either form of Fig. 4.8, with the required three tap sets listed in both the g and h forms (as in Table 4.3). The column A_f is the multiplicity of the nearest neighbors, averaged over the binary-code trellis states.[19]

An important element in this coding method is the mapping from binary 3-tuples at the convolutional encoder output to 8PSK points. The PSK mapping most commonly used is the so-called *natural mapping*, in which the 3-tuples count up in binaries around the PSK circle, from 0 to $M_c - 1$. For example, the 8PSK natural map is $\{000, 001, 010, \ldots, 111\} \rightarrow \{0°, 45°, 90°, \ldots, 315°\}$. (Here, 0° is the rightmost 8PSK point.) This map, with the 3-tuple taken as the values $z_2 z_1 z_o$ in Fig. 4.8, is the one used in Table 4.8. It is rare that another map improves d_f but improvements by as much as 0.5 dB have been made to the BER in 8PSK codes [29,30].

As the state space grows in these 8PSK codes, the free distance continuously grows, with no apparent limit from a same subset bound. By comparison, the best rate 2 PSK-type code at this bandwidth which has a true set partition has $d_f^2 = 4$ and is the top 8PSK one in Table 4.3. Some codes have BER considerably worse than the theoretical prediction $Q(\sqrt{d_f^2 E_b/N_0})$, but the gap can often be closed to some extent by increasing the encoder state space. The baseband positive-frequency bandwidth of all these schemes in NBW terms is $(2T/2)[(1 + \alpha)/2T]$ Hz-s/data bit, where α is the pulse excess bandwidth factor. For a 30% root RC pulse and the 99% power criterion, this is $(1 + \alpha/2)/2 = 0.575$ Hz-s/bit.

Little is known about PSK coding at 16PSK and higher without subsets. In any case, these have a relatively weak master constellation.

16QAM Systems

Table 4.8 also lists some best-known convolutional code–16QAM combinations at rate 3 bits/2-dim. These are due mostly to Zhang [30], and all use a rate 3/4 binary encoder (the encoder is the extension of Fig. 4.8 to three inputs). The mapping here is a special one given in Table 4.9. Standard normalized square distances are given. By comparison with true set-partition codes in Table 4.3, it can be seen that the move to a rate 3/4 convolutional encoder does not improve d_f, but multiplicities (A_f) are improved in all cases except the 8 and 16 state encoders. BER improves in every case. The NBW calculation is

---

[18] That is, $d_f^2 = D^2/2E_b$, where D is the closest Euclidean approach of two codeword signals and E_b is the average energy/data bit measured with the same signals. The measures of distance used in the literature vary widely.

[19] As computed by Schlegel [1].

Table 4.9 Zhang's map from/to rate 3/4 convolutional encoder output bits $z_3 z_2 z_1 z_0$ to 16QAM constellation points. Encoder as defined in Fig. 4.8; constellation (I, Q) points located as in Fig. 2.18

QAM point location	Binary 4-tuple	QAM point location	Binary 4-tuple
$(1, -1)$	0000	$(-1, -3)$	1000
$(1, -3)$	0001	$(-1, -1)$	1001
$(3, -3)$	0010	$(-3, -1)$	1010
$(3, -1)$	0011	$(-3, -3)$	1011
$(1, 3)$	0100	$(-1, 1)$	1100
$(1, 1)$	0101	$(-1, 3)$	1101
$(3, 1)$	0110	$(-3, 3)$	1110
$(3, 3)$	0111	$(-3, 1)$	1111

$(2T/3)[(1+\alpha)/2T]$ Hz-s/bit, which for 30% root RC and 99% power bandwidth gives $(0.575)2/3 = 0.383$ Hz-s/bit.

Other Systems

A method studied at some length is to implement Fig. 4.23 with a block code instead of a convolutional one, a method called block coded modulation (BCM). A selection of BCM results appears in [1,36]. One way to construct a receiver is to express the block code in a trellis form, in this case a rather irregular one. A variety of other methods exist to form Euclidean code words in a block manner.

4.6. Problems

1. Repeat the calculation of the inter-subset square minimum distance that was performed in Section 4.2.1 for the subset selection machine (7, 5), but instead of this machine use the 8-state (64,74) convolutional encoder. Let the symbol energy and subset assignments continue to be $E_s = 1$ and those of Fig. 4.1. Show that the inter-subset square distance must be of the form $k(2) + \ell(0.586)$, k and ℓ integers. What is d_f^2 for this TCM code?

2. Derive Eq. (4.2-3); that is, show that the average energy of a rectangular grid of M_c points with unit spacing is $(1/6)(M_c - 1)$.

3. Let $\bar{E}_s$ be the average energy of points on a unit-spaced rectangular grid that lie within a circle. Show that $2\pi \bar{E}_s / M_c$ tends to 1 as M_c grows, where M_c is the number of points in the circle.

4. (a) Show that the average energy $\bar{E}_s$ of standard rectangular 16QAM is 2.5 if its minimum distance is 1. (b) Show that standard 32CR has $\bar{E}_s = 5$.

Set-partition Coding

5. Consider the values of free distance shown in Table 4.3 for rate 2 TCM based on the 8PSK master constellation. Show that these distances are restricted to a certain pattern of values. Do this as follows. Scale the constellation to radius 1; then note that the worst-case inter-subset distance is achieved between two sequences of 8PSK constellation points; list all possible square distances between two 8PSK constellation points; normalize these components by $R/2\bar{E}_s = 1$. Finally, work out what combination of components sum up to each square free distance value.

6. Repeat the previous problem for the 16PSK codes in Table 4.3.

7. The first entry under "Rectangular Family" in Table 4.3 is the 1-for-2 4-state subset selector (5, 2), which gives a rate 3 bits/QAM symbol TCM code. Consider the code based on the standard 16QAM master constellation. Use the standard symbol designations $u_3 u_2 u_1$ and $z_1 z_0$ in Fig. 4.4 and the standard set partitioning in Fig. 4.6.

   (a) Give a trellis section of this subset selector. On each branch show $z_1 z_0$ and the u_1 that causes the branch.

   (b) Give an acceptable mapping of $u_3 u_2 z_1 z_0$ to the 16QAM points.

   (c) Show that d_f^2 is the value in Table 4.3.

   (d) Give a one-term asymptotic estimate of the form $A_f Q(\sqrt{d_f^2 E_b/N_0})$ for the first event error probability.

8. Consider the introductory rate 2 8PSK set-partition code in Section 4.2.1 and assume that the allzero data sequence is transmitted. The probability that the first data bit pair is decoded in error has the asymptotic form $Q(\sqrt{d_f^2 E_b/N_0})$, where $d_f^2 = 4$.

   (a) By counting the number of codeword neighbors that lie at square distance 4, refine this estimate to the form $C_f Q(\sqrt{d_f^2 E_b/N_0})$. (This is similar to the first term of Eq. (4.3-6), but now the allzero input is assumed.)

   (b) Find the number of next-to-nearest neighbors, and further refine the estimate to the form $C_f Q(\sqrt{d_f^2 E_b/N_0}) + C_1 Q(\sqrt{d_1^2 E_b/N_0})$.

   (c) Will these forms change with the transmission of other data?

9. Show that the generator matrix G_Λ for the 2-dimensional sphere packing lattice A_2 is Eq. (4.4-5), and find the fundamental lattice volume $V(A_2)$. Find the average energy $\bar{E}_s$ for the set of 16 least-energy points of A_2.

10. Use Stirling's approximation to show that the shape gain γ_s given by Eq. (4.4-15) tends to $\pi e/6$ as $N \to \infty$.

11. Find the energy-minimizing 64-point constellation whose points are taken from a rectangular array. Or equivalently, find c such that a 64-point subset of the coset $\mathbf{Z}^2 + c$ has minimum energy. Compare to the average energy of standard 64QAM. (Assume throughout that $d_{\min} = 1$). *Hint:* A useful fact is that for any particular subset of $\mathbf{Z}^2$, the energy-minimizing offset c is the centroid.

12. Show that the energy-minimizing 16-point constellation whose points are taken from a rectangular array is standard 16QAM.

13. Find the energy-minimizing 16-point constellation whose points have a sphere packing arrangement. That is, find c such that a 16-point subset of the coset $\mathbf{A}_2 + c$ has minimum energy. Compare to the average energy of the energy-minimizing 16 points of the unshifted $\mathbf{A}_2$, given in Table 4.5. What is the total dB energy gain of your solution over standard 16QAM? (Assume throughout that $d_{\min} = 1$.)

14. The checkerboard lattice $\mathbf{D}_4$ in 4 dimensions consists of the points of $\mathbf{Z}^4$ for which $\sum_{j=1}^{4} i_j$ is even. Show that $d_\Lambda = \sqrt{2}$ and $\kappa = 24$.

15. Consider the first rate 2/3 convolutional code–8PSK modulation system in Table 4.8.

    (a) Verify that its square free distance is 4.59, as shown in the table.

    (b) Now change the mapping from the natural map given in Section 4.5 to the *Gray code* mapping $\{000, 001, 011, 010, 110, 111, 101, 100\} \rightarrow \{0°, 45°, 90°, \ldots, 315°\}$. What is the free distance now?

16. Consider again the first rate 2/3 convolutional code–8PSK modulation system in Table 4.8.

    (a) Construct an 8-state trellis section. Proceed as follows. Define the encoder state as the register contents, left to right. Use a notation similar to Fig. 4.4, in which the convolutional encoder input is $u_2 u_1$ and the output is $z_2 z_1 z_0$. Use the natural mapping to 8PSK points that is given in Section 4.5. On each trellis branch, show the 8PSK point phase that will occur and the data $u_2 u_1$ that cause the branch.

    (b) In the absence of noise the receiver is such that the 8PSK points form a unit circle in the I/Q plane. There is AWGN noise, however, and the first three I/Q points received are $1\angle 22.5°$, $1.2\angle 0°$, $1.2\angle 10°$. Find the nearest and second-nearest paths through the trellis after three sections, in terms of Euclidean distance.

Bibliography

References marked with an asterix are recommended as supplementary reading.

[1] *C. Schlegel, *Trellis Coding*. IEEE Press, New York, 1997.
[2] E. Biglieri, D. Divsalar, P. J. McLane, and M. Simon, *Introduction to Trellis-Coded Modulation with Applications*. Macmillan, New York, 1991.
[3] L. H. C. Lee, *Convolutional Coding*. Artech House, Boston, 1997.
[4] G. Ungerboeck, "Channel coding with multilevel/phase signals," *IEEE Trans. Inf. Theory*, **IT-28**, 55–67, Jan. 1982.
[5] *G. Ungerboeck, "Trellis-coded modulation with redundant signal sets," Parts I–II, *IEEE Commun. Mag.*, **25**, 5–21, Feb. 1987.
[6] D. P. Taylor and H. C. Chan, "A simulation study of two bandwidth-efficient modulation techniques," *IEEE Trans. Commun.*, **COM-29**, 267–275, Mar. 1981.
[7] L.-F. Wei, "Rotationally invariant convolutional channel coding with expanded signal space," Parts I–II, *IEEE J. Sel. Areas Communs.*, **SAC-2**, 659–686, Sept. 1984.
[8] L.-F. Wei, "Trellis-coded modulation with multidimensional constellations," *IEEE Trans. Inf. Theory*, **IT-33**, 483–501, July 1987.
[9] S. S. Pietrobon, et al., "Rotationally invariant nonlinear trellis codes for two-dimensional modulation," *IEEE Trans. Inf. Theory*, **IT-40**, 1773–1791, Nov. 1994.

Set-partition Coding

[10] S. S. Pietrobon and D. J. Costello, Jr., "Trellis coding with multidimensional QAM signal sets," *IEEE Trans. Inf. Theory*, **IT-39**, 325–336, Mar. 1990.

[11] S. S. Pietrobon, *et al.*, "Trellis-coded multidimensional phase modulation," *IEEE Trans. Inf. Theory*, **IT-36**, 63–89, Jan. 1990.

[12] E. J. Rossin, N. T. Sindhushayana, and C. D. Heegard, "Trellis group codes for the Gaussian channel," *IEEE Trans. Inf. Theory*, **IT-41**, 1217–1245, Sept. 1990.

[13] L. W. Couch, Jr., *Digital and Analog Communication Systems*, 6th Edn. Prentice-Hall, Upper Saddle River, NJ, 2000.

[14] *R. Johannesson and K. Zigangirov, *Fundamentals of Convolutional Coding*. IEEE Press, New York, 1999.

[15] J. B. Anderson, T. Aulin, and C.-E. Sundberg, *Digital Phase Modulation*, Plenum, New York, 1986.

[16] J. B. Anderson and D. P. Taylor, "A bandwidth-efficient class of signal space codes," *IEEE Trans. Inf. Theory*, **IT-24**, 703–712, Nov. 1978.

[17] J. B. Anderson and R. deBuda, "Better phase-modulation error performance using trellis phase codes," *Electronics (Lett.)*, **12**(22), 587–588, 28 Oct. 1976.

[18] *M. M. Mulligan and S. G. Wilson, "An improved algorithm for evaluating trellis phase codes," *IEEE Trans. Inf. Theory*, **IT-30**, 846–851, Nov. 1984.

[19] C. E. Shannon, "Communication in the presence of noise," *Proc. IRE*, **37**, 10–21, 1949; reprinted in *Claude Elwood Shannon: Collected Papers*, Sloane and Wyner, (eds), IEEE Press, New York, 1993.

[20] C. E. Shannon, "Probability of error for optimal codes in a Gaussian channel," *Bell Syst. Tech. J.*, **38**, 611–656, 1959; see Sloane and Wyner, *ibid.*

[21] R. de Buda, "The upper error bound of a new near-optimal code," *IEEE Trans. Inf. Theory*, **IT-21**, 441–445, 1975.

[22] R. de Buda, "Some optimal codes have structure," *IEEE J. Sel. Areas Communs.*, **SAC-7**, 893–899, Aug. 1989.

[23] J. Conway and N. J. A. Sloane, *Sphere Packings, Lattices and Groups*, Springer, New York, 1988.

[24] G. D. Forney, Jr., "Coset codes – Part I, II," *IEEE Trans. Inf. Theory*, **IT-34**, 1123–1187, Sept. 1988.

[25] G. D. Forney, Jr. and L.-F. Wei, "Multidimensional constellations – Part I," *IEEE J. Sel. Areas Communs.*, **SAC-7**, 877–892, Aug. 1989; G. D. Forney, Jr., "Multidimensional constellation – Part II," *ibid.*, 941–958.

[26] A. R. Calderbank and N. J. A. Sloane, "New trellis codes based on lattices and cosets," *IEEE Trans. Inf. Theory*, **IT-33**, 177–195, Mar. 1987.

[27] A. R. Calderbank and N. J. A. Sloane, "An eight-dimensional trellis code," *Proc. IEEE*, **74**, 757–759, May 1986.

[28] G. Lindell, "On coded continuous phase modulation," PhD Thesis, Telecommunication Theory Dept., University of Lund, Lund, Sweden, 1985.

[29] J. Du and M. Kasahara, "Improvements in the information-bit error rate of trellis code modulation systems," *IEICE Japan*, **E72**, 609–614, 1989.

[30] W. Zhang, "Finite-state machines in communications," PhD Thesis, Digital Communications Group, University of South Australia, Adelaide, Australia, 1995.

[31] J.-E. Porath, "On trellis coded modulation for Gaussian and bandlimited channels," PhD Thesis, Computer Engineering Department, Chalmers University of Technology, Göteborg, Sweden, Dec. 1991; *see also* "Fast algorithmic construction of mostly optimal trellis codes," Tech. Rpt. No. 5, Division of Information Theory, Electrical and Computer Engineering, Chalmers University of Technology, Göteborg, Sweden, 1987.

[32] L. H. Zetterberg, "A class of codes for polyphase signals on a bandlimited Gaussian channel," *IEEE Trans. Inf. Theory*, **IT-11**, 385–395, July 1965.

[33] D. Slepian, "Group codes for the Gaussian channel," *Bell Syst. Tech. J.*, **47**, 575–602, April 1968.

[34] G. R. Welti and S. L. Lee, "Digital transmission with coherent four-dimensional modulation," *IEEE Trans. Inf. Theory*, **IT-20**, 497–502, July 1974.

[35] L. Zetterberg and H. Brändström, "Codes for combined phase and amplitude modulated signals in four-dimensional space," *IEEE Trans. Commun.*, **COM-25**, 943–950, Sept. 1977.

[36] E. Biglieri and M. Luise (eds), *Coded Modulation and Bandwidth-Efficient Transmission*. Elsevier, Amsterdam, 1992 (chapter 4, Block Coded Modulation).

5

Continuous-phase Modulation Coding

5.1. Introduction

The second class of coded modulations in this book is a method of coding for carrier signals with constant envelope. It works by convolution, as do many coding systems, but this time a generator signal $q(t)$ is convolved with the symbols $a_0, a_1, \ldots$ to form a *phase* signal, which drives a phase modulator. In spirit this is not different from ordinary binary convolutional coding or the real-number partial response signaling (PRS) code convolution in Chapter 6, but it produces a nonlinearly modulated signal. The signal bandwidth and distance depend in different, subtle ways on the generator signal. Both need to be explored rather carefully in order to find the best codes. This is in contrast to trellis coded modulation (TCM) coding, where bandwidth is fixed by the orthogonal linear modulation pulse, and to PRS coding, where the optimal codes are solutions of a linear program.

In this first section we give an overview, define ideas, and review some history. Continuous-phase modulation (CPM) was the first class of coded modulations to receive extensive study, and it was the source of the term coded modulation.[1] CPM grew from the works of a number of researchers through the 1970s. Its direct ancestors can perhaps be said to be Doelz and Heald's MSK [6], patented in 1961, and a related scheme by de Buda [7], called Fast FSK, which he published in 1972. These were both FSK schemes with modulation index 1/2 and piecewise-linear continuous phase. Consequently, transmissions in adjacent symbol intervals were of necessity *dependent*; there was memory between symbols and therefore a crude kind of coding. De Buda pointed out to colleagues that Fast FSK in fact had a trellis structure, the trellis notion having recently been proposed.[2] During 1971–1976,

---

[1] The authors recall a meeting in 1979 in which three names were proposed – coded modulation, modulation coding, and codulation. The first eventually won the day, although the second is heard from time to time. Until the early 1980s, constant envelope coding schemes were most often called phase coding.

[2] By Forney, around 1970. See Section 3.3.2.

there appeared important papers [8]–[10] on continuous-phase FSK (CPFSK) by Schonhoff and Pelchat, Osborne *et al.* These extended the MSK idea to general modulation indices, found attractive distance properties, and showed that optimal detection required MLSE-like procedures and a signal observation considerably wider than one symbol. Papers in 1975–1978 by Miyakawa *et al.* [11], Anderson and de Buda [13], and Anderson and Taylor [12] finally introduced a formal coding theory foundation. They introduced encoders with memories longer than one (first in the form of multi-h codes), cut-off rates, formal free distance calculation, and outright trellis decoding (the last three in [12,13]).

CPM reached full development as a coded modulation method – and also acquired the name continuous-phase modulation – in 1979 with the thesis of T. Aulin [14] and the subsequent 1981 classic papers of Aulin and Sundberg *et al.* [15,16]. Phase continuity is critical to efficient constant-envelope coded modulation; without the 'CP' in CPM, bandwidth is unacceptably wide, and the combined energy–bandwidth performance of a scheme can never be competitive. Continuity also guarantees at least a crude level of coding.

In parallel with all this, the late 1970s also saw intense discussion of the role of energy and bandwidth in coding. This discussion was driven by CPM. As discussed in Section 3.6, traditional Shannon theory certainly never excluded the bandwidth dimension, but the everyday wisdom of practitioners in the mid-1970s was that coding increased bandwidth. CPM showed concrete schemes that could achieve coding gains without bandwidth expansion. By the end of the 1970s, the papers that appeared suggested that coding could reduce energy or bandwidth, or even *both*.

CPM researchers originally had satellite communication in mind, but the focus soon shifted to mobile radio. In both, constant-envelope modulation is important because the transmitter electronics are much more power-efficient. At the same time, bandwidth efficiency is important. In satellites, the standard traveling-wave tube class C final RF amplifier is strongly nonlinear and at the same time, 2–4 dB more efficient than a linear class A or B amplifier. This is a corresponding gain in E_b/N_0, or a large saving in vehicle weight. In mobile telephony, the 2–4 dB class C advantage translates to 50–150% more battery life. Were TCM to be used in these systems instead of a constant-envelope method, the 2–4 dB advantage would be lost. In such systems, 2–4 dB must be deducted from the TCM coding gain. This nearly cancels the energy advantage of TCM at the bandwidths where both TCM and CPM can be applied.

When the transmission system presents a nonlinear class C channel, there is little choice but to employ constant-envelope coding. When there is a choice, we can summarize CPM by saying that it works in a middle range of bandwidth and energy. More specifically, it achieves an error rate near 10^{-5} in bandwidth 0.5–1.5 Hz-s/data bit and an E_b/N_0 of 4–12 dB, depending on the CPM parameters and complexity. This is more narrowband than simple modulation and parity-check coding, but wider band/lower energy than TCM.

Formal Definition of CPM

We need to define a general signal form and some terminology that will be used throughout the chapter. The general form of a CPM signal is Eq. (2.5-27), which we repeat now as

$$s(t, \boldsymbol{a}) = \sqrt{2E_s/T} \cos[\omega_0 t + \phi(t, \boldsymbol{a})], \qquad (5.1\text{-}1)$$

in which ω_0 equals $2\pi f_0$, f_0 the carrier frequency, T is the symbol time, E_s is the energy per symbol time, and $\boldsymbol{a}$ is a sequence $\ldots, a[n], a[n+1], \ldots$ of M-ary data symbols, each taking one of the values $\pm 1, \pm 3, \ldots, \pm(M-1)$. The *excess phase* $\phi(t, \boldsymbol{a})$ is the phase in excess of the carrier at time t caused by the data, and it is defined by convention to be

$$\phi(t, \boldsymbol{a}) = 2\pi \sum_n a[n] h_n q(t - nT). \qquad (5.1\text{-}2)$$

Here h_n is the *modulation index* operating in interval n. If there are H different indices applied in cyclical fashion, the CPM is called *multi-h*. If the index is constant, we shall simply denote it as h; this is the case in almost all CPM schemes. The function $q(t)$ is the *phase response* of the CPM code, and its *frequency pulse* $g(t)$ is the derivative of $q(t)$. It is assumed that $g(t)$ is of finite duration, occupying the interval $[0, LT]$. If $g(t)$ has support $[0, T]$ or less, the scheme is *full response* ($L = 1$); otherwise, it is a *partial response* scheme. For normalization purposes, $g(t)$ integrates to $1/2$ whenever it has nonzero integral. Equivalently, $q(t)$ satisfies $q(t) = 0, t < 0$ and $q(t) = 1/2, t \geq LT$. The CPM signal can be thought of as either phase or frequency modulated, and we will adopt whichever is convenient at the moment.

We will take the term "standard CPM" to mean the case where all the following hold: M is a power of 2, the symbols are equiprobable, h is fixed and noninteger, $q(t)$ is continuous, and $q(LT) = 1/2$. Most of the analysis applies without any of these restrictions, but the CPMs obtained without them are seldom used, or they are well-known simple modulations. The requirement that $q(t)$ be continuous is what makes CPM signals continuous phase.

Like any signal with form (5.1-1) and (5.1-2), a CPM signal is defined by its in-phase and quadrature signals, which we can denote as $I(t, \boldsymbol{a})$ and $Q(t, \boldsymbol{a})$. I/Q signal formulations are reviewed in Section 2.5. The most convenient way to work with CPM signals, either in analysis or in practical receivers, is usually with their excess phase $\phi(t, \boldsymbol{a})$ or their I/Q signals. From Eq. (2.5-20), the normalized square Euclidean distance between two CPM signals may be expressed directly as

$$d^2 = \frac{\log_2 M}{2T} \int [I(t, \boldsymbol{a})^2 + Q(t, \boldsymbol{a})^2] \, dt, \qquad (5.1\text{-}3)$$

where the integral is over the time the signals differ. It is possible, of course, to form a signal space representation (see Section 2.3) and sum the square distance components in each dimension, as was natural for TCM signals. But CPM signals tend to project onto many dimensions, while having significant activity in only a few; quaternary CPFSK requires, for example, eight dimensions per symbol interval in order to be expressed in signal space. Bandwidth is only roughly related to signal space dimension, and what is important in transmission is physical bandwidth and its rolloff. It is also possible to represent CPM signals – and most other coded modulation – as a symbolic logic device that drives a discrete memoryless channel or a memoryless modulator. See, for example, [19]. But to do so can complicate the representation and take attention away from the physical reality.

In this chapter, the signal $s(t, \boldsymbol{a})$ will always be transmitted over an additive white Gaussian noise (AWGN) channel. The detector will be coherent with perfect synchronization, and the symbol boundaries $\ldots, -T, 0, T, \ldots$ are known exactly. Since $q(t)$ has infinite positive support, the transmitted signal in any particular interval depends in general on the present and all previous data symbols. This, however, is simply a statement that CPM signals are differential phase modulations: it can generally be said that for a signal that has reached time nT there will be a time $N'T$ in the past such that all phase activity before time $(n - N')T$ can be represented by a single offset (assume $N' \geq L$)

$$\phi_0 = 2\pi \sum_{i < n - N'} ha[i]/2.$$

If symbols up to $a[n - N' - 1]$ are correctly detected, then the correct ϕ_0 is known and may be subtracted from all signals.

The most common CPM frequency pulses are listed in Table 5.1. CPM schemes are denoted by an abbreviation followed by a prefix 'L' that denotes the length of the pulse. In two cases, TFM and GMSK, the pulses have infinite support and no L is given. The meaning of the abbreviations is:

REC – rectangular frequency pulse (also denoted *CPFSK*)

RC – raised cosine

SRC – spectral raised cosine

TFM – tamed frequency modulation

GMSK – Gaussian-shaped MSK

The frequency pulses $g(t)$ are the derivatives of the phase pulses. Figure 5.1 shows three representative frequency pulses (on the left) together with their phase pulses. 1REC–3REC and 1RC–3RC are shown in standard form, with support $[0, LT]$. GMSK is offered as an example of pulses with infinite support, and here it is convenient to center the responses at time 0. In a practical scheme, this kind of pulse must be truncated in time. SRC is another infinite-support pulse; when $\beta = 1$, LSRC is almost the same as LRC, so no separate plot of it is given.

Continuous-phase Modulation Coding

Table 5.1 Basic frequency pulses of CPM. RC, REC, and SRC pulses are parameterized by the pulse width L, GMSK by the bandwidth parameter B_b. GMSK, TFM, and SRC have infinite support

LREC
$$g(t) = \begin{cases} \dfrac{1}{2LT}, & 0 \le t \le LT; \\ 0, & \text{otherwise.} \end{cases}$$

LRC
$$g(t) = \begin{cases} \dfrac{1}{2LT}\left[1 - \cos\left(\dfrac{2\pi t}{LT}\right)\right], & 0 \le t \le LT; \\ 0, & \text{otherwise} \end{cases}$$

TFM
$$g(t) = \tfrac{1}{8}[g_0(t-T) + 2g_0(t) + g_0(t+T)]$$

$$g_0(t) \approx \sin\left(\dfrac{\pi t}{T}\right)\left[\dfrac{1}{\pi t} - \dfrac{2 - (2\pi t/T)\cot(\pi t/T) - (\pi^2 t^2/T^2)}{\pi t^3/T^2}\right]$$

LSRC
$$g(t) = \dfrac{1}{LT}\dfrac{\sin(2\pi t/LT)}{2\pi t/LT}\dfrac{\cos(\beta(2\pi t/LT))}{1 - ((4\beta/LT)t)^2}, \quad 0 \le \beta \le 1.$$

GMSK
$$g(t) = \dfrac{1}{2T}[Q\left(2\pi B_b \dfrac{t-T/2}{\sqrt{\ln 2}}\right) - Q\left(2\pi B_b \dfrac{t+T/2}{\sqrt{\ln 2}}\right)], \quad 0 \le B_b T \le 1$$

In Section 4.1 the concept of normalized bandwidth (NBW) was introduced as a measure for bandwidth comparison across chapters in this book. In TCM coding, the NBW is TWN_{dim}/R Hz-s/data bit, where T is the symbol time, W is a baseband positive-frequency measure of bandwidth, $N_{\text{dim}} = 2$ dimensions are used each symbol time, and the code rate is R data bits per symbol time. In CPM coding, N_{dim} continues to be 2, since CPM signals comprise an in-phase and quadrature component. The rate is $\log_2 M$; sometimes we will express the ratio T/R as simply $T_b = T/\log_2 M$. W will normally be the 99% or 99.9% power bandwidth (as defined in Section 2.7.1) of the baseband spectrum, positive frequencies only. This baseband quantity will be denoted B. The NBW is thus given by

$$\text{NBW} = \dfrac{TWN_{\text{dim}}}{R} = 2BT_b. \tag{5.1-4}$$

However, the basic spectra in Section 5.3.2 are simply plotted against fT_b.

The chapter that follows summarizes the fundamentals of CPM, as the subject developed during 1975–1990. The plan of the chapter is to study distance and spectrum in Sections 5.2 and 5.3, and then receivers and transmitters in Sections 5.4 and 5.5. Basic references are given, but not an inclusive list. More recent developments and more specialized work are found elsewhere in the book. For many older details and contributions, pages in *Digital Phase Modulation*, [1], are cited throughout. Other textbooks with at least a chapter coverage of CPM are Proakis [3], Wilson [4], and especially Xiong [5].

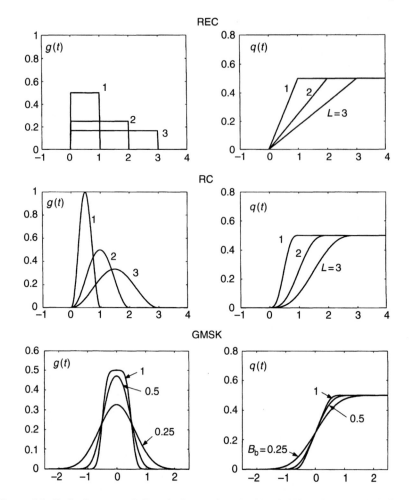

Figure 5.1 Basic frequency (*left*) and phase pulses (*right*) of CPM, showing 1REC–3REC, 1RC–3RC, and GMSK with bandwidth parameter $B_b = 0.25, 0.5, 1$. GMSK has infinite support. $T = 1$ here.

How to Find Good CPM Codes

What constitutes a "best" CPM code depends in a complex way on both its bandwidth and distance, and of course on the complexity needed to obtain these. For those who want to adopt a CPM code, we can attempt a brief summary of how to find one in the pages that follow.

1. Roughly speaking, there are two kinds of phase pulses, the linear ones (the REC family) and the smoothed ones (e.g. the RC, SRC families); the second group have better spectral sidelobes.

2. If large minimum distance is the object, the symbol alphabet size should be taken as large as possible, and good indices h can be read off the plots in Section 5.2.
3. If narrow bandwidth is the object, the insight into good codes comes from the spectra in Section 5.3.2.
4. If the code is to be best in a combined energy–bandwidth sense, then the best code should be read off the energy–bandwidth plots of Section 5.3.3. M and the pulse width L act to improve both energy and bandwidth.
5. h acts to trade off energy and bandwidth.
6. The best M is probably 4.

5.2. CPM Distances

As introduced in Chapter 2, the asymptotic form $P_e \sim Q(\sqrt{d_{\min}^2 E_b/N_0})$ for the error probability of a coded modulation greatly simplifies the search for good schemes because it reduces their energy efficiency to the single parameter $d_{\min}$, the minimum distance. This section deals with the properties of $d_{\min}^2$ in CPM and its outcomes. The plan is first to define an upper bound d_B^2, which is easier to compute, but which is nonetheless tight and shows the overall dependence of distance on the CPM parameters. This is done in Section 5.2.1. The bound development there was first devised by Aulin and Sundberg [14,15]. Next we will calculate in Section 5.2.2 some actual $d_{\min}^2$ values; it will turn out that $d_{\min}^2$ is the same as d_B^2, except at certain "weak" combinations of parameters, so long as the observation time of the signals is long enough. Distance calculation is the natural time to define a CPM state trellis, and this is done in Section 5.2.3.

5.2.1. Bounds on Minimum Euclidean Distance

To begin, we will consider standard CPM signals of the full response type; that is, $g(t)$ occupies a single interval.

The *phase tree*, introduced briefly in Section 2.5, is an important tool for studying CPM distances. This tree is formed by all phase trajectories $\phi(t, \boldsymbol{a})$ having a common starting excess phase of zero at time $t = 0$. Figure 5.2 shows a phase tree for a 4-ary 1REC scheme over $[0, 3T]$. This can be compared to Fig. 2.20 in Chapter 2, which is for binary 1REC. The modulation index h is arbitrary, but it is important to keep in mind that the phase must always be viewed mod-2π, and so the tree actually wraps around itself along the ϕ axis. A good way to visualize this wraparound is to draw the tree on a transparency and fold

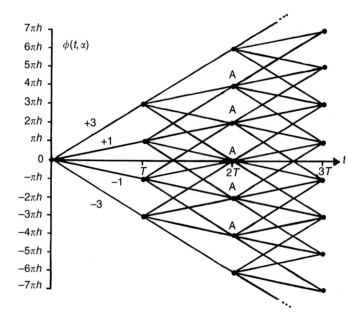

Figure 5.2 Phase tree of quaternary 1REC CPM, for paths that begin from phase 0. 'A' denotes a first phase merger.

this into a cylinder with the time axis down the cylinder.[3] By varying the radius of the cylinder, different modulation indices are modeled. Trajectories which seem far apart in the phase tree may actually be close or even coincide when viewed mod-2π, and this phenomenon is controlled in turn by h. The distance between two phase signals over some interval is the integral of the ordinary Euclidean distance through the body of the cylinder.

A projection of Fig. 5.2 with $h = 1/4$ onto a cylinder gives Fig. 5.3. The dotted rings mark the time axis, and the heavy dots in both figures mark corresponding branch points. At time T, 1REC CPM can have phases $\{\pm\pi/4, \pm 3\pi/4\}$; at $2T$ it can have $\{\pm\pi/2, 0, \pi\}$; at $T = 3$ the phases are back to the first set.

Figure 5.4 is a complete phase cylinder, showing all transitions of quaternary 1REC when $h = 1/4$. A phase transition at time 0 can begin from any multiple of $\pi/2$. This picture gives a good intuition about general CPM signal patterns: standard CPM has phases that form a pattern that repeats each interval down the cylinder and some integer number of times around the cylinder. However, it is easier to calculate distances with the aid of the earlier two pictures.

---

[3] This was suggested by R. de Buda in 1976; his first cylinder was a water glass. Cylindrical intuition is not easily obtained without de Buda's water glass.

Continuous-phase Modulation Coding

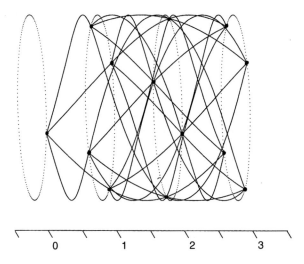

Figure 5.3 Phase tree of Fig. 5.2 with $h = 1/4$, projected onto a cylinder. Dotted circles mark times $0, T, 2T, 3T$.

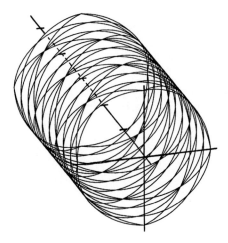

Figure 5.4 The complete phase cylinder diagram of $h = 1/4$ 1REC CPM, showing three symbol intervals. Trajectories can begin at phase $0, \pi/2, \pi, 3\pi/2$.

Suppose that two CPM signals differ over the interval $[0, NT]$. From Chapter 2, Eq. (2.5-25), and the linearity of $\phi(t, a)$, the normalized square Euclidean distance between them may be taken as

$$d^2 = \frac{\log_2 M}{T} \int_0^{NT} [1 - \cos \phi(t, \Delta a)] \, dt, \qquad (5.2\text{-}1)$$

where Δa is the difference sequence $a - \tilde{a}$ between the transmitted data sequence a and a sequence $\tilde{a}$ viewed by the detector. Note that several pairs of such sequences can correspond to the same Δa. Each element in the difference sequence can take one of the values $0, \pm 2, \pm 4, \ldots, \pm 2(M-1)$. To calculate $d_{\min}^2$ for an observation width of N intervals, all differences in a phase tree like Fig. 5.2 must be plugged into Eq. (5.2-1). Because of the symmetries in the CPM phase pattern, it is sufficient to start all signals at $t = 0$ and from a single root at phase $\phi = 0$. $\Delta a[0]$ must not equal 0. The minimum over all these distances is $d_{\min}$ for width N.

Applying these ideas first to the binary 1REC tree (Fig. 2.20), we see that if two data sequences are zero before time 0 and thereafter have the form

$$a = +1, -1, \alpha_2, \alpha_3, \ldots,$$
$$\tilde{a} = -1, +1, \alpha_2, \alpha_3, \ldots, \tag{5.2-2}$$

with $\alpha_2, \alpha_3, \ldots$ the same in both, then the two phase trajectories form a diamond and coincide for all $t \geq 2T$. They have a *phase merge* at time $2T$ and phase 0. Δa is $2, -2, 0, 0, \ldots$ for this pair, and the square distance for this pair, calculated from Eq. (5.2-1), forms the upper bound on $d_{\min}^2$ given by

$$d_{\min}^2 \leq 2 - (1/T) \int_0^{2T} \cos\{2\pi h[2q(t) - 2q(t-T)]\} \, dt. \tag{5.2-3}$$

For binary 1REC, the right-hand side reduces to the bound expression

$$d_B^2 = 2\left(1 - \frac{\sin 2\pi h}{2\pi h}\right). \tag{5.2-4}$$

Before continuing, we observe that $-\Delta a$ will give the same result in Eq. (5.2-1), since $\cos(\ )$ is an even function, and so $\Delta a[0]$ may always be chosen positive. Furthermore, when both trajectories are linear over an interval, the square distance contribution from that interval is

$$d^2 = \begin{cases} \left(1 - \dfrac{\sin \Delta\phi_b - \sin \Delta\phi_e}{\Delta\phi_b - \Delta\phi_e}\right) \log_2 M, & \Delta\phi_b \neq \Delta\phi_e; \\ (1 - \cos \Delta\phi_b) \log_2 M, & \Delta\phi_b = \Delta\phi_e, \end{cases} \tag{5.2-5}$$

where $\Delta\phi_b$ is the beginning phase difference and $\Delta\phi_e$ is the ending difference. (See Problems.)

We turn now to the quaternary case of 1REC and Fig. 5.2. Several pairs of phase trajectories may be found that all merge at $t = 2T$, and these merge points are all labeled A in the figure. It is even possible for two pairs of trajectories to have the same merge point. However, we see from Eq. (5.2-5) that only the phase difference at $t = T$ matters, those at 0 and $2T$ being zero. These are $2\pi h, 4\pi h,$

Continuous-phase Modulation Coding

and $6\pi h$, when the sign symmetry is removed. The argument is easily extended to the M-ary case and to other full response phase pulses. An upper bound on $d_{\min}^2$ for the general case is the minimum of Eq. (5.2-3) over $M-1$ difference sequences $\delta_0, -\delta_0, 0, 0, \ldots$, with $\delta_0 = 2, 4, \ldots, 2(M-1)$. This is

$$d_B^2 = \min_{1 \le k \le M-1} \log_2 M$$
$$\times \left(2 - (1/T) \int_0^{2T} \cos\{2\pi hk[2q(t) - 2q(t-T)]\} \, dt \right). \quad (5.2\text{-}6)$$

For M-ary 1REC, this specializes to

$$d_B^2 = \min_{1 \le k \le M-1} \log_2 M \; 2\left(1 - \frac{\sin k 2\pi h}{k 2\pi h}\right). \quad (5.2\text{-}7)$$

Figure 5.5 shows the upper bounds d_B^2 as a function of the modulation index h for M-ary 1REC schemes with $M = 2, 4, 8, 16$. The dotted lines show the different terms in the minimization of Eq. (5.2-7) for the $M = 4$ and 8 cases. For a fixed h the bound grows with M, and grows actually without limit.[4] It will

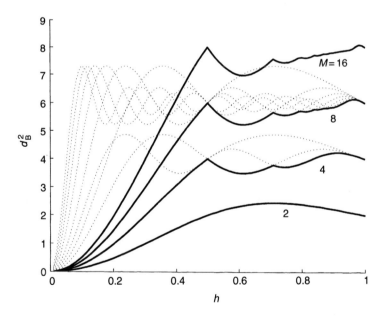

Figure 5.5 The upper bound d_B^2 as a function of h for M-ary 1REC. The dotted curves show the component curves for $M = 4, 8$ in the minimization of Eq. (5.2-7).

---

[4] But the spectrum grows as well, and in Section 5.3 it will turn out that $M > 8$ are not useful in a joint energy–bandwidth sense.

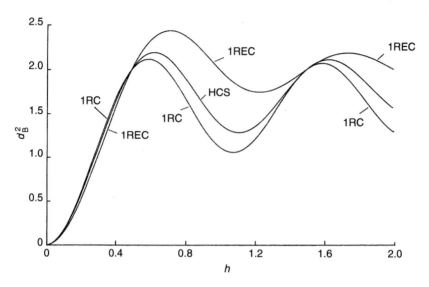

Figure 5.6 Upper bounds d_B^2 vs h for binary schemes with the three pulses 1REC, HCS, 1RC. (From Anderson, Aulin, and Sundberg [1].)

be shown in the next subsection that in fact $d_{\min}^2 = d_B^2$ if the observation width is long enough, except at a few h values. Furthermore, for h small enough, generally $h < 0.5$, difference sequence $+2, -2, 0, 0, \ldots$ directly yields $d_{\min}^2$ (this is the $k = 2$ term in Eq. (5.2-7)); these h will turn out to be the most effective ones in a joint energy–bandwidth sense.

The discussion so far may be repeated for other width-1 frequency pulses, and the bound d_B^2 that results is plotted against h in Figs 5.6–5.8 for the cases $M = 2, 4, 8$, respectively. The pulses here are REC (or CPFSK), RC, and HCS; the first two were defined in Table 5.1 and "half-cycle sinusoid" (HCS) consists of a positive sine half-cycle centered in $[0, T]$. All of these pulses are symmetrical about $T/2$, and for pulses with this property, the bound d_B^2 reduces to

$$d_B^2 = \log_2 M \min_{1 \le k \le M-1} \left(2 - \frac{2}{T} \int_0^T \cos 4\pi hk q(t)\, dt \right). \qquad (5.2\text{-}8)$$

It is clear from this form that d_B^2 cannot exceed $4 \log_2 M$. For example, binary systems can have square distance atmost 4, a 3 dB gain over QPSK or MSK.

There has been some attempt to find a full response pulse $g(t)$ that optimizes distance; see, for example, [1], Section 3.2. However, there is little to be gained in the important range $h < 0.5$. This is evident from the figures. There is, however, a difference in the spectral properties of these pulses.

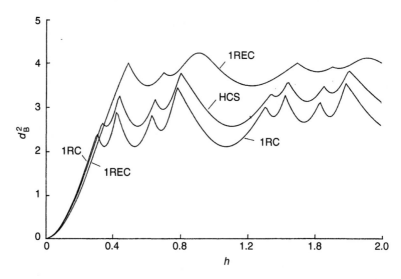

Figure 5.7 d_B^2 vs h for quaternary schemes with pulses 1REC, HCS, 1RC. (From [1].)

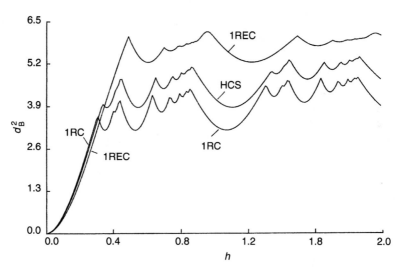

Figure 5.8 d_B^2 vs h for octal schemes with pulses 1REC, HCS, 1RC. (From [1].)

The Partial Response Case

Now we extend the distance bound argument to the case when the frequency pulse covers more than one interval, the case $L > 1$. We will begin the discussion with a 3REC example and then move on to other pulses and pulse lengths.

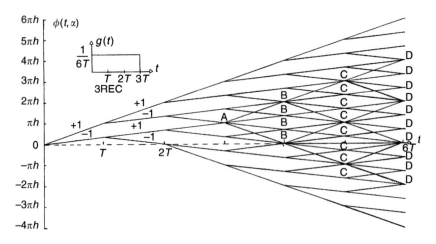

Figure 5.9 The phase tree for binary 3REC CPM. Before $t = 0$, data are $+1$. B, C, and D are first–third mergers; A is a crossing, not a merger. (From Aulin and Sundberg [2], copyright 1984 by John Wiley Ltd., used with permission.)

Figure 5.9 shows the example of a phase tree for binary 3REC, a rectangular frequency pulse extending across three symbol intervals. The tree is drawn for the case where all the data symbols before time zero are $+1$. A major difference from the full response trees appears: the slopes of the trajectories forward of a node depend on the data history up to the node. To make this more clear, data symbols are attached to some early branches. The slopes in front of $+1, +1$ and $-1, -1$ are different; the slopes in front of $+1, -1$ and $-1, +1$ are the same, but not the same as the first two slopes. The slopes are in fact determined by the present and the earlier $L - 1$ data symbols. These $L - 1$ are called the *prehistory* of the node.

Our interest is in distance calculation, and fortunately the prehistory does not affect the calculation of distances forward of a node. This is seen by considering Eq. (5.2-1). Since the data symbols prior to the node are the same, it follows that components of Δa before the node are all 0, and make no contribution. We have only to consider differences with components satisfying

$$\Delta a[n] = 0, \qquad n < 0,$$
$$\Delta a[0] = 2, 4, \ldots, 2(M - 1), \qquad (5.2\text{-}9)$$
$$\Delta a[n] = 0, \pm 2, \pm 4, \ldots, \pm 2(M - 1), \quad n \geq 1.$$

A phase tree formed by using only the differences of Eq. (5.2-9) in Eq. (5.1-2), instead of the data values a, is called a *phase difference* tree. Figure 5.10 shows such a tree for the binary 3REC code in Fig. 5.9. The tree is a way of keeping the books during the computation of $d^2_{\min}$. Every distinct initially positive difference

Continuous-phase Modulation Coding

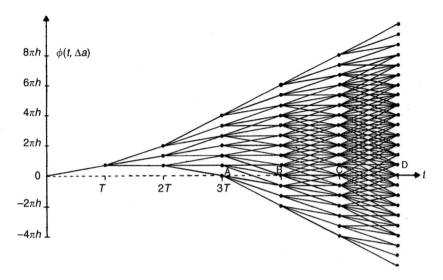

Figure 5.10 The phase difference tree for binary 3REC CPM. B,C,D,A as in Fig. 5.9. (From [2], copyright 1984 by John Wiley Ltd., used with permission.)

that can occur is in the tree; if a difference begins in a positive sense, but later reverses sense, it will switch to negative values.

The construction of an upper bound on the infinite-N minimum distance is similar to the full response case. In Fig. 5.9, mergers between two phase paths are identified by the property that the phase coincides at some time t_m and ever after. In Fig. 5.10, the mergers are marked by trajectories that are zero for all $t \geq t_m$. The first such merger is marked in both figures by B. The point marked A is not a merger, but only a place where trajectories happen to cross, or in the case of Fig. 5.10, happen to cross the t axis. In calculating the distance of a phase path pair, Eq. (5.2-7) may be used, and it need only be applied during $0 \leq t \leq t_m$, even if the observation width is longer.

In general, the first time a merger can occur is $t = (L+1)T$, L being the frequency pulse width. It is called a *first merger*, and the data difference sequences must be of the form $\Delta a[0], -\Delta a[0], 0, 0, \ldots$. The phase difference trajectories that result for any full response $q(t)$ are

$$\phi(t, \Delta a) = \begin{cases} 0, & t \leq 0; \\ 2\Delta a[0]\pi h q(t), & 0 \leq t \leq T; \\ 2\Delta a[0]\pi h[q(t) - q(t-T)], & T \leq t \leq (L+1)T; \\ 0, & t \geq (L+1)T. \end{cases} \quad (5.2\text{-}10)$$

A bound d_B is obtained for an h, M, and $q(t)$ by calculating the minimum of distances associated with difference of Eq. (5.2-10) via Eq. (5.2-1).

One can also consider *second mergers*, that is, phase difference trajectories which merge at $t = (L+2)T$. These merge at C in the figures. They are defined by the difference sequences that sum to zero and also satisfy

$$\Delta a[n] = \begin{cases} 2, 4, \ldots, 2(M-1), & n = 0; \\ 0, \pm 2, \pm 4, \ldots, \pm 2(M-1), & n = 1; \\ \pm 2, \pm 4, \ldots, \pm 2(M-1), & n = 2; \\ 0, & \text{otherwise.} \end{cases}$$

These mergers also occur in full response CPM, but it can be shown that these and all later mergers always have larger distance than the first merger. With partial response this is not true, and the upper bound can be tightened by taking into account second and later mergers. An mth merger is one for which the conditions $\sum_0^J \Delta a[n] = 0$; $\Delta a[n] = 0, n > J$ happen for the mth time as J grows.

In Fig. 5.11 the Euclidean distances are plotted against h which correspond to the first through third mergers in 3REC CPM. For first and second mergers there is only one upper bounding component, but there are three third

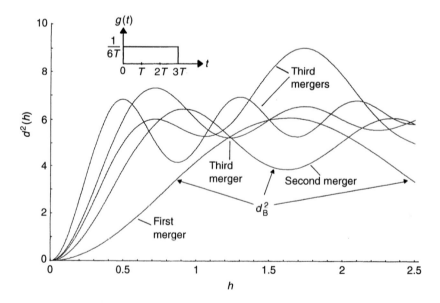

Figure 5.11 Construction of the upper bound for binary 3REC in Fig. 5.10. Only one third merger briefly contributes. (From [2], copyright 1984 by John Wiley Ltd., used with permission.)

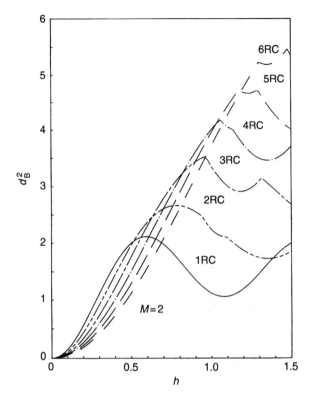

Figure 5.12 Upper bound d_B^2 vs h for binary schemes 1RC–6RC. (From [2], copyright 1984 by John Wiley Ltd., used with permission.)

merger bounds. Over the range $0 \leq h \leq 2.5$, only one of these three actually tightens the overall bound. The final upper bound d_B^2 for binary 3REC is taken as the minimum of all the curves in Fig. 5.11.

The final bounds d_B^2 for the binary schemes 1RC–6RC are shown in Fig. 5.12. It is clear that d_B^2 is built up of segments from several component functions, and this means that different types of mergers lead to d_B^2 at different h. The first merger, however, determines d_B^2 over the initial range of small h, in every case. With regard to pulse width, the trend in Fig. 5.12 is that increasing L leads to smaller d_B^2 at low indices but larger d_B^2 in the range $0.5 < h < 1.5$.

Bounding the quaternary minimum distance is the same in principle, although there are more cases in each type of merger. Figure 5.13 shows d_B^2 against h for quaternary 1RC–3RC. The bounds are much larger than in the previous figure. For full response schemes it was found that M larger than two led to much larger bounds for the same modulation index, and here we see the same for partial

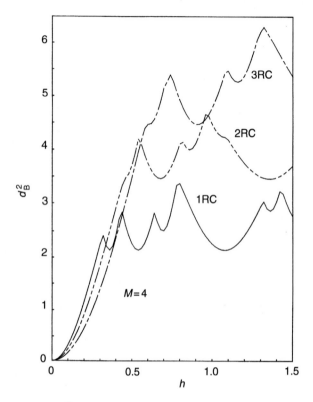

Figure 5.13 Upper bound d_B^2 vs h for quaternary schemes 1RC–3RC. (From [2], copyright 1984 by John Wiley Ltd., used with permission.)

response schemes. d_B^2 for octal 1RC–3RC appear in Fig. 5.14, and now the bound values are even larger. Good distances are obtained even at small h values, which means that higher-M schemes will have attractive bandwidth properties. In both the 4-ary and 8-ary cases, the closest first merger again sets d_B^2 in the initial h range.

A phase tree for quaternary 2RC over the interval $0 \leq t \leq 4T$ is shown in Fig. 5.15. This may be roughly compared with the 1REC and 3REC trees in Figs 5.2 and 5.9. Some study shows several trends. In the smoother RC tree, neighbor paths split more slowly and later merge more slowly, but this does not much affect the distance when the pulse width L is the same; the chief effect is to reduce the spectrum, as we will see in Section 5.3. With wider pulses of either type, path pairs remain apart longer but they lie closer; the net effect at lower modulation indices (when distance through the phase cylinder is comparable to distance in the plane of the figure) is to lower distance, but to raise it at higher indices. To judge whether a better overall coded modulation exists at some h, one must compare systems over both distance and spectrum.

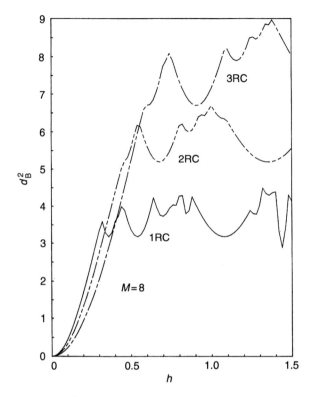

Figure 5.14 Upper bound d_B^2 vs h for octal schemes 1RC–3RC. (From [2], copyright 1984 by John Wiley Ltd., used with permission.)

Although distance bounds for smooth pulses have been given here only for RC pulses, the general trends carry over to other smoothed pulse shapes, so long as $g(\ )$ is positive. Another interesting smooth class is the SRC pulses, which have a g main lobe of width L and temporal sidelobes to either side. SRC trees with $\beta = 1$ are almost identical to RC trees, and their distance properties must consequently be the same. The *spectral* sidelobe behavior of SRC, however, is superior.

Weak indices and weak pulse shapes. The phase mergers used to construct d_B^2 are the inevitable ones that occur when two symbol sequences differ for only a finite time. They occur for any h. In Section 5.2.3 we will call them state merges. There are other kinds of phase mergers, too, that depend on the specific index h. These occur because there is a phase difference trajectory that equals a nonzero multiple of 2π over a certain time. The distance contribution is zero even though the phase difference is not. The h leading to this condition is called a *weak index*.

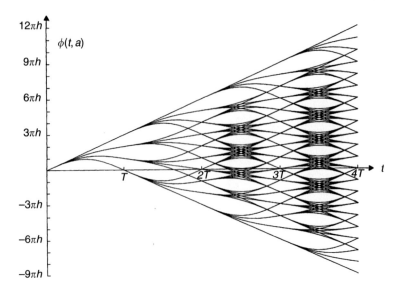

Figure 5.15 The phase tree for quaternary 2RC CPM.

One condition for a weak index can be derived by setting the data difference $\Delta a[0]$ equal $2, 4, \ldots, 2(M-1)$, and $\Delta a[n] = 0$, $n > 0$. The phase difference will be zero mod-2π at all times $t > LT$ whenever $2\pi h \Delta a[0] q(LT) = k 2\pi$, $k = 1, 2, \ldots$. This phase merger occurs one interval earlier than a standard first merger, and in the simple 1REC case, for example, it reduces distance by half. This kind of weak index satisfies

$$h_W = \frac{k}{\Delta a[0] q(LT)}, \quad k = 1, 2, \ldots; \quad \Delta a[0] = 2, 4, \ldots, 2(M-1). \tag{5.2-11}$$

(Assume that $q(LT) \neq 0$.) Other conditions can be found (see [1], section 3.3), but Eq. (5.2-11) defines the most important weak indices.

A CPM scheme can also be weak if the pulse shape is such that the first phase merge occurs before $(L+1)T$, independent of h. 4REC is such a pulse shape. 7RC is weak and 8RC is almost so, and for this reason these pulses were not shown in Fig. 5.12. We can conclude that $L = 6$ is a limit to practical pulse width for both RC and SRC pulses.

Multi-h CPM

CPFSK schemes with more than one index were proposed in [11] and analyzed from a coding point of view in [12]. The idea was extended to smoother and longer phase pulses in [14,15]. A CPM scheme with H indices in a cyclic rotation is called a *multi-h* scheme. Figure 5.16 shows the phase tree for a binary

Continuous-phase Modulation Coding

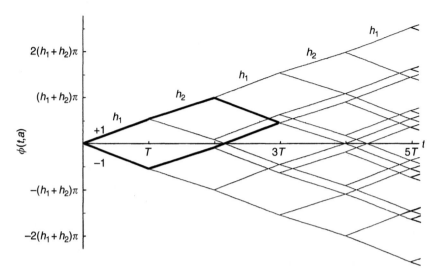

Figure 5.16 Phase tree for binary 2-h scheme based on 1REC pulse. Index $h_2 = (5/6)h_1$. Heavy lines show a first merger. (From Aulin and Sundberg [17], copyright 1982 IEEE; used with permission).

1REC two-h scheme in which $h_2 = 5h_1/6$; the h that applies to each interval is indicated. By using H indices in a cyclic manner with a 1REC pulse, the first merger is delayed from $2T$ to $(H+1)T$. Careful study of the figure will show that it occurs at $t = 3T$ there. The result is a potentially larger minimum distance.

Upper bounds of the type in this section may be computed for multi-h codes, and we can take the simplest case, the binary 1REC codes, as an example. Distances in a multi-h phase tree depend on which index comes first, but in either case in Fig. 5.16 the first merger stems from data difference $\Delta a = 2, 0, -2, 0, \ldots$. A first-merger phase path pair is outlined in the figure. A first-merger d_B^2 may be written as

$$d_B^2 = \min\{d_{B,1}^2, d_{B,2}^2\}, \quad \text{with} \qquad (5.2\text{-}12)$$

$$d_{B,i}^2 = [1 - \cos 2\pi h_i] + 2\left[1 - \frac{\sin 2\pi h_i}{2\pi h_i}\right], \quad i = 1, 2.$$

The first term here stems from the middle of the three intervals. Unfortunately, such first-merger bounds are not in general tight for multi-h codes. A careful study with more bound components shows that the maximum value of a tight d_B^2 in the range $0 \leq h_1, h_2 \leq 1$ is 4, attained when both h_i are near 1/2. For three and four indices, it is 4.9 and 5.7, respectively. These are gains of 3.4 and 4.5 dB over MSK.

With 4-ary and 8-ary data, these numbers grow significantly: for octal multi-h 1REC, maximum values of 7.5 and 8.4 have been reported [1] for d_B^2 when there are 3 indices. Figure 5.17 shows d_B^2 as a function of h_1, h_2, and the

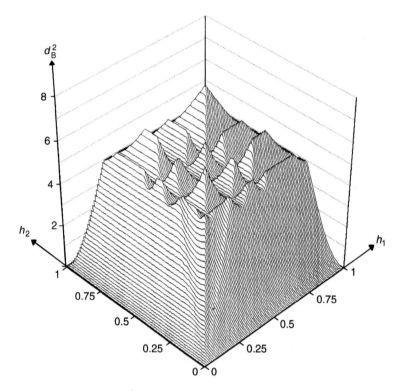

Figure 5.17 The upper bound d_B^2 for quaternary 2-h 1REC schemes when both hs lie in the interval (0, 1). The maximum occurs when $h_1 \approx h_2 \approx 0.77$. (From [17], copyright 1982 IEEE; used with permission.)

complexity of the bound components is obvious there. The peak of d_B^2 is around 6.5 and occurs near the back at $h_1 = h_2 = 0.772$. Bounds have also been constructed for partial response multi-h CPM, but they are quite complicated; see [1,17,18].

A similar idea to multi-h coding is multi-T coding, proposed by Holubowicz and Szulakiwicz [25]. Here the symbol time varies cyclically instead of h. Occasionally, better distances are obtained for the same state size, compared to multi-h.

Conclusion: Distances

We have introduced useful tools for understanding CPM schemes, such as the phase tree, symbol differences, and the difference tree, and we have used them to derive simple bounds on the minimum distance of CPM schemes. For the most part these bounds are tight, and for a range of moderate h the bound obtained from the closest first merger is in fact usually $d_{\min}$. These bounds are a powerful

Continuous-phase Modulation Coding

tool for the classification and comparison of CPM schemes. They show that data alphabet size has a powerful effect on minimum distance, and they show for what indices a wide pulse width improves distance. They show that CPM offers major energy gains over simple constant-envelope modulations such as MSK, without the use of parity checks, and without large bandwidth gain. d_B will also be a crucial aid to finding the actual $d_{\min}$ in what follows.

5.2.2. Calculation of Minimum Euclidean Distance

We turn now to numerical results for the actual minimum distance of CPM schemes. The results will be presented as plots of $d_{\min}^2$ against modulation index h, obtained with different observation widths N. As N grows, $d_{\min}^2$ tends in general to the upper bounds d_B^2 derived in the previous subsection. The exception is at weak modulation indices, where at no N will the bound be reached, and near weak indices, where N must be larger than otherwise. Whatever limit $d_{\min}$ reaches with N, it is the free distance d_f of the CPM code. N here is length of path memory in the decoder; if the transmission system is to perform at the potential promised by d_f, N must satisfy $N \geq L_{\text{dec}}$, where L_{dec} is the code decision depth introduced in Section 3.4.1.

CPM minimum distances have most often been found by the branch and bound technique that is discussed in Section 6.3.2; the technique was introduced to CPM by Aulin [14]. At first [12], distances were found by the double dynamic programming method of Section 4.3.3, but this method proves in most cases to be more time consuming. Among other reasons, the branch and bound method is more suited to h that are not simple fractions. Both methods are also discussed in [1], Appendix A.

Full Response Schemes

The first results are for binary 1REC schemes. Figure 5.18 shows $d_{\min}^2$ vs h for observation width $N = 1-5$, with the upper bound d_B^2 from Eq. (5.2-7) for comparison. For this scheme, Eq. (5.2-11) says that weak indices are to be expected at $h_W = 1, 2, \ldots$. The weak index $h_W = 1$ is evident in the figure. MSK is binary 1REC with $h = 1/2$: its $d_{\min}^2$ of 2 lies on the upper bound and an observation width of $N = 2$ is needed to achieve it.[5] The bound curve peaks at $h \approx 0.715$ and distance 2.43, a gain of 0.85 dB over MSK (see Problems); $N = 3$ is required. For all $h \leq 0.5$, $N = 2$ is sufficient. Single-interval detection, the case of $N = 1$, incurs significant loss at all h except 1.

---

[5] The observation width concept here and throughout assumes correct decoding until the present time, with the $Q(\ )$ expression estimating the probability of the initial error event. Without this assumption, a width of 3 is required if MSK is to have asymptotic error probability $Q(\sqrt{2E_b/N_0})$.

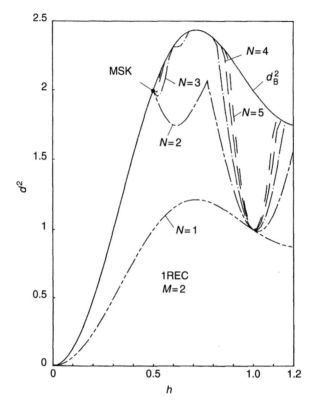

Figure 5.18 Comparison of actual d_{min}^2 to upper bound d_B^2, plotted against h, for binary 1REC CPM. $N = 1, \ldots, 5$ observation intervals. (From [2], copyright 1984 by John Wiley Ltd., used with permission.)

The next results are for quaternary 1REC schemes, in Fig. 5.19. Equation (5.2-11) now predicts weak indices at $h_W = 1/3, 1/2, 2/3, 1, 4/3, \ldots$. All of these except the first have a strong effect on the plot. Once again, $N = 2$ is sufficient detection width for small indices ($h < 0.3$). To reach the highest d_{min}^2 values of 4 or more, a width of 8 is required. The distances at small h hardly differ from those in the previous binary figure, but Section 5.3 will show that the quaternary REC bandwidth is much better, and that consequently quaternary REC is an important CPM scheme. It is a good example of a coding scheme that works to reduce bandwidth more than increase distance.

Figure 5.20 shows octal 1REC schemes. There is a potential for large distance gain, but there are many weak indices and in only a few index ranges does a width 5 detector achieve these. Still, indices near 0.45, 0.6, and 0.87 can achieve d_{min}^2 in the range 5–6.

Continuous-phase Modulation Coding

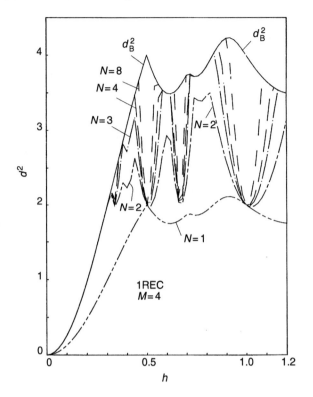

Figure 5.19 d_{min}^2 vs h for quaternary 1REC. 1–8 observation intervals. (From [2], copyright 1984 by John Wiley Ltd., used with permission.)

It can be shown that except at the weak indices defined by Eq. (5.2-11), all M-ary full response schemes achieve d_{min}^2 equal d_B^2 for a wide enough observation interval [1, Section 3.3.1]. For h less than about 0.3, $N = 2$ is sufficient, regardless of M, but with larger h, and in particular with the h that optimizes d_B^2, very large widths are needed at large alphabet sizes. An idea of this is given by Table 5.2, which lists the optimal h at each M, and the width required for it. Observe that d_B^2 grows linearly with log M, apparently without limit, in a manner similar to the distance growth in TCM. But the distance is probably not worth the complexity of the scheme.

Partial Response Schemes

A great many partial response schemes have been studied, and rather than cover all of these, we take as representative the classes RC and REC. SRC results are very close to RC, and the TFM pulse is approximately RC with a width 3.7 T.

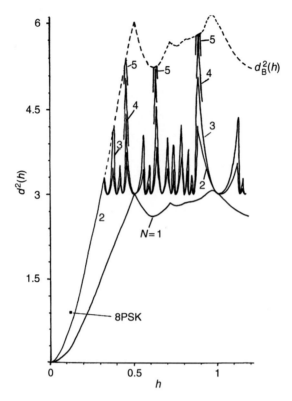

Figure 5.20 d^2_{min} vs h for octal 1REC. 1–5 observation intervals. (From [2], copyright 1984 by John Wiley Ltd., used with permission.)

Table 5.2 Optimum indices h and corresponding square minimum distances for M-ary 1REC schemes (reproduced from Anderson, Aulin and Sundberg [1])

M	Optimum $h(h_{opt})$	$d^2_B(h_{opt})$	N_B
2	0.715	2.434	3
4	0.914	4.232	9
8	0.964	6.141	41
16	0.983	8.088	178
32	0.992	10.050	777

Continuous-phase Modulation Coding

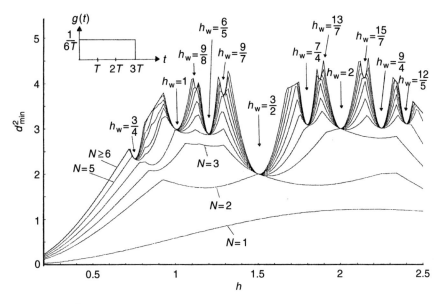

Figure 5.21 d_{min}^2 vs h for binary 3REC. 1–10 observation intervals. Weak indices are marked with arrows. (From [2], copyright 1984 by John Wiley Ltd., used with permission.)

Once again, the d_B^2 obtained in Section 5.2.1 will be a tight bound on d_{min}^2 except at certain weak indices.

Figure 5.21 shows d_{min}^2 vs h for binary 3REC schemes and detection widths 1–10. The upper bound function is shown in Fig. 5.11. There is a bewildering number of weak modulation indices. Some of these are predicted by Eq. (5.2-11), but most come from more subtle weak-index phenomena, and some stem from the shape of $g(t)$ itself. Some study of Fig. 5.10 shows that there are a number of instances of constant phase difference with 3REC, and with the wrong h, these lead to a CPM scheme with poor distance. This kind of behavior is typical of partial response CPM. Aside from its weak index problem, binary 3REC does show considerable distance improvement over 1REC, typically 3 dB, with a reasonable detection width.

Figures 5.22–5.24 show the minimum distance of 3RC with binary, quaternary, and octal alphabets, and a selection of detector widths N up to 15. d_B^2 in these pictures is computed from first through third mergers. In all cases, d_{min}^2 is very poor when $N = 1$, but it increases rapidly with each increase in N. For binary 3RC, the upper bound is reached over the range $h < 0.6$ when $N = 4$. d_{min}^2 as large as ≈ 3.5 can be reached with binary 3RC, a 2.4 dB improvement over MSK. When the RC pulse is widened, d_{min}^2 shrinks at h values below $1/2$ but grows at larger h, and the detection width also grows; d_{min}^2 more or less follows the bounds

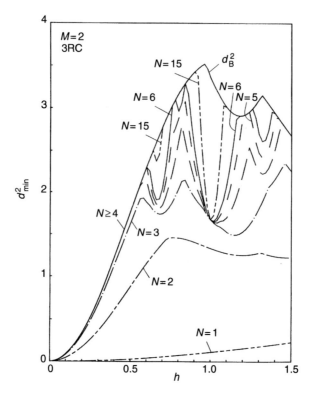

Figure 5.22 d_{min}^2 vs h for binary 3RC. 1–15 observation intervals. (From [2], copyright 1984 by John Wiley Ltd., used with permission.)

shown in Fig. 5.12. Binary 6RC achieves d_{min}^2 as high as 5 at width $N = 30$, a gain of 4 dB over MSK. Longer pulses than 6RC require very large N.

Figures 5.22–5.23 show large distance gains for $M = 4, 8$ over binary RC. Quaternary d_{min}^2 exceeds 6 at $h = 1.3$ and $N = 15$ symbol observation; this is a 5 dB gain over MSK. A complicated weak index picture begins to appear in the octal case, but d_{min}^2 can exceed 9 with 12 symbol observation (6.5 dB gain).

Multi-h Distances

We give a selection of largest distance 1REC multi-h codes and their distances in Table 5.3. The average h value, $\bar{h}$, is a rough spectrum measure, and the normalized 99% baseband bandwidth BT_b, computed by Lereim [24], is given where available. $h_1, \ldots, h_H$ are each multiples of $1/c$, c an integer, and the table shows the largest distance code for each c and H. c must be at least 2^H if a code with an early merge is to be avoided, and c sets the complexity of the multi-h phase

Continuous-phase Modulation Coding

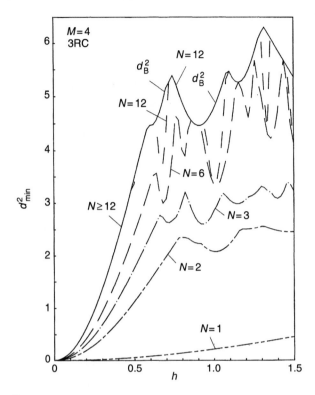

Figure 5.23 d_{min}^2 vs h for quaternary 3RC. 1–12 observation intervals. (From [2], copyright 1984 by John Wiley Ltd., used with permission.)

trellis. Even the very simple $(h_1, h_2) = (2/4, 3/4)$ code has a gain of 1.4 dB over MSK. As the 2-h phase structure becomes more complicated, gains of up to 3 dB are possible, but a larger c does not automatically imply a higher distance. The simplest good 3-h code $(4/8, 5/8, 6/8)$ has a gain of 2.8 dB, compared to an upper limit of 3.9 dB for 3-h systems. The simplest 4-h code has a gain of 3.7 dB. Because of particular phase trellis paths, it can happen that a code with fewer indices has a larger distance than one with more. d_{min}^2 for these codes is best found by the double dynamic programming (DDP) approach, with the h_i limited to ratios of relatively small integers. The distance search must be repeated with each h_i as the starting index. Upper bound analysis gives quite a tight upper bound for full response multi-h cases, and d_{min}^2 tends to follow a relatively sensible pattern as the h_i vary.

With partial response, behavior is much more chaotic. The DDP method requires a bound of the d_B type in order to be efficient, and an exhaustive search for good indices may be more reasonable. Some results for partial response are given in [1, Chapter 3]. An extensive treatment of multi-h coding is Xiong [5, Chapter 7].

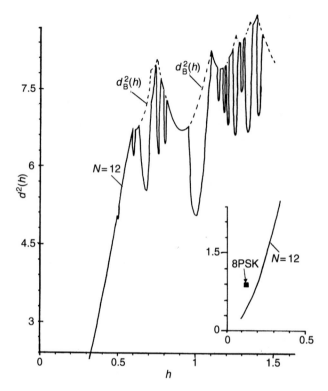

Figure 5.24 d^2_{min} vs h for octal 3RC at 12 observation intervals. (From [1].)

5.2.3. Trellis Structure and Error Estimates

A finite-state trellis description lies at the heart of the Viterbi algorithm (VA) and most other decoders that one would use with CPM signals. We present here the most common of several trellis descriptions that have evolved for CPM. We make no claim that it is in any sense minimal; rather, it should be taken as a simple, intuitive way for a decoder to keep the books. Several other trellis representations will be mentioned briefly. The section concludes with the traditional trellis-based error estimation.

From the foregoing sections, it is clear that a straightforward definition of the state of a CPM signal is its recent data symbol history and the signal's phase offset. If the modulation index h is a ratio of integers, the evolution of phase settles into a repeating pattern like that in Fig. 5.4, and then there are only a finite number of possible offsets. If the pulse width is L, then $L - 1$ previous symbols affect the future phase shape. The phase component of the state that applies to symbol interval $[nT, (n + 1)T]$ is called the *phase state* and denoted θ_n; the data

Continuous-phase Modulation Coding

Table 5.3 Largest distance multi-h 1REC CPM schemes with 2–4 indices, with h_i a multiple of $1/c$, c an integer. Bandwidth is 99% normalized baseband bandwidth BT_b (positive frequencies only)

Number of phases, c	Number of indices, H	h_1	h_2	h_3	h_4	Square min. distance	Baseband bandwidth
4	2	2/4	3/4			2.79	
5	2	3/5	4/5			3.07	
6	2	3/6	4/6			3.45	
7	2	3/7	5/7			3.32	
8	2	4/8	5/8			3.55	
	3	4/8	5/8	6/8		3.79	0.84
9	2	5/9	6/9			3.46	
	3	4/9	7/9	8/9		2.70	
10	2	5/10	6/10			3.62	
	3	5/10	6/10	7/10		3.82	0.78
11	3	6/11	7/11	8/11		3.73	
12	3	6/12	7/12	8/12		3.81	0.66
13	3	8/13	9/13	10/13		4.17	0.90
14	3	8/14	9/14	10/14		4.12	0.84
15	3	8/15	9/15	11/15		3.88	
16	3	10/16	11/16	12/16		4.34	0.89
	4	8/16	11/16	10/16	12/16	4.65	0.85
20	4	10/20	14/20	12/20	15/20	4.78	
22	4	12/22	15/22	14/22	16/22	4.89	

symbol component is $(a[n-1], a[n-2], \ldots, a[n-L+1])$ and is called the *correlative state*.

We can now define precisely the relation of the excess phase $\phi(t, a)$ to the state. First, break up Eq. (5.1-2) into two terms

$$\phi(t, a) = 2\pi \sum_{i=n-L+1}^{n} a[i] h_i q(t - iT) + \pi \sum_{i=-\infty}^{n-L} a[i] h_i, \quad nT \leq t \leq (n+1)T.$$

Here the second term follows from the fact that $q(t) = 1/2, t > LT$. Take as the phase state the second term mod-2π. The modulation indices take one of the H values

$$\frac{2k_\ell}{p}, \quad \ell = 1, \ldots, H; \; k_\ell \text{ integer}, \qquad (5.2\text{-}13)$$

in which the k_ℓ and p are relatively prime. There are thus p phase states, and we can write

$$\theta_n \triangleq \left[\pi \sum_{i=-\infty}^{n-L} a[i]h_i \right], \quad \text{mod-}2\pi, \qquad (5.2\text{-}14)$$

$$\theta_n \in \left\{ 0, \frac{2\pi}{p}, \frac{4\pi}{p}, \ldots, \frac{2(p-1)\pi}{p} \right\}.$$

The total state of the transmitted phase is the L-tuple

$$\sigma_n = (\theta_n, a[n-1], \ldots, a[n-L+1]). \qquad (5.2\text{-}15)$$

and there are pM^{L-1} distinct states. The current symbol $a[n]$ directs the transition from state σ_n to σ_{n+1}, and $a[n]$ together with σ_n define $\phi(t, a)$ during $[nT, (n+1)T]$.

Some simple examples are given by CPM with a simple 1REC pulse. With $h = 1/4$, there will be 8 phase states and no correlative states, for a total of 8. When there are no correlative states, it is convenient to draw the state trellis directly as a phase trellis. This will be the phase trellis of Fig. 5.2 or Fig. 2.20 completely grown, with transitions around the back of the phase cylinder accounted for. Figure 5.25(a) shows an example for $h = 1/2$ (MSK). Observe that the use of balanced data symbols (here $+1, -1$) means that the trellis alternates between two section types, one ending at phases $(\pi/2, 3\pi/2)$ and one at $(0, \pi)$. It can be argued that the number of states here is artificially doubled, and since the beginning, CPM trellises have sometimes been drawn "biased," with data symbols taken as $0, 2, \ldots, 2(M-1)$. An example of a biased phase state trellis is the $(h_1, h_2) = (2/4, 3/4)$ multi-h trellis shown in Fig. 2.25(b). Despite the fact that the indices are multiples of $1/4$, there are 4, not 8, phase positions. Biasing phases does not change the distances calculated from them. In actuality, doubling the phase states affects only the bookkeeping in a practical decoder, not the complexity,[6] and so the more natural unbiased trellises have become the standard.

Figure 5.26 shows a more complicated trellis, one for single-index binary 3RC with $h = 4/5$. Because the h numerator is even, there is no doubling of the phase states. There are a total $pM^2 = 5 \cdot 4 = 20$ states. A trellis for the same 3RC with $h = 2/3$ was given before in Fig. 3.8; now there are 12 states. Figure 3.8 also shows a method of locating the states on a phase tree. It can be helpful to imagine these trellises wrapped around a phase cylinder, but this is irrelevant to a Viterbi decoder, which has no need of such intuition.

An open problem is what the minimal trellis representation of CPM might be, and a closely related problem, what is the precise role of the phase state θ_n.

---

[6] See Section 5.4.1; the Viterbi decoder deals with the smaller number of states in either case, and the M algorithm is completely unaffected.

Continuous-phase Modulation Coding

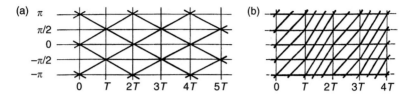

Figure 5.25 (a) Balanced phase and state trellis for MSK; the trellis alternates between two section types. (b) Biased phase and state trellis for multi-h 1REC CPM with $(h_1, h_2) = (2/4, 3/4)$; phase transitions are biased upward by $t\pi/2T$ in odd intervals and $3t\pi/4T$ in even intervals, compared to a standard phase tree.

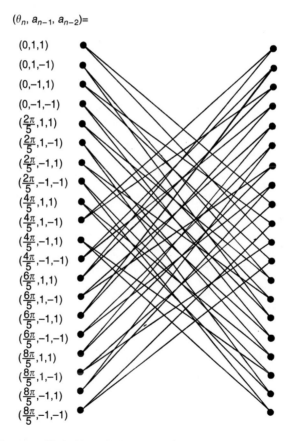

Figure 5.26 The state trellis for binary 3RC CPM with $h = 4/5$. Compare with Fig. 3.8. (From [2], copyright 1984 by John Wiley Ltd., used with permission.)

It is known, for example, that detection is asymptotically unaffected by a constant phase offset, which means that θ_n in effect need not be known. Practical MLSE receivers have been proposed that work at RF and ignore the phase offset [20].

CPM Error Estimates

As with almost all coded modulations, computation of exact error event and bit error probabilities is impractical. But with a trellis description and $d_{\min}$ in hand, the methods of Section 3.4.1 lead to reasonable estimates. To review briefly, an error event begins when the decoded trellis path splits from the correct one, and it ends when the decoded path merges again. An upper bound to the probability P_{ev} that an error event begins at nT, given correct decoding until that time, has the form

$$P_{\text{ev}} \leq \sum_{i \geq 0} A_i Q\left(\sqrt{d_i^2 E_b/N_0}\right). \tag{5.2-16}$$

For a sufficiently wide, but finite, trellis observation, P_{ev} has asymptotic form $\sim A_0 Q(\sqrt{d_f^2 E_b/N_0})$. When the decoder observation N is narrower than L_{dec}, the decision depth of the code, d_f here must be replaced with $d_{\min}$ computed for trellis depth N; the computation of $d_{\min}$ must take account of unmerged as well as merged trellis paths. With many convolutional and TCM codes, the coefficient A_0 is 1, but CPM codes are not *regular*, meaning that the distance structure varies forward of different trellis nodes, and consequently A_0 is seldom as large as 1.

Several methods exist to compute some of the A_i, and thus obtain a tighter estimate than simply $Q(\sqrt{d_{\min}^2 E_b/N_0})$. The transfer function bound method was applied to CPM by Aulin; see [1,21]. The method is complex to apply, but it does give accurate estimates at practical error rates. Figure 5.27 plots a quaternary 3RC, $h = 1/2$ P_{ev} calculation, repeated for observation widths 1–8, together with a lower bound. The last is the transfer function calculation of $A_0 Q(\sqrt{d_f^2 E_b/N_0})$ when $N \geq L_{\text{dec}}$, which can be shown to be a lower bound to P_{ev}. The method gives a tight estimate for $P_{\text{ev}} < 0.01$. It shows clearly the effect of growing observation width.

Balachandran [22] and Macdonald [23] developed a much simpler method that also gives accurate estimates. In their approach, the event probability is computed given that a certain, small number of symbols were in fact transmitted after nT. We can take for example the first three symbols $a[n], a[n+1], a[n+2]$ and for brevity denote them abc. Then

$$P_{\text{ev}} = \sum P[abc] P_{\text{ev}\,|\,abc} \geq \sum \frac{1}{8} Q\left(\sqrt{d_{\min\,|\,abc}^2 E_b/N_0}\right). \tag{5.2-17}$$

Here we use the fact that each abc has the same probability, $1/8$, and $d_{\min\,|\,abc}$ is the outcome of a minimum distance calculation that assumes abc. The distance is easy to find, and in fact can often be found by hand. Since each $Q(\sqrt{d_{\min\,|\,abc}^2 E_b/N_0})$

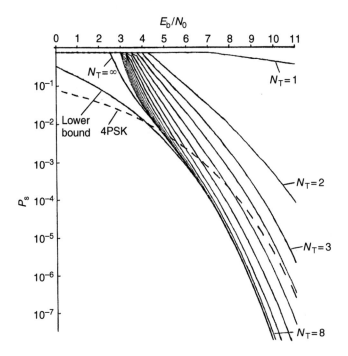

Figure 5.27 Upper bounds on the symbol error probability vs E_b/N_0 for quaternary 3RC with $h = 1/2$, when the decoder is limited to path memory $N_T = 1$–10. The decision depth of the code is 11 symbols. (From Aulin [21], copyright 1981 IEEE, used with permission.)

stems from only one of the possible error events given abc, Eq. (5.2-17) gives a lower bound. The lower bound is asymptotically tight, and has the correct A_0. As an example, binary 1REC with $h = 2/3$ yields

$$P_{ev} \geq 0.5\, Q(\sqrt{2.41E_b/N_0}) + 0.25\, Q(\sqrt{2.93E_b/N_0}) + 0.25\, Q(\sqrt{3.84E_b/N_0}).$$

Binary 1REC, $h = 1/3$ yields $0.5\, Q(\sqrt{1.17E_b/N_0}) + 0.5\, Q(\sqrt{2.67E_b/N_0})$, where we have conditioned now on two symbols. Tests [23] show that CPM decoding performs very near these two estimates over a wide range of E_b/N_0. Above low E_b/N_0, all but first terms here can be ignored, and the important contribution of the method is its calculation of A_0.

5.3. CPM Spectra

The second fundamental parameter of a CPM scheme is its bandwidth. Unlike the case with TCM and PRS codes, CPM coding is a nonlinear process and the linear modulation calculation in Theorem 2.7.1 cannot provide a quick and easy computation of the spectrum. A few closed-form expressions are available, notably for CPFSK; these were given in Section 2.7.2 as Eqs (2.7-12)–(2.7-13), with representative binary spectra drawn in Fig. 2.27. MSK may be viewed as linear modulation, and so its spectrum follows from Theorem 2.7.1. Otherwise, it is necessary to calculate the complex baseband autocorrelation $R(\tau)$ in Eq. (2.7-10), and Fourier transform it to obtain the baseband PSD $G(f)$ in Eq. (2.7-11). The logic behind the quantities $R(\tau)$ and $G(f)$ is discussed in Section 2.7.2. The forms (2.7-10)–(2.7-11), which comprise Theorem 2.7-2, apply when there are no discrete spectral components and the CPM symbols are IID. Calculation of standard CPM spectra consists of numerical evaluation of these, a relatively straightforward programming exercise.

Section 5.3.1 proves Theorem 2.7.2 and shows how to extend it to some more general cases. Those without the need of these mathematical details can skip directly to Section 5.3.2, which calculates example spectra. With information on spectra in hand, Section 5.3.3 plots the energy–bandwidth performance of important CPM schemes.

The calculation method of Section 5.3.1 has good stability and computes accurate spectra typically 80 dB or more below the PSD main lobe peak. This is usually sufficient to detail the complete sidelobe structure of the spectrum. As discussed in Section 5.1, the NBW measure for CPM is $2BT_b$ Hz-s/data bit, where $T_b = T/\log_2 M$ is the bit time and B is some baseband frequency measure, here the 99% or 99.9% power bandwidth of the baseband spectrum, positive frequencies only. Spectra in Sections 5.3.1 and 5.3.2 are plotted against fT_b, which is bit-normalized baseband frequency, and at the same time, half the NBW. In the comparison of schemes in Section 5.3.3, the NBW is used.

Extreme sidelobe rolloff in CPM follows Baker's Rule, Eq. (2.7-14), which states that the baseband spectrum rolls off asymptotically as $\sim|f|^{-2n+4}$, where n is the number of continuous derivatives in the excess phase $\phi(t)$. Alternately, $n-1$ is the number of derivatives in the frequency pulse. A modern proof is given in [26]. The rule means that phase-discontinuous schemes behave as $\sim|f|^{-2}$, all REC schemes behave as $\sim|f|^{-4}$, and RC schemes as $\sim|f|^{-8}$ (see Problems).

5.3.1. A General Numerical Spectral Calculation

Numerical spectral calculation methods developed during the period 1965–1985. In the beginning spectra were simply computed by averaging the transforms of a list of data signals, but this list grows exponentially with signal length. Some of the history is recounted in [1, Chapter 4]. In brief, researchers sought a method whose computation grows linearly instead of exponentially with the length of $q(t)$, which is based on numerical integration rather than evaluation of series, and which accurately computes sidelobes. The last consideration eliminates, for example, methods based on piecewise linear approximation of signals. The development

Continuous-phase Modulation Coding

that follows was published by Aulin and Sundberg in [26,27]. To insure continuity with [1], it closely follows Chapter 4 of that volume.

We take the CPM signal in standard form (5.1-1) and (5.1-2), with M-ary data symbols $a[n], a[n+1], \ldots$ that are independent and identically distributed with *a priori* probabilities

$$P_k = P[a[n] = k], \quad k = \pm 1, \pm 3, \ldots, \pm(M-1), \text{ all } n, \quad (5.3\text{-}1)$$

where $\sum P_k = 1$. It will save notation to work with complex numbers, so that the CPM signal with data a becomes

$$s(t, a) = \sqrt{2E/T}\, \Re\{e^{j\phi(t,a)} e^{j2\pi f_0 t}\}. \quad (5.3\text{-}2)$$

The average PSD corresponding to $s(t, a)$, expressed as a complex baseband spectrum, will be denoted $S(f)$ in what follows.

The time-average autocorrelation of $s(t, a)$ is the complex baseband autocorrelation function $R(\tau)$, given by[7] $\sqrt{E/T}\, \Re\{R(\tau)e^{-j2\pi f_0 \tau}\}$, in the limit $f_0 T \to \infty$. For the standard CPM signal, $R(\tau)$ works out to be

$$R(\tau) = \mathcal{E}_{\delta,a}\left\{e^{j\phi(t+\delta+\tau,a)} e^{-j\phi(t+\delta,a)}\right\}$$

$$= \mathcal{E}_{\delta,a}\left\{e^{j2\pi h \sum_i a_i [q(t+\delta+\tau-iT) - q(t+\delta-iT)]}\right\}. \quad (5.3\text{-}3)$$

The sum in the exponent can be written as a product of exponentials

$$R(\tau) = \mathcal{E}_{\delta,a}\left\{\prod_i e^{j2\pi h a_i [q(t+\delta+\tau-iT) - q(t+\delta-iT)]}\right\},$$

and since the symbols in a are independent, the averaging with respect to a is simply performed by

$$R(\tau) = \mathcal{E}_\delta\left\{\prod_i \left[\sum_{\substack{k \text{ odd} \\ |k| \leq M-1}} P_k\, e^{j2\pi h k [q(t+\delta+\tau-iT) - q(t+\delta-iT)]}\right]\right\}. \quad (5.3\text{-}4)$$

The product contains infinitely many factors, but the phase pulse $q(t)$ is constant outside the interval $[0, LT]$, which makes the exponent in Eq. (5.3-4) cancel to zero for all but a finite number of factors. By assuming that $\tau \geq 0$ and writing

$$\tau = \tau' + mT, \quad 0 \leq \tau' < T, \quad m = 0, 1, 2, \ldots$$

---

[7] See Eqs (2.7-6)–(2.7-8). Recall from Section 2.7 that the spectrum denoted $S(f)$ in these pages is always for a CPM signal that has been randomized in time.

it can be shown that (see [27])

$$R(\tau) = R(\tau' + mT),$$

$$= \frac{1}{T} \int_0^T \prod_{i=1-L}^{m+1} \left\{ \sum_{\substack{k \text{ odd} \\ |k| \leq M-1}} P_k e^{j2\pi hk[q(t+\tau'-(i-m)T)-q(t-iT)]} \right\} dt, \quad (5.3\text{-}5)$$

where the expectation over δ has now been computed explicitly. $R(\tau)$ for $\tau < 0$ follows directly because all autocorrelations satisfy

$$R(\tau) = R_c(\tau) + jR_s(\tau), \quad R_c(\tau), R_s(\tau) \text{ real} \quad (5.3\text{-}6)$$

with

$$R_c(-\tau) = R_c(\tau); \quad R_s(-\tau) = -R_s(\tau). \quad (5.3\text{-}7)$$

Before proceeding with the Fourier transformation, we need to rewrite $R(\tau)$ in a form that shows a special exponential character. Consider Eq. (5.3-5) for $R(\tau)$ when $\tau = \tau' + mT \geq LT$. We can write

$$R(\tau) = R(\tau' + mT),$$

$$= [C_\alpha]^{m-L} \frac{1}{T} \int_0^T \prod_{i=1-L}^{0} \left\{ \sum_{\substack{k \text{ odd} \\ |k| \leq M-1}} P_k e^{j2\pi hk[q(LT)-q(t-iT)]} \right\} dt$$

$$\times \prod_{i=1-L}^{1} \left\{ \sum_{\substack{k \text{ odd} \\ |k| \leq M-1}} P_k e^{j2\pi hkq(t+\tau'-iT)} \right\} dt, \quad m \geq L, \quad (5.3\text{-}8)$$

where

$$C_\alpha \triangleq \sum_{\substack{k \text{ odd} \\ |k| \leq M-1}} P_k e^{j2\pi hkq(LT)} \quad (5.3\text{-}9)$$

is a constant, independent of τ. It is clear that $|C_\alpha| \leq 1$. The most important fact about Eq. (5.3-8) is that it is independent of m. We can conclude that when $\tau \geq LT$, $R(\ )$ has the form

$$R(\tau) = R(\tau' + mT) = [C_\alpha]^{m-L} \Psi(\tau'), \quad m \geq L, \quad 0 \leq \tau' < T; \quad (5.3\text{-}10)$$

that is, it is *separable in the arguments m and* τ'. The function $\Psi(\tau')$ is the integral in Eq. (5.3-8) and has only to be calculated in the interval $0 \leq \tau' < T$. $R(\tau)$ then follows from successive multiplications by C_α and has therefore an overall *geometric decay* from symbol interval to symbol interval.

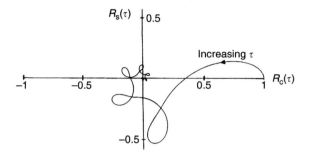

Figure 5.28 Plot of time-average autocorrelation function $R(\tau)$ in the complex plane. Quaternary 1SRC CPM with $h = 5/8$; data distribution (0.05, 0.45, 0.05, 0.45). (From [1].)

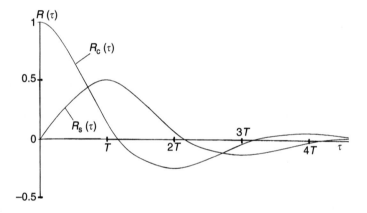

Figure 5.29 Real and imaginary components of autocorrelation $R(\tau)$. Binary 1RC CPM, $h = 1/2$, distribution B2. (From [1].)

We can pause to look at some example autocorrelations. In Fig. 5.28 is shown a plot of $R(\tau)$ in the complex plane for the quaternary scheme 1SRC with $h = 5/8$. The probabilities of the data symbols are $(P_{-3}, P_{-1}, P_1, P_3) = (0.05, 0.45, 0.05, 0.45)$. The trajectory begins with $(R_c(\tau), R_s(\tau)) = (1, 0)$ for $\tau = 0$ and then collapses inward to the origin as τ grows. Once τ has increased somewhat, the form of the trajectory is rotating, roughly repetitive and contracting; this is a consequence of the complex exponential decay factor C_α in Eq. (5.3-10). The interval $0 \leq \tau \leq 15T$ elapses in the figure. Another illustration of $R(\tau)$ is shown in Fig. 5.29. Here the functions $R_c(\tau)$ and $R_s(\tau)$ are plotted for the binary 1RC scheme with index $h = 1/2$. The data symbols have probabilities $(P_{-1}, P_1) = (1/4, 3/4)$ and these give $C_\alpha = j/2$. The autocorrelation here is completely determined by its values over the interval $0 \leq \tau \leq 2$. Using C_α, we find that $R_c(\tau)$ over the interval $2T \leq \tau \leq 3T$ equals $-R_s(\tau)/2$ over the interval

$T \leq \tau \leq 2T$, and similarly $R_s(\tau)$ over the interval $2T \leq \tau \leq 3T$ equals $R_c(\tau)/2$ over the interval $T \leq \tau \leq 2T$.

What remains is to find the complex baseband spectrum. This is

$$S(f) = \mathcal{F}\{R(\tau)\} = \int_{-\infty}^{\infty} R(\tau)e^{-j2\pi f\tau}\,d\tau = 2\Re\left\{\int_0^{\infty} R(\tau)e^{-j2\pi f\tau}\,d\tau\right\}. \tag{5.3-11}$$

The most important case is $|C_\alpha| < 1$, since this will be the case when $S(f)$ lacks discrete components. Assuming this, then, the integral in Eq. (5.3-11) can be divided into

$$\int_0^{\infty} R(\tau)e^{-j2\pi f\tau}\,d\tau = \int_0^{LT} R(\tau)e^{-j2\pi f\tau}\,d\tau + \int_{LT}^{\infty} R(\tau)e^{-j2\pi f\tau}\,d\tau,$$

where the last term can be written as

$$\int_{LT}^{\infty} R(\tau)e^{-j2\pi f\tau}\,d\tau = \sum_{m=L}^{\infty} \int_{mT}^{(m+1)T} [C_\alpha]^{m-L}\Psi(\tau')e^{-j2\pi f(\tau'+mT)}\,d\tau'.$$

The term can be further reduced as follows. From Eq. (5.3-10), $\Psi(\tau') = R(\tau' + LT), 0 \leq \tau' < T$. Thus,

$$\int_{LT}^{\infty} R(\tau)e^{-j2\pi f\tau}\,d\tau$$

$$= e^{-j2\pi fLT}\int_0^T R(\tau' + LT)e^{-j2\pi f\tau'}\,d\tau'\sum_{m=0}^{\infty}[C_\alpha]^m e^{-j2\pi fmT}. \tag{5.3-12}$$

The sum in Eq. (5.3-12) is geometric and adds to $1/[1 - C_\alpha e^{-j2\pi fT}]$. The last term therefore reduces to

$$\int_{LT}^{\infty} R(\tau)e^{-j2\pi f\tau}\,d\tau = \frac{e^{-j2\pi fLT}}{1 - C_\alpha e^{-j2\pi fT}}\int_0^T R(\tau + LT)e^{-j2\pi f\tau}\,d\tau. \tag{5.3-13}$$

The baseband spectrum has finally the expression

$$S(f) = 2\Re\left\{\int_0^{LT} R(\tau)e^{-j2\pi f\tau}\,d\tau + \frac{e^{-j2\pi fLT}}{1 - C_\alpha e^{-j2\pi fT}}\right.$$

$$\left. \times \int_0^T R(\tau + LT)e^{-j2\pi f\tau}\,d\tau\right\}. \tag{5.3-14}$$

It can be seen that $R(\tau)$ only needs to be computed over the interval $[0, (L+1)T]$ and that the baseband spectrum is a straightforward numerical integration of it. $R(\tau)$ itself is computed from Eq. (5.3-8).

Continuous-phase Modulation Coding

Equal symbol probabilities in standard CPM. When the symbol probabilities P_k are all $1/M$, $q(LT)$ has the standard value $1/2$, and h is non-integer, we have standard CPM. It can be shown that

$$C_\alpha = \frac{1}{M} \frac{\sin M\pi h}{\sin \pi h}. \tag{5.3-15}$$

$R(\tau)$ is real and takes the values

$$R(\tau) = \frac{1}{T} \int_0^T \prod_{i=1-L}^{m+1} \frac{\sin 2\pi h M[q(u+\tau-iT) - q(u-iT)]}{M \sin 2\pi h[q(u+\tau-iT) - q(u-iT)]} du, \tag{5.3-16}$$

for $mT \leq \tau \leq (m+1)T$.

The baseband spectrum is[8]

$$S(f) = 2\int_0^{LT} R(\tau) \cos 2\pi f\tau \, d\tau + \frac{2 - 2C_\alpha \cos 2\pi fT}{1 + C_\alpha^2 - 2C_\alpha \cos 2\pi fT}$$

$$\times \int_{LT}^{(L+1)T} R(\tau) \cos 2\pi f\tau \, d\tau$$

$$- \frac{2C_\alpha \sin 2\pi fT}{1 + C_\alpha^2 - 2C_\alpha \cos 2\pi fT} \int_{LT}^{(L+1)T} R(\tau) \sin 2\pi f\tau \, d\tau. \tag{5.3-17}$$

The last two equations are (2.7-10) and (2.7-11) in Chapter 2.

Spectral lines. The case $|C_\alpha| = 1$. For this case one can put

$$C_\alpha = e^{j2\pi\beta}, \quad 0 \leq \beta < 1. \tag{5.3-18}$$

From Eq. (5.3-10) it follows that for $|\tau| \geq LT$, $R(\tau)$ is periodic in τ with period T. Extend the periodic part to cover also the interval $[-LT, LT]$; call this part $R_{\text{dis}}(\tau)$ and what remains $R_{\text{con}}(\tau)$, so that

$$R(\tau) = R_{\text{con}}(\tau) + R_{\text{dis}}(\tau). \tag{5.3-19}$$

Since $R_{\text{dis}}(\tau)$ is periodic with period T, it follows that its Fourier transform contains only impulses, located at frequencies $f_{\text{dis}} = \pm(\beta + k)/T$, k an integer, with amplitudes that can be found from the coefficients of the Fourier series expansion of $R_{\text{dis}}(\tau)$. On the other hand, the function $R_{\text{con}}(\tau)$ is zero outside the interval $|\tau| \geq LT$, cannot lead to spectral lines, and yields the continuous part of the PSD. This part may be found separately by the usual method for the case $|C_\alpha| < 1$.

---
[8] Due to a misprint, Eq. (4.35) of [1] omits the third term in Eq. (5.3-17). This error is repeated by authors of several later books.

We can conclude that the spectrum has spectral lines if and only if $|C_\alpha| = 1$. Lines are not a desirable property since they carry no information and their transmitter power is wasted.[9] It is easy to check whether $|C_\alpha| = 1$. An obvious condition for this is that $q(LT) = 0$; this means that all CPM signals having frequency pulse integrating to zero will have a PSD containing discrete spectral components. Another condition is that one data symbol value, say i, has probability one, with $q(LT) \neq 0$; then

$$f_{\text{dis}} = \pm \frac{ihq(LT) + k}{T}, \quad k \text{ integer}. \qquad (5.3\text{-}20)$$

These cases, however, are not standard CPM. When $q(LT)$ takes the standard value 1/2, a condition for $|C_\alpha| = 1$ is that the modulation index h is an integer. Then,

$$f_{\text{dis}} = \begin{cases} \pm 1/2T, \pm 3/2T, \pm 5/2T, \ldots & h \text{ odd}, \\ 0, \pm 1/T, \pm 2/T, \ldots & h \text{ even}. \end{cases} \qquad (5.3\text{-}21)$$

Multi-h CPM spectra. The introduction of H different modulation indices in a cyclic pattern does not change the cyclostationary property of the process $s(t, a)$, but it does change the period of the cycle from T to HT. By changing the time averaging in Eq. (5.3-3) from $[0, T]$ to the interval $[0, HT]$, the correct baseband spectrum will be found. Now the time-average autocorrelation is completely determined by its values over the interval $0 \leq \tau \leq (L + H)T$. In Eq. (5.3-5) we write $\tau = \tau' + mHT$, with $0 \leq \tau' \leq HT$. The autocorrelation still decays exponentially, but now over blocks of H intervals. The expression for C_α will differ and will now be a function of $h_1, h_2, \ldots, h_H$. Just as for $H = 1$, $|C_\alpha| = 1$ indicates discrete spectral components. It is easiest to think of the multi-h case as CPM with a M^H-ary alphabet.

Filtered CPM; amplitude limiting; correlated data. Filtering and correlation in the data both introduce new dependences from interval to interval in the signal, and this modifies $R(\tau)$. Amplitude limitation changes, for example, C_α, but not the basic method. In principle, the new $R(\tau)$ may be computed. References [1, Section 4.5], and [28,29] show how to do this in certain cases; [1] discusses amplitude limitation and finite impulse response filtering and [28,29] treat the correlated data case. However, these methods are not very useful when strong, infinite response filters are applied to CPM. A better approach to distance and bandwidth in filtered CPM is the method given in Section 6.5.

5.3.2. Some Numerical Results

This section contains examples of PSDs computed by the method just given. We will see in these examples how main lobe width and sidelobe decay

---

[9] However, some transmission systems actually require lines, for example, synchronization.

Continuous-phase Modulation Coding

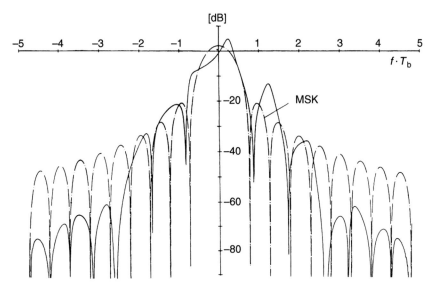

Figure 5.30 PSD for binary 1RC, $h = 1/2$, distribution B2. MSK spectrum shown for comparison (dashed). (From [1].)

depend on the alphabet size and phase pulse shape. The PSDs here are plotted as a function of baseband fT_b in Hs-s/data bit, where $T_b = T/\log_2 M$. They integrate to unity.

The first example appears in Fig. 5.30 and is the PSD corresponding to the autocorrelation in Fig. 5.29. This was the full response scheme 1RC, with $M = 2$, $h = 1/2$, $(P_{-1}, P_1) = (1/4, 3/4)$. The spectrum of MSK is shown for comparison; this scheme has a different phase response shape and equiprobable data, but is otherwise identical. Because of its unbalanced data, the 1RC spectrum is not symmetrical, but the main lobe is about the same width as the MSK main lobe. This is because the alphabet and phase response length L are the same in both schemes. The sidelobe decay is much faster with 1RC; this is because its phase response has more derivatives.

Figure 5.31 shows two schemes with a longer phase response. These are the binary schemes TFM and 4RC with $h = 1/2$, both with equiprobable data symbols. They have similar free distances. The TFM scheme has by definition $h = 1/2$ and a frequency pulse $g(t)$ with Fourier transform

$$G(f) = \mathcal{F}\{g(t)\} = (1/2) \cos^2 \pi f T \frac{\pi f T}{\sin \pi f T}. \qquad (5.3\text{-}22)$$

As discussed in Section 5.2, no analytical expression has been found for $g(t)$; rather, one can invert $G(f)$ numerically to find $g(t)$, and the result is a frequency

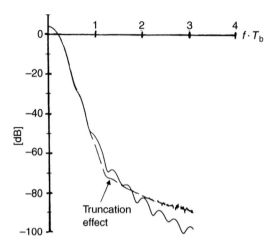

Figure 5.31 PSDs for binary 4RC, $h = 1/2$ (solid) and TFM with truncation to $7T$ (dashed). Equiprobable symbols. (Adapted from [1].)

pulse that resembles the 4RC pulse. The two spectra are almost identical down to -50 dB. (The TFM spectrum has a knee around -75 dB, but this was caused by time-truncation of $q(t)$ to $[0,7T]$.) These longer response spectra are a little narrower in the main lobe than the full response spectra in Fig. 5.30. However, their near-in sidelobe decay rate is much faster, and this is primarily due to the long phase response.

The next PSD examples are for 6RC and 6SRC, both with $M = 2$ and $h = 1/4$. The symbol probabilities are now unbalanced according to the following notation:

$$(P_{-1}, P_1) = \begin{cases} (1/2, 1/2) & \text{B1}, \\ (1/4, 3/4) & \text{B2}, \\ (1/10, 9/10) & \text{B3}. \end{cases} \quad (5.3\text{-}23)$$

Figure 5.32(a) shows the PSD for 6RC and Fig. 5.32(b) for 6SRC. The two frequency pulses are very similar, and two sets of PSDs are also similar, but 6RC shows small sidelobes below -50 db. This is because the SRC frequency pulses have derivatives to a much higher order.[10] The sidelobe structure is reduced primarily by the long length-6 pulse, but to remove all vestiges, the smoother 6SRC pulse is required. PSDs for the data symbol probabilities B1–B3 are shown in both figures. The balanced case B1 gives a symmetrical PSD; as data $+1$ becomes more

---

[10] In theory, the SRC frequency pulse has infinite support and has all derivatives; the pulse in the figure was truncated to width $16T$.

Continuous-phase Modulation Coding

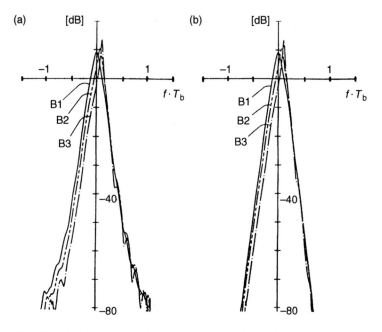

Figure 5.32 PSDs for binary 6-interval smoothing with $h = 1/4$ and symbol distributions B1–B3. (a) 6RC and (b) 6SRC. (Adapted from [1].)

likely, a peak starts to develop at frequency $f = h/2T_b$. This is consistent with Eq. (5.3-20).

Figure 5.33(a) and (b) is for the same conditions as Fig. 5.32(a) and (b), except that the modulation index h has been changed from 1/4 to 1/2. This approximately doubles the PSD widths above -10 dB (the part roughly corresponding to a main lobe), but there is less widening in the lower regions.

Next, two quaternary schemes will be considered. The frequency pulses are 3RC and 3SRC, both with $h = 1/4$, in Fig. 5.34(a) and (b). The symbol probabilities are now unbalanced according to

$$(P_{-3}, P_{-1}, P_1, P_3) = \begin{cases} (1/4, 1/4, 1/4, 1/4) & Q1, \\ (0.1, 0.2, 0.3, 0.4) & Q2, \\ (0.05, 0.10, 0.15, 0.70) & Q3. \end{cases} \quad (5.3\text{-}24)$$

Just as in the previous binary figure, the small sidelobes in the 3RC PSD disappear in the 3SRC plot. A similar result can be observed at other h and at higher alphabet sizes. Another observation is that the upper spectral width is 30% narrower in this $M = 4$, $L = 3$ case, compared to the $M = 2$, $L = 6$ case in the previous figure. In fact, h has a roughly linear effect on main lobe width, as will become more clear

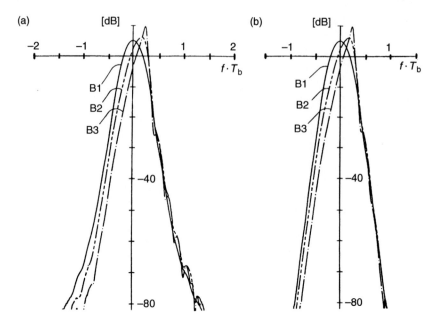

Figure 5.33 PSDs as Fig. 5.32, but $h = 1/2$. (a) 6RC and (b) 6SRC. (Adapted from [1].)

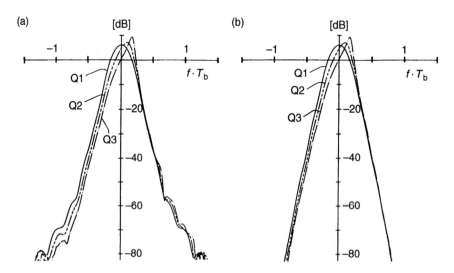

Figure 5.34 PSDs for quaternary 3-interval smoothing with $h = 1/4$ and symbol distributions Q1–Q3. (a) 3RC and (b) 3SRC. (Adapted from [1].)

Continuous-phase Modulation Coding

presently. The effect is somewhat counteracted here by the shorter 3RC/3SRC frequency pulse.

Plots of Extreme Sidelobes

By taking the PSD plots to extreme fT_b, it is easier to see how the different CPM parameters act to reduce main lobe width and to increase sidelobe decay. The general spectrum calculation method with single precision gives reliable results for $S(f)$ only down to 80–90 dB below the spectrum peak (100–110 dB with double precision), and below this a specialization of the method is available that extends the calculation to any fT_b. The calculation is based on successive integration by parts; details may be found in [1], Section 4.4, or in [27]. The following plots use this method. Data are equiprobable throughout.

Figure 5.35 compares power out of band (POB) for the binary RC family from 1RC to 5RC. h is 1/2 here. What the picture therefore shows is the effect of increasing the pulse width parameter L. The main lobe steadily narrows with growing L, even though the effect is not very rapid; second, the sidelobes are kept under steadily better control as L grows, although the asymptotic rate of decay is the same in all cases.

Next, Fig. 5.36 shows binary 4RC with $h = 0.3, 0.5, 0.8$; that is, L is now fixed, but the index varies. Above -20 dB the PSDs are roughly linearly proportional in width to the index h. The linear relation holds much less well

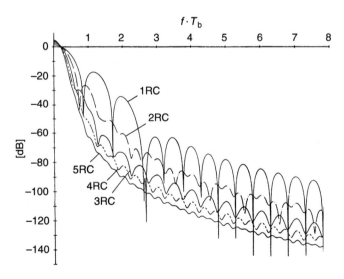

Figure 5.35 PSD comparison for binary 1RC–5RC when $h = 1/2$. Equiprobable data. (From [1].)

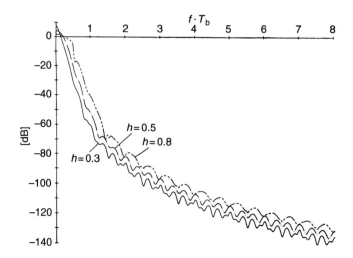

Figure 5.36 PSD comparison for binary 4RC with $h = 0.3, 0.5, 0.8$. Equiprobable data. (From [1].)

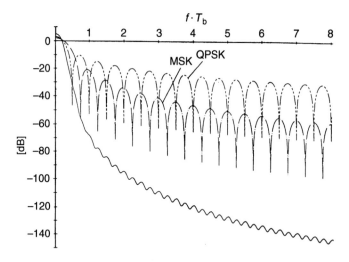

Figure 5.37 PSD comparison of QPSK (square phase pulse), MSK, and quaternary 3RC with $h = 0.3$. All have the same d_{min}. (From [1].)

at higher fT_b, and has little effect at extreme fT_b, but a smaller h always reduces $S(fT_b)$.

Finally, Fig. 5.37 shows the PSD of quaternary 3RC with $h = 0.3$, with a comparison to the modulations MSK and QPSK. All three schemes have the same free distance, and hence the same asymptotic error performance, but it is

Continuous-phase Modulation Coding

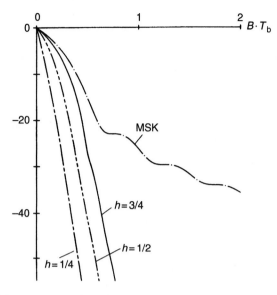

Figure 5.38 POB plot for binary 6SRC with equiprobable symbols. $h = 0.25, 0.5, 0.75$. MSK shown for comparison. (From [1].)

clear that quaternary 3RC is spectrally superior in every way to the other two. Here, coding has dramatically reduced bandwidth, while keeping distance fixed.

Power out-of band plots. The POB function $P_{OB}(f)$ was defined in Eq. (2.7-5). Figure 5.38 shows a POB plot for binary 6SRC with $h = 1/4, 1/2, 3/4$, together with the POB plot for MSK. The picture thus highlights the effect of h. Sidelobe structure is virtually absent in this SRC plot, even if the plot is extended much further. Above -20 dB the POBs for 6SRC have width linearly proportional to the index h, with a weaker proportionality at higher fT_b.

Conclusions: CPM Spectrum Properties

We should summarize what we have learned about this complex subject.

1. *Effect of L.* Increasing the frequency pulse width mildly reduces main lobe width and strongly controls and reduces sidelobe excursions; it does not in general change the asymptotic sidelobe decay rate.
2. *Effect of h.* Main lobe width is roughly proportional to modulation index; h has much less effect on sidelobe width.
3. *Effect of M.* Spectrum is roughly proportional to the inverse of the alphabet size M. This will be more clearly demonstrated in Section 5.3.3.

4. *Pulse derivatives.* Asymptotic sidelobe decay rate is proportional to the number of derivatives in the frequency pulse; pulses exist with all derivatives, and these have no sidelobe structure.

5.3.3. Energy–Bandwidth Performance

Neither energy nor bandwidth alone is a sufficient measure of a coded modulation system, and this is particularly true of CPM coding, where minimum distance and bandwidth are different, complicated functions of the CPM parameters. A redesign to achieve better distance can have an unpredictable effect on bandwidth consumption, and vice versa. A small distance gain is not worth a large bandwidth loss. In order to see overall parameter effects, it is necessary to plot the position of a large number of systems in the energy–bandwidth plane. The object is to clarify which parameter changes improve the overall energy and bandwidth in the coding system, and which primarily trade one off against the other.

Section 5.1 gives the standard way to plot energy and bandwidth, and we will follow that now. The energy consumption of a coding scheme will be its square minimum distance, expressed in dB above or below the MSK/QPSK benchmark 2; that is, *Energy Consumption* $= 10\log_{10}(d^2_{\min}/2)$. Since $P_e \sim Q(\sqrt{d^2_{\min} E_b/N_0})$, this measure directly expresses the asymptotic energy the system needs, relative to MSK or QPSK.[11] Distances come from Section 5.2. The bandwidth consumption of a scheme will be its NBW as defined in Eq. (5.1-4), with a POB criterion; that is, *Bandwidth Consumption* $= 2BT_b$. B will be either the 99% or 99.9% power bandwidth, taken from the positive-frequency complex baseband spectra in Sections 5.3.1 and 5.3.2. The 99% bandwidth more or less measures the main lobe width and the 99.9% measures the width of the significant sidelobes.

Figure 5.39 plots energy consumption vs bandwidth consumption for a variety of 1REC (i.e. CPFSK) schemes with different modulation indices h and alphabet sizes M. As h varies for each M, the location of a 1REC scheme roughly traces the path shown. As one moves down a trajectory, smaller h are employed, which cause the scheme to use less bandwidth but more energy. There are two families of lines, one for 99% power and one for 99.9%; the 99.9% bandwidth of a scheme must always lie to the right of the 99% bandwidth. At a fixed distance, better schemes are to the left; at a fixed bandwidth, better schemes lie above. The overall best schemes thus lie to the upper left. MSK is shown here and in the next two figures as a reference.

It is clear from Fig. 5.39 that $M = 2$ is a poor alphabet size to use at either power bandwidth, since its paths lie well to the right of the others. An alphabet size of 4 is almost as good as the best alphabet size, which seems to be close to 8. A size of 16 is worse than 8. We can conclude that $M = 4$ is a good size to use for CPFSK

---

[11] This measure, expressed as a loss, is used again to compare PRS codes in Chapter 6; see Section 6.4.

Continuous-phase Modulation Coding

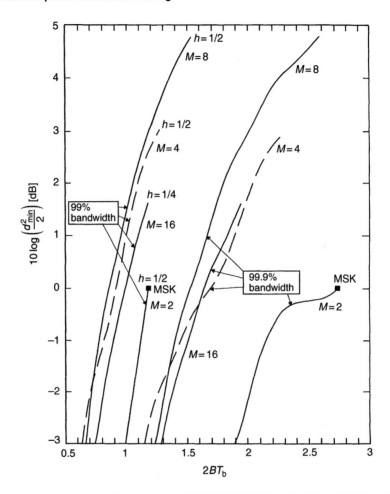

Figure 5.39 Energy–bandwidth tradeoffs in 1REC CPM (CPFSK), with 99% and 99.9% in-band power bandwidth B. h decreases down each curve. (From [2], copyright 1984 by John Wiley Ltd., used with permission.)

schemes, since it leads to a simple system and it performs essentially as well as any other. The index h primarily trades off energy for bandwidth, since the paths for different M do not move significantly away from each other as h changes. Large distance gains are achievable, but only at the expense of bandwidth expansion. It is interesting to note that information-theoretic arguments based on the cut-off rate for both AWGN and CPM channel models predict roughly the same outcome for the best M (see [1], Chapter 5, and Wozencraft and Jacobs [30]). The weakness of $M = 2$ has been noticed in practical applications.

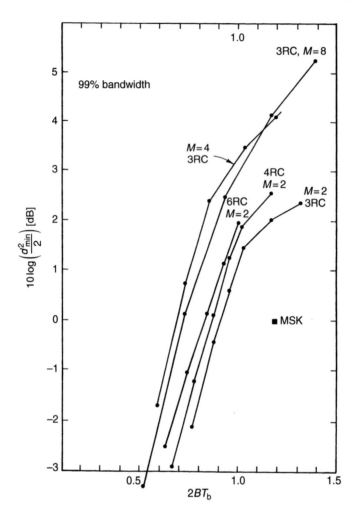

Figure 5.40 Energy–bandwidth tradeoffs in RC-pulse CPM, with 99% in-band power bandwidth B. h decreases down each curve; straight lines connect points denoting specific schemes. (From [2], copyright 1984 by John Wiley Ltd., used with permission.)

Next, Fig. 5.40 uses the RC family to show how longer pulses improve energy–bandwidth performance. Once again, h grows up a path, and the bandwidth measure is 99%. There is a small improvement in the performance as the pulse widens from 2RC to 6RC, but alphabet size has the greater influence. Once again, $M = 4$ appears to be a good size. The RC systems here are only marginally better than the CPFSK systems of Fig. 5.40. This reflects the fact that the 99% bandwidth measures the main lobe, and the main lobes are not much different in the systems compared, when the effect of h is factored out.

Continuous-phase Modulation Coding

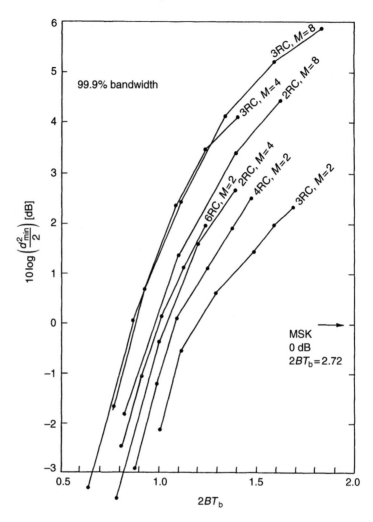

Figure 5.41 RC energy–bandwidth tradeoff as in Fig. 5.40, but 99.9% in-band power bandwidth. (From [2], copyright 1984 by John Wiley Ltd., used with permission.)

Figure 5.41 shows RC systems again, this time in terms of 99.9% bandwidth. This criterion stresses sidelobes and gives a good indication of adjacent-channel interference. Now an increase in pulse width has a strong effect and produces a much better coded modulation. This is true at all alphabet sizes, but $M = 4$ is always close to the best size, except in the very wide bandwidth/low energy region. The 99.9% curves of Fig. 5.39 are much worse than the curves of Fig. 5.41.

The examples here indicate that longer, smoother phase responses dramatically improve energy–bandwidth performance only if the bandwidth criterion takes some account of spectral sidelobes. Otherwise, the improvement is moderate, if all systems are allowed to use their best h. A second feature of these plots is the strong influence of alphabet size, up to a size of about 8. The behavior here reflects the basic difficulty of narrowing the spectral main lobe in a constant-envelope coded modulation. Unlike the case with TCM, one cannot, in theory, endlessly increase the underlying constellation.

All of these results are for specific schemes that we can now imagine and construct, and the ultimate abilities of phase coding are still a matter for speculation. The Shannon theory is helpful here, but it is not well developed.

5.4. Receivers and Transmitters

In this section we turn to receivers, transmitters, and related subsystems, such as synchronizers, for CPM signals. Because CPM is nonlinear, constant envelope signaling, it is more challenging to implement than TCM or PRS. The most difficult subject is that of receivers. In this section we will focus on receivers that are optimal or essentially so. The background for the discussion is the signal space theory of Section 2.3, the basic optimal receivers of Section 2.4, and the trellis coding and VA in Section 3.4. Reduced-complexity receivers are an important subject for any coded modulation, and many approaches exist for CPM signals; these appear in a separate Section 5.5.

Synchronizer circuits are a major part of any receiver. In CPM as in other coded modulations, the rest of the receiver needs in the best of worlds to know the carrier phase (actually, the offset ψ_o in Section 2.6.2) and the symbol timing (the times nT in Eq. (5.1-2)). Reception actually comprises the estimation of both these and the data. An optimal receiver would estimate all of them simultaneously, subject to some suitable criterion such as the maximum likelihood (ML) principle, but with simple quadrature amplitude modulation (QAM) or PSK modulation it is sufficient in practice to estimate them with independent circuits. With partial response CPM, this is not true, and we will look at combined circuits for that case.

In an imperfect world the CPM phase reference is often not fully known. A more fruitful point of view is to characterize the carrier phase as random to some degree, and we will investigate optimal receivers for several such cases. Let ψ_o and $\hat{\psi}_o$ be the carrier phase offset at the transmitter and receiver, respectively, and let the carrier phase error be $\Delta \psi_o$. If $\Delta \psi_o$ is fixed but is otherwise unknown at the receiver, we can take $\Delta \psi_o = \psi_o - \hat{\psi}_o$ to be a uniform random variable, with density function

$$f_{\Delta\psi}(\Delta\psi_o) = \begin{cases} 1/2\pi, & |\Delta\psi_o| \leq \pi; \\ 0, & |\Delta\psi_o| > \pi. \end{cases}$$

Continuous-phase Modulation Coding

This reception is referred to as *noncoherent*. An important result is that most often the optimal noncoherent receiver has the same asymptotic error probability as the coherent one. It often happens that a synchronizer works to extract statistical information about $\Delta\psi_o$ through a device like a phase-lock loop. Once the loop is in a stable operating mode, the reference that it provides differs from the true one by a slowly varying random variable $\Delta\psi_o$. A standard assumption is that an observation of $\Delta\psi_o$ has the *Tikhonov* density

$$f_{\Delta\psi}(\Delta\psi_o) = \begin{cases} \dfrac{e^{\alpha \cos \Delta\psi_o}}{2\pi I_0(\alpha)}, & |\Delta\psi_o| \leq \pi; \\ 0, & |\Delta\psi_o| > \pi. \end{cases}$$

Here $I_0(\ )$ is the modified Bessel function of zero order and first kind. Some Tikhonov distributions are plotted in Fig. 5.42. The parameter α is the loop SNR, and it generates a whole class of models, ranging from the noncoherent case ($\alpha = 0$) to the fully coherent case ($\alpha = \infty$). The cases in between are *partially coherent* detection.

The section ends with a review of several CPM transmitter implementations.

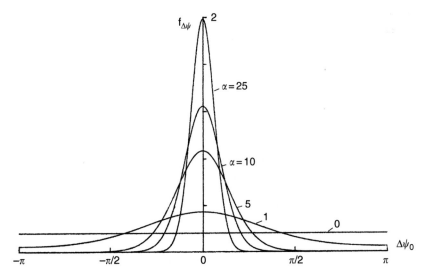

Figure 5.42 The Tikhonov density function for random phase error $\Delta\psi_0$, for $\alpha = 0, 1, 5, 10, 25$.

5.4.1. Optimal Coherent Receivers

We will develop two optimal receivers. The first, the Osborne–Luntz receiver, makes an ML decision about one symbol only, based on observation of a sequence of successive symbols. The second, the VA, forms an ML estimate of the entire sequence. The VA is the reduction of the correlator or matched-filter receiver, Fig. 2.9 or Fig. 2.10, to trellis coding.

The Osborne–Luntz Receiver: Coherent Case

The derivation of a one-symbol-only detector roughly follows Section 2.3, but it is sufficiently different and the principle is used often enough in the rest of the chapter and book, to justify a review of the derivation here. For simplicity, we will focus on CPFSK and equiprobable data, but the derivation extends easily to general CPM and general probabilities. After it decides the first symbol, the receiver can decide the second symbol by starting from scratch again, sliding the observation interval one forward. Both the coherent and the succeeding noncoherent derivations here follow the original work of Osborne and Luntz [9] and Schonhoff [10].

The received signal is $s(t, \boldsymbol{a}) + \eta(t)$, where $s(t, \boldsymbol{a})$ is a CPM signal (5.1-1) and (5.1-2) over time $[0, NT]$ and $\eta(t)$ is AWGN. To derive the receiver, we modify Eq. (2.3-2), asking what initial data symbol $a[1]$ has maximum *a posteriori* probability, instead of what whole signal. Define $\mathcal{A}_k$ to be a set of particular symbols $a[2], \ldots, a[N]$, the kth of M^{N-1} such sets. Denote by $s(t, a[1], \mathcal{A}_k)$ the signal that carries all N symbols. The *a posteriori* probability of the event $\{a[1] = i\}, i = \pm 1, \ldots, \pm(M-1)$, given that $r(t)$ is received, is formally

$$P[a[1] = i | r(t)] = \frac{\sum_k P[r(t) \text{received} | \mathcal{A}_k \text{sent}] P[\mathcal{A}_k \text{sent}]}{P[r(t) \text{received}]}. \quad (5.4\text{-}1)$$

By a series of steps like those in Section 2.3.1, we can reduce Eq. (5.4-1) to the likelihood value

$$\sum_k \exp[\|\boldsymbol{r} - \boldsymbol{s}_k\|^2 / N_0] P[\mathcal{A}_k], \quad (5.4\text{-}2)$$

where $\boldsymbol{s}_k$ is the vector representation of the signal formed from $a[1]$ and $\mathcal{A}_k$. This is a form similar to Eq. (2.3-13). The receiver works by forming the M likelihood values for $a[1] = \pm 1, \ldots, \pm(M-1)$, and deciding $a[1]$ as the one with highest value.

If the symbols are equiprobable, $P[\mathcal{A}_k]$ may be dropped. Furthermore, the Euclidean distance in Eq. (5.4-2) may be replaced by the correlation integral $\int r(t) s(t, a[1], \mathcal{A}_k) dt$, together with suitable scale factors which may be factored out. This leads to an equivalent form for the likelihood of the event $\{a[1] = i\}$ given by

$$\lambda_i \triangleq \sum_k \exp\left[\frac{2}{N_0} \int_0^{NT} r(t) s(t, a[1], \mathcal{A}_k) dt \right]. \quad (5.4\text{-}3)$$

Continuous-phase Modulation Coding

This is the form favored by Osborne and Luntz. The receiver decides the data symbol corresponding to the largest of λ_i. In its physical implementation, the receiver selects the largest of M subsystem outputs; each of these comprises M^{N-1} actions consisting of a mixing $r(t)s(t, a[1], \mathcal{A}_k)$, an integration, and an exponentiation.

This Osborne–Luntz receiver rapidly grows with N and it is not amenable to precise analysis. However, in the limit of high E_b/N_0 it is clear that in each of the M likelihoods (5.4-2), the term having s_k closest to r dominates, and consequently the decision on $a[1]$ is in favor of the $s(t, a)$ that lies closest in Euclidean distance to r. From signal space theory, then, the asymptotic error probability is $\sim Q(\sqrt{d_{\min}^2 E_b/N_0})$, with $d_{\min}$ the minimum distance of the N-interval signal set. Thus the distance analysis of Section 5.2 gives the Osborne–Luntz asymptotic probability.

The Viterbi Receiver

The VA of Section 3.4 is an efficient, recursive implementation for trellis codes of the basic ML sequence detector in Section 2.4. We will begin with some notes on how the VA is applied to CPM and then give a few test results.

The VA is based on an encoder state description, and we will use the one given in Section 5.2.3. The output of the VA is the ML estimate of the entire *sequence* of encoder states, and from these can be derived the ML sequence of data symbols. The encoder begins and ends at known states. The pure form VA thus does not estimate a single symbol like the Osborne–Luntz receivers do, but in most applications, including ours here, a first-symbol "sliding window" modification is employed that is in fact a single-symbol estimator. It observes a window $[0, NT]$ of the signal and decides the first symbol; this symbol is the initial symbol on the path closest in distance to the received signal. The window then moves forward to $[T, (N+1)T]$, the algorithm repeats the process to decide the second symbol, and so on indefinitely. Symbol $n+1$ is decided by observing interval $[nT, (n+N)T]$ and surviving paths are saved over only their N most forward trellis branches.[12] N is the decoding delay and is the same N as the one in the Section 5.2.2 distance calculations. Strictly speaking, the output stream of the decoder here is no longer the ML symbol sequence, but the asymptotic first-event error probability[13] is $\sim Q(\sqrt{d_{\min,N}^2 E_b/N_0})$ nonetheless, with $d_{\min,N}$ the minimum distance of the CPM code observed to depth N. When N equals or exceeds the decision depth L_{dec} of the code, $d_{\min}$ in this expression becomes the full free distance d_f of the code. None of

---

[12] The precise manner of dropping paths can follow several strategies; also, a means of storing paths, or alternately, a "traceback" procedure, must be devised. A cell of distance and path history information is associated with each trellis state. These and other implementation details appear in applied coding texts such as [33].

[13] Defined in Section 4.3.3.

this discussion differs in principle from the VA application to binary convolutional codes.

With both TCM and PRS coding, it is natural to model and handle signals in discrete time, there being one (possibly complex) signal space dimension associated with each symbol. The VA is that of ordinary convolutional coding, but with real-number arithmetic and Euclidean distances. With CPM, it is clumsy to handle the signals as signal space vectors. The dimension of the space is usually much larger than the data alphabet size, and the components in some dimensions can be very small. A better way is as follows. (i) Define paths by their state history; this is equivalent to recording their data symbol history over $N + L - 1$ intervals, plus one phase offset. This so-called "path map" for a path beginning at time nT is $(\theta_n, a[n+N], a[n+N-1], \ldots, a[n+1], a[n], \ldots, a[n-L+2])$, adapting the notation in definition (5.2-15). (ii) Do not perform matched filtering or signal correlation. Instead, compute Euclidean distance directly from the continuous-time signals, by means of a stairstep approximation. These signals can be the carrier $r(t)$ and the carrier CPM signals $s(t, a)$, but in practicality they would be the baseband I and Q signals, after down-conversion to baseband in the receiver. The full Euclidean distance is obtained from the baseband signal square differences as given in Eq. (2.5-20).

To illustrate the stairstep method, consider the Euclidean square distance $\int |r_I(t) - I(t, a)|^2 dt$ between the in-phase received signal and an in-phase CPM signal. We can approximate the integral by the time-discrete approximation

$$\int |r_I(t) - I(t, a)|^2 dt \approx \frac{T}{\kappa} \sum_{j=1}^{J} |r_I(j\Delta t) - I(j\Delta t, a)|^2, \qquad (5.4\text{-}4)$$

where the integral covers k intervals, the signals are sampled κ times each symbol interval, and $J = k\kappa + 1$. As $\kappa \to \infty$, the discrete approximation tends to the continuous integral for any bandlimited signals. Four to sixteen samples per interval give a good approximation, depending on the smoothness of the signals. The proper number is best found by experiment during the receiver design; as a rule, receiver distances need only be roughly calculated and κ is often surprisingly small.

Figures 5.43 and 5.44 show some typical VA decoding performances for CPM over a simulated AWGN channel. P_e in these plots is the observed event error rate, that is, the rate that trellis error events begin at an interval, given that an event is not already in progress (see Section 3.4.1). Figure 5.43 shows $h = 1/2$ binary 3RC VA decoding, implemented by the stairstep method with $\kappa = 10$ and N much longer than L_{dec}. There are 16 trellis states, counted by the Section 5.2.3 method. The solid curve is an error estimate of the form (5.2-17) based on one symbol, given by $0.5Q(\sqrt{1.77E_b/N_0}) + 0.5Q(\sqrt{2.28E_b/N_0})$. d_f^2 is $1.77 - 0.5$ dB below MSK – (see Fig. 5.22), and the event error rate essentially achieves this; the bit error rate (BER) will be somewhat higher. The 99% bandwidth of this

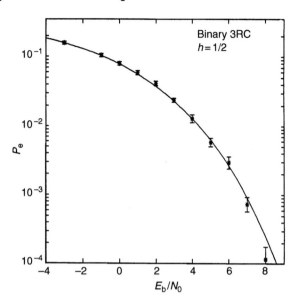

Figure 5.43 Experimental event error rate for Viterbi decoded binary $h = 1/2$ 3RC, with comparison to estimate of form (5.2-17). (From [23], copyright 1991 by A. Macdonald, used with permission.)

scheme is 71% of MSK, and the 99.9% bandwidth is only 41% of the MSK 99.9% bandwidth.[14]

Figure 5.44 plots BER for octal CPFSK with h equal to 1/4 and 5/11. For the first scheme, $d_f^2 = 2.18$ (0.4 dB above MSK), $L_{dec} = 2$ and there are 8 trellis states; for the second, $d_f^2 = 5.41$ (4.3 dB above MSK), $L_{dec} = 5$ and there are 22 states. The VA is implemented with $\kappa = 20$ sampling and the path memory is $N = 50$. The mapping from data bits to octal CPM symbols is Gray coded in such a way that any error to a closest neighbor phase tree branch causes only a single data bit error. The solid curve estimates in this figure are of the single form $2/3 Q(\sqrt{d_f^2 E_b/N_0})$. At BER $= 10^{-4}$, the results in this figure are close to the estimate for $h = 1/4$ and about 1 dB off for $h = 5/11$. The 99% bandwidth of the $h = 1/4$ scheme is 75% of MSK, and the 99.9% is 60%; the 99% and 99.9% bandwidths of the $h = 5/11$ scheme, with its large coding gain, are 115% and 90% of MSK, respectively.

Notes on Simulation Using the Stairstep Approximation

Since most software VA receiver implementations are based on the stairstep method, it is worth pausing to show how the method is applied during

---

[14] The NBW 99% bandwidth of MSK is 1.2 Hz-s/bit; the 99.9% bandwidth is 2.72 Hz-s/bit.

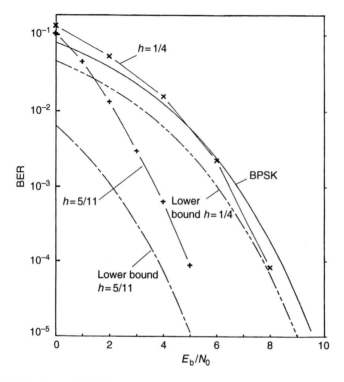

Figure 5.44 Experimental BER for Viterbi decoded octal CPFSK with $h = 1/4$ and $5/11$, with comparison to lower bound. (Adapted from [1].)

simulation of an AWGN channel. A baseband transmission component $I(t, a_o)$ that has been converted to stairsteps can be viewed as a succession of 1-dimensional square pulse QAM mini-transmissions, with amplitudes $I(j\Delta t, a_o), j = 1, \ldots, J$ in AWGN. Each mini-transmission may be modeled as an independent use of the discrete-time Gaussian channel (3.5-4), the output of which is the value $r_j = \sqrt{2E_s/T} I(j\Delta t, a_o) + \eta_j$, where η_j is zero-mean IID Gaussian with some variance σ_m. r_j lies at Euclidean square distance $|\eta_j|^2$ from $\sqrt{2E_s/T} I(j\Delta t, a_o)$. Now consider the whole transmission. The total signal space square distance between $r = (r_1, \ldots, r_J)$ and $I = \sqrt{2E_s/T}[I(\Delta t, a_o), \ldots, I(J\Delta t, a_o)]$ is $\Delta t \sum_1^J |\eta_j|^2$, the sum of J independent contributions. Similarly, the signal space square distance between two successions $\sqrt{2E_s/T} I(j\Delta t, a_1)$ and $\sqrt{2E_s/T} I(j\Delta t, a_2), j = 1, \ldots, J$ is $\Delta t 2E_s/T \sum |I(j\Delta t, a_1) - I(j\Delta t, a_2)|^2$.

Now turn to the distance expression (5.4-4), and assume that κ is large enough so that the error in the calculation is negligible. We wish to derive the mini-transmission noise variance σ_m that leads to a stairstep distance with the same statistics as the true one. To do this, start with the fact that the distance satisfies $\int |r(t) - s(t)|^2 dt = \int r(t)^2 dt - 2\int r(t)s(t)dt + \int s(t)^2 dt$. The last term is a constant offset; the first term will be a constant during comparison of the various $s(t)$ to $r(t)$. Therefore, it suffices to consider only the middle correlation term, which has stairstep approximation

$$\int r(t)s(t)\,dt \approx \frac{T}{K}\sum_{1}^{J} r(j\Delta t)s(j\Delta t). \qquad (5.4\text{-}5)$$

In either guise, the term is a Gaussian variable, and as such, is defined statistically by its mean and variance. If $\sqrt{2E_s/T}I(t, a_1)$ is transmitted and $s(t)$ is $\sqrt{2E_s/T}I(t, a_2)$, the mean of either side of Eq. (5.4-5) is approximately $2E_s/T \int I(t, a_1)I(t, a_2)dt$. A calculation shows that the variances approximate when

$$\sigma_m^2 = \kappa N_0/2T. \qquad (5.4\text{-}6)$$

A detailed calculation appears in [1], Section 7.4.1. We conclude that the variance of the noise added to each signal sample in the stairstep method is effectively scaled up by κ.[15]

5.4.2. Partially Coherent and Noncoherent Receivers

When the signal reception is offset by a constant but unknown phase, there are many receiver possibilities, some simple but suboptimal in performance and some complex but optimal. Two of the simple alternatives are diagrammed in Fig. 5.45. Both function independently of a carrier phase offset.

A *differential* detector is one that multiplies the input carrier signal by a T-delayed version of the same signal. The version in Fig. 5.45 also shifts the

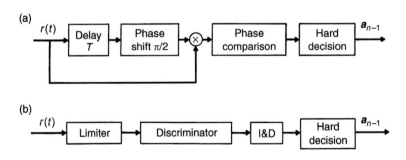

Figure 5.45 (a) Block diagram of differential detector. (b) Block diagram of discriminator detector.

---

[15] The scale factor is κ when $T = 1$; changing to an arbitrary T scales all samples by $1/\sqrt{T}$ irrespective of the symbol energy – see Eq. (5.1-1) – and the noise added to the waveform must be scaled by the same factor.

delayed version by $+\pi/2$, which creates in the absence of noise the multiplier output

$$(1/2) \sin[\omega_0 T + \Delta\phi(t)], \qquad (5.4\text{-}7)$$

in which the excess phase difference over T seconds is

$$\Delta\phi = \phi(t, a) - \phi(t - T, a) \mod 2\pi.$$

A lowpass filter has removed a double-frequency term here, and the term $\omega_0 T$ can be tuned out. There follows a comparison to one or more phase thresholds and a decision. Figure 5.46 shows the ensemble of differences $\Delta\phi(t)$ for the example binary 2RC. Noise will disturb the measured $\Delta\phi(t)$. It is clear that one wants to sample $\Delta\phi(t)$ at a position with an open eye. Two such positions are indicated, and once a set of thresholds is selected, the various regions need to be associated with data symbols.

Many variations on this theme are possible. Further results and an error analysis appear in [1], Chapter 7, and references therein.

The second alternative in Fig. 5.45 is the *discriminator* detector. A discriminator is a device that extracts the derivative of the carrier phase. Consequently, it does not respond to a constant phase offset. An interesting and practical example

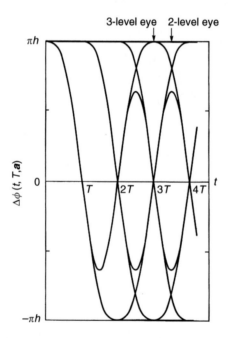

Figure 5.46 The ensemble of differences between phase paths in binary 2RC, when one path is delayed T. Two possible observation positions are shown. (From [1].)

Continuous-phase Modulation Coding

of discriminator detection occurs when the CPM is CPFSK. Since the excess phase consists of a symmetrical set of M linear slopes, its derivative is a square pulse waveform with M symmetrical amplitudes. Except for the fact that the noise is rather different, the remaining detection apparatus is simply that of NRZ-pulse, 1-dimensional QAM in AWGN. Figure 5.45 thus shows an integrate-and-dump (I and D) detector, which is the optimal one at least for AWGN. For more complex CPMs, the discriminator output is just that of a linear modulator whose pulse is the frequency pulse $g(t)$; that is, it is $\sum a_n g(t - nT)$. A favorite scheme is Gaussian MSK, $g(t)$ for which is given in Table 5.1. All the usual linear modulation insights, such as eye diagrams, apply, and when $g(t)$ has response longer than T seconds, one can even apply the VA to decode the discriminator output. Because the pulse modulation here frequency-modulates a carrier, the name "digital FM" is sometimes applied to the combination of a simple CPM scheme with discriminator detection.

Although everyday circuitry like discriminators and I and Ds are used in digital FM, they are by no means optimal. The noise is far from Gaussian and contains "clicks" that occur when the phase rotates 2π radians. Some noise analysis and error probability results were given in [1], Chapter 7, and the references therein. One standard technique is to place a prefilter in front of the discriminator and then search for a good one; this is described in [1], and a more modern reference is [39]. As a rule, one can say that the discriminator detection is 3–5 dB less energy efficient than optimal detection.

The Osborne–Luntz Receiver: Noncoherent Case

For noncoherent detection, the initial phase offset $\Delta\psi_0$ takes a value uniformly distributed in $[0, 2\pi]$ during the detection process for a symbol. What follows is a detector that finds the ML estimate of one symbol, due originally to Osborne and Luntz. Following again [9,10], we observe the signal for $N = 2\ell + 1$ symbols. The receiver decides the middle symbol $a[\ell + 1]$, and the remaining symbols $\{a[1] \cdots a[\ell]a[\ell + 2] \cdots a[2\ell + 1]\}$, which take on $J = M^{2\ell}$ values, form the set $\mathcal{A}_k$. As in the coherent Osborne–Luntz receiver, there are M likelihood values, but now the probability evaluation in Eq. (5.4-1) must include an integration over the distribution of $\Delta\psi_0$. For equiprobable symbols, the eventual likelihood parameters are

$$\int_{-\pi}^{\pi} \sum_k \exp\left[\frac{2}{N_0} \int_0^{NT} r(t) s(t, a[\ell+1], \mathcal{A}_k, \Delta\psi_0) \, dt\right] f_\Delta \, d\Delta\psi_0,$$

for $a[\ell + 1] = \pm 1, \ldots, \pm(M - 1)$,

where $s(t, a, \mathcal{A}_k, \Delta\psi_0)$ is a CPM signal offset by $\Delta\psi_0$, f_Δ is the density of $\Delta\psi_0$, and factors that affect all likelihoods the same are removed. A linear density f_Δ

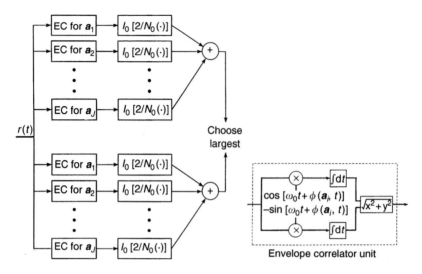

Figure 5.47 The Osborne and Luntz optimal one-symbol noncoherent receiver for binary CPM. Envelope correlator unit defined at right.

leads to

$$\lambda_i \triangleq \sum_k I_0(2z_{i,k}/N_0), \quad i = \pm 1, \ldots, \pm(M-1), i \text{ odd}, \quad (5.4\text{-}8)$$

in which

$$z_{i,k}^2 = \left[\int_0^{NT} r(t)s(t, i, \mathcal{A}_k, 0)\, dt \right]^2 + \left[\int_0^{NT} r(t)s(t, i, \mathcal{A}_k, \pi/2)\, dt \right]^2. \quad (5.4\text{-}9)$$

$z_{i,k}$ is a variable with a Rician distribution (see [32], Sections 8.2 and 8.3). $I_0(\cdot)$ is the zeroth-order modified Bessel function. The receiver decides $a[\ell + 1]$ to be the index of the largest λ.

The receiver structure is shown in Fig. 5.47 for binary CPM. The operation in Eq. (5.4-9) is called an *envelope correlation* and it is denoted EC in the figure. There are two sets of ECs, an upper set for middle symbol $+1$ and a lower set for -1. Each EC feeds a Bessel operation $I_0[(\cdot)2/N_0]$; the function can be stored as some simple approximation. The Bessel output sets are summed, and $a[\ell + 1]$ is set to the symbol with the largest sum. The structure is optimum at all E_b/N_0.

The receiver can be simplified somewhat at high E_b/N_0 by using the facts that the Bessel function is monotone and at high SNR (low N_0), $\sum_k I_0(2z_{i,k}/N_0) \approx I_0(2\max_k\{z_{i,k}\}/N_0)$ holds. Effectively, this eliminates the Bessel operation, because the receiver can simply set $a[\ell + 1]$ equal to the i leading to the max of all the $z_{i,k}$. As a rule, for high SNR and a wide enough observation window, the noncoherent Osborne–Luntz receiver has asymptotic symbol error probability $\sim Q(\sqrt{d_{min}^2 E_b/N_0})$, in which d_{min} is the minimum distance of Section 5.2. This is true despite the unknown phase offset, and it shows that what is really important at reasonable SNR is a stable phase, not a known one. The noncoherent probability will now be derived as a special case of partial coherence.

Optimal Partially Coherent Detection

We now extend the Osborne–Luntz idea to partial coherence and to an arbitrary observation interval $[-N_1 T, N_2 T]$. The symbol to be detected is $a[1]$, corresponding to interval $[0, T]$. The error probability depends asymptotically on a new minimum distance d_e, which by means of a wider observation can usually be driven to d_f for the CPM code.

Represent the CPM signal by

$$s(t, \boldsymbol{a}, \hat{\psi}_0) = \sqrt{2E_s/T} \cos[\omega_0 t + \phi(t, \boldsymbol{a}) + \hat{\psi}_0], \quad (5.4\text{-}10)$$

in which $\hat{\psi}_0$, obtained from the synchronizer loop, is the receiver estimate of the transmitter carrier phase offset ψ_0. The receiver phase error is thus $\Delta\psi_0 = \psi_0 - \hat{\psi}_0$. The receiver knows the mean value of ψ_0, which it uses to force the equality $\mathcal{E}[\hat{\psi}_0] = \psi_0$; thus the average of ψ_0 and $\hat{\psi}_0$ can be taken as zero. The phase error is assumed to be Tikhonov distributed. The receiver knows the parameter α, which depends on the SNR in the synchronizer loop.

Now form the likelihood function [38]

$$\int_{-\pi}^{\pi} P_{\Delta\psi}(\Delta\psi_0) \exp\left[(2/N_0) \int_{-N_1 T}^{N_2 T} r(t, \boldsymbol{a}, \psi_0) s(t, \tilde{\boldsymbol{a}}, \hat{\psi}_0) \, dt\right] d\Delta\psi_0 \quad (5.4\text{-}11)$$

for all possible received sequences $\tilde{a}[-N_1 + 1], \ldots, \tilde{a}[N_2]$, and find the maximizing one. The value $\tilde{a}[1]$ in this sequence is taken as the ML estimate of $a[1]$.

Some analysis [35] shows that the integrand in the exponential is only a function of $\Delta\psi_0$, assuming that $\omega_0 T \gg 1$. Equation (5.4-11) can be integrated analytically [36,37]. A monotone transformation of the result is the set of likelihood parameters

$$\lambda(\boldsymbol{a}, \tilde{\boldsymbol{a}}) \triangleq C(\boldsymbol{a}, \tilde{\boldsymbol{a}}) + \sqrt{T/E_s} n_c(\tilde{\boldsymbol{a}}) + \frac{\alpha}{2E_s/N_0} + [S(\boldsymbol{a}, \tilde{\boldsymbol{a}}) + \sqrt{T/E_s} n_s(\tilde{\boldsymbol{a}})]^2. \quad (5.4\text{-}12)$$

Here the following are defined:

$$C(a, \tilde{a}) = \frac{1}{T} \int_{-N_1 T}^{N_2 T} \cos[\phi(t, a) - \phi(t, \tilde{a})] \, dt,$$

$$S(a, \tilde{a}) = \frac{1}{T} \int_{-N_1 T}^{N_2 T} \sin[\phi(t, a) - \phi(t, \tilde{a})] \, dt,$$

$$n_c(\tilde{a}) = \frac{1}{T} \int_{-N_1 T}^{N_2 T} n_c(t) \cos \phi(t, \tilde{a}) \, dt + \frac{1}{T} \int_{-N_1 T}^{N_2 T} n_s(t) \sin \phi(t, \tilde{a}) \, dt,$$

$$n_s(\tilde{a}) = \frac{1}{T} \int_{-N_1 T}^{N_2 T} n_c(t) \sin \phi(t, \tilde{a}) \, dt - \frac{1}{T} \int_{-N_1 T}^{N_2 T} n_s(t) \cos \phi(t, \tilde{a}) \, dt.$$

The last two are independent white Gaussian random variables with zero mean, which are defined through the formula $n(t) = \sqrt{2} n_c(t) \cos \omega_0 t - \sqrt{2} n_s(t) \sin \omega_0 t$.

In receiver (5.4-11), an error occurs if the $\tilde{a}$ that maximizes $\lambda(a, \tilde{a})$ satisfies $\tilde{a}[1] \neq a[1]$. There may be many sequences $\tilde{a}$ giving correct estimates of $a[1]$. For large E_b/N_0, only the $\tilde{a}$, among those with $\tilde{a}[1] \neq a[1]$, which gives a value $\lambda(a, \tilde{a})$ closest to the value $\lambda(a, a)$, significantly contributes to the probability of error. For large SNR we thus have

$$p_e \sim P[\lambda(a, a) < \lambda(a, \tilde{a})], \tag{5.4-13}$$

where a and $\tilde{a}$ satisfy $\tilde{a}[1] \neq a[1]$ and are chosen to minimize the noiseless value of $\lambda(a, a) - \lambda(a, \tilde{a})$.

Both $\lambda(a, a)$ and $\lambda(a, \tilde{a})$ are Rician random variables [38], and in Eq. (5.4-13) we have the problem of calculating the probability of one Rician variable exceeding another. This probability is found with the aid of the Marcum Q function, and after considerable calculation (summarized in [35] and [1], pp. 391–392), one arrives at the following. The asymptotic probability p_e has form

$$p_e \sim Q\left(\sqrt{d_e^2 E_b/N_0}\right) \tag{5.4-14}$$

in which

$$d_e^2 = \log_2 M \; \frac{\zeta_N^2}{\xi_N + (\xi_N^2 - \xi_D \zeta_N^2)^{1/2}}. \tag{5.4-15}$$

Here the following are defined:

$$\xi_N = (N_1 + N_2 + \beta)[(N_1 + N_2)^2 - Z^2(a, \tilde{a})]$$
$$+ (\beta^2/2)[N_1 + N_2 - C(a, a)],$$
$$\xi_D = (N_1 + N_2)^2 - Z^2(a, \tilde{a}),$$
$$\zeta_N = (N_1 + N_2)^2 - Z^2(a, \tilde{a}) + \beta[N_1 + N_2 - C(a, \tilde{a})],$$
$$Z^2(a, \tilde{a}) = C^2(a, \tilde{a}) + S^2(a, \tilde{a}); \quad \beta = \alpha/(E_s/N_0).$$

Continuous-phase Modulation Coding

d_e is only a function of $\Delta a = a - \tilde{a}$, since $C(a, \tilde{a})$ and $S(a, \tilde{a})$ are so. In the limiting cases of coherent and noncoherent detection, d_e^2 becomes

$$\lim_{\alpha \to \infty} d_e^2 = [N_1 + N_2 - C(a, \tilde{a})] \log_2 M \quad \text{(coherent)},$$
$$\lim_{\alpha \to 0} d_e^2 = [N_1 + N_2 - Z(a, \tilde{a})] \log_2 M \quad \text{(noncoherent)},$$
(5.4-16)

and it can be seen from Eq. (5.2-1) and the definition of $C(a, \tilde{a})$ that d_e coincides with the ordinary Euclidean normalized distance d in the coherent case. An interesting property of the noncoherent case is that $d_e = 0$ when two phase trajectories are parallel and offset; the two are indistinguishable and need to differ in shape somewhere in $[-N_1 T, N_2 T]$ to give $d_e > 0$.

The receiver here is quite complex, and there have been studies of simpler approximations to it. A VA-like adaptation with good performance was given in [1, pp. 410–411].

Properties of the Partial Coherence Distance d_e

For large SNR we can thus conclude that error probabilty under partial coherence has the asymptotic form

$$p_e \sim Q\left(\sqrt{d_{e,\min}^2 E_b / N_0}\right)$$

in which the distance-like parameter $d_{e,\min}$ obtained for the interval $[-N_1 T, N_2 T]$ plays the role of normalized Euclidean minimum distance $d_{\min}$ in coherent CPM. A large number of results are known for $d_{e,\min}$ and a summary appears in [1], Sections 10.2–10.4. We can list some main points here and then give a few details. The CPM scheme is M-ary, full or partial response, but is fixed-h.

1. An upper bound $d_{e,B}^2$ on the normalized $d_{e,\min}^2$ may be constructed by using Δa associated with mergers, just as was done in Section 5.2.1. $d_{e,B}^2$ depends on N_1, N_2 and the normalized Tikhonov parameter β.

2. Weak modulation indices force valleys in the bound plots $d_{e,B}^2$ vs h, just as in coherent CPM. As the Tikhonov $\beta \to \infty$ and $N_1, N_2 \to \infty$, it is generally true that $d_{e,B}$ tends to $d_{c,B}$ except at weak indices, where the subscript c denotes the bound for the coherent CPM scheme.

3. For large enough N_1 and N_2, the actual $d_{e,\min}$ in general also approaches $d_{c,B}$, regardless of β. Again, the result is not always true at weak indices.

4. For finite β and $N_1 = 0$ (partial coherence and no observation before the detected symbol), $d_{e,\min}$ often cannot be driven to $d_{c,B}$ at some or all h. For example, with binary CPFSK, $d_{e,\min}^2$ cannot exceed $d_{c,B}^2/2$, for any β or N_2. Thus pre-observation is a hallmark of optimal partially coherent detection.

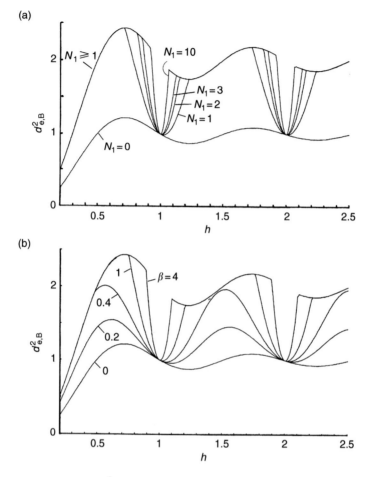

Figure 5.48 (a) Upper bound $d_{e,B}^2$ on the normalized squared equivalent distance for partially coherent binary 1REC, plotted against modulation index h; $N_1 = 0\text{--}3, 10$, $N_2 = \infty$. (b) The bound when $N_1 = 0$, $N_2 = \infty$, and coherence parameter $\beta = 0\text{--}4$. (Adapted from [1].)

First, some further detail on point 2. The convergence of bound $d_{e,B}$ to $d_{c,B}$ for binary CPFSK is shown as a function of h in Fig. 5.48(a) and (b). In Fig. 5.48(a), $\beta = 0$ (noncoherence), total observation is 10 intervals, and pre-observation is $N_1 = 0\text{--}3$. $d_{e,B} = d_{c,B}/2$ holds always when $N_1 = 0$; otherwise, $d_{e,B} = d_{c,B}$ for some h, with the valleys around the weak indices narrowing as N_1 grows. In the limit $N_1 \to \infty$, the two bounds are equal except at weak indices. The same behavior occurs for $M = 4, 8$ except that there are more weak indices and a larger N_1 is needed to force the bounds equal. In Fig. 5.48(b), $N_1 = 0$ (detection of the first observed symbol) and the degree of coherence is varied over $\beta = 0, 0.2, 0.4, 1, 4$.

Continuous-phase Modulation Coding

As the coherence grows, $d_{e,B}$ rises toward $d_{c,B}$, and when $\beta = 0.4$ there are regions where $d_{e,B} = d_{c,B}$. With enough coherence, the bounds are the same except near weak indices. The actual minimum distances are the same as well, and this shows that with CPFSK and enough coherence, pre-observation is unnecessary.

Some further detail on the relation between the actual $d_{e,\min}^2$ and the $d_{e,B}$ and $d_{c,B}$ bounds, points 3 and 4, are shown in Figs 5.49(a) and (b). The CPM scheme is again binary CPFSK, this time with noncoherent detection ($\beta = 0$) and post-observation of 1–6 intervals. In Fig. 5.49(a), the pre-observation is $N_1 = 0$;

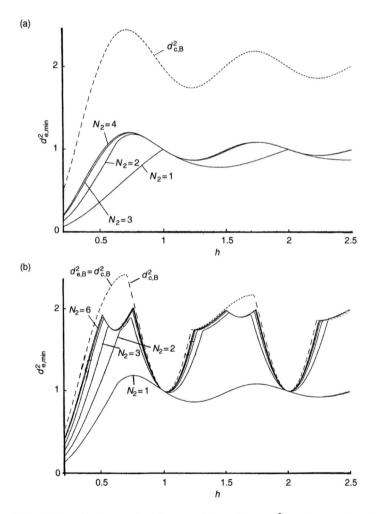

Figure 5.49 (a) Normalized squared minimum equivalent distance $d_{e,\min}^2$ for noncoherent binary 1REC ($\beta = 0$); $N_1 = 0$, $N_2 = 1$–4. Coherent upper bound shown for comparison. (b) The same, when N_1 is increased to 1. (Adapted from [1].)

the actual $d^2_{e,\min}$ grows with N_2, saturating at $N_2 = 4$, but it never exceeds half of $d^2_{c,B}$. In Fig. 5.49(b), the pre-observation is increased to one interval; $d^2_{e,\min}$ is much improved and in fact both $d_{e,B}$ and the actual $d^2_{e,\min}$ for $N_2 > 3$ are very close to $d_{c,B}$ in the region $h < 0.52$.[16] The weak indices still have a considerable negative effect. Distance continues to improve when N_1 increases to 2. The behavior shown here for binary CPFSK is similar with higher alphabet sizes, except that there are more weak indices. It is interesting from Fig. 5.49(b) that $d^2_{e,B} \approx 2.43$ at $h \approx 0.715$; the same h optimizes distance for binary coherent CPFSK and the same distance is achieved. In fact, noncoherence leads to little distance loss when N_1 grows, and the scheme that results far outperforms coherent MSK and QPSK.

These results for CPFSK more or less repeat for partial response CPM. Figure 5.50 shows $d^2_{e,\min}$ vs h for binary 3RC with pre-observation $N_1 = 0$, $\beta = 0$

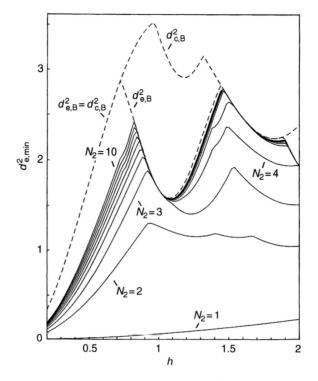

Figure 5.50 Normalized squared minimum equivalent distance $d^2_{e,\min}$ for noncoherent binary 3RC; $N_1 = 0$, $N_2 = 1$–10, $\beta = 0$. Upper bounds $d^2_{e,B}$ and $d^2_{c,B}$ shown for comparison. (From [1].)

---

[16] The shortfall between $d_{e,B}$ and $d_{c,B}$ in the h region [0.52,0.8] can be mostly removed by considering more Δa in the bound process.

(noncoherence), and $N_2 = 1$–10. The bounds $d_{e,B}$ and $d_{c,B}$ are shown, and unlike the case with CPFSK, $d_{e,B} = d_{c,B}$ holds over some ranges of h; $d_{c,B}$ can actually be achieved in these regions with large enough post-observation. This shows that performance under partial coherence depends on CPM pulse shape in a way that it does not under full coherence.

For a given total observation width, it is interesting to study how to best allocate the width between pre- and post-observation. This is done in [1], Section 10.4. Osborne and Luntz [9] state that it is ML-optimal to detect the middle data symbol. When the detection measure is the distance parameter d_e, it is relatively easy to compute the best N_1 and N_2, given a fixed $N_1 + N_2$. As a rule, it does not matter much from a d_e point of view which is the detected symbol position, so long as it is not very near the beginning or end. As the coherency drops, the best position for the detection moves from the beginning toward the middle.

5.4.3. CPM Phase Synchronization

The basics of synchronization were reviewed in Section 2.6. For coherent detection, the carrier with correct frequency and phase as well as the data symbol transition instants must be available. In many systems, these are provided in the form of training sequences that are transmitted before or during the main transmission. Otherwise, one of the circuits like those in Section 2.6 must be used. We will concentrate here on the carrier phase problem. For present purposes, we can divide phase synchronizers into three loose categories. In the first, a power loop generates a phase reference that is provided to the detector "open loop"; that is, the results of demodulation with this reference are not fed back to help in the phase estimation, although the phase estimator itself may be closed loop. A fourth power loop example is shown in Fig. 2.23. In the second category, detected symbols are used to generate a CPM signal whose phase is compared to the received signal, and this error signal helps drive the synch loop. These systems are called *data aided*. In the sense just defined, they are closed loop synchronizer systems. An example is the remodulation loop in Fig. 2.24. The third category comprises joint estimators that form an ML or MAP estimate of both the phase and data. To date, they have been only a theoretical topic.

An exhaustive treatment of CPM phase synchronization, including a systematic exposition of most known methods with comparison to the ML solution, has been given in the thesis by Macdonald [23].

Open Loop Reception with Raised Power Synchronizers

These schemes work by producing a carrier phase estimate and passing it on to the detector, without provision for feeding back the data outcome. The usual way of producing the carrier estimate is a nonlinearity, normally a simple raising to a power, followed by a standard phase-lock loop (PLL). The PLL requires a spectral line, but CPM signals almost never contain discrete spectral components. The condition for such a PSD component is Eq. (5.3-18), namely $|C_\alpha| = 1$, and the only reasonable condition leading to this is that the index h be an integer. For a rational index of the form q/p, q and p integers, raising the CPM signal to the pth power will create a new harmonic CPM signal with a p-fold increased index and p times the carrier offset. It will have the desired line. There will be other harmonics of the CPM signal, which can be filtered away, and a marked increase in the in-band channel noise, which cannot. This noise is generically called "squaring noise" and grows worse with p. In addition to the random noise, there will be in-band CPM signals other than the line. This is the "data noise." All this interference needs to be removed as much as possible by some narrowband apparatus. Most often it is a PLL, but with burst transmission, such as occurs in TDMA, it can be advantageous to use a straightforward, narrow bandpass filter and take a carrier reference tone directly from its output.

Before considering the overall synchronizer, we focus on the PLL element. In the analog version of Fig. 2.21, the circuit tries to lock a VCO to a target discrete spectral component. Once near lock, the PLL generates a phase reference whose jitter has variance

$$\sigma_\psi \approx \frac{B_N T}{E'_s/N'_0}, \qquad (5.4\text{-}17)$$

where B_N is the loop noise-equivalent bandwidth, N'_0 is the PSD of the random noise after the raising to a power, and E'_s is the new symbol energy (this equation is derived from (2.6-7) with the substitution $A^2 = 2E'_s/T$). The symbol-normalized noise equivalent bandwidth $B_N T$ is the most important parameter of the PLL, and it is effectively the reduction in the post-nonlinearity SNR afforded by the PLL. The size of $B_N T$ depends on the stability of the channel: a rapidly varying carrier reference requires a large $B_N T$ if the PLL is to track properly and a stable carrier allows a small one. Thus the viability of the raised power method depends on the interaction of three factors, the integer p, the signal SNR E_s/N_0, and the channel stability.

Very often data noise dominates random noise, and then the ratio E'_s/N'_0 which appears at the PLL input can be replaced approximately by the measure

$$\text{CNR} \triangleq 10\log_{10} \frac{b_k}{\int_\mathcal{I} T S_{\text{con}}(f_k + f)\mathrm{d}f}. \qquad (5.4\text{-}18)$$

Here $S_{\text{con}}(f)$ is the continuous part of the baseband PSD, b_k is the square magnitude of a target spectral line at $hk/2T$ Hz, k an integer, $\mathcal{I}$ is the interval $[f_k - W/2, f_k + W/2]$, and W is the positive-frequency bandwidth of a bandpass filter that comes before the PLL. If there is no PLL, then Eq. (5.4-18) is the SNR of the output carrier reference tone $\cos[(\omega_0 + 2\pi hk/2T)t]$.[17] A spectrum calculation

---

[17] The locations of discrete PSD lines are given in Eq. (5.3-21). If h in the pth power signal is odd, lines appear at odd multiples of $1/2T$, and for even h, at even multiples. See [1, Section 9.1].

program can plot Eq. (5.4-18) as a function of WT for a CPM scheme and show what synchronizer bandwidths are practical. The conclusions are approximate but instructive. Figure 5.51 shows a plot of CNR for binary 3RC when h has been raised to the value 2. In this example there are spectral lines at offsets $fT = 0$ and ± 1 from the carrier. It appears that the $1/T$ tone is usable with a filter bandwidth as large as $WT = 0.1$ (corresponding to a PLL $B_N T$ of about 0.05), and the 0 tone is usable to about twice this.

Now to circuits. The first such CPM synchronizer was proposed for MSK in the form of de Buda's FFSK [7] in 1972, and in fact he viewed the synchronizer as the central contribution of the paper. The idea has evolved [40,41] to a form applicable to any CPM scheme, which is shown in Fig. 5.52. Let the original CPM have index k_1/k_2. After raising to a power k_2, the signal is filtered by bandpasses centered at $k_2(f_0 \pm f')$ Hz, $f_0 = \omega_0/2\pi$; a multiplication then produces sum

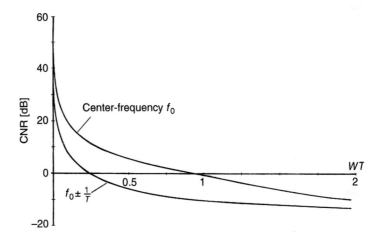

Figure 5.51 CNR measure vs WT for binary 3RC, $h = 2$; CNR shown for spectral lines at carrier f_0 and $f_0 \pm 1/T$. (Adapted from [1].)

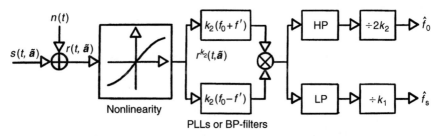

Figure 5.52 General form of a raised power synchronizer for CPM, without data feedback. The power is k_2, where $h = k_1/k_2$. Outputs are the transmitter carrier estimate $\hat{f}_0$ and $\hat{f}_s$, the estimate of $1/T$. (From [1].)

and difference tones at $2k_2 f_0$ and $2k_2 f'$ Hz. In most cases the strongest PSD line occurs at $f_1 = h/2T = k_1/2k_2T$ Hz. For this line and f' set to f_1, the latter multiplier tone has frequency k_1/T. Dividing the first multiplier tone by $2k_2$ and the second by k_1 produces respectively the carrier and the symbol rate. There is a π/k_2-fold ambiguity in the carrier phase, but this is usually of no consequence, because the CPM is impervious to such shifts.

Data-aided Synchronizers

A stronger approach is to regenerate the clean constant-envelope CPM signal and multiply it against the delayed received $r(t)$, as sketched in the remodulation loop of Fig. 2.24. This eliminates the squaring noise, and more importantly, the data noise, and allows synchronization at a much lower E_b/N_0. There is, however, the signal delay, and effectively the data detection/CPM remodulation must project the synchronizer phase estimate at time nT into the future at least as far as $(n+1)T$ in order to regenerate the signal. In reality the delay is worse, because CPM is a coded modulation and its detection requires more than one interval. This extra delay can be as small as one symbol ($2T$ total) for CPFSK, but can grow to 10 or more with more complex CPM. Excess delay slows the synchronizer response to phase transients. Fortunately, the detection can be arranged so that an early, less reliable data estimate for the synchronizer can be tapped off at extra delay D, with the true customer symbol output taking place much later. The synchronizer can accept less reliable data because it must contend with other, more serious disruptions. Macdonald [23] has thoroughly investigated the early-tap remodulation idea. As a general rule, he finds that this kind of CPM synchronization works best at $D = 1$, that is, a phase projection of $2T$ into the future.

The first data-aided CPM synchronizer proposed seems to be the multi-h method of Mazur and Taylor [42] (alternately, see [1], Section 9.1.5). It contained a Viterbi detector whose outputs drove a digital PLL; the extra delay D was zero (immediate data feedback). The authors assume perfect symbol and index ("superbaud") timing and report simulations for the case when the only channel disturbance is Gaussian noise. Simulations for the 1REC 2-h scheme (3/8, 4/8) and 4-h scheme (12/16, 10/16, 11/16, 8/16) show a BER that is little degraded over most of its E_b/N_0 range. The authors also give an analysis of the phase jitter.

Later, Premji and Taylor [45,46] proposed the general structure in Fig. 5.53 for CPM synchronization (the figure actually shows a slight rearrangement due to Macdonald). The reconstructed signal is multiplied against a one-symbol delayed $r(t)$ to produce the usual phase error signal $A \sin(\Delta \psi)$. This signal is integrated over one symbol interval and the value held for one interval in a zero-order hold. After this comes an integrator (equivalent to a VCO) and the usual loop filter $F(s)$. The Premji method has advantages in a digital implementation, which is the means of choice at any but the highest speeds, but the projection forward of the phase reference is as much as two intervals, even with

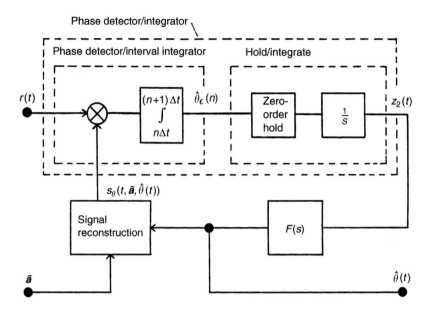

Figure 5.53 Macdonald's version of the Premji/Taylor data-aided CPM synchronizer. (From [23], copyright 1991 by A. Macdonald, used with permission.)

no extra delay in the detection. Fortunately, many applications allow and require a slow-acting synch loop. Premji [45] also thoroughly investigates simultaneous phase and symbol time acquisition in multi-h CPM.

Another type of data-aided synchronizer was proposed by Ascheid *et al.* [43,44]. According to Macdonald [23, Chapter 5], this family of methods is equivalent to Premji, with no-extra-delay detection and a zero-order hold, except for the phase detector, which is designed from Kalman and ML principles, rather than taken simply as the multiplier/integrator in Fig. 5.53. The method produces a sequence of phase estimates, one for use in each interval. The basic operation in the ML phase detector is to produce an $(n+1)$st estimate of the transmitter phase offset

$$\tan^{-1} \frac{\int r(t) \sin[\omega_0 t + \phi(t, \tilde{a}) + \hat{\psi}_0] \, dt}{\int r(t) \cos[\omega_0 t + \phi(t, \tilde{a}) + \hat{\psi}_0] \, dt}. \qquad (5.4\text{-}19)$$

Here $\tilde{a}$ is the decoded symbol stream to date, $\hat{\psi}_0$ is the nth estimate of the transmitter offset, and the integrals are over $[nT, (n+1)T]$. Ascheid *et al.* also study symbol time extraction and how often to sample $r(t)$ and the CPM signals; they find that rates as slow as $4/T$ are acceptable for synchronizers.

Yet another data-aided approach is that of Simmons and McLane [47]. Here two tentative Viterbi decoders run simultaneously with phase estimates that differ by $\Delta\theta$; after some time the reference in the VA with the best metric is declared to be the new primary reference, and the two tentative VAs take new phases relative to this. This idea of pursuing multiple phase assumptions is extended further in Macdonald and Anderson [48], where a synchronizer is associated with *each* VA state. Each data-dependent synchronizer operates with zero delay, using the best data path leading to its node. Digital PLLs are cheap to replicate, but the method can be simplified by supplying only the 2–3 best nodes with synchronizers. Tests of 3RC CPM under challenging phase transients show that the 2-synchronizer method has BER performance that lacks only a fraction of a dB compared to a data-aided synchronizer with known data [23, Chapter 9].

5.4.4. Transmitters

For all but the fastest applications, CPM signals are probably best generated from sampled and quantized I and Q waveforms stored in binary ROMs. A library of one-interval signal pieces can be maintained, and the ongoing signal constructed an interval at a time. The theory and technology of signal sampling is covered in digital signal processing texts, and we will leave most of the subject to them. A few guidelines will suffice. (i) As a rule, the sampling rate in a receiver can be low, because the approximation does not much affect the distance calculation, but sampling must be considerably faster in a transmitter, since otherwise the signal spectral sidelobe properties will be destroyed. When a burst begins and ends, one must be careful not to truncate signals in time, for the same reason. (ii) Once a digital sequence is down-converted to an analog waveform, it must be lowpass (bandpass if at passband) filtered in order to set the final spectral sidelobes.[18] This step cannot usually be avoided, and can in fact be turned to an asset, since with careful final filtering, signals can be synthesized with less digital precision. (iii) Since CPM signals are constant envelope, the remainder of the transmitter can be standard class C amplifier components.

Guidelines (i) and (ii) apply equally to generation of TCM and PRS signals. We will give no separate treatment of transmitters under those headings. However, it should be emphasized that with all linear pulse modulation systems, time truncation is especially damaging to the sidelobe spectrum. This can mean that quite a long data symbol history is required in order to generate the waveform for an interval. This is a major factor in ROM size.

For ROM-based CPM transmitters, 4–8 bits per I and Q waveform sample and a sampling rate of $4/T$–$8/T$ are usually enough. Further details about this appear in [1, Chapter 6]. From Section 5.2.3, Eq. (5.2-14), there are in theory p phase shifts of $2M^{L-1}$ baseband interval waveforms to store in the ROM. Often

---

[18] Of course, the repeating spectral replicas in the reconstruction from discrete time must be filtered out as well.

Continuous-phase Modulation Coding

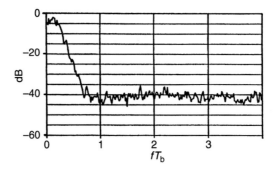

Figure 5.54 Measured power spectrum for $h = 1/3$ quaternary 2RC realized from stored I and Q samples, with sampling rate $16/T$ and 4 bits/sample. Sample quantization produces a noise floor. (Adapted from [1].)

this can be reduced by various symmetries, and in particular a storage of $2M^{L-1}$ waveform pieces can be combined with a separate p-fold phase shifting circuit. Assuming this is done, we get a total ROM size of 24–64 × $2M^{L-1}$ bits, or around 2 kbits for quaternary 3RC. Figure 5.54 shows a typical reconstruction spectrum, in this case for quaternary 2RC, $h = 1/3$, 16-fold sampling and 4 bits/sample. There is a spectral noise floor at about 35 dB below the mainlobe peak, caused by the 4-bit precision. A rule of thumb states that 6 dB reduction in the floor is gained for each bit/sample in the reconstruction. Thus 8 bits lowers the floor another 24 dB. These floor positions change little for different CPMs.

High-speed Analog Implementations

A straightforward frequency modulation of a carrier by the CPM excess frequency signal $2\pi h \sum a[n]g(t - nT)$ is unsuccessful because drift makes the CPM phase trellis framework difficult to maintain. A transmitter structure based on an ordinary PSK modulator followed by a PLL with a certain loop filter [1, Section 6.2, 49] provides a reasonably simple and accurate high-speed analog implementation. The building blocks are shown in Fig. 5.55. A rectangular-pulse M-ary PSK modulator is the input to the loop phase detector. A VCO tries to track this PSK phase, but misses (hopefully) by an amount that makes its output the desired CPM signal; the filter $F(s)$ is tuned to optimize this outcome. The phase detector can be an RF mixer, and all the loop components need to have a wide enough phase response for the h.[19] A derivation [1, Section 6.2] shows that $F(s)$

---

[19] For the basic circuit in Fig. 5.55, the limits are around $h \approx 0.5$ for binary and $h \approx 0.1$ for quaternary CPM; but CPM at higher indices can be synthesized at a lower index and then multiplied up k-fold, k an integer.

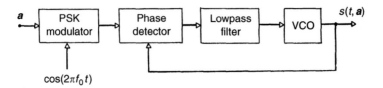

Figure 5.55 PSK followed by PLL generation of CPM. The PLL smoothes rectangular PSK transitions to form CPM.

should satisfy the relationship

$$\mathcal{L}\{q(t)\} \approx \frac{F(s)/2}{s[s + F(s)]}, \qquad (5.4\text{-}20)$$

where $q(t)$ is the CPM phase response pulse and $\mathcal{L}$ indicates the Laplace transform operation.

Another approach is based on a PSK modulator or a simple CPM (e.g. CPFSK) modulator, followed by a bandpass filter and a hard limiter. It can be shown that the bandpass filter to a first approximation acts as a filter on the baseband phase when h is small. Whatever h is, the hard limiter ensures that this implementation produces some sort of CPM signal. In any case, it is only necessary to approximate the CPM signal, and this is aided by a rough method for computing the spectral effect of a filter/limiter combination. That analysis and details of the transmitter appear in [1, Sections 4.5 and 6.3].

Several special transmitter implementations exist for the case of MSK.

5.5. Simplified Receivers

The previous section has introduced optimal CPM receivers of several types. These receivers are sometimes rather large and our aim now is to explore ways to reduce this complexity. Throughout this section coherent detection is assumed, with perfect carrier and timing recovery, an AWGN channel, and IID data symbols. The problem is then equivalent to reducing CPM tree and trellis searching, and we can break that into two approaches, reducing the *size* of the trellis that is searched, or reducing the *search* of a trellis of given size. As an example of the first, the transmitter state description may contain symbols that affect the CPM signal in insignificant ways as far as detection is concerned; the receiver can utilize a trellis description that lacks these symbols. The phase variable θ_n in the standard CPM state description (5.2-15) is one such variable. In some coded modulations, older symbols affect the present-interval signal less than newer ones, a decaying of code memory over time. This effect is more pronounced in PRS coding than in CPM and TCM; we will give a CPM discussion but delay

Continuous-phase Modulation Coding

most treatment of the idea until Section 6.6. Examples of the second approach are standard sequential and reduced-search decoding, which were introduced in Section 3.4. Here, only a part of a large trellis is searched, the part where the minimum distance path is most likely to be, and one wants the smallest search that yields approximately ML error performance.

Sections 5.5.1 and 5.5.2 focus on two exemplars of the first approach, simplifying the description of the phase pulse at the receiver and retaining a compromise description of earlier signal parts. Section 5.5.3 shows some results of reduced-search decoding, using the M-algorithm. With CPM, this simple algorithm greatly reduces complexity, and it makes practical large-alphabet, long-response schemes at any index. A last section reviews special receivers for the case $h = 1/2$.

5.5.1. Pulse Simplification at the Receiver

The state complexity of a CPM scheme depends exponentially on the length of its phase response pulse $q(t)$. A long response is desirable because it tends to reduce bandwidth without much loss of distance; the joint energy–bandwidth of the scheme is thus improved. But only the transmitter needs to use the complex narrowband pulse. If a way can be found to base the receiver on a shorter, simpler pulse, without costing much distance, then the receiver trellis will be reduced to a fraction of the transmitter trellis. This idea will be the main topic of this subsection.

The phase tree in Fig. 5.56 shows an example of such pulse simplification. The transmitted CPM is standard binary 3RC (solid tree), and superposed on it is the phase tree created by the shorter 2REC pulse (dashed). The trees are offset to achieve the best alignment and then offset slightly more so that they are easier to distinguish. It is clear that a 2REC receiver tree with the proper phase and time offset (zero degrees and $T/2$ s in this case) approximates the 3RC scheme very well. Receiver trellis state complexity, counting correlative states, drops from M^2 to M, a reduction by 1/2. Receiver error performance cannot have dropped much, because the phase trellises hardly differ, although the association of data symbols with paths does differ in the two trellises.

The two signal sets in Fig. 5.56 are said to be mismatched, and the new error performance can be analyzed with the aid of the *mismatched Euclidean distance*. The notion appears from time to time in CPM analysis and elsewhere, for example, in studying filtered or distorted signals [1, Section 3.5.2], and this is a good place to review it. Let the transmitter send the signal set

$$s_T(t, a) = \sqrt{2E_s/T} \, \cos[\omega_0 t + \phi_T(t, a)], \qquad (5.5\text{-}1)$$

generated by phase pulse $q_T(t)$, but the receiver compares $r(t)$ to the signal set

$$s_R(t, a) = \sqrt{2E_s/T} \, \cos[\omega_0 t + \phi_R(t, a)], \qquad (5.5\text{-}2)$$

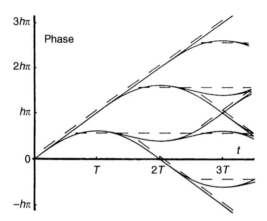

Figure 5.56 Comparison of the phase trees of binary 3RC (*solid*) and 2REC shifted $T/2$ (*dashed*), showing a close fit. (From [1].)

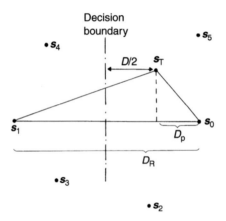

Figure 5.57 Receiver signals plus s_T in signal space. s_T becomes s_0 in the receiver signal set. (From [1].)

generated by pulse $q_R(t)$. Both observe a time interval of length NT. Presumably, q_R is a shorter pulse than q_T. We view all signals in the same signal vector space. The receiver selects the signal alternative s_R that lies closest in Euclidean distance to r. Figure 5.57 shows the projection of some of these signals onto the 2-dimensional plane formed by the two receiver-based signals s_0 and s_1. Here s_0 is the receiver version of the transmitted signal s_T, and s_1 is some other receiver-version signal; other receiver alternatives are $s_2, \ldots, s_5$. These are noise-free signals, and under noise, a noise vector η is added to s_T. If the resultant lies to the left of the dashed decision boundary, an error occurs. This particular error,

Continuous-phase Modulation Coding

confusing s_1 for s_0, can occur if $\|\eta\| \geq D/2$, where $D/2$ is the perpendicular distance parameter defined on the figure.

The minimum of the D parameter, over both the receiver signals s_i and the transmitter signals s_T, sets the asymptotic error performance. Some manipulations express D in terms of the signals. Simple geometry gives that $D/2 = D_R/2 - D_P$, where

$$D_R^2 = \|s_1 - s_0\|^2, \qquad D_P = \frac{\langle s_T - s_0, s_1 - s_0 \rangle}{\|s_1 - s_0\|}.$$

Here $\langle \cdot \rangle$ is the inner product and D_P is the projection of $s_T - s_0$ onto $s_1 - s_0$. Consequently, D is given by

$$D = \frac{\|s_1 - s_T\|^2 - \|s_0 - s_T\|^2}{\|s_1 - s_0\|}. \tag{5.5-3}$$

The minimum of Eq. (5.5-3) over all s_T and receiver pairs s_1, s_0 yields a worst case distance D_m. We normalize D_m as usual to form $d_m = D_m/2E_b$, and the asymptotic error probability is then $\sim Q(\sqrt{d_m^2 E_b/N_0})$. It can be shown that $d_m^2 \leq D_{\min,T}^2/2E_b$, where $D_{\min,T}$ is the minimum distance among T-signals; this simply says that asymptotic performance cannot be improved by misforming the receiver signals. It can also be shown that $d_m^2 \geq D_{\min,R}^2/2E_b$, in which $D_{\min,R}$ is the minimum distance among signals seen at the receiver. This means that changing the transmitter while keeping the receiver the same can possibly improve performance. Some detailed examples that illustrate these facts are given in [52].

Solving for the best mismatched receiver of a given pulse length is relatively time consuming, since the optimization of D in Eq. (5.5-3) depends on the shape of the pulse $g_R(t)$ and can have a different outcome for each index h. One must also test different time and phase offsets for the receiver set. Details appear in [1], Sections 8.1.4–8.1.6, and in original references [50–53]. Table 5.4 gives an idea of the savings possible with different CPM schemes and shortenings of the phase pulse. In general, one can say that reductions of the correlative state size to 1/2–1/4 can be achieved without significant error performance loss.

Table 5.4 Asymptotic E_b/N_0 degradation vs state complexity reduction for pulse CPM shortening, at modulation index 3/4. L_T and L_R denote transmitter and receiver pulse lengths. (From [1].)

Transmitted scheme	$M=2$ 2RC	$M=2$ 3RC	$M=2$ 4RC	$M=2$ 3RC	$M=2$ 4RC	$M=4$ 2RC	$M=2$ 4RC
$L_R - 1$	1	2	3	1	2	1	1
Degradation [dB]	0.13	0.02	0.02	0.86	0.36	1.28	2.40
$M^{L_T - L_R}$	2	2	2	4	4	4	8

There are several other effective ways of simplifying the state description of CPM, which are based on reducing its signal space description. They work because even though CPM signals can occupy many signal space dimensions, most of the signal energy is in just a few; in the same spirit as above, the extra complexity can be ignored at the receiver but not at the transmitter. Huber and Liu [54] introduced the idea of detecting with only a few dominant signal space dimensions. Simmons [56] proposes a method composed of a simple bandwidth-B lowpass filter on the signal I and Q lines, followed by a sampler at rate $2B$. The trellis decoding is performed on the discrete-time sample stream rather than on the original I and Q signals. Nearly optimal detection is achieved over the range of useful h with sampling rates in the range $4/T$–$8/T$, even for CPM as complex as 3RC. The method combines easily with the reduced-search methods in Section 5.5.3.

5.5.2. The Average-matched Filter Receiver

An average-matched filter (AMF) is a filter matched to the average of a set of signals rather than to a particular one. The concept was introduced by Osborne and Luntz [9] for CPFSK schemes as a way to simplify their receiver, the one we introduced at the beginning of Section 5.4.1. The matched filterings before and/or after the detected symbol position are matched to average signals. The AMF that results is the ML symbol estimate at low E_b/N_0 and it is of course much simpler. One can hope that the simplicity is gained without much error probability loss at high E_b/N_0. The loss can be assessed through computation of an equivalent distance parameter d, which through an expression of the form $Q(\sqrt{d^2 E_b/N_0})$ gives the usual asymptotic error probability. A generalization of the AMF idea is to match the receiver to an average over part of the signal set or part of the observation time, and have a complete filter ensemble otherwise.

We will now briefly review the AMF idea. Many further details, especially the extension beyond CPFSK and an exposition on the d parameter, are available in section 8.4 of [1].

Suppose that the receiver observes the signal over N intervals spanning $[-NT, 0]$ and seeks to decide the first transmitted symbol $a[0]$. The pure AMF receiver is a bank of M filters, the first matched to the average of all signals with $a[0] = -(M-1)$, the second to the average of all with $a[0] = -(M-3)$, and so on to $a[0] = (M-1)$. The filter outputs are sampled each symbol interval and the estimate of $a[0]$ taken from the filter with highest correlation. For the extension to partial response schemes, experience has shown that there should be M^L filters, matched to the correlative variables $a[-L+1], \ldots, a[0]$, where LT is the frequency pulse width and the variables $a[-L+1], \ldots, a[-1]$ are the result of previous symbol decisions; the averaged signal segment is the average of the signals leading to these variables.

The receiver can be further generalized to a *multiple-symbol* AMF receiver. Instead of M filters in the CPFSK case, M^{N_A} can be used, where M^{N_A-1} of these assume the same value of $a[0]$ and with each having a precise match to a set of N_A symbols. $a[0]$ may be located before, after, or during the N_A symbols. Note that some average-matched responses may be phase rotations of or even identical to others, a fact that can further simplify the receiver. Sometimes a nonuniform weighting of the responses in the averaging can improve performance, although it is difficult to identify such weightings except by trial and error.

A careful analysis (section 8.4.2 of [1]) under AWGN gives the d parameter. With binary CPFSK and the minimum precise-match width ($N_A = 1$), d can be brought reasonably near the standard upper distance bound $d_{c,B}$ when $h < 0.5$ and the total observation width N lies in a certain middle range, typically 2–3. Longer observation, hence longer averaging, can dramatically reduce d. For $N_A > 1$, d approaches $d_{c,B}$ much more closely. The behavior is roughly similar for binary partial response systems, for example, 3RC, but it is more difficult to drive d toward $d_{c,B}$; the AMF idea does not work well with quaternary partial response. Tests of actual receivers show error rates reasonably close to $Q(\sqrt{d^2 E_b/N_0})$ when $N_A > 1$. In general, it can be said that AMF receivers work well for short pulses, binary transmission, low h, and an averaging that extends over only a few intervals.

5.5.3. Reduced-search Receivers via the M-algorithm

A reduced-search receiver examines only a part of the code trellis, the part that hopefully lies closest in distance to the transmitted trellis path. It is usually the case that the error performance of such a receiver is substantially that of a full Viterbi receiver, since the probability is very small that the correct path wanders far away from the apparent best path at the receiver. Almost all the tests of this idea with CPM codes have been performed with the M-algorithm or its near relatives. This scheme and some background are discussed in Section 3.4.1. In brief, the algorithm proceeds without backtracking through the trellis and maintains constant M paths in contention. The tests reported in this section are for CPM without filtering; reduced-search decoding of filtered CPM is well studied, but this subject is covered in Section 6.6.3.

Figure 5.58 shows one of a series of tests by Aulin [55], which though unpublished, were among the first to be performed (1985). It shows M-algorithm decoding of quite a sophisticated CPM, 4-ary 3RC with $h = 1/3$. Seven retained paths are enough to achieve VA symbol error performance. This can be compared to the standard VA state description (5.2-15), which leads to $6 \cdot 4^2 = 96$ states, or if only correlative states are counted, 16 states. At about the same time, Seshadri [59] published M-algorithm tests for CPFSK and 2RC CPM schemes that were filtered. His tests with weak filtering give an indication of M-algorithm CPM performance

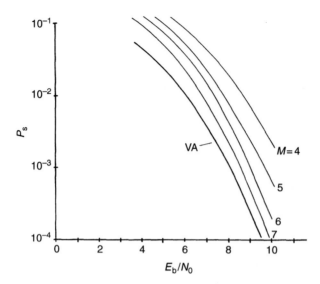

Figure 5.58 Observed symbol error performance vs E_b/N_0 for M-algorithm decoding of quaternary $h = 1/3$ 3RC CPM over the AWGN channel. Number of paths kept, M, as shown. A VA estimate is based on code minimum distance.

when there is no filtering, and these showed that 4–8 retained paths were generally enough to attain MLSE error performance for CPFSK, 1RC and 2RC over a range of modulation indices. This outcome roughly confirms Fig. 5.58. (For M-algorithm results for CPM under severe filtering, see Section 6.6.3.)

A series of papers by Svensson, Palenius and Simmons [56–58] in the early 1990s showed that for CPFSK schemes with a variety of indices and alphabet sizes, the M-algorithm needs to retain only *two* paths.[20] This holds for $h \lesssim 0.5$ for 2-ary, $h \lesssim 0.35$ for 4-ary, and $h \lesssim 0.28$ for 8-ary CPFSK [57]. These authors also look at more complex CPMs, with the same outcome as in the previous paragraph. As well, they apply the M-algorithm and its relatives to simplified trellis descriptions of CPM, such as those at the end of Section 5.5.1, with a similar outcome.

An implicit conclusion in all these results is that the needed number of retained paths is not very sensitive to the modulation index h. The reduced search in a sense removes most of the dependence of the CPM state size on the phase states. With CPFSK and small, practical h, the retained number is 2, the smallest it can be if there is to be true multi-interval decoding; more complex CPM schemes

---

[20] The Svensson–Palenius work actually employed a trellis reduction procedure called reduced state sequence detection (RSSD), which more properly belongs in Section 5.5.1, since it reduces the trellis to a smaller structure, which is then fully searched. However, it has been shown that for the same asymptotic performance, the M-algorithm on the original trellis never needs more survivors than RSSD; RSSD conclusions thus overbound the M-algorithm.

Continuous-phase Modulation Coding

seem to need a larger number, in rough proportion to their correlative states. At large h, there is a mild increase in the needed paths, but these h are bandwidth-inefficient. Discounting phase states is only reasonable, since the information encoded in a CPM signal depends on its phase differences, not its absolute phase. Reduced-search decoders are a means to do this.

5.5.4. MSK-type Receivers

When the modulation index h is 1/2, some special, simple receiver structures work well for CPM with pulses out to width 3–4. Two of these are shown in Fig. 5.59, the *parallel* and *series MSK* receivers. The first consists of an I/Q mix-down, and a special filter with response $f(t)$, and samplings that occur every $2T$ in alternate quadrature arms. A decision logic box contains differential decoding and time demultiplexing. The serial MSK receiver consists of the I/Q mix-down, two special filters $h_1(t)$ and $h_2(t)$ in the I and Q arms, a summing junction, a sampling each nT, and some decision logic.

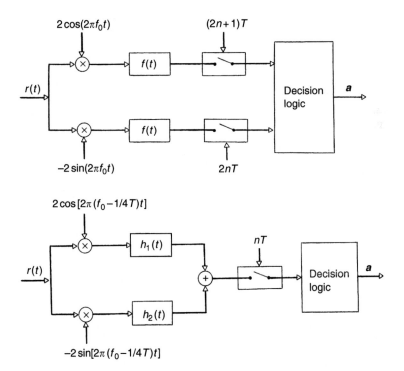

Figure 5.59 Parallel MSK (*top*) and serial MSK (*bottom*) receiver structures. (Adapted from [1].)

MSK itself can be viewed as a linear modulation, for which the standard quadrature receiver is an ML-optimal detector (see Section 2.5.3 and Fig. 2.13). Both the parallel and serial MSK receivers can be viewed as quadrature receivers which are optimal for MSK.[21] The filters in the two types can be shown to satisfy the relation

$$h_1(t) = f(t)\cos(\pi t/2T),$$
$$h_2(t) = -f(t)\sin(\pi t/2T). \quad (5.5\text{-}4)$$

When the CPM scheme is partial response or full response without the REC shape, the two receivers are not in general optimal detectors. In the same spirit as Section 5.5.2, a distance parameter d may be derived in the AWGN case, where $d \leq d_{\min}$ for the CPM scheme, and the asymptotic error probability has the form $Q(\sqrt{d^2 E_b/N_0})$. We can look for filters that yield d close to $d_{\min}$.

To give a flavor of MSK-receiver analysis, we review the high-SNR parallel MSK optimum receiver filter of Galko [60], who gave the first such analysis in 1983. We will consider just one quadrature arm and its sampler. The sampler output is a noisy ± 1 symbol, which, in the usual way, is compared to 0 in order to decide what the noise-free value was; these symbols may require differential decoding and interleaving with the other arm's outputs in order to produce the original data sequence. Denote the noisefree signals that appear in the arm as $\tilde{s}(t, \boldsymbol{a})$. The ensemble of such signals can be divided into two subsets. If t_0 is the sampling time, the subset having $\tilde{s}(t_0, \boldsymbol{a}) > 0$ will be denoted by $\mathcal{S}^+$, while the subset having $\tilde{s}(t_0, \boldsymbol{a}) \leq 0$ will be $\mathcal{S}^-$. The sets of corresponding data sequences can be denoted $\mathcal{A}^+$ and $\mathcal{A}^-$. The ith of the "+" signals in signal space is denoted by the vector s_i^+, and similarly for s_i^-, where $i = 1, \ldots, I$. The total number of signals is $2I$ owing to the fact that with $h = 1/2$ the signals in $\mathcal{S}^+$ are the negatives of those in $\mathcal{S}^-$. Gaussian noise is added to each signal component. Because of symmetry, the best detection occurs if the sampler output is compared to 0. The object is to decide whether the received $r(t)$ stems from an $\boldsymbol{a}$ in $\mathcal{A}^+$ or one in $\mathcal{A}^-$.

Galko gave the optimal receiver in vector form as

$$\boldsymbol{f} = \sum_{i=1}^{I} c_i \boldsymbol{s}_i^+, \quad (5.5\text{-}5)$$

or as an impulse response, $f(t) = \sum c_i \tilde{s}(t, \boldsymbol{a}_i)$. The coefficients $\{c_i\}$ are chosen to minimize the average error probabilty, and they depend on the SNR. Define a normalized squared distance, given that s_i^+ is transmitted, as

$$d_i^2 = \frac{2\langle s_i^+, f\rangle^2}{T\|f\|^2} = \frac{2\left[\int_{\mathcal{I}} f(-t)\tilde{s}(t, \boldsymbol{a}_i)\,dt\right]^2}{T\int_{\mathcal{I}} f^2(t)\,dt}, \quad (5.5\text{-}6)$$

---

[21] The filters $f(t)$, $h_1(t)$, and $h_2(t)$ for the MSK case are given in [1, pp. 315–317].

Continuous-phase Modulation Coding

where $\mathcal{I}$ denotes the signal observation interval. Define the parameter d by $d = \min_i d_i$. Then it can be shown that at high E_b/N_0, the symbol probability of error is $\sim Q(\sqrt{d^2 E_b/N_0})$.

Galko [60] gave a quadratic programming algorithm to solve for the optimum filter $f(t)$ at high SNR. Later, references [53,62] presented a simpler algorithm, gave a numerical algorithm to find the optimal filter at a specified SNR, and extended the solution to the Rayleigh channel (a review appears in [1, Section 8.2.2]). In these references and others, a number of optimal filters have been computed, for binary schemes such as 3RC–4RC, TFM, and GMSK with parameter B_b set so that the effective frequency pulse width is 3–5. For a wide enough filter response width, it can be said that the d parameter comes reasonably close to the CPM d_f. For widths in the range 2–4, losses of perhaps 1–3 dB occur, depending on the width. When SNR is very low, the filter solution tends to the AMF solution. At higher SNR, there is not much change in the filter solution as the SNR varies.

The serial MSK receiver can be derived from the parallel one, and we will not discuss it in detail. In a practical application, one receiver can be preferred over the other. For example, parallel behaves worse than serial when the carrier phase recovery is not exact. If the bit timing is imprecise, the opposite is true. For burst transmission, the serial receiver is said to be better. There are also electronic implementation pros and cons to consider.

5.6. Problems

1. Starting from Eq. (2.5-25), show that when two CPM excess phases are linear over an interval, the square distance contribution from that interval is Eq. (5.2-5),

$$d^2 = \begin{cases} \left(1 - \dfrac{\sin \Delta\phi_b - \sin \Delta\phi_e}{\Delta\phi_b - \Delta\phi_e}\right) \log_2 M, & \Delta\phi_b \neq \Delta\phi_e; \\ (1 - \cos \Delta\phi_b) \log_2 M, & \Delta\phi_b = \Delta\phi_e, \end{cases}$$

where $\Delta\phi_b$ is the beginning phase difference and $\Delta\phi_e$ is the ending difference. Using this formula, verify bound (5.2-4).

2. Find the h that maximizes the binary 1REC bound formula (5.2-4). What is the maximum value of d_B^2? Is it in fact $d_{\min}^2$? (This is de Buda's energy-optimal CPFSK scheme from the early 1970s.)

3. Repeat Problem 2 for quaternary ($M = 4$) 1REC CPM, by using Eq. (5.2-7); that is, find the maximizing h and the maximum value of d_B^2. (Do not analyze whether the bound is reached.)

4. Except at certain weak modulation indices, $d_{\min}^2$ for the important quaternary CPFSK class of CPM schemes is equal to the bound d_B^2 given by Eq. (5.2-7).

(a) Find the range of indices h, for $h < 2$, over which the data difference sequences $\Delta a = 2, -2, 0, \ldots, 4, -4, 0, \ldots,$ and $6, -6, 0, \ldots$ each lead to d_{min}^2. (See Fig. 5.5.)

(b) Using Eq. (5.2-6) and numerical integration, repeat for quaternary 1RC.

5. Derive an upper bound expression on the normalized Euclidean minimum distance for binary 2REC CPM, as a function of h.

6. Derive an upper bound expression on the normalized Euclidean minimum distance for $M = 2$ 3REC CPM, as a function of h.

7. Derive an upper bound expression on the normalized Euclidean minimum distance for multi-h binary 1REC CPM with two cycling indices $h_1, h_2, h_1, h_2, \ldots$. The bound should be an analytical expression in h_1 and h_2.

8. (a) List the weak modulation indices for binary ($M = 2$) 1REC CPM, for $h \leq 4$, as predicted by formula (5.2-11); (b) List the indices for quaternary 1REC CPM, for $h \leq 2$; (c) Repeat (b) for octal 1REC.

9. List the values of the weak modulation indices for binary 3REC CPM.

10. Derive the error probability estimate of the type (5.2-17), Section 5.2.3, for quaternary 1REC CPM, $h = 2/3$. Compare it with the result given after Eq. (5.2-17) for binary $h = 2/3$ 1REC.

11. Using MATLAB[22] or similar software package, make a CPM baseband spectrum calculator for your toolbox by programming Eqs (5.3-15)–(5.3-17). Assume standard CPM with equiprobable data symbols and $q(LT) = 1/2$. Your program should accept the alphabet size, index, and phase response $M, h,$ and $q(t)$, and put out the baseband spectrum $S(f)$ at a set of positive frequencies. (*Hint:* If your package works with complex variables, it will be simpler to work from Eq. (5.3-14) than from Eq. (5.3-17).)

12. Demonstrate Eq. (5.3-15); that is, show that the spectrum factor C_α for standard CPM with equiprobable data is equal to $\sin M\pi h / M \sin \pi h$.

13. The baseband spectrum of CPFSK with equiprobable data symbols may be expressed as the closed-form Eqs (2.7-12) and (2.7-13). Derive these equations for the binary case ($M = 2$) from the general standard-CPM spectrum Eqs (5.3-16) and (5.3-17).

14. (a) Consider the CPM family LRC. Show that the phase response $q(t)$ for LRC has two continuous derivatives, regardless of L. Using this information and Baker's formula (2.7-14), write an expression for the asymptotic rate of side lobe decay in this CPM family; (b) Repeat for the REC family, finding first the number of continuous derivatives of $q(t)$, then the asymptotic expression.

15. Draw a standard state trellis section for binary 3RC, $h = 4/5$. Describe in detail the operation of a standard Viterbi decoder that utilizes discrete-time sampling of I and Q waveforms. How many states are there? What sampling rate do you recommend?

---

[22] MATLAB is a trademark of the MathWorks, Inc. Further information is available from The MathWorks, Inc., Cochituate Place, 24 Prime Park Way, Natick, MA 01760, USA.

Continuous-phase Modulation Coding

16. By a series of steps like those in Section 2.3.1, derive the Osborne–Luntz optimal coherent receiver from first principles. Start from the probabilities in Eq. (5.4-1) and work to the likelihood values in Eq. (5.4-2).

17. (a) Demonstrate Eq. (5.4-16); that is, find the partial-coherence equivalent square distance $d^2_{e,\min}$ for observation $[-N_1 T, N_2 T]$, in the limit that the Tikhonov parameter α tends to 0 and to ∞; (b) Show that $\lim_{\alpha \to \infty} d^2_{e,\min} = d^2_{\min}$, the square minimum distance for the coherent CPM scheme; (c) Show that $\lim_{\alpha \to 0} d^2_e = 0$ for any $[-N_1 T, N_2 T]$, when d_e is the distance between two phase trajectories that are equal except for a constant phase offset.

Bibliography

References marked with an asterix are recommended as supplementary reading.

[1] *J. B. Anderson, T. Aulin, and C.-E. Sundberg, *Digital Phase Modulation*. Plenum, New York, 1986.
[2] *T. Aulin and C.-E. Sundberg, "CPM – An efficient constant amplitude modulation scheme," *Int. J. Satellite Commun.*, **2**, 161–186, 1984.
[3] *J. Proakis, *Digital Communications*, 3rd edn. McGraw-Hill, New York, 1995.
[4] *S. G. Wilson, *Digital Modulation and Coding*. Prentice-Hall, Upper Saddle River, NJ, 1996.
[5] *F. Xiong, *Digital Modulation Techniques*. Artech House, Boston, 2000.
[6] M. L. Doelz and E. H. Heald, *Minimum Shift Data Communication System*, US Patent No. 2917417, 28 Mar. 1961.
[7] R. de Buda, "Coherent demodulation of frequency-shift keying with low deviation ratio," *IEEE Trans. Commun.*, **COM-20**, 429–436, June 1972.
[8] M. G. Pelchat, R. C. Davis, and M. B. Luntz, "Coherent demodulation of continuous phase binary FSK signals," *Proc. Int Telemetering Conf.*, Washington, 181–190, Nov. 1971.
[9] *W. P. Osborne and M. B. Luntz, "Coherent and noncoherent detection of CPFSK," *IEEE Trans. Commun.*, **COM-22**, 1023–1036, Aug. 1974.
[10] T. A. Schonhoff, "Symbol error probabilities for M-ary CPFSK: coherent and noncoherent detection," *IEEE Trans. Commun.*, **COM-24**, 644–652, June 1976.
[11] H. Miyakawa, H. Harashima, and Y. Tanaka, "A new digital modulation scheme – multimode binary CPFSK," *Proc. Third Int. Conf. Digital Satellite Commun.*, 105–112, Kyoto, Japan, Nov. 1975.
[12] J. B. Anderson and D. P. Taylor, "A bandwidth-efficient class of signal space codes," *IEEE Trans. Inf. Theory*, **IT-24**, 703–712, Nov. 1978.
[13] J. B. Anderson and R. de Buda, "Better phase-modulation error performance using trellis phase codes," *Electronics* (Lett.), **12**(22), 587–588, 28 Oct. 1976.
[14] T. Aulin, "CPM – A power and bandwidth efficient digital constant envelope modulation scheme," PhD Thesis, Telecommunication Theory Department, Lund University, Lund, Sweden, Nov. 1979.
[15] *T. Aulin, N. Rydbeck, and C.-E. Sundberg, "Continuous phase modulation," Parts I–II, *IEEE Trans. Commun.*, **COM-29**, 196–225, March 1981.
[16] J. B. Anderson, C.-E. Sundberg, T. Aulin, and N. Rydbeck, "Power-bandwidth performance of smoothed phase modulation codes," *IEEE Trans. Commun.*, **COM-29**, 187–195, March 1981.
[17] T. Aulin and C.-E. Sundberg, "On the minimum Euclidean distance for a class of signal space codes," *IEEE Trans. Inf. Theory*, **IT-28**, 43–55, Jan. 1982.

[18] T. Aulin and C.-E. Sundberg, "Minimum Euclidean distance and power spectrum for a class of smoother phase modulation codes with constant envelope," *IEEE Trans. Commun.*, **COM-30**, 1721–1729, 1982.

[19] B. E. Rimoldi, "A decomposition approach to CPM," *IEEE Trans. Inf. Theory*, **IT-34**, 260–270, March 1988.

[20] S. Nanayakkara and J. B. Anderson, "High speed receiver designs based on surface acoustic wave devices," *Int. J. Satellite Commun.*, **2**, 121–128, 1984.

[21] T. Aulin, "Symbol error probability bounds for coherently Viterbi detected continuous phase modulated signals," *IEEE Trans. Commun.*, **COM-29**, 1707–1715, 1981.

[22] K. Balachandran and J. B. Anderson, "A representation for CPM error event probability," *Conf. Rec. 24th Conf. Information Sciences and Systems*, Princeton, NJ, March 1990.

[23] A. Macdonald, "Characterization and development of ML-based synchronization structures for trellis coded modulation," PhD Thesis, Department of Electrical, Computer and Systems Engineering, Rensselaer Poly. Inst., Troy, NY, May 1991.

[24] A. T. Lereim, "Spectral properties of multi-h codes," MEng Thesis, Department of Electrical and Computer Engineering, McMaster University, Hamilton, Canada, July 1978.

[25] W. Holubowicz and P. Szulakiwicz, "Multi-T realization of multi-h phase codes," *IEEE Trans. Inf. Theory*, **IT-31**, 528–529, July 1985; see also P. Szulakiwicz, "M-ary linear phase multi-T codes," *IEEE Trans. Commun.*, **COM-37**, 197–199, Mar. 1989.

[26] T. Aulin and C.-E. Sundberg, "Exact asymptotic behavior of digital FM spectra," *IEEE Trans. Commun.*, **COM-30**, 2438–2449, Nov. 1982.

[27] T. Aulin and C.-E. Sundberg, "An easy way to calculate power spectrum for digital FM," *Proc. IEE*, Part F, **130**, 519–526, Oct. 1983.

[28] G. L. Pietrobon, S. G. Pupolin, and G. P. Tronca, "Power spectrum of angle modulated correlated digital signals," *IEEE Trans. Commun.*, **COM-30**, 389–395, Feb. 1982.

[29] P. K. M. Ho and P. J. McLane, "The power spectral density of digital continuous phase modulation with correlated data symbols," Parts I–II, *Proc. IEE*, Part F, **133**, 95–114, Feb. 1986.

[30] J. M. Wozencraft and I. M. Jacobs, *Principles of Communication Engineering*. Wiley, New York, 1965.

[31] G. Lindell, "On coded continuous phase modulation," PhD Thesis, Telecommunication Theory Department, University of Lund, Sweden, 1985.

[32] M. Schwartz, W. R. Bennett, and S. Stein, *Communication Systems and Techniques*. McGraw-Hill, New York, 1966 (re-issued by IEEE Press, New York, 1995).

[33] J. B. Anderson and S. Mohan, *Source and Channel Coding*. Kluwer, Boston, 1991.

[34] *M. K. Simon, S. M. Hinedi, and W. C. Lindsey, *Digital Communication Techniques*. Prentice-Hall, Englewood Cliffs, NJ, 1995.

[35] T. Aulin and C.-E. Sundberg, "Differential detection of partial response continuous phase modulated signals," *Proc. Int. Conf. Commun.*, 56.1.1–56.1.6, Denver, June 1981.

[36] T. Aulin and C.-E. Sundberg, "Partially coherent detection of digital full response continuous phase modulated signals," *IEEE Trans. Commun.*, **COM-30**, 1096–1117, May 1982.

[37] A. J. Viterbi, "Optimum detection and signal selection for partially coherent binary communication," *IEEE Trans. Inf. Theory*, **IT-11**, 239–246, Mar. 1965.

[38] H. L. van Trees, *Detection, Estimation and Modulation Theory*, Part I. Wiley, New York, 1968.

[39] T. Andersson and A. Svensson, "Error probability for a discriminator detector with decision feedback for continuous phase modulation," *Electronics* (Lett.), **24**(12), 753–753, 9 June 1988.

[40] W. U. Lee, "Carrier synchronization of CPFSK signals," *Conf. Rec. Nat. Telecommun. Conf.*, 30.2.1–30.2.4, Los Angeles, Dec. 1977.

[41] T. Aulin and C-E. Sundberg, "Synchronization properties of continuous phase modulation," *Conf. Rec. Global Telecommun. Conf.*, D7.1.1–D7.1.7, Miami, Nov. 1982.

[42] B. A. Mazur and D. P. Taylor, "Demodulation and carrier synchronization of multi-h phase codes," *IEEE Trans. Commun.*, **COM-29**, 259–266, March 1981.

[43] G. Ascheid, Y. Chen, and H. Meyr, "Synchronization, demodulation und dekodierung bei bandbreiteneffizienter übertragung, *NTZ Archiv*, **4**(12), 355–363, 1982.

[44] G. Ascheid, M. J. Stahl, and H. Meyr, "An all-digital receiver architecture for bandwidth efficient transmission at high data rates," *IEEE Trans. Commun.*, **COM-37**, 804–813, Aug. 1981.

[45] A. Premji, "Receiver structures for M-ary multi-h phase codes," PhD Thesis, Electrical and Computer Engineering Department, McMaster University, Hamilton, Canada, 1986.

[46] A. Premji and D. P. Taylor, "Receiver structures for multi-h signaling formats," *IEEE Trans. Commun.*, **COM-35**, 439–451, April 1987.

[47] S. J. Simmons and P. J. McLane, "Low-complexity carrier phase tracking decoders for continuous phase modulations," *IEEE Trans. Commun.*, **COM-33**, 1285–1290, Dec. 1985.

[48] A. J. Macdonald and J. B. Anderson, "Performance of CPM receivers with phase synchronizers that utilize tentative data estimates," *Proc. 24th Conf. Inf. Sciences and Systems*, Princeton, NJ, 41–46, Mar. 1990; A. J. Macdonald and J. B. Anderson, "PLL synchronization for coded modulation," *Proc. Int. Conf. Commun.*, Denver, 1708–1712, June 1991.

[49] K. Honma, E. Murate, and Y. Rikou, "On a method of constant-envelope modulation for digital mobile radio communication," *Conf. Rec. Intern. Conf. Commun.*, Seattle, 24.1.1–24.1.5, June 1980.

[50] A. Svensson, C-E. Sundberg, and T. Aulin, "A class of reduced complexity Viterbi detectors for partial response continuous phase modulation," *IEEE Trans. Commun.*, **COM-32**, 1079–1087, Oct. 1984.

[51] A. Svensson and C-E. Sundberg, "Optimized reduced-complexity Viterbi detectors for CPM," *Conf. Rec. Global Commun. Conf.*, San Diego, 22.1.1–22.1.8, Nov. 1983.

[52] A. Svensson, C.-E. Sundberg, and T. Aulin, "A distance measure for simplified receivers," *Electronics* (Lett.), **19**(23), 953–954, 1983.

[53] A. Svensson, "Receivers for CPM," PhD Thesis, Telecommunication Theory Department, University Lund, Sweden, May 1984.

[54] J. Huber and W. Liu, "An alternative approach to reduced complexity CPM receivers," *IEEE. J. Sel. Areas Commun.*, **SAC-7**, 1437–1449, Dec. 1989.

[55] T. Aulin, "Study of a new trellis decoding algorithm and its applications," Final Report, ESTEC Contract 6039/84/NL/DG, European Space Agency, Noordwijk, Netherlands, Dec. 1985.

[56] S. J. Simmons, "Simplified coherent detection of bandwidth-efficient CPFSK/CPM modulations," *Proc. IEEE Pacific Rim Conf.*, Victoria, Canada, 174–177, May 1991; *see also* S. J. Simmons, "Simplified coherent detection of CPM," *IEEE Trans. Commun.*, **COM-43**, 726–728, Feb–Apr. 1995.

[57] A. Svensson, "Reduced state sequence detection of full response continuous phase modulation," *Electronics* (Lett.), **26**(9), 652–654, 10 May 1990; *see also* A. Svensson, "Reduced state sequence detection of partial response continuous phase modulation," *IEE Proc.*, Part I, **138**, 256–268, Aug. 1991.

[58] T. Palenius and A. Svensson, "Reduced complexity detectors for continuous phase modulation based on a signal space approach," *European Trans. Telecommun.*, **4**(3), 285–297, May 1993.

[59] N. Seshadri, "Error performance of trellis modulation codes on channels with severe intersymbol interference," PhD Thesis, Electrical, Computer and Systems Engineering Department, Rensselaer Poly. Inst., Troy, NY, Sept. 1986.

[60] P. Galko, "Generalized MSK," PhD Thesis, Department of Electrical Engineering, University Toronto, Canada, Aug. 1982.

[61] P. Galko and S. Pasupathy, "Optimal linear receiver filters for binary digital signals," *Proc. Int. Conf. Commun.*, 1H.6.1–1H.6.5, Philadelphia, June 1982.

[62] A. Svensson and C.-E. Sundberg, "Optimum MSK-type receivers for CPM on Gaussian and Rayleigh fading channels," *IEE Proc.*, Part F, **131**(8), 480–490, Aug. 1984; *see also* A. Svensson and C.-E. Sundberg, "Optimum MSK-type receivers for partial response CPM on Rayleigh fading channels," *Conf. Rec. Int. Conf. Commun.*, Amsterdam, 933–936, May 1984.

6

PRS Coded Modulation

6.1. Introduction

Partial response signaling, (PRS), refers in this book to a collection of methods that intentionally or unintentionally use linear ISI to create signals that save bandwidth or energy or both. As we introduced in Chapter 1, this kind of coded modulation tends to work in the high-energy/narrow bandwidth region of the energy–bandwidth plane, and it requires a linear channel. One can think of PRS coding as a fixed, small alphabet of symbols such as $\{-1, +1\}$ that linearly modulate a standard pulse train, followed by a strong filtering of the pulse train. The result is a narrowband signal. Trellis coded modulation (TCM), in contrast, achieves bandwidth efficiency with subsets of a large alphabet and a standard orthogonal pulse train, with no filtering. Truly narrowband signaling cannot be built up from orthogonal pulses modulated by independent small-alphabet transmission values, since the combination of these must run at bandwidths greater than 0.5 Hz-s/data bit, to take again the binary case. Either a large alphabet of transmission values must be used, or something akin to filtering must take place. Significant envelope variation must occur in any consequent carrier modulated signal.

All PRS coding is linear filtering. It must also therefore be based on the convolution operation. A notation for PRS coding in terms of a discrete-time convolution is set up in Section 6.2. Just as in traditional binary-symbol convolutional coding, there is feedforward and feedback codeword generation and there are trellises, states, and minimum distances. But now all these are expressed in terms of ordinary real numbers. Consequently, the mathematical methods are sometimes quite different. For example, it is possible to *solve* for optimal PRS codes, rather than search through lists of generators. This is discussed in Section 6.4. Other techniques remain the same: just as with the other trellis codes in this book, the Viterbi algorithm (VA)/MLSE receiver provides a benchmark PRS maximum likelihood decoder, whose error rate stems from a minimum distance. The distance is discussed in Sections 6.2 and 6.3.

PRS signaling traces back to the introduction in the 1960s by Lender of duobinary signaling (see e.g. [2]). The method was extended by Kretzmer in several papers, including [3]. The symbols to be transmitted were convolved with a generator sequence via ordinary arithmetic, but the generator contained only integer values. The duobinary generator, for example, was the sequence $\{1, 1\}$; this scheme offered no real improvement in energy or bandwidth in the modern sense, but had more tractable pulse characteristics. The generator sequences $\{1, 0, -1\}$ and $\{-1, 0, 2, 0, -1\}$ create baseband spectral zeros at DC, an important property for undersea cables and carrier systems with pilot tones. A readable summary of these integer schemes is Pasupathy [4]. In contrast, the generators in this chapter contain unrestricted real or complex numbers. PRS signaling began to be viewed this way in the 1980s.

A major development that occurred in parallel was the invention of equalizers, beginning in the late 1960s, together with the trellis decoding view of ISI, which stems from the early 1970s. We will review the array of equalizer techniques in Section 6.6.1. They can be viewed as simplified PRS decoders. They have an array of error performances that can be distinctly different from the MLSE one. A slightly more complex decoder is the M-algorithm, and this one actually performs near to the MLSE limit, at a small fraction of the cost of the full MLSE receiver. This decoder is discussed in Section 6.6.2.

Another major development that began in the 1970s was of course TCM and continuous-phase modulation (CPM). By the 1990s, PRS signaling and the related ISI problems were considered by many to be part of one larger coded modulation view. A closely related modulation subject is the optimal detection of signals that have encountered heavy filtering. An example is heavily filtered QPSK modulation: this is in spirit PRS coding where the "encoder" is the filter and the filter could be, for example, a severe cutoff multi-pole Butterworth filter. The chapter's discussion of this subject begins with generation and properties of such coded signals in Section 6.5 and continues with decoder designs in Section 6.6.3. Special receiver designs and new analytical methods are needed, because the traditional trellis state description grows unwieldy and because very narrowband error events are qualitatively different from those in wider band systems. The result is a coded modulation that performs surprisingly well and is fully competitive with more ordinary types.

Finally, other categories of PRS codes are those used as line codes over wirelines and for magnetic recording. These are not emphasized here.

We begin by setting up a signal theory framework for PRS coding and ISI. The signals can be expressed in a natural way as components of a vector in Euclidean signal space, and it is often (but not always) natural to handle them in this way. Like TCM signals, PRS-type signals can directly be seen as patterns on the component values. Seen this way, they are another concrete implementation of the Chapter 1 Shannon concept of coding explicitly with signal space components.

6.2. Modeling and MLSE for ISI and Linear Coded Modulation

Communication ultimately takes place with real numbers and continuous functions. In the kind of coding in Sections 6.3 and 6.4 and in the equalization

PRS Coded Modulation

discussion, we will work directly with patterned time-discrete sequences of reals, where each number stands for the amplitude of a standard continuous pulse. In this section, we will set down the time-discrete modeling framework and derive the MLSE receiver from it. By contrast, with heavily filtered modulation in Section 6.5, it will be more convenient to work directly with the continuous signals.

6.2.1. A Modeling Framework for PRS Coding and ISI

An implicit assumption throughout this chapter is that the signal generation works by linear modulation; that is, signals are produced from data symbols $\{a_n\}$ by

$$s(t) = \sqrt{E_s} \sum_{n=1}^{N} a_n h(t - nT), \qquad (6.2\text{-}1)$$

subject to

$$\mathcal{E}\{|a_n|^2\} = 1,$$

$$\int |h(t)|^2 \, dt = 1.$$

When $a_1, a_2, \ldots$ are uncorrelated, the conditions imply that $\mathcal{E}\{\int |s(t)|^2 \, dt\} = NE_s$, that is, that E_s is the average symbol energy. Furthermore, we will discuss only baseband signals and assume that passband signals are created from two (6.2-1) signals in the usual way given in Section 2.5.

The basics of modeling linear modulation signals like these were given in Section 2.8. Except for a few side excursions, we will be able to base the chapter on the additive white Gaussian noise (AWGN) PAM receiver model given there. That is to say, a transmitter and ML receiver system that works with certain discrete-time reals will make the same decisions and have the same error performance as an ML receiver that works with the actual continuous signal functions. The time-discrete receiver has as its input the transmitted sequence of reals with IID Gaussians η_n added to each element. η_n is zero mean with variance $N_0/2$.

Here are three cases of the signal generation in Eq. (6.2-1) which will prove important in the chapter. Each is more general than the previous one. The first will serve as a formal definition of PRS coding.

Case (i). $h(\,)$ is a sum of orthogonal pulses. Formally, let

$$h(t) = \sum_{\ell \geq 0} b_\ell v(t - \ell T) \qquad (6.2\text{-}2)$$

with $\sum |b_\ell|^2 = 1$ and $v(t)$ orthonormal. Each transmission symbol generates a sequence of orthogonal pulses weighted by $b_0, b_1, \ldots$. We will take the form (6.2-2) as defining a *PRS encoding* with generator sequence $\{b_\ell\}$. If the sequence has $m+1$ elements $b_0, \ldots, b_m$, the PRS is finite response with memory m; otherwise it is infinite response. Note that the requirement $\int |h(t)|^2\, dt = 1$ in Eq. (6.2-1) is still met. By inserting Eq. (6.2-2) into the linear signal model Eq. (6.2-1), we find that the total signal is

$$s(t) = \sum c_n v(t-nT), \quad \text{with } c_n = \sum b_\ell a_{n-\ell}, \qquad (6.2\text{-}3)$$

which has already been given in Section 2.8 as Eq. (2.8-3). That is, $s(t)$ is just an orthogonal pulse modulation by the sequence $\{c_n\}$, which are the values of the convolution $a*b$. The PAM Model outputs in Fig. 2.29 are $y_n = c_n + \eta_n$, $n = 1, \ldots$.

Equation (6.2-2) also models a kind of multipath interference where the paths have weights $b_0, b_1, \ldots$, but the fact that the path delays are integer multiples of T makes it a rather limited model.

Case (ii). $h(\ )$ is a filtering of simple linear modulation. Now $h = g*v$, where $g(t)$ is the filter impulse response. Such an h arises in several natural ways. For v an orthogonal pulse in pulse amplitude modulation (PAM) or quadrature amplitude modulation (QAM), g can represent ISI in the transmission medium or a bandlimitation filter in the transmitter. g could also be an intentional coding; that is, $g*v$ can synthesize coded modulation. ML detection of $\{a_n\}$ from the signal $\sum a_n h(t-nT) + \eta(t)$ via a discrete-time model and receiver requires the Matched Filter Model in Section 2.8. The receive filter that leads to this model is matched to response $g*v$ and its $1/T$-rate samples are the discrete-time received sequence $\{y_n\}$. Included in $\{y_n\}$ are noise variates that are not IID Gaussians and significant further processing of $\{y_n\}$ may be needed for an ML detection of the $\{a_n\}$.

Case (iii). $h(\ )$ is a general bandlimited pulse. By this case is meant general linear modulations that are not based on orthogonal pulses. A symbol a simply produces a response $ah(t)$. An interesting example is "Faster than Nyquist" (FTN) signaling, a method that sends T-shifted orthogonal pulses, but sends them faster than the $1/T$ rate. These signals have some intriguing properties which generalize to other coded modulations and they make excellent coded modulations. We will take them up in Section 6.5. Other Case *(iii)* signals occur when the pulse is the impulse response of some bandlimited natural system, such as the recording head in magnetic recording. Case *(iii)* signals in general require the Matched Filter Model.

Each of these cases includes the previous ones. An obvious question we can ask is when is the simpler PAM Model sufficient for ML detection in Cases *(ii)* and *(iii)*. Or a similar and more direct question: when can h be safely represented by the form $\sum c_n v(t-nT)$ in Eq. (6.2-3)? The simplified answer is that the bandwidth of $h(t)$ should be $<0.5/T$. A refinement of this is the following theorem.

PRS Coded Modulation

THEOREM 6.2-1. *For h continuous, let $H(f) = 0, |f| > W$, with $W < 0.5/T$. Let the real T-orthogonal pulse v have transform $V(f)$ that satisfies*

$$V(f) = \begin{cases} C_0, \text{ a constant}, & |f| \leq W \\ 0, & |f| \geq 1/T - W. \end{cases} \quad (6.2\text{-}4)$$

Then $h(t)$ may be expressed as $\sum h_n v(t - nT)$, where $\{h_n\}$ are the generalized Fourier coefficients given by

$$h_n = \int_{-\infty}^{\infty} h(t)v(t - nT)\,dt, \quad \text{integer } n. \quad (6.2\text{-}5)$$

Furthermore, $\{h_n\}$ are $\{(T/C_0)h(nT)\}$, the scaled samples of $h(t)$.

The proof is an exercise in Fourier theory and it will be given after some important remarks.

1. The hypothesis on $v(t)$ in the theorem closely fits the standard root RC orthogonal pulse in Section 2.2.2. Figure 2.1 shows $|V(f)|^2$ for some root-RC pulses; Fig. 6.1 compares the spectra of $v(t)$ and $h(t)$ in the theorem when $W = 1/3T$. It is clear that root-RC pulses with excess bandwidth α up to 1/3 satisfy the theorem. When $\alpha = 0$, $v(t)$ is the sinc pulse, and $W = 0.5/T$, its widest value.

2. A $v(t)$ that satisfies the hypothesis may be thought of as a function that interpolates $h(t)$ from its samples every T seconds. It is interesting to observe that the *same* $h(t)$ is created from the samples $\{h_n\}$ with any v that satisfies the hypothesis. Using sinc( ) for v replicates the traditional sampling theorem.

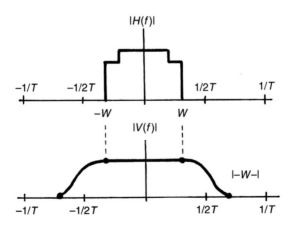

Figure 6.1 Illustration of the spectral condition on $v(t)$ and $h(t)$ in Theorem 6.2-1. $\alpha = 1/3$ case shown.

3. Equation (2.8-4), which generates the coefficients γ_n of the PAM model $H_{PAM}(z)$, and Eq. (6.2-5) are formally identical. However, the PAM calculation applies to any h, not just the bandlimited one in the theorem. When h fails to meet the theorem, the PAM model coefficients are not the samples $h(nT)$, Eq. (6.2-2) does not exactly produce the true signals (6.2-1), and MLSE based on the PAM model outputs may not be optimal. When h and v meet the theorem, the model coefficients are $h(nT)$, Eqs (6.2-3) and (6.2-1) produce the same signals, and MLSE with the true analog signals or the PAM outputs produces the same outcome.

4. When h and v do not quite fit the theorem – when $H(f)$ is only approximately 0, $|f| > 0.5/T$, and $V(f)$ does not quite satisfy Eq. (6.2-4) – the h_n in Eq. (6.2-5) are in general still very close to the sample train $h(nT)$. An example of this will be given later in Example 6.2-2.

5. For $h(t)$ with bandwidth greater than $0.5/T$ Hz, it is in general possible to *detect* $\{a_n\}$ in an ML fashion from samples every T seconds; it is not possible, of course, to *recreate* $h(t)$ from such samples. The basic result in Section 6.3, that distance and bandwidth stem solely from the autocorrelation of h, continues to hold when h has the wider bandwidth. A full discussion of the wideband case appears in [5].

Here is the proof of the theorem, as given by Balachandran [8].

Proof. Sample $h(t)$ at the rate $1/T$ to obtain the sequence $\{h(nT)\}$. The Fourier transform of these samples is $H_{\text{sam}}(f) = \sum h(nT) e^{-j2\pi f nT}$. It will then be true that

$$H(f) = T H_{\text{sam}}(f) = \sum_n T h(nT) e^{-j2\pi f nT}, \quad |f| \leq 1/T - W.$$

At the same time, $V(f)$ is constant over the support of $H(f)$, so that

$$h(t) \triangleq \int H(f) e^{j2\pi ft} \, df = \frac{1}{C_0} \int H(f) V(f) e^{j2\pi ft} \, df. \quad (6.2\text{-}6)$$

We may substitute $T H_{\text{sam}}(f)$ directly for $H(f)$ in Eq. (6.2-6); since $V(f) = 0$ for $|f| \geq 1/T - W$, this gives

$$h(t) = \frac{1}{C_0} \int \left[\sum_{n=-\infty}^{\infty} T h(nT) e^{-j2\pi f nT} \right] V(f) e^{j2\pi ft} \, df,$$

$$= \sum_n \left[\frac{T h(nT)}{C_0} \right] \int V(f) e^{j2\pi f(t-nT)} \, df,$$

$$= \sum_n h_n v(t - nT),$$

where $h_n = T h(nT)/C_0$, all n. □

6.2.2. Maximum Likelihood Reception and Minimum Distance

For any of signal types (i)–(iii) and at any bandwidth, it is always possible to build an ML receiver for the AWGN channel in the manner of Section 2.4 and Figs 2.9 and 2.10. Fundamentally, one computes the Euclidean distance $\int |r(t) - s(t)|^2 dt$ between the received signal $r(t)$ and all the possible signals of the form $s(t) = \sqrt{E_s} \sum a_n h(t - nT)$, and chooses the closest $s(t)$. This is the MLSE process. But for Case (i) signals and Case (ii) and (iii) signals that satisfy Theorem 6.2-1, the processing can equally well be carried out with the discrete-time sequences obtained from a PAM model of the transmission. From Eq. (6.2-3), the PRS signal of length N intervals is $s(t) = \sum_1^N c_n v(t - nT)$ as a continuous signal; N terms can represent all of the signal or some part of it. The PAM model outputs are $y_n = c_n + \eta_n$; the IID noise variates are obtained from $\eta_n = \int \eta(t) v^*(t - nT) dt$ (See Fig. 2.29). Each $\{s_i(t)\}$ is in one-to-one correspondence with some sequence $c_i[1], \ldots, c_i[N]$. (For clarity we switch now to the bracket sequence notation.) Compute the square Euclidean distance between sequence $y[1], \ldots, y[N]$ and some $\sqrt{E_s}(c_i[1], \ldots, c_i[N])$ as $\sum_1^N |y[n] - \sqrt{E_s} c_i[n]|^2$. It is easy to show that $\int |r(t) - s_i(t)|^2 dt = \sum_1^N |y[n] - \sqrt{E_s} c_i[n]|^2$; we leave the proof of this as one of the exercises. Thus, it is sufficient to consider only the sequences $y[1], \ldots, y[N]$ and $c_i[1], \ldots, c_i[N]$ in order to carry out the same ML decoding.

What is shown here is that for the cases considered, partial response signals in AWGN can be directly represented as *components of a vector in Euclidean signal space*. It is often but not always natural to do so. With TCM coding it was almost always natural.

If the PRS generation in Eq. (6.2-3) has memory m, then $c_i[1], \ldots$ are generated by a finite state machine operation (namely, $\sum b_\ell a_{n-\ell}$) with the m state variables $a_{n-m}, \ldots, a_{n-1}$. A general introduction to this was given in Sections 3.3 and 3.4. With a finite state trellis code, the MLSE process is efficiently performed by the VA. The VA works with pieces of signals and their distances, and what we have just shown is that the VA, working instead on the PAM model output $y_i[1], \ldots$ the various $c_i[1], \ldots$ sequences, can perform MLSE as if it were working with the continuous signals.

For any linear signaling, the asymptotic error probability as E_b/N_0 grows is, from Eq. (2.3-29), $\sim Q(\sqrt{d_{\min}^2 E_b/N_0})$, neglecting constant multipliers, and it is set by the *normalized minimum distance* of the signal set, $d_{\min}$. It is often easier to find this critical parameter by reducing the signals to time-discrete form and by using properties of h. But it needs to be said that straightforward numerical calculation of $\int |s_i(t) - s_j(t)|^2 dt$ is always an option, no matter how strange h may be. $d_{\min}$ tells us the asymptotic ML receiver error performance. The various modelings, PAM, Matched Filter, or otherwise, also have an asymptotic performance; so also do suboptimal receivers such as equalizers. If these various measures do not reduce the asymptotic performance, we know that asymptotically at least, they are optimal models or receivers.

The computation of $d_{\min}$ thus leads to an important benchmark. Without normalization, the square minimum distance is

$$D^2_{\min} = \min_{i,j} E_s \int |s_i(t) - s_j(t)|^2 dt. \qquad (6.2\text{-}7)$$

Recall some notation from Section 2.5: $\Delta \boldsymbol{a} = \Delta a[1], \Delta a[2], \ldots$ denotes the difference $\boldsymbol{a}_i - \boldsymbol{a}_j$ between two data sequences $\boldsymbol{a}_i$ and $\boldsymbol{a}_j$; $s_i(t)$ and $s_j(t)$ are the corresponding signals. If the data are M-ary, then $E_s = \log_2 M E_b$. From Eq. (6.2-1), $s_i(t) - s_j(t) = \sum_n \Delta a[n] h(t - nT)$. Normalizing Eq. (6.2-7) by the standard $2E_b$ thus gives the normalized PRS square minimum distance

$$d^2_{\min} = \min_{i,j} \tfrac{1}{2} \log_2 M \int \left| \sum_n \Delta a[n] h(t - nT) \right|^2 dt. \qquad (6.2\text{-}8)$$

By minimizing over all legal $\Delta \boldsymbol{a}$, the minimum distance is obtained. We have introduced the concept of trellis error event in Section 3.4, along with the idea of free distance, which is the minimum distance of a trellis code with unbounded length. It is sufficient to define a PRS error event in terms of the data symbol differences $\Delta \boldsymbol{a}$ in the following convenient way. The components of $\Delta \boldsymbol{a}$ are zero up to some time which we may as well take as time 0, and zero again after some future time. We suppress the zero components of $\Delta \boldsymbol{a}$ outside this range and write the error event in terms of the remaining components,[1] with the requirement of Eq. (6.2-1) that $\mathcal{E}[|a_n|^2] = 1$. A binary data symbol alphabet becomes $\{\pm 1\}$ and a quaternary equi-spaced one becomes $\{\pm 1/\sqrt{5}, \pm 3/\sqrt{5}\}$; the alphabet for $\Delta a[n]$ thus becomes $\Delta \mathcal{A} = \{\pm 2, 0\}$ and $\Delta \mathcal{A} = \{\pm 6/\sqrt{5}, \pm 4/\sqrt{5}, \pm 2/\sqrt{5}, 0\}$ in these two cases. From the form of Eq. (6.2-8) it is clear that event $-\Delta \boldsymbol{a}$ leads to the same distance as $\Delta \boldsymbol{a}$. In a search for minimum distance, then, $\Delta a[1]$ may be restricted to positive values. Other properties may help reduce the searching further.

A major property of the linear coded modulation $d_{\min}$ is the following lemma, which is due to Said [5]. It has important coding implications.

LEMMA 6.2-1. *The minimum distance of a linear coded modulation is upper bounded by Eq. (6.2-8) with $\Delta \boldsymbol{a}$ taken as $\Delta a[1], 0, \ldots, 0$, in which $\Delta a[1]$ takes*

---

[1] Note that in terms of the time function $\sum a_n h(t - nT)$ the event may begin long before $t = 0$, since there is no requirement that $h(t)$ be causal.

PRS Coded Modulation

the least nonzero magnitude $\Delta a^\ddagger$ in the symbol difference alphabet $\Delta \mathcal{A}$. That is,

$$\begin{aligned} d_{\min}^2 &\leq \min_{|\Delta a[1]| \in \mathcal{A}} \frac{1}{2} \log_2 M \, \|\Delta a[1]\|^2 \int |h(t)|^2 \, dt, \\ &= \tfrac{1}{2} \log_2 M \, \min_{\mathcal{A}} |\Delta a[1]|^2, \\ &= \tfrac{1}{2} \log_2 M |\Delta a^\ddagger|^2 \triangleq d_{\text{MF}}^2. \end{aligned} \qquad (6.2\text{-}9)$$

The proof of the lemma follows by substituting all of the single-component difference sequences into Eq. (6.2-8). Since $\int |h(t)|^2 \, dt = 1$, the least of these is Eq. (6.2-9). Because the minimum distance cannot be greater than these particular cases, Eq. (6.2-9) is an overbound.

The value in Eq. (6.2-9) is called the *matched filter bound* or sometimes the *antipodal signal distance*, because it applies with equality when the signals are a single pulse at equal and opposite amplitudes and the detector is a matched filter/sampler. For a binary antipodal symbol alphabet, the lemma states that $d_{\text{MF}}^2 \leq 2$; for the equi-spaced quaternary alphabet, $d_{\text{MF}}^2 \leq 0.8$. A limit to distance like this is in stark contrast to other kinds of coded modulation, which exhibit arbitrarily large minimum distances. Some thought shows that such a limit is not necessarily in conflict with Shannon theory. When linear coded modulation is at its best, it is a method of narrow bandwidth/high energy signaling; it should logically then have small distances. In fact, at narrow bandwidths a larger distance than d_{MF} would soon contradict bandwidth–energy capacity formulas like those in Section 3.6.

The fact is that many $h(t)$ lead to a minimum distance that achieves the upper bound d_{MF}. Furthermore, we will find in Section 6.5 that $d_{\min}$ for heavily filtered modulations continues to achieve d_{MF} at surprisingly narrow bandwidths. A relatively well-known example of this is the FTN signaling in Section 6.5. Such schemes are of obvious interest: since coded modulation of the form (6.2-1) cannot do better than d_{MF}, we would like to find the most narrowband scheme with this maximal value. Eventually, $d_{\min}$ must fall below the bound, since otherwise signaling at a set rate and E_b could take place reliably at arbitrarily narrow bandwidth; this, again, would violate the capacity formulas in Section 3.6.

Even when $d_{\min} < d_{\text{MF}}$, the lemma is still a useful fact, because the easiest way to estimate $d_{\min}$ is often simply to try out a set of sequences Δa. As the Δa lengthen, the distance of the error events grows in general, although the growth may not be monotone. Through this process, the lemma assures that only the few events with distances less than d_{MF} are of interest. Eventually, one can be reasonably sure that $d_{\min}$ has been observed.[2] Another use of the lemma is in evaluating simplified discrete-time models and suboptimal receivers such as equalizers: minimum distances and asymptotic error probabilities of the form $Q(\sqrt{d^2 E_b/N_0})$ may be found for these, and if the distance is d_{MF} then the model

---

[2] Methods based on this idea are discussed in Section 6.3.

or receiver is automatically optimal in the sense of asymptotic error, since no better probability can exist.

Here are several illustrations of $d_{\min}$ for coded linear modulation. The first coding system is based directly on orthogonal pulses, the second on a Butterworth filter response.

Example 6.2-1 *(Free Distance of a Simple Finite-Length PRS Code).* Consider the memory 2 PRS code with unnormalized generator sequence $b = \{1, 0.9, 0.9\}$. This is Case (i) in Section 6.2.1. We can attempt to find the free distance by substituting a number of symbol difference sequences Δa into Eq. (6.2-8); after normalizing, $h(t)$ is $(1/\sqrt{2.62})[v(t) + 0.9v(t-T) + 0.9v(t-2T)]$, in which $v(t)$ is an orthonormal pulse.[3] The single symbol difference $\{2\}$ leads to the matched filter bound distance, $d_{\rm MF}^2 = 2$. The length-2 difference sequence $\Delta a = \{2, -2\}$ leads to square distance 1.39. All other Δa lead to larger outcomes, although some lead to numbers that increase only slowly away from 1.39; for example, $\Delta a = \{2, -2, 0, 2, -2\}$ leads to square distance 1.40. The free distance is therefore 1.39, and the MLSE benchmark is this number. It represents a 1.6 dB energy loss compared to transmission with h that is a single orthogonal pulse. The largest loss of any three-tap PRS code is 2.3 dB and it occurs with normalized taps $\{0.503, 0.704, 0.503\}$; this was discovered by Magee and Proakis [13], who also gave an optimization procedure to solve for the position (see Section 6.4). Most three-tap generators in fact meet the $d_{\rm MF}$ bound. It can be shown that *all* two-tap generators meet it.

Example 6.2-2 *(PRS Code Derived from Narrowband Butterworth Filter).* Consider the 4-pole Butterworth lowpass filter[4] with cut-off $0.4/T$ and impulse response $g(t)$ and a root-RC orthogonal pulse $v(t)$ with 30% excess bandwidth. This is Case (iii). It could be binary orthogonal linear modulation, to which has been added a baseband transmitter bandlimitation filter with the Butterworth response. The central part of the combined impulse response $h = v * g$ is shown in Fig. 6.2 (the response has doubly infinite support in theory). $h(t)$ is a mildly distorted $v(t)$, delayed about $1.2T$. The circles in the figure denote the PAM model solution (2.8-4). The response h cannot quite be represented as a Case (i) weighted sum of orthogonal pulses, because the spectrum of v is not quite flat over the support of $H(f)$ and $H(f)$ has some small response outside $0.5/T$ Hz. Nonetheless, the γ_n of the model lie very close to the samples $h(nT)$; this is because v and h nearly satisfy Theorem 6.2-1. We can adopt h as a near-member of Case (i) and take the set $\{\gamma_n\}$ as a PRS code tap set that mimics the Butterworth filter action. But first we calculate the true free distance by means of Eq. (6.2-8), without reference to the PAM model, by using the true response $h(t)$. As in the previous example,

---

[3] A simple MATLAB program to carry out the calculation appears in [1], Chapter 7.

[4] The Butterworth filter response is created by the MATLAB statement $[b, a] = {\rm butter}(4, 0.4 * 2/f_s)$, followed by filter($b, a,$[1, zeros(1, 200)]), where f_s is the number of samples per T interval. See also Example 2.8-1, which calculates discrete-time models for the same filter with cutoff $0.5/T$.

PRS Coded Modulation

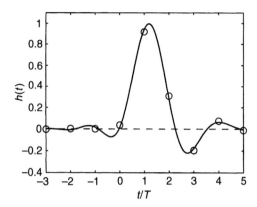

Figure 6.2 Total response of the channel $h = v * g$, where v is a 30% root-RC pulse and g is impulse response of a 4-pole Butterworth filter with cut-off $0.4/T$. Circles denote PAM discrete-time model of this channel.

difference sequence $\{2\}$ leads to $d_{MF}^2 = 2$; this is the square difference between the pulse in Fig. 6.2 and its negative. But this time a search for difference sequences with a lower distance outcome turns up nothing, and the square free distance and MLSE benchmark are 2. The next smallest square distance outcome is 3.1 with $\Delta a = \{2, -2\}$. Alternatively, we can find the free distance of the PRS code with generator $\{\gamma_n\}$. Ignoring taps smaller than 0.013 and renormalizing yields the tap set $b_0, \ldots, b_4 = \{0.035, 0.927, 0.309, -0.196, 0.077\}$. Repeating the free distance search, this time using $h(t) = \sum b_n v(t - nT)$, we get an almost identical set of distances, and the same end result, that the square free distance is 2.

6.3. Distance and Spectrum in PRS Codes

Our purpose in this section is to calculate the distance – hence the energy requirement – and the spectrum of PRS codes. We take these as imposed by taps $\{b_\ell\}$ in the total response $h(t) = \sum b_\ell v(t - \ell T)$ to a data symbol, as defined in Eq. (6.2-2). This is Case (*i*) signaling. The basic minimum distance calculation in Eq. (6.2-8) is not different in principle from that for binary-symbol codes, but the switch to real-number arithmetic leads eventually to a few conclusions that are very different. We have seen one of these already in Lemma 6.2-1, which states that d_{min} in PRS coding cannot exceed a bound d_{MF}. The basic spectrum calculation differs from parity-check and TCM coding in that it is not simply a modulator spectrum scaled by the code rate: the spectrum is also a function of the generator taps. In this way, PRS coding resembles CPM coding.

The central fact about distance and spectrum in PRS codes is that both stem solely from the autocorrelation function of the response $h(t)$. From this comes the whole structure of the theory, and ultimately, the fact that optimal encoders are the solution of a linear program. Here is a major departure from the other code classes, which essentially require an exhaustive search to find the best codes. We will begin with some basic definitions and notations for PRS transforms and autocorrelations.

6.3.1. Basic PRS Transforms

Consider independent and identically distributed data symbols $a_n \in \mathcal{A}$, where $\mathcal{A}$ is a possibly complex alphabet of symbols that are symmetric about the origin of the complex plane, scaled so that $\mathcal{E}[|a_n|]^2 = 1$. Coded signals are generated by Eq. (6.2-1), namely $s(t) = \sqrt{E_s} \sum_{n=1}^{N} a_n h(t - nT)$, in which the possibly complex response $h(t)$ is Eq. (6.2-2), namely, $h(t) = \sum_{\ell \geq 0} b_\ell v(t - \ell T)$. The Fourier transform of $h(t)$ is

$$H(f) = \int_{-\infty}^{\infty} h(t) e^{-j2\pi f t} \, dt \quad \text{all } f. \tag{6.3-1}$$

Here $v(\ )$ is real but the PRS taps $\{b_\ell\}$ may be complex. As mentioned in Section 6.2, the conditions here imply that E_s is the energy per symbol. By allowing complex a_n and $h(\ )$ we extend the discussion to passband signals with I and Q components, and to simultaneous encoding of I and Q signals, but the focus throughout Chapter 6 will mainly be on real, baseband signaling.

To define an autocorrelation of $s(t)$, we utilize the random offset function $\bar{s}(t) = s(t + \delta)$, with δ selected at random in $(0, T)$, which was defined in Section 2.7, and compute the time-average autocorrelation of $s(t)$ given in Eq. (2.7-6). The required expectation over the symbols $\{a_n\}$ is

$$\mathcal{E}_a\{s(t+\delta+\tau)s^*(t+\delta)\} = E_s \sum_k \sum_n h(t+\delta+\tau-kT)h^*(t+\delta-nT)\mathcal{E}[a_k a_n^*],$$

$$= E_s \sum_n h(t+\delta+\tau-nT)h^*(t+\delta-nT),$$

because of the standard symbol conditions (6.2-1). The time-average autocorrelation of $s(t)$ at τ is the expectation of this over δ, which is

$$\frac{1}{T} \int_0^T E_s \left[\sum_n h(t+\delta+\tau-nT)h^*(t+\delta-nT) \right] d\delta,$$

which rearranges to

$$\frac{E_s}{T} \int_{-\infty}^{\infty} h(t+\tau) h^*(t) \, dt. \tag{6.3-2}$$

As usual with an autocorrelation, we normalize by the value at $\tau = 0$; the result will be denoted as $\rho(\tau)$, and we have

$$\rho(\tau) = \int_{-\infty}^{\infty} h(t+\tau)h^*(t)\,dt. \qquad (6.3\text{-}3)$$

We notice immediately that at $\tau = kT$ this is the same calculation as Eq. (2.8-7), which computed the coefficient set $\{\rho_k\}$ of the matched filter discrete-time model. We will retain the notation $\rho[k]$ to mean $\rho(kT)$.

The autocorrelation function $\rho(\tau)$ has many useful properties. These, with the following notations, will figure prominently in what follows.

1. Like all autocorrelations, $\rho(\tau)$ has Hermitian symmetry: $\rho(\tau) = \rho^*(-\tau)$ and $\rho[k] = \rho^*[-k]$.

2. The z-transform of the double-sided discrete-time sequence $\ldots, \rho[-1], \rho[0], \rho[1], \ldots$, represented in vector notation by $\boldsymbol{\rho}$, is given by

$$R_Z(z) \triangleq \sum_{k=-\infty}^{\infty} \rho[k] z^{-k}. \qquad (6.3\text{-}4)$$

We assume that there are no poles on the unit circle.

3. The Fourier transform of the sequence $\boldsymbol{\rho}$ is given by

$$R_D(f) \triangleq \sum_k \rho[k] e^{-j2\pi f k} = R_Z(e^{j2\pi f}) \qquad (6.3\text{-}5)$$

and is always real.

4. The continuous transform of $\rho(\tau)$ is

$$R(f) \triangleq \int_{-\infty}^{\infty} \rho(\tau) e^{-j2\pi f \tau}\,d\tau. \qquad (6.3\text{-}6)$$

5. It follows that

$$R(f) = |H(f)|^2. \qquad (6.3\text{-}7)$$

6. The *folded power spectrum*, also called the aliased power spectrum, of $\rho(\tau)$ is defined by

$$R_{\text{fold}}(f) \triangleq \sum_{n=-\infty}^{\infty} R_D(f + n/T),$$

$$= \sum_{n=-\infty}^{\infty} |H(f + n/T)|^2. \qquad (6.3\text{-}8)$$

We note especially that if $H(f) = 0, |f| \geq 1/2T$, then $R_{\text{fold}} = |H(f)|^2, -1/2T < f < 1/2T$; otherwise, it does not. We can expect such bandlimited $H(f)$ to have special behavior.

7. The ρ sequence Fourier transform satisfies

$$R_D(f) = R_Z(e^{j2\pi f}) = (1/T)R_{\text{fold}}(f/T), \qquad (6.3\text{-}9)$$

$$= (1/T)\sum_n \left| H\left(\frac{f+n}{T}\right)\right|^2. \qquad (6.3\text{-}10)$$

8. It follows from the Hermitian symmetry property that the z-transform $R_Z(z)$ satisfies

$$R_Z(z) = R_Z(1/z^*). \qquad (6.3\text{-}11)$$

Consequently, when $R_Z(\ )$ is rational, if z_0 is a zero of $R_Z(z)$ then so is $1/z_0$, and similarly for the poles. Furthermore, if $h(t)$ is real, poles and zeros must both occur in conjugate pairs. Therefore, if z_0 is a zero, so are z_0^*, $1/z_0$ and $1/z_0^*$, and similarly for the poles.

In fact, many $h(t)$ lead to the same $R_Z(z)$. The concept of *spectral factorization* develops from the last property. It follows from Eq. (6.3-11) that R_Z of the type here may be expressed in the form

$$R_Z(z) = H_{\text{SF}}(z)H_{\text{SF}}(1/z^*), \qquad (6.3\text{-}12)$$

where

$$H_{\text{SF}}(z) = \frac{\prod_{i=1}^{N_Q}(1 - \zeta_i z^{-1})}{\prod_{j=1}^{N_P}(1 - p_j z^{-1})}.$$

Here $\{\zeta_i\}$ are the positions of the N_Q zeros and $\{p_j\}$ are the positions of the N_P poles. The process of obtaining an H_{SF} from an R_Z in this way is called spectral factorization. Because of the special properties of poles and zeros, many factorizations are possible, and each such $H_{\text{SF}}(z)$ maps back via the inverse z-transform to some time sequence $h_{\text{SF}}[n], n = \ldots, -1, 0, 1, 2, \ldots$. In effect, each of these sequences $h_{\text{SF}}[n]$ represents a discrete-time channel model impulse response. Since $\mathcal{Z}\{g^*[-n]\} = G^*(1/z^*)$, it follows that

$$\rho(kT) = \rho[k] = \sum_{\ell=-\infty}^{\infty} h_{\text{SF}}[\ell + k]h_{\text{SF}}^*[\ell], \qquad (6.3\text{-}13)$$

that is, all such sequences $h_{\text{SF}}[k]$ lead to the same autocorrelation.

PRS Coded Modulation

Now let $h(t)$ equal a sum of orthogonal pulses $\sum b_\ell v(t - \ell T)$, as in Eq. (6.2-2). The autocorrelation (6.3-2) at $\tau = kT$ becomes

$$\rho(kT) = \int_{-\infty}^{\infty} \left[\sum_{\ell'} b_{\ell'} v(t - kT - \ell'T) \right] \left[\sum_{\ell} b_\ell^* v(t - \ell T) \right] dt$$

$$= \sum_{\ell} b_{\ell+k} b_\ell^*. \qquad (6.3\text{-}14)$$

$\rho(kT)$ has z-transform $R_Z(z)$. For this $R_Z(z)$, then, one of the spectral-factorization-derived sequences is the actual PRS tap set $\{b_\ell\}$. There are potentially many others as well, all derived from the same $R_Z(z)$ and all creating the same autocorrelation. These may be displayed by performing all the possible factorizations. Since by Eq. (6.3-11) all the poles and zeros of $R_Z(z)$ occur in reciprocal pairs, there is a special unique one for which $H_{SF}(z)$ in Eq. (6.3-12) has $|\zeta_i| < 1$ and $|p_j| < 1$, all i and j. This is called the *minimum phase* factorization of $R_Z(z)$. It can be shown that $h_{SF}[n]$ corresponding to it is stable and that $H_{SF}(e^{j2\pi f}) = \sum h_{SFE} e^{-j2\pi kf}$ has the least phase of any sequence leading to $R_Z(z)$. The opposite case, with $|\zeta_i| > 1$ and $|p_j| > 1$, all i and j, is the maximum phase $H_{SF}(z)$; the corresponding sequence in this case is the minimum phase $h_{SF}[n]$ reversed in time. The many other factorizations have phases in between. Since we have assumed that $h(t)$ is $\sum b_\ell v(t - \ell T)$, then $h_{SF}[k]$ is directly $b_0, \ldots, b_m$ and all these factorizations correspond to different PRS encoder tap sets.

The phase of a tap set has a major effect on decoding complexity and drives other behavior as well, and so it is worth pausing for a closer look at it. Roughly speaking, sequences with lower phase respond sooner than those with the same autocorrelation but a higher phase. A useful measure of this responsiveness is the *partial energy function* $E(n)$ given by

$$E(n) = \sum_{\ell=0}^{n} |b_\ell|^2, \quad n = 0, 1, \ldots, m. \qquad (6.3\text{-}15)$$

Since $\sum |b_\ell|^2$ is normalized to unit energy, the value $E(m)$ is 1. $E(n)$ shows the growth in energy of the response with time. It can be shown that

$$E_{\max}(n) \leq E(n) \leq E_{\min}(n), \qquad (6.3\text{-}16)$$

where $E_{\max}(n)$ and $E_{\min}(n)$ are the partial energies of the max and min phase response [9,10]. Thus the min phase response in a sense occurs fastest. Partial energy also is related to phase, group delay, and the column distance function of the PRS code [9], but $E(n)$ is the most important phase-type measure in the sequel.

An example of phase in a PRS tap set is given by the tap set $\{b_\ell\} = \{0.035, 0.927, 0.309, -0.196, 0.077\}$ derived in Example 6.2-2. It stems from the

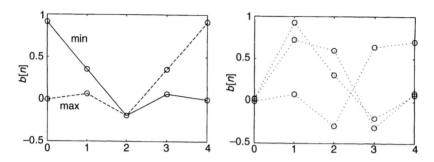

Figure 6.3 Tap sets having the autocorrelation $0.003z^4 + 0.064z^3 - 0.147z^2 + 0.243z + 1 + 0.243z^{-1} + \cdots$ of the tap set in Example 6.2-2. (*Left*) Min and max phase tap sets; (*right*) remaining tap sets (reverse sets not shown).

response $h = g * v$, where $g(t)$ is the impulse response of a 4-pole Butterworth filter with cut-off $0.4/T$ Hz and $v(t)$ is a 30% root-RC T-orthogonal pulse; the $\{b_\ell\}$ are almost identical to the response time samples $h(nT), n = 0, \ldots, 4$. The autocorrelation z-transform is $R_Z(z) = 0.003z^4 + 0.064z^3 - 0.147z^2 + 0.243z + 1 + 0.243z^{-1} - 0.147z^{-2} + 0.064z^{-3} + 0.003z^{-4}$, which is $\{b[n]\} * \{b[-n]\}$. $R_Z(z)$ has no poles and four reciprocal pairs of zeros, and listing only the zero locations inside the unit circle, the locations are $\{\zeta_i\} = \{-0.0386, -0.771, 0.212 \pm 0.252j\}$. The minimum phase factorization leads to $H_{SF}(z) = 1 + 0.385z^{-1} - 0.205z^{-2} + 0.075z^{-3} + 0.003z^{-4}$, whose coefficients after normalizing represent the tap set $\{b_\ell\}_{\min} = \{0.914, 0.352, -0.187, 0.069, 0.003\}$. The set is plotted on the left in Fig. 6.3 together with the max phase tap set, which is this set time reversed. Observe that the set is not the tap set in Example 6.2-2. This shows that the modeling process in the example does not necessarily lead to the minimum phase tap set. There are eight allowed factorizations in all, assuming that $h(t)$ is real; these include the max and min phase ones just discussed, together with the ones on the right in Fig. 6.3 plus their time reversals. All lead to the same autocorrelation $R_Z(z)$. Figure 6.4 shows the partial energy functions of all the responses.

6.3.2. Autocorrelation and Euclidean Distance

In Section 6.2 we derived the normalized Euclidean distance between two signals $s_i(t)$ and $s_j(t)$ that are produced by the general linear modulation process $\sum a[n]h(t - nT)$. It had the form

$$d^2(s_i(t), s_j(t)) = \frac{1}{2}\log_2 M \int \left|\sum_n \Delta a[n]h(t - nT)\right|^2 dt. \qquad (6.3\text{-}17)$$

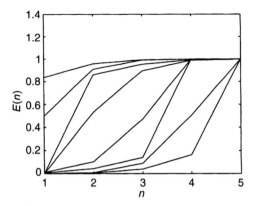

Figure 6.4 All 8 partial energy functions of the tap sets in Fig. 6.3.

Here, Δa are the differences between the data symbols in $s_i(t)$ and $s_j(t)$, and these differences are the sole aspect of the signals other than $h(\ )$ that is needed to find the distance. We will now show that Eq. (6.3-17) is in fact a *linear* form if $h(\ )$ is replaced by the autocorrelation ρ. $d^2(\ ,\ )$ has other useful properties as well those that stem from the autocorrelation.

A rewriting in terms of ρ follows easily by expanding Eq. (6.3-17) as

$$d^2 = \tfrac{1}{2}\log_2 M \sum_n \sum_k \int \Delta a[k] h(t-kT) \Delta a[n]^* h^*(t-nT)\, dt,$$

which is

$$d^2 = \tfrac{1}{2}\log_2 M \sum_n \sum_k \Delta a[k] \rho[n-k] \Delta a[n]^*, \qquad (6.3\text{-}18)$$

using definition (6.3-3). Furthermore, when $h(t)$ is a *PRS code* impulse response $\sum_\ell b[\ell] v(t-\ell T)$, v orthogonal, Eq. (6.3-14) gives $\rho[n-k]$ directly in terms of the taps as $\sum_m b[m+n-k] b[m]^*$. In summary, we have shown that given the data symbol differences Δa, distance depends on $h(\ )$ or on the generator taps only through their autocorrelation. Two generators with the same autocorrelation lead to identical code distance structures.

Euclidean distance may also be computed in the frequency domain, by Fourier-transforming the autocorrelation. This fact will play a major role in Section 6.5, and we delay discussion of it until then.

Equation (6.3-18) is a quadratic form in Δa, but the equation can be recast as a fully linear form by taking the autocorrelation of the Δa sequence as

$$\rho_{\Delta a}[k] \triangleq \sum_{n=-\infty}^{\infty} \Delta a[n+k] \Delta a[n]^*. \qquad (6.3\text{-}19)$$

Then the distance in Eq. (6.3-18) becomes

$$d^2 = \tfrac{1}{2} \log_2 M \sum_{k=-\infty}^{\infty} \rho[n]\rho^*_{\Delta a}[n]. \qquad (6.3\text{-}20)$$

We see that we can find d_f for $h(\ )$ by *optimizing a linear function* over the allowed $\rho_{\Delta a}$ sequences. This will be done in Section 6.4. $\rho_{\Delta a}$ has Hermitian symmetry and consequently

$$d^2 = \tfrac{1}{2} \log_2 M \left[\rho_{\Delta a}[0] + 2\sum_{k=1}^{\infty} \Re\{\rho[k]\rho_{\Delta a}[k]\} \right]. \qquad (6.3\text{-}21)$$

Note that when Δa has one nonzero component, the first term leads to d_{MF}^2 in Lemma 6.2-1; the tap set cases where $d_f^2 < d_{\text{MF}}^2$ are the ones for which the second term is negative.

It will be useful in Section 6.4 to have upper bounds on d_f in terms of ρ, since these will act to restrict the search over candidate Δa (or $\rho_{\Delta a}$). One such bound from Said [5] now follows.

LEMMA 6.3-1. d_f^2 is upper bounded by

$$d_f^2 \le \min_{k<0}\, \{2d_{\text{MF}}^2[1 - |\Re\{\rho[k]\}|]\}. \qquad (6.3\text{-}22)$$

Proof. The bound follows from Eq. (6.3-21) by restricting attention to data symbol differences of the form $\Delta a = (\alpha, 0, \ldots, 0, \pm\alpha^*, 0, \ldots)$, in which there are only two nonzero differences and $\pm\alpha^*$ appears in place k. The form $\rho_{\Delta a}$ contains the nonzero element $\pm|\alpha|^2$ in places $-k$ and k and $|\alpha|^2$ in place 0. The bracket in Eq. (6.3-21) is thus $2|\alpha|^2 + 2|\alpha|^2 \Re\{\rho[k]\} = 2|\alpha|^2[1 \pm \Re\{\rho[k]\}]$. For any k the minimum of this is the least of the allowed $|\alpha|^2$, which is $(2/\log_2 M)d_{\text{MF}}^2$ in Lemma 6.2-1, times twice $1 - |\Re\{\rho[k]\}|$. It remains to minimize over k, and from that follows Eq. (6.3-22). □

A restatement of this lemma is that

$$d_f^2 \le 2d_{\text{MF}}^2[1 - |\rho^\dagger|], \qquad (6.3\text{-}23)$$

where $|\rho^\dagger|$ is the second largest real part of the autocorrelation. For real-tap PRS encoders, the conclusion is particularly simple to state: the second largest autocorrelation sets an upper bound on d_f.[5] In the PRS tap set of Example 6.2-1, the lemma directly gives the free distance. The normalized taps are $1/\sqrt{2.62}\{1, 0.9, 0.9\}$. Their autocorrelation is $(1/2.62)\{0.9, 1.71, 2.62, 1.71, 0.9\}$

---

[5] Similar results appear elsewhere in signal processing theory; for example, the time to acquire a radar range or a timing mark in noise depends on the second highest autocorrelation of the signal.

PRS Coded Modulation

and $d_{MF}^2 = 2$. Equation (6.3-23) thus states that $d_f^2 \leq 2 \cdot 2[1 - 0.652] = 1.39$. 1.39 is indeed the square free distance for this code. In Example 6.2-2, the second highest autocorrelation is .243, so that $d_f^2 \leq 3.03$; but we already know that $d_f^2 \leq d_{MF}^2 = 2$.

Searching for Free Distance

PRS codes are trellis codes and free distance may therefore be found by a dynamic program over the trellis. As with TCM and CPM codes, the distance to the nearest neighbor depends on the path transmitted and in particular, on its starting state. A "double" dynamic program over both the transmitted and neighbor path symbols must in general be performed, and some details of this were given in Sections 4.3.3 and 5.2.2. With PRS codes and often with CPM codes, a faster search procedure is based on the so-called branch and bound idea. In addition, the search can be performed over the data symbol differences rather than the joint transmitted and neighbor symbol pairs; with binary and quaternary alphabets, there are 3 and 7 differences, respectively, compared to 4 and 16 joint pairs. We pause briefly to illustrate the branch and bound search with a PRS application.

In finding the PRS free distance, it is sufficient to search over all data symbol difference sequences that are nonzero beginning with a certain trellis stage, which may as well be stage 1. We can furthermore limit the search to sequences with $\Delta a[1] > 0$. The PRS code impulse response is $h(t) = \sum b_\ell v(t - \ell T)$, in which $v(t)$ is an orthogonal pulse. In general, v is noncausal and even if time-truncated, its main response occurs only after significant delay. The delay seriously lengthens the branch and bound search.

Fortunately, the causality problem with v is finessed as follows. Parallel with Eq. (6.2-3), we can write the difference signal that stems from some Δa as

$$\Delta s(t) = \sum_n \Delta c[n] v(t - nT), \quad \text{all } t, \qquad (6.3\text{-}24)$$

where $\Delta c[n]$ is given by the convolution $\Delta c[n] = \sum_{\ell=0}^{m} b_\ell \Delta a[n - \ell]$, $n = 1, 2, \ldots$. The square distance generated by $\Delta s(t)$ is then

$$D^2(\Delta a) = \int |\Delta s(t)|^2 \, dt$$

$$= \int \left[\sum_n \Delta c[n] v(t - nT) \right] \left[\sum_{n'} \Delta c^*[n'] v^*(t - n'T) \right] dt$$

$$= \sum_n |\Delta c[n]|^2, \qquad (6.3\text{-}25)$$

because v is orthonormal. Now consider the contribution to $D^2(\Delta a)$ from just the first N differences $\Delta a[1], \ldots, \Delta a[N]$ in Δa. Since $\Delta c[n]$ is not a function of future $\Delta a[N + 1], \ldots$, the contribution $\sum_1^N |\Delta c[n]|^2$ to $D^2(\Delta a)$ from these

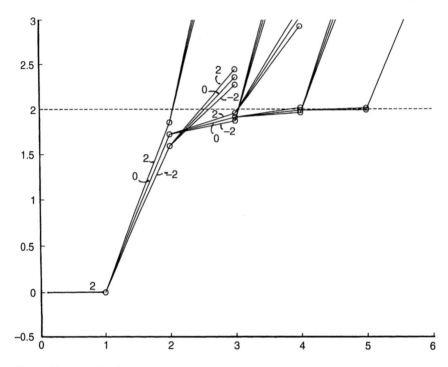

Figure 6.5 Tree of underbounds to normalized square distances for the tap set of Example 6.2-2. Each path of length N corresponds to a possible error difference sequence of this length. Bounds only increase with path length. Search for d_f may stop when all exceed the global bound (2 here). All difference sequences start with $\Delta a[1] = 2$; since $b_0 = 0.035$, bounds after one difference are ≈ 0. Small figures indicate $\Delta a[n]$: Leftmost single 2 is $\Delta a[1]$; next set are $\Delta a[2]$; rightmost sets are $\Delta a[3]$.

first N differences does not depend on the future differences. It is true that some of $\Delta a[1], \ldots, \Delta a[N]$ can affect $\Delta c[N+1], \ldots$ and certainly $\Delta a[N+1], \ldots$ affects it, but the distance contributions $|\Delta c[N + 1]|^2, \ldots$ are all nonnegative. Therefore $\Delta a[1], \ldots, \Delta a[N]$ creates an underbound $\sum_1^N |\Delta c[n]|^2$ to D^2 that holds for any continuation $\Delta a[N + 1], \ldots$ that may occur.

We can keep track of the bounds that correspond to initial segments of difference sequences in a tree structure like that of Fig. 6.5. The example there is the binary PRS code in Example 6.2-2, so that the tree begins with one branch corresponding to the one allowed value $\Delta a[1] = 2$, and is thereafter ternary branching, corresponding to $\Delta a[n]$ being 2, 0, or -2. Each node at depth N in the tree corresponds to an allowed sequence $\Delta a[1], \ldots, \Delta a[N]$, and associated with the node is the underbound $\sum_1^N |\Delta c[n]|^2$ on the distances of all sequences with this prefix. The vertical axis in the plot is the normalized value of the underbound, $(1/2) \sum_1^N |\Delta c[n]|^2$. The difference sequence of the form

PRS Coded Modulation

$\Delta a[1], \ldots, \Delta a[N], 0, 0, \ldots$ has a special significance because it represents a merged trellis path pair and its distance is a candidate for d_f.

The discussion here motivates the following branch and bound procedure. For clarity, we will call the bound that drives the procedure the global bound. Begin with a general upper bound of the form (6.3-22) for the global bound. Paths forward of a node for which $\sum_1^N |\Delta c[n]|^2$ exceeds the global bound *need not be explored*; they cannot yield the free distance. At every node form the full merged square distance $\sum_1^\infty |\Delta c[n]|^2$ corresponding to $\Delta a[1], \ldots, \Delta a[N], 0, 0, \ldots$.[6] If this distance is less than the global bound, it becomes the new global bound for the search. The global bound thus continually tightens. The search stops when all paths in the tree have hit the global bound.

In Fig. 6.5 the initial global bound from Eq. (6.3-22) is square distance 2; the bound is never reduced since 2 is in fact d_f^2. One path corresponds to $\Delta a = 2, 0, 0, \ldots$ and some values of $\Delta a[n]$ along paths are indicated. The tree search stops whenever a path's $(1/2) \sum_1^N |\Delta c[n]|^2$ hits or exceeds 2.

In finding d_f for a generator $b_0, \ldots, b_m$ with a given autocorrelation, it is important to apply the branch and bound procedure to the minimum phase tap set that leads to ρ. We have seen that the energy response to a unit isolated symbol difference grows fastest to 1 when the tap set is minimum phase. Some contenders for d_f may be quite complicated Δa, and there is no guarantee that the minimum phase tap set causes these to be evaluated most rapidly, but such a tap set usually does lead to the shortest branch and bound search. Note finally that the Δa leading to d_f is the same for all tap set phase versions; this is implied by Eq. (6.3-20).

6.3.3. Bandwidth and Autocorrelation

We now calculate the bandwidth of a PRS code and find that bandwidth, like distance, depends linearly on the tap set autocorrelation. The subsection concludes with some relationships between distance and bandwidth as bandwidth becomes small.

Since PRS coding is linear modulation, its bandwidth follows directly from Theorem 2.7-1: this states that for the signal $s(t) = \sqrt{E_s} \sum a_n h(t - nT)$ in Eq. (6.2-1), with the conditions on a_n given there, the power spectral density (PSD) is taken as $(1/T)|H(f)|^2$. For $h(t) = \sum_{\ell=0}^m b_\ell v(t - \ell T)$, the transform here is

$$|H(f)|^2 = \left| \mathcal{F} \left\{ \sum_{\ell=0}^m b_\ell v(t - \ell T) \right\} \right|^2 = |V(f)|^2 \left| \sum_0^m b_\ell e^{-j2\pi \ell f T} \right|^2. \quad (6.3\text{-}26)$$

---

[6] Note that the sum is in fact limited to $\sum_1^{N+m} |\Delta c[n]|^2$.

After some manipulation (see Problems), it can be shown that[7]

$$|H(f)|^2 = |V(f)|^2 \sum_{k=-m}^{m} \rho[k] e^{-j2\pi kfT} = |V(f)|^2 R_Z(e^{j2\pi kfT}), \quad (6.3\text{-}27)$$

in which $R_Z(z)$ is the autocorrelation z-transform in Eqs (6.3-4) and (6.3-5). Equation (6.3-27) shows that for a given pulse v, the PSD is determined by the autocorrelation. Potentially, very many tap sets may lead to the same autocorrelation and therefore the same PSD.

We would eventually prefer to work with bandwidth as a single parameter instead of a function $H(f)$, and this will bring out the correlation dependency in a different way. In Chapter 2 was introduced the concept of power out of band at baseband bandwidth W, the function $P_{OB}(W)$, the fraction of the PSD that lies outside the bandwidth $[-W, W]$ Hz. By setting $1 - P_{OB}(W) = \mathcal{C}$, we define a single-parameter measure $W_\mathcal{C}$, inside which lies the fraction $\mathcal{C}$ of the PSD. Most of the time we will take $\mathcal{C} = 0.99$, so that W is the 99% power bandwidth, denoted $W_{0.99}$. Further, we will usually consider real, baseband signals; then Eq. (2.7-5) set to 0.99 defines $W_{0.99}$. With the unit normalizations of v and $\{b_\ell\}$, Eq. (2.7-5) becomes

$$P_{OB}(W) = 1 - \mathcal{C} = \frac{\int_{|f|>W} (1/T)|H(f)|^2 \, df}{\int_{-\infty}^{\infty} (1/T)|H(f)|^2 \, df} = \int_{|f|>W} (1/T)|H(f)|^2 \, df.$$

Thus,

$$\mathcal{C} = \int_{-W}^{W} |V(f)|^2 \sum_{k=-m}^{m} \rho[k] e^{-j2\pi kfT} \, df,$$

$$= \sum_{k=-m}^{m} \rho[k] \int_{-W}^{W} |V(f)|^2 e^{-j2\pi kfT} \, df. \quad (6.3\text{-}28)$$

This defines W as a function of $\mathcal{C}$. The autocorrelation sequence ρ explicitly appears. In fact for a given W we may gather the remaining material in Eq. (6.3-28) together into the special discrete variable[8]

$$\chi[k] \triangleq \int_{-W}^{W} |V(f)|^2 e^{-j2\pi kfT} \, df, \quad (6.3\text{-}29)$$

---

[7] Note that although we assume that v is an orthonormal pulse, Eq. (6.3-27) holds for any v.
[8] Here we follow Said [5], Chapter 6.

PRS Coded Modulation

so that the power out of band constraint becomes the simple form

$$C = \sum_{k=-m}^{m} \rho[k]\chi^*[k]. \qquad (6.3\text{-}30)$$

This expression will be used in the optimization of Appendix 6B.

Said has called the quantity C the *spectral concentration* of the PRS code; it is the reverse of the power out of band. We will set it to 99% unless otherwise stated. For every tap set and v, there is a W that achieves the concentration. This W is our bandwidth measure.

Maximal Spectral Concentration

All tap sets with the same autocorrelation ρ lead to the same spectrum, but Eqs (6.3-27) and (6.3-28) show that the spectrum still depends on $V(f)$. In a PRS code, v is an orthogonal pulse. As $V(f)$ becomes more narrowband, the bandwidth W_C for concentration C in general falls. If the pulse v leads to the least W_C, the PRS total response $H(f)$ is said to have *maximum spectral concentration* (MSC). In fact, a whole range of v can lead to an $H(f)$ with MSC. Consider the integral $(1/T)\int_{-W}^{W} \sum_{-m}^{m} \rho[k] e^{-j2\pi k f T} df$, which is the concentration of power inside $[-W, W]$ of the tap set alone; it can be shown that when $W = 1/2T$, the value of the integral is unity. For concentration C there is thus some $W_C < 1/2T$ inside which lies fraction C. Now suppose that $|V(f)|^2 = T$ for f in the range $[-W_C, W_C]$; the root-RC pulses with parameter $\alpha < 1 - 2W_C T$ are such a class (see Section 2.2.2). Then bandwidth $[-W_C, W_C]$ also has concentration C of the power in $H(f)$, since the integral (6.3-28) will be unchanged. This holds true for all pulses with $0 < \alpha < 1 - 2W_C T$: all have the same spectral concentration. We will show next that it is maximal.

A condition for maximal concentration is given in the next proposition. To prepare the way for it, let us back up and consider *all* the total responses $h(t)$ which have autocorrelation sequence ρ. All of them have the same distance structure, since this depends only on ρ, but many power spectra are possible. If $h(t) = \sum_{\ell=0}^{m} b_\ell v(t - \ell T)$, with v orthonormal, as it does in Eq. (6.3-26), many spectra are possible depending on v. Some of these are shown in Fig. 6.6, which shows some spectra (6.3-27) for the tap set of Example 6.2-2 and the root-RC pulse v with $\alpha = 0.1, 0.5, 1$; the dotted curve is the discrete transform $R_Z(e^{j2\pi f T})$ alone. These spectra are quite different. The bandwidths at spectral concentration 0.99 are $W = 0.46/T, 0.63/T$, and $0.83/T$ Hz, for $\alpha = 0.1, 0.5, 1$. Many other h with this ρ can be demonstrated, other than those created by root-RC pulses, and Said explores these [5].

To investigate further, imagine a function $b(t)$ with transform $B(f)$, whose samples $b(0), b(T), b(2T), \ldots$ happen to be the tap sequence

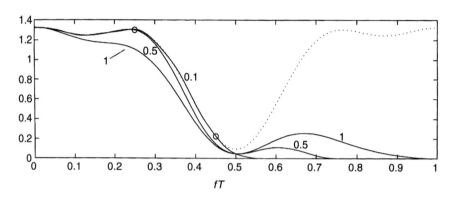

Figure 6.6 Power spectra of Butterworth $h(t)$ in Example 6.2-2, with root-RC pulses v having excess bandwidth $\alpha = 0.1, 0.5, 1$ (solid curves). Dotted curve shows $R_Z(e^{j2\pi fT})$ derived directly from tap set. $|H(f)|^2$ equals $R_Z(e^{j2\pi fT})$ out to 0.45 when $\alpha = 0.1$ and out to 0.25 when $\alpha = 0.5$ (shown by circles).

$b[0], b[1], b[2], \ldots$. The Fourier transform of this sequence is $R_Z(e^{j2\pi fT}) = \sum_k \rho[k] e^{-j2\pi k fT}$. From Property (6.3-10), $R_Z(e^{j2\pi fT})$ is the folded power spectrum $(1/T) \sum_n |B(f + n/T)|^2$. If $b(t)$ is bandlimited to $[-1/2T, 1/2T]$ – which means that the sequence $b[0], b[1], \ldots$ is infinite – then $R_Z(e^{j2\pi fT}) = |B(f)|^2$, $f \in (-1/2T, 1/2T)$; there is no aliasing in the formation of $R_Z(e^{j2\pi fT})$. With a finite tap set, R_Z contains an aliasing contribution. The spectral concentration in fact depends on how many $n \neq 0$ terms distort the central spectral term $|B(f)|^2$. Depending on the frequency range $[-W, W]$, $V(f)$ in the PRS scheme may modify $H(f)$ further.

Now we are prepared for two propositions. They are both due to Said [5].

Proposition 6.3-1. *Consider $W < 1/2T$ and a unit-energy total response function $h(t)$ having autocorrelation sequence ρ, which in turn has z-transform $R_Z(z)$. The responses h with MSC in $[-W, W]$ are those for which*

$$|H(f)|^2 = T R_D(fT) = T R_Z(e^{j2\pi fT}), \quad f \in [-W, W]. \qquad (6.3\text{-}31)$$

Proof. From Eqs (6.3-9) and (6.3-10), we have $R_{\text{fold}}(f) = T R_Z(e^{j2\pi fT}) = \sum_n |H(f + n/T)|^2$. Thus

$$|H(f)|^2 = T R_Z(e^{j2\pi fT}) - \Delta(f),$$

where $\Delta(f) = \sum_{n \neq 0} |H(f + n/T)|^2 > 0$, all f. The concentration C in $[-W, W]$ is

$$C = \int_{-W}^{W} |H(f)|^2 \, df = \int_{-W}^{W} T R_Z(e^{j2\pi fT}) \, df - \int_{-W}^{W} \Delta(f) \, df,$$

an expression that holds for any h with autocorrelation ρ. Clearly, C is largest when $\Delta(f) = 0$, $f \in [-W, W]$, from which follows Eq. (6.3-31). □

PRS Coded Modulation

Which $h(t)$ satisfy Eq. (6.3-31)? As it turns out, these include ones that stem from PRS codes with a certain range of v pulses. This is shown in the next proposition. Effectively, it implies that all useful PRS responses have the MSC property.

Proposition 6.3-2. *Suppose an orthonormal pulse v satisfies $|V(f)|^2 = T$ over the interval $(-W, W)$. Then for any tap set $\{b_\ell\}$, the response $h(t) = \sum b_\ell v(t - \ell T)$ has MSC in $(-W, W)$ and in any smaller interval.*

Proof. From Eq. (6.3-27), $|H(f)|^2 = |V(f)|^2 R_Z(e^{j2\pi fT})$. By hypothesis, $|V(f)|^2 = T$, $f \in [-W, W]$, and so the result follows from Proposition 6.3-1. □

The proposition is well illustrated by the $h(t)$ responses in Fig. 6.6. The root-RC pulse with $\alpha = 0.1$ is equal to T for $f \in (-0.45/T, 0.45/T)$. $0.45/T$ is marked with 'o' on the created $h(t)$; $|H(f)|^2$ equals $R_Z(e^{j2\pi fT})$ out to this mark. The $\alpha = 0.5$ root-RC pulse leads to the same property over $(-0.25/T, 0.25/T)$; $0.25/T$ is also marked. The $\alpha = 1$ pulse leads to an h that is not MSC in any bandwidth.

The tap set power spectra in Fig. 6.6 are not really typical of good PRS codes. Figure 6.7 shows spectra for the tap set $\{0.325, 0.656, 0.625, 0.271\}$, which is taken from an optimal codes list in the next section; it has the best free distance of any 4-tap code with 99% bandwidth $0.26/T$. The spectra created by root-RC $v(t)$ with either $\alpha = 0.1$ or 0.5 are indistinguishable from $R_Z(e^{j2\pi fT})$ out to $fT = 0.5$, and virtually zero thereafter. Since the concentration is 0.99, Proposition 6.3-2 implies that $v(t)$ with the condition $(1 - \alpha)/2T > 0.26/T$ leads to an $h(t)$ with MSC at the corresponding bandwidth; this means that $\alpha < 0.52$. v with $\alpha = 1$ leads to a slightly reduced $|H(f)|^2$ which has a small bump centered at $fT \approx 0.85$.

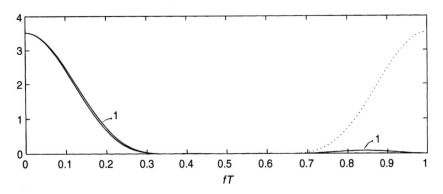

Figure 6.7 Power spectra of bandwidth 0.26 optimal PRS code with taps $\{0.325, 0.656, 0.625, 0.271\}$ and root-RC pulses having $\alpha = 0.1, 0.5, 1$. Dotted curve shows $R_Z(e^{j2\pi fT})$. Upper line is $\alpha = 1$ spectrum; lower line is $\alpha = 0.1, 0.5$ spectra.

Distance Bounds Based on Spectrum

It is only sensible that the free distance of a PRS code drops as bandwidth drops, since otherwise transmission would seem to be possible at error rate $\sim Q(\sqrt{d_f^2 E_b/N_0})$ in a channel whose Shannon capacity is dropping toward zero. In fact it is possible to demonstrate several useful bounds on d_f as a function of bandwidth. These depend in a subtle way on the bandwidth definition, and they all depend somehow on the concentration of power in the bandwidth in question.

The following propositions apply to any $h(t)$, PRS or not.

Proposition 6.3-3. *(Said [5]). Consider a unit energy total response $h(t)$ with spectral concentration $0 \leq C \leq 1$ in bandwidth $[-W, W]$, $W \leq 1/2T$. Then over the range $(-1/2T, 1/2T)$ Hz*

$$d_f^2 \leq 2 d_{MF}^2 [2 - C[1 + \cos(2\pi WT)]]. \tag{6.3-32}$$

Furthermore,

$$d_f^2 \leq 2 d_{MF}^2 [2(1 - C) + C 2\pi^2 (WT)^2], \tag{6.3-33}$$

which is a tight bound to Eq. (6.3-32) at small WT.

Proof. From Property (6.3-10),

$$T R_Z(e^{j2\pi fT}) \geq |H(f)|^2.$$

From the definition of C, therefore,

$$C \triangleq \int_{-W}^{W} |H(f)|^2 \, df \leq T \int_{-W}^{W} R_Z(e^{j2\pi fT}) \, df = \int_{-WT}^{WT} R_Z(e^{j2\pi f}) \, df. \tag{6.3-34}$$

$R_Z(e^{j2\pi f})$ is the discrete Fourier transform of ρ and from the properties of transforms,

$$\rho[k] = \int_{-1/2}^{1/2} R_Z(e^{j2\pi f}) e^{j2\pi fk} \, df.$$

Therefore $\Re\{\rho[1]\} = \int_{-1/2}^{1/2} R_Z(e^{j2\pi f}) \cos 2\pi f \, df$, since R_Z is real. It can be shown that

$$\Re\{\rho[1]\} \geq [1 + \cos 2\pi u] \int_{-u}^{u} R_Z(e^{j2\pi f}) \, df - 1, \quad \text{all } u \in [0, 1/2].$$

From this and Eq. (6.3-33), it follows that

$$\Re\{\rho[1]\} \geq [1 + \cos 2\pi WT] C - 1,$$

when $WT < 1/2$. Equation (6.3-32) now follows from Lemma 6.2-1. Equation (6.3-33) follows by use of the bound $\cos 2\pi WT \geq 1 - (2\pi WT)^2/2$. □

For a strictly bandlimited $h(t)$ with $\mathcal{C} = 1$ in $[-W, W]$, bound (6.3-33) for small WT becomes

$$d_f^2 \leq 4d_{MF}^2(\pi WT)^2. \qquad (6.3-35)$$

This has an important implication: For any such $h(t)$, $d_f^2 \to 0$ at least as fast as $(WT)^2$. It is true that the proposition is only an upper bound, but we will see in the next section that d_f^2 for optimal codes does in fact obey a quadratic law.

A second interesting point is that when $\mathcal{C} < 1$, bound (6.3-33) does *not* tend to zero as $(WT)^2$, but instead tends to $4d_{MF}^2(1 - \mathcal{C})$. This too is not solely due to a weak bounding technique. Said gives the following sharper result.

Proposition 6.3-4. *For any W, $0 < W < 1/2T$, and spectral concentration $\mathcal{C}$ in $(-W, W)$, there is at least one $h(t)$ for which*

$$d_f^2 \geq [1 - \mathcal{C}]d_{MF}^2. \qquad (6.3-36)$$

Some of this distance may lie outside the $(-1/2T, 1/2T)$ Hz bandwidth.

The reason for Eq. (6.3-36) is that with a $(-W, W)$ and concentration $\mathcal{C} < 1$, some distance escapes to bandwidths outside $(-W, W)$ and resides there no matter how narrow $(-W, W)$ becomes. The distance contribution inside $(-W, W)$ tends to zero as WT drops. If there is interference outside $(-W, W)$ which destroys the signal there, then the detection will act as if the distance is the smaller, inside value. Phenomena like this have been observed when bandwidth is very narrow for several code types, including CPM and PRS codes and the modulation plus filter coding in Section 6.5.

6.4. Optimal PRS Codes

By now we have seen that both distance and spectral concentration are simple linear functions of the tap autocorrelation sequence ρ of a PRS code. The functions are given by Eqs (6.3-20) and (6.3-30). After some earlier attempts at nonlinear optimization directly from the taps had appeared in the literature, it was Said [5] who suggested a linear programming approach based directly on ρ, which simultaneously optimized signal energy and bandwidth. He expressed the program in terms of constraints on the spectral concentration and the distances of a set of symbol difference sequences; a few other constraints arise as well. The method applies to multidimensional codes, and especially to the common practical case of coding the I and Q components in a bandpass signal. The details of Said's method appear in Appendix 6B. In this section we will explore the optimal solutions themselves.

Energy–Bandwidth Performance

We first define specific energy and bandwidth measures. These will make it easier to compare coded and uncoded systems, baseband and bandpass systems, and systems with differing alphabets.

The measure of distance, which becomes a measure of energy through the relation $p_e \sim Q(\sqrt{d_f^2 E_b/N_0})$, will be the *minimum distance loss* (MDL) defined by

$$\text{MDL} = -10\log_{10}(d_f^2/2) \text{ (in dB)}. \tag{6.4-1}$$

d_f is the data bit normalized minimum distance in Eq. (6.2-8) or free distance if appropriate. The MDL for a scheme is its loss in energy in dB, asymptotically as E_b/N_0 grows, compared to binary antipodal signaling, a scheme with $d_f^2 = 2$. It is the binary matched filter bound of Eq. (6.2-9). Note that most nonbinary antipodal schemes will show a loss when measured in this way; for example, 4-ary equi-spaced PAM has $d_{\text{MF}}^2 = 0.8$ and the MDL is $-10\log(0.8/2) = 3.98 \text{ dB}$.

The measure of bandwidth will be the *normalized bandwidth* (NBW) given by

$$\text{NBW} = \frac{TWN_{\text{dim}}}{\log_2 M} \text{ (Hz-s/data bit)}, \tag{6.4-2}$$

in which T is the coded modulation symbol time, M is its alphabet size, W is some single parameter baseband measure of bandwidth like 99% power bandwidth, and N_{dim} is the number of signal space dimensions used by the scheme per symbol time. For simple baseband transmission, $N_{\text{dim}} = 1$; for bandpass schemes with I and Q components, $N_{\text{dim}} = 2$. This states that an I and Q scheme uses twice the bandwidth of a baseband system. It reflects the fact that two independent baseband systems can be sent on the same carrier, or alternately, that jointly encoded I and Q signals can as well be carried as two baseband transmissions. In Eq. (6.4-2), W can be baseband bandwidth or the RF bandwidth of a corresponding bandpass system; the NBW is respectively baseband or RF.

The MDL and NBW are effectively per-data-bit normalized measures of energy and bandwidth. It is important to emphasize that they are a joint measure: An MDL of a certain size cannot be pronounced "good" without reference to the NBW, and vice versa. It is true, for example, that any 4-ary equispaced PAM has loss 3.98 dB compared to binary, but it is also true that its NBW is half the binary value. We will see that binary schemes may attain such an NBW only via significant PRS coding, and their distance loss may exceed 3.98 dB; they are thus "worse" in the joint sense.

Figures in this section show the NBW and MDL for optimum PRS codes with 4–15 taps. A selection of the actual tap sets is given in Tables 6A.1–6A.6 in Appendix 6A. The symbol alphabets are the three most important ones in practice, namely:

$$2\text{PAM}: \mathcal{A} = \{\pm 1\},$$
$$4\text{PAM}: \mathcal{A} = \{\pm 1/\sqrt{5}, \pm 3/\sqrt{5}\},$$
$$4\text{QAM}: \mathcal{A} = \{\pm 1/\sqrt{2}, \pm j/\sqrt{2}\}.$$

The last represents joint I and Q coding, which at memory 0 (single tap $b_0 = 1$) becomes ordinary QPSK. The formalism in the chapter analyzes the 4QAM case through letting all quantities become complex; details of this appear in [5,6]. The bandwidth criterion is the range $(-WT, WT)$ achieving 99% and 99.9% concentration ($P_{\text{OB}}(W) = 0.01$ and 0.001). A summary of the optimal code solution for each number of taps is as follows: choose $(-WT, WT)$; set constraint that $(-WT, WT)$ contains concentration 99% or 99.9%; find autocorrelation ρ with the largest d_f; find roots of ρ and take the minimum phase tap set.[9]

Binary Transmission

Figure 6.8 shows bandwidth (as NBW) against distance (MDL) for 2PAM, 99% concentration, and 4, 8, and 15 taps. Each symbol in the plot is a different set of taps; a selection of these tap sets is given in Appendix A. The first conclusion to draw from the figure is that by switching to a more complex code, energy can be saved (i.e. distance reduced) without bandwidth expansion, or bandwidth can be reduced without loss of distance. For each number of taps, there is a minimum possible NBW: at 4 taps it is about 0.26 Hz-s/data bit, at 8 taps it is 0.14, and at 15 taps it is 0.09. Each tap number also has a minimum bandwidth at which 0 dB distance loss is possible: at 4, 8, 15 taps, these bandwidths are 0.385, 0.37 and 0.355 Hz-s/data bit. By comparison, a single tap $b_0 = 1$ leads to 0.495 Hz-s/data bit. As the tap numbers increase there is a steady improvement in both distance and bandwidth. For all three tap numbers, the fall in d_f^2 as bandwidth drops closely follows the law $d_f^2 \sim (WT)^2$ suggested by Proposition 6.3-3 (the slope of the dashed line represents this law).

To focus on one set of optimal codes, Fig. 6.9 shows the optimal tap sets at bandwidth $WT = 0.26$ and 4, 6, 8, 10, 15 taps. The tap sets settle toward the 15 tap solution, which is roughly a lowpass filter impulse response. Figure 6.10(a) shows some power spectra (actually $|\sum b_\ell e^{-j2\pi \ell f T}|^2$), with the effect of the pulse $V(f)$ removed. It is clear that all are lowpass filters with cutoff ≈ 0.26. However, these are special lowpasses with maximal free distance, and it is also clear that the power concentration constraint $\mathcal{C}$ does not force down the stopband response any more than is exactly needed to achieve $\mathcal{C} = 0.99$. This pattern repeats generally

---

[9] This last is easily done, for example, by MATLAB routines; apply 'roots' to ρ, then 'poly' to the zeros inside the unit circle.

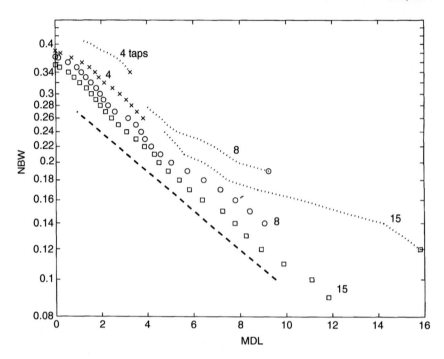

Figure 6.8 NBW against MDL for 2PAM codes with 99% spectral concentration and 4, 8, and 15 taps. Dotted trajectories show same cases with 99.9% concentration. Heavy dashed line shows the law $d_f^2 \sim (WT)^2$.

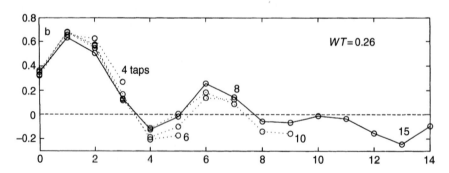

Figure 6.9 Optimal 2PAM tap sets for 4–15 tap codes having NBW 0.26 and 99% concentration.

for most tap sets. Figure 6.10(b) shows some power spectra for optimal 99.9% tap sets at the same bandwidth; these show much more stopband attenuation.

Figure 6.8 also shows as dotted lines the location of the optimal 2PAM tap set performances with the same tap numbers, but for the 99.9% concentration.

PRS Coded Modulation

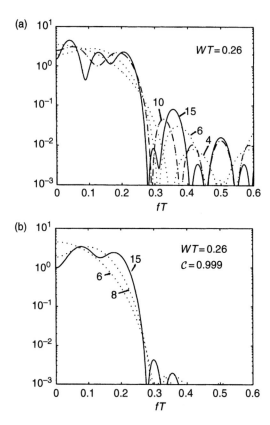

Figure 6.10 Power spectra of bandwidth 0.26 optimal codes in Fig. 6.9, plotted against normalized frequency. (a) 99% concentration codes, (b) 99.9% concentration 2PAM codes at bandwidth 0.26, for comparison; 6, 8, 15 taps.

These codes need 20–25% higher bandwidth than the 99% codes, or viewed along the distance axis, about 2 dB more energy. It is also true that these trajectories pull away from the $d_f^2 \sim (WT)^2$ law at narrow bandwidths. They need a wider $(-WT, WT)$ to achieve a given d_f^2.

Escaped Free Distance

The qualitative difference between the 99% and 99.9% bandwidth–energy plots in Fig. 6.8 – that the 99.9% results trend away from a straight line – can be explained in terms of an "escape" of free distance from the operating band $(-WT, WT)$. A similar phenomenon has been observed in very narrowband CPM, and we will see it again in Section 6.5.

The effect is easily quantified by the *difference power spectrum*. Heretofore, we have conceived of Euclidean distance as taken between time domain signals $s_1(t)$ and $s_2(t)$, but by Parseval's Theorem, it may equally well be taken between the frequency domain signals as

$$d^2(s_1(t), s_2(t)) = \int_{-\infty}^{\infty} |s_1(t) - s_2(t)|^2 \, dt = \int_{-\infty}^{\infty} |S_1(f) - S_2(f)|^2 \, df, \quad (6.4\text{-}3)$$

where $S_1(f)$ and $S_2(f)$ are the Fourier transforms of $s_1(t)$ and $s_2(t)$. It is thus natural to think of distance as residing in different regions of the spectrum, as having a distribution over bandwidth. Spectral distance analysis of the difference $\Delta S(f) = S_1(f) - S_2(f)$ often yields great insight.

The application to PRS signaling works as follows. Suppose a PRS code has spectral concentration $\mathcal{C}$ and free distance d_f; furthermore, d_f is achieved by the data difference sequence Δa. A difference power spectrum $|\Delta S(f)|^2$ may be defined for Δa as

$$|\Delta S(f)|^2 = \left| \mathcal{F}\left\{ \sum_\ell \Delta a_\ell v(t - \ell T) \right\} \right|^2,$$

$$= |V(f)|^2 \left| \sum \Delta a_\ell e^{-j2\pi \ell f T} \right|^2, \quad (6.4\text{-}4)$$

and it must be that $d_f^2 = \int_{-\infty}^{\infty} |\Delta S(f)|^2 \, df$.

Figure 6.11 plots this $|\Delta S(f)|^2$ for three of the optimal 2PAM codes with 99% concentration in $(-0.26, 0.26)$ that were just discussed. The pulse is 30% root RC. The difference spectra depend strongly on the tap number; this is because

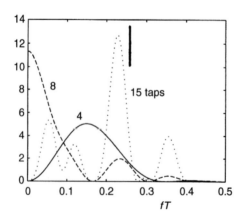

Figure 6.11 Illustration of escaped free distance: difference spectrum plot for error difference sequence leading to free distance, for NBW $= 0.26$ 99% 2PAM codes with 4, 8, 15 taps. An 18% escape occurs with 15 taps. Code taps and spectra shown in Figs 6.9 and 6.10.

PRS Coded Modulation

Table 6.1 Escape of free distance in 2PAM codes: total square distance d_f^2 and portion of this inside bandwidth $(-0.26, 0.26)$ Hz-s/data bit for 99% and 99.9% concentrations. Modulation uses 30% root-RC pulse. Error difference sequence shown achieves d_f^2 in the full bandwidth. Error spectra appear in Fig. 6.11 for 99% codes

	99% Concentration			99.9% Concentration		
$m+1$	Difference sequence	Sq. free distance	Sq. dist. in BW	Difference sequence	Sq. free distance	Sq. dist. in BW
4	2, −2	0.83	0.78	No code exists		
6	2, −2, 2	0.95	0.88	2, −2	0.55	0.54
8	2, −2, 2	0.97	0.92	2, −2, 0, 2, 0, −2, 2	0.69	0.67
10	2, −2, 2, −2, 2, −2, 2	1.09	0.99	2, −2, 0, 2, 0, −2, 2	0.70	0.68
15	2, −2, 0, 2, −2,	1.14	0.94	2, −2, 0, 2, −2, 0, 2, −2	0.84	0.82

the Δa that lead to d_f are very different in each case (details appear in Table 6.1). The phenomenon of escaped distance is visible in the 15 tap spectrum: 18% of the difference energy that makes up d_f^2 is in a region $0.3 < |fT| < 0.4$ that lies entirely outside the design bandwidth $(-0.26, 0.26)$. If the detection operates only inside $(-0.26, 0.26)$ Hz-s/data bit – for example, because of interference outside this band – then the distance of the 15 tap code must be taken as 18% less, 0.94 instead of 1.14. The escape of distance appears to be less serious in the 4 and 8 tap cases, and when the spectral concentration is 99.9%, the table shows that only 1–2% of d_f^2 escapes outside the design bandwidth.

The outward escape of free distance worsens as signal design bandwidth drops. A rather large escape occurs at $WT = 0.12$. This is shown in Fig. 6.12 for the 99% (dashed curve) and the 99.9% concentrations (solid curve), taking the 15 tap optimal code in both cases. The 99% code has $d_f^2 = 0.255$, but only 41% of this lies inside bandwidth $(-0.12, 0.12)$; the rest escapes outside and the signal components that create it are part of the 1% of signal power allowed out of band. The 99.9% code has d_f^2 only 0.053, but 74% of this lies in $(-0.12, 0.12)$.

Propositions 6.3-3 and 6.3-4 are not conclusive, but they do raise the possibility that some free distance, possibly as much as $4d_{MF}^2(1-C)$, escapes out of band, in the limit $WT \to 0$. In any case, a serious escape of distance is an observed fact as WT becomes small. It means that codes optimized for a 99.9% or higher concentration are the important ones to consider when the detector is truly limited to the design bandwidth. Optimizations at very high concentrations can be affected by numerical instability.[10]

---

[10] Said [5], Chapter 6, gives a modification of the method in Appendix 6B that finds a class of codes with 100% concentration. Note that any such code must be infinite response.

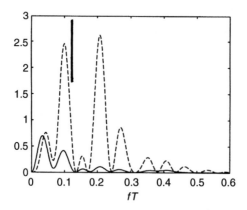

Figure 6.12 Escaped free distance: comparison of 99% (dashed) and 99.9% (solid) concentration 2PAM codes with NBW = 0.12 and 15 taps. Difference spectrum plots for error difference leading to free distance in the unrestricted bandwidth. 99% code has 41% loss.

Higher Alphabet Sizes

Figure 6.13 shows the bandwidth–energy performance of the optimal 15 tap 4PAM and 4QAM partial response codes, along with the 2PAM case for comparison. Star, diamond, and square symbols, respectively, show individual codes at 99% concentration, while dotted lines show only a trajectory for the 99.9% codes. Here are some observations about these codes.

1. Larger alphabet sizes are an important way to reduce bandwidth at a fixed d_f: 4PAM and 4QAM are uniformly 15% more narrowband than 2PAM. Alternately, 4PAM and 4QAM codes save 1–1.5 dB in energy at the same bandwidth. It is thus important to use these when the link allows.

2. Furthermore, 4PAM codes can work in bandwidths below $WT = 0.09$ which are unavailable to 15 tap binary codes; but there is an upper limit to their usefulness in terms of bandwidth and d_f. The lesson here applies elsewhere in coding as well: the type of coding and its alphabet size need to be tuned to the energy and bandwidth resources that are available.

3. 4QAM, which implies cross coding in two signal space dimensions, has a definite range of usefulness. In a middle range of energy and bandwidth, 4QAM with two binary alphabets has performance that matches 4PAM with its more difficult to implement quaternary alphabet. If we view 4QAM as two jointly coded 2PAMs, the conclusion is clear that such multidimensional coding leads to a definite performance improvement.

4. The 99.9% concentration codes show some departure from the $\sim (WT)^2$ rule in the 99% codes, just as in the 2PAM case.

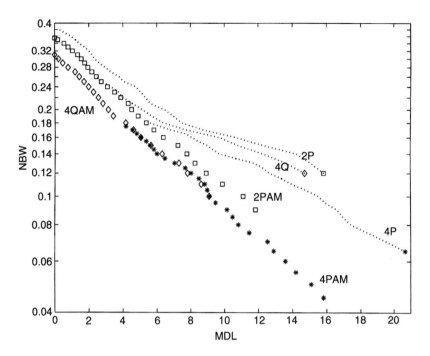

Figure 6.13 Comparison of 2PAM (squares), 4PAM (stars) and 4QAM (diamonds) codes. NBW against free distance loss for optimal codes with 99% concentration and 15 taps. Dotted trajectories show same cases with 99.9% concentration.

The overall performance of PRS codes was compared to TCM and CPM in Chapter 1. There is a distinct region, given approximately by $WT < 0.3$ and $E_b/N_0 > 12$ dB, where they perform better than the other code types.

4QAM codes, with their complex taps, have some interesting special properties that follow from the fact that the autocorrelation ρ is no longer real. One consequence is that the power spectrum $|H(f)|^2$ (see Eqs (6.3-26) and (6.3-27)) is in general *asymmetric*. An example, taken from [6], is shown in Fig. 6.14; the code here is the optimal 15 tap 4QAM code for 99.9% concentration in NBW 0.25 Hz-s/data bit. In a bandpass QAM implementation, this code would have positive-frequency RF bandwidth twice 0.25, or 0.50 Hz-s/QAM symbol, and in finding the very best tap set, the asymmetry means that it is necessary to let this bandwidth move up and down the spectrum. The power out of band definition in Eq. (6.3-28) needs to change from

$$P_{OB}(W) = 1 - C = \int_{|f|>W} (1/T)|H(f)|^2 \, df$$

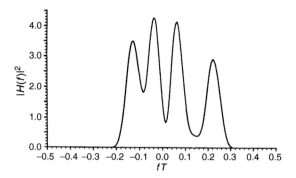

Figure 6.14 Example of asymmetric power spectrum for an optimal 4QAM code. 15 taps, NBW = 0.25, 99.9% concentration. (From Said and Anderson [6], copyright 1998 IEEE; used with permission.)

to

$$P_{\text{OB}}(W) = 1 - C = \int_{-W+\varphi/T}^{W+\varphi/T} (1/T)|H(f)|^2 \, df, \qquad (6.4\text{-}5)$$

in which φ, $-W < \varphi < W$, is a frequency offset. The optimization now finds the least d_f^2 over both the tap set and φ. For the case in Fig. 6.14, the best offset turns out to be $\varphi/T = 0.045$. The power spectrum is asymmetric both in its shape and its offset. In 4QAM, Tables 6A.5 and 6A.6, optimal offsets are given for each code.

Another property evident in Fig. 6.14 is that the spectrum is by no means flat. This property extends to the symmetrical 2PAM and 4PAM spectra as well.

The Worst Case ISI Problem

Magee and Proakis [13] posed the following problem for ISI with a discrete-time linear model: given a model with $(m+1)$ taps and unit energy, what is the *worst* d_f the model can have? The classic application is a mobile communication link with time-varying ISI which has an MLSE receiver; as the channel changes, the worst asymptotic error rate that can occur is $\sim Q(\sqrt{d_w^2 E_b/N_0})$, where d_w is this worst free distance. Earlier in this section, the object has been to maximize d_f with a constraint on bandwidth. Now instead we seek to minimize it, with no constraint on bandwidth.

Reference [13] expresses the square distance from a given error difference sequence as the quadratic form bCb', where b is the model tap vector and C is the correlation matrix of the error sequence (i.e. $C_{ij} = \rho_{\Delta a}[|i - j|]$, with $\rho_{\Delta a}$ from Eq. (6.3-19)). Then the worst distance for this error sequence is the minimum eigenvalue of C and the taps are the corresponding eigenvalue. The procedure must be repeated for each error sequence and a global minimum taken. Table 6.2 lists

PRS Coded Modulation

Table 6.2 Square free distance losses in dB below 2 for worst-case ISI discrete-time channel models of memory m. Unique worst-case channels (except $m + 1 = 2$); all are symmetric. Worst-case error difference sequence may not be unique

$m+1$	Loss (dB)	Worst channel taps	Worst error sequence
1	0	1	2
2	0	Any unit energy set	2
3	2.3	0.50,0.71,0.50	2, −2
4	4.2	0.38,0.60,0.60,0.38	2, −2
5	5.7	0.29,0.50,0.58,0.50,0.29	2, −2
6	7.0	0.23,0.42,0.52,0.52,0.42, …	2, −2
7	8.2	0.19,0.35,0.46,0.50,0.46, …	2, −2
8	11.5	0.16,0.24,0.43,0.49,0.49,0.43, …	2, −2, −2, 2, 2, −2
9	13.2	0.11,0.21,0.35,0.46,0.48,0.46, …	2, −2, −2, 2, 2, −2
10	14.7	0.09,0.16,0.31,0.39,0.47,0.47,0.39, …	2, −2, −2, 2, 2, −2

the worst case taps, sequences, and distance losses up to 10 taps.[11] It is interesting that there is no loss at all when MLSE is applied to 2 tap models. Thereafter the loss increases 1.3–2 dB with each additional tap. The worst case taps are always symmetric.

6.5. Coded Modulation by Outright Filtering

Obtaining a coded modulation through some form of linear filtering is the theme of Chapter 6. Until now it has meant PRS coding, which is discrete-time finite-response encoding, and most of the baseband signal bandwidths are less than $0.5/T$ Hz. We turn now to Cases (ii) and (iii) in Section 6.2. The fact is that simply filtering an ordinary modulation, such as PSK or FSK, with a common analog filter, for example, a Butterworth, will produce an admirable coded modulation. Even modulation of an impulse train to produce $\sum a_n \delta(t - nT)$, when applied to a lowpass filter, can generate signals with good bandwidth–energy properties. This section looks at properties of such filtered signals. Some of these properties resemble those of FTN signaling, an extension of orthogonal pulse theory that arose in the 1970s, and so the story will begin there.

In fact, finding good codes and constructing this kind of encoder is not difficult, and it is the decoding that is more subtle, largely because the encoders are infinite state. Fortunately, effective schemes have been discovered, but the discussion of these will be delayed until Section 6.6.3.

---

[11] There are a number of errors in the original table of [13].

6.5.1. Faster-than-Nyquist Signaling

We have seen in Section 2.2 that T-orthogonal pulses exist with bandwidth as small as $(-0.5/T, 0.5/T)$ Hz (Theorem 2.2.2); after matched filtering in the linear receiver the pulse train consists of Nyquist criterion pulses, from which in the absence of noise the symbols $a_0, a_1, \ldots$ can be obtained from the samples at times $0, 1, \ldots$ without intersymbol interference. The most narrowband transmission pulse is $\text{sinc}((t - nT)/T)$, sent at times $n = 0, T, 2T, \ldots$. As shown in Sections 2.2 and 2.3, the linear receiver, working from matched filter samples at times $0, T, \ldots$, performs maximum likelihood detection of $a_0, a_1, \ldots$. The symbol error probability is $\sim Q(\sqrt{d_{\text{MF}}^2 E_b/N_0})$, where d_{MF} is effectively the symbol alphabet minimum distance, and in the binary antipodal case, the error probability is precisely $Q(\sqrt{2E_b/N_0})$.

If T-orthogonal pulses are transmitted faster than rate $1/T$, there must be ISI and each a_n cannot be ML-detected simply from the nth sampling of the matched filter. The transmission is said to be FTN. It will be easiest to explore what now happens if we restrict to the binary case and pulses $v(t)$ whose transform is flat in $(-W, W)$ Hz and zero otherwise. This is Case (iii) signaling. ISI in the filter samples will grow worse as $1/T$ grows beyond $2W$, or alternately, as the bandwidth $(-W, W)$ falls below $(-0.5/T, 0.5/T)$ for a fixed T. For concreteness, take the latter case and set the pulse spectrum to the unit energy expression

$$|V(f)|^2 = \begin{cases} T/\theta, & -\theta/2T \leq f \leq \theta/2T, \\ 0, & \text{otherwise,} \end{cases} \quad (6.5\text{-}1)$$

in which $0 < \theta < 1$. The pulse is $v(t) = \sqrt{\theta/T} \text{sinc}(\theta t/T)$. The linearly modulated pulse train is $\sqrt{E_s} \sum a_n v(t - nT)$, but v is no longer T-orthogonal.

Signaling is still possible, but the detector eye narrows and the asymptotic error probability is $\sim Q(\sqrt{x E_b/N_0})$, with $x \leq 2$. The precise value of x may be found by checking the worst case ISI. However, if a full ML *sequence* detection is performed, which compares in principle all 2^K K-symbol signals to the received signal, then a surprising thing happens: $p_e \sim Q(\sqrt{2E_b/N_0})$, the same as with $\sqrt{1/T} \text{sinc}(t/T)$ pulses, until the band of $V(f)$ falls below $(-0.401/T, 0.401/T)$ Hz. This phenomenon was announced by Mazo in 1975 [16]; proofs of the 0.401 factor appeared later [17,18]. To summarize, the FTN phenomenon is that for brickwall filtered linear signaling the minimum distance remains effectively at d_{MF} until quite a narrow bandwidth, and in particular, until considerably narrower bandwidth than the limit to Nyquist criterion pulses.

The underlying reason for the FTN phenomenon is that the antipodal pair of signals leads always to square distance 2, whereas the next-to-nearest pair is forced steadily closer by the narrowing bandwidth. At bandwidth $(-0.401/T, 0.401/T)$ Hz its distance drops below 2 and thus so must d_{min}^2; in fact, the offending difference sequence in the present case is $2, -2, 2, -2, 2, -2, 2, -2$.

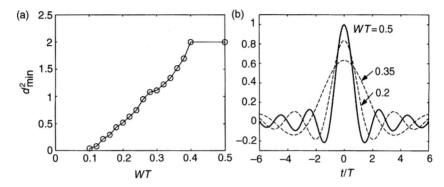

Figure 6.15 FTN signaling. (a) d_{min}^2 against NBW WT for pulses with brickwall $(-WT, WT)$ spectrum. (b) Some time pulses for $WT = 0.2, 0.35, 0.5$ Hz-s/data bit.

Figure 6.15(a) plots d_{min}^2 for the brickwall signaling case in Eq. (6.5-1) as a function of the signal bandwidth normalized to T; Fig. 6.15(b) plots some of the "too slow" v pulses.

We are interested in the FTN phenomenon because its pattern repeats with many linearly filtered coded modulations. Also, we see here for the first time a linear encoding system with *infinite response*: the pulse $v(t)$ can equally be generated by passing $\sqrt{1/T}\,\mathrm{sinc}(t/T)$ or another standard T-orthogonal pulse with flat spectrum in $(-\theta/2T, \theta/2T)$ through a lowpass filter realized in discrete time (see Problems). To realize the spectrum in Eq. (6.5-1), the filter must have infinite response.

6.5.2. Euclidean Distance of Filtered CPM Signals

The next three subsections continue the theme of infinite response filtering. They explore severe filtering of a modulation, which is Case (ii) signaling. We will first catch up with a topic related to Chapter 5, the filtering of CPM signals.

In Section 6.4 we introduced the difference power spectrum as a tool for analyzing the relationship between bandwidth and Euclidean distance. It will now provide much insight into what happens to the distance of phase modulations, or of CPM coded modulations, when they are filtered. We will find a FTN-like tendency: no energy loss down to a surprisingly narrow bandwidth, provided that means is available to detect the more complicated signals. The attractive bandwidth–energy performance of these signals makes them important "coded" modulations – if we accept filtering as a means of encoding.

The difference spectrum expression (6.4-3) applies just as well to signals after filtering. If all transmissions $s(t)$ are filtered by some $G(f)$, then the square distance between $s_1(t)$ and $s_2(t)$ after filtering has a particularly simple form:

$$\int_{-\infty}^{\infty} |s_1(t)*g(t) - s_2(t)*g(t)|^2\,dt = \int_{-\infty}^{\infty} |[s_1(t) - s_2(t)]*g(t)|^2\,dt,$$

$$= \int_{-\infty}^{\infty} |S_1(f) - S_2(f)|^2 |G(f)|^2\,df,$$

$$= \int_{-\infty}^{\infty} |\Delta S(f)|^2 |G(f)|^2\,df; \quad (6.5\text{-}2)$$

that is, the difference spectrum at the input is simply filtered to produce the one at the output. For the special case where $G(f)$ is an ideal lowpass filter with unit-amplitude passband $(-W, W)$ Hz, Eq. (6.5-2) yields the portion of distance inside $(-W, W)$.

If we view $G(f)$ as synthesizing a new coded modulation, then distances like these should be properly normalized. For baseband linear modulation, a straightforward argument goes as follows. Let a modulation $\sqrt{E_s}\sum a_n v(t-nT)$ with unit orthogonal pulses v and symbol energy E_s be filtered by $G(f)$ to produce $s(t) = \sqrt{E_s}\sum a_n h(t-nT)$, where $h = v*g$. With only a little loss of generality, let $V(f)$ take constant value $\sqrt{T}$ over the support of $G(f)$; that is, $G(f)$ is narrower band than the flat spectrum of the basic pulse. From Theorem 2.7-1, the "encoded" average energy spectrum is in general

$$E_s \mathcal{E}[|a_n|^2]|H(f)|^2 = E_s |V(f)|^2 |G(f)|^2$$

on a per-symbol basis, and the average energy is the integral

$$E_s \int_{-\infty}^{\infty} |V(f)|^2 |G(f)|^2\,df,$$

which is $E_s \int T|G(f)|^2\,df$ under the assumptions just made. After the standard normalization, then, the signal distance in Eq. (6.5-2) becomes

$$d(s_1(t), s_2(t)) \triangleq \frac{\int |s_1(t) - s_2(t)|^2\,dt}{2E_b} = \frac{\int |\Delta S(f)|^2 |G(f)|^2\,df}{(2E_s/R) \int T|G(f)|^2\,df} \quad (6.5\text{-}3)$$

for signaling at rate R bits/symbol interval.

For a clarifying example, we can take Mazo's basic binary FTN scheme. We can construct the system with bandwidth $(-\theta/2T, \theta/2T)$, $0 < \theta \leq 1$, by taking $v(t)$ as the T-orthogonal pulse $\sqrt{1/T}\,\text{sinc}(t/T)$, setting $s(t) = \sqrt{E_s}\sum a_n v(t-nT)$, and taking $G(f)$ as the filter in Eq. (6.5-1). Consider the error difference $\Delta a = 2$. $\Delta s(t)$ is then $2\sqrt{E_s/T}\,\text{sinc}(t/T)$, and $|\Delta S(f)|^2$ takes value $4E_s T$ for $-0.5/T \leq f \leq 0.5/T$. The second integral in Eq. (6.5-3) yields 2, no matter what θ is. The square minimum distance of the FTN scheme will be 2 so long as

PRS Coded Modulation

no other signal pairs have square distance less than 2. $d_{\min}^2$ indeed remains 2 until the bandwidth falls below $(-0.401/T, 0.401/T)$, as discussed in Section 6.5.1.

In order to work with CPM signals, we extend Eq. (6.5-2) to signals in I/Q form. Let the standard CPM signal be $s(t) = \sqrt{2E_s/T}\cos(\omega_0 t + \phi(t))$. From Eq. (2.5-20), $\int |\Delta s(t)|^2\,dt$ for any bandpass signal may just as well be taken as $E_s/T \int [|\Delta I(t)|^2 + |\Delta Q(t)|^2]\,dt$. The effect of a bandpass filter $g(t)$ on a bandpass signal may be broken into a baseband equivalent circuit, in which equivalent filters g^I and g^Q both act on the I and Q components of $s(t)$. We simplify the discussion, while still including most cases of practical interest, by assuming that $G(f)$ is symmetrical about the signal center frequency f_0; this[12] sets $g^Q(t) = 0$. The bandpass $g(t)$ and baseband g^I are related by

$$g(t) = g^I(t)\cos(\omega_0 t), \quad \omega_0 = 2\pi f_0, \tag{6.5-4}$$

and the filtered I and Q baseband signals are $g^I(t) * I(t)$ and $g^I(t) * Q(t)$. By Parseval once again, then, the distance between two filtered CPM signals becomes

$$\int_{-\infty}^{\infty} |\Delta s(t)|^2\,dt = (E_s/T)\int_{-\infty}^{\infty} |G^I(f)|^2 [|\Delta I(f)|^2 + |\Delta Q(f)|^2]\,df. \tag{6.5-5}$$

Equation (6.5-5) may be used to analyze the effect of band limitation on CPM, but it is more interesting now to treat the filter as synthesizing a new coded modulation. The proper transmitted symbol energy is then the average energy per symbol *after* filtering, hereafter denoted $E_s^{(g)}$. The normalization to square distance is $2E_s^{(g)}/R$. Let the notation $|\tilde{S}(f)|^2$ denote the baseband per-symbol energy density of the modulated signal $s(t)$ (baseband spectra are discussed in Section 2.7 and CPM spectra in Section 5.3). After the filtering by $G(f)$, the density of $s(t)$ becomes $|G^I(f)|^2|\tilde{S}(f)|^2$, and the average symbol energy can be taken as

$$E_s^{(g)} = \int_{-\infty}^{\infty} |G^I(f)|^2|\tilde{S}(f)|^2\,df. \tag{6.5-6}$$

Note that the new coded modulation is almost never constant envelope and that it is still nonlinear modulation. Furthermore, the trellis of these codes has *infinitely many states* whenever $G(f)$ has infinite response. For this reason we will use the notation $d_{\min}$ for minimum distance, reserving d_f for traditional finite state trellis codes. Despite the infinite response, effective decoding algorithms exist (in Section 6.6.3).

CPM Minimum Distance under Ideal Lowpass Filtering

When $G(f)$ is a brickwall filter with $G(f) = 1$, $f_0 - W < |f| < f_0 + W$, and 0 otherwise, $d_{\min}$ has a behavior as bandwidth falls that mimics the FTN

---

[12] For a derivation of the full equivalence and of this symmetry property, see [1,22].

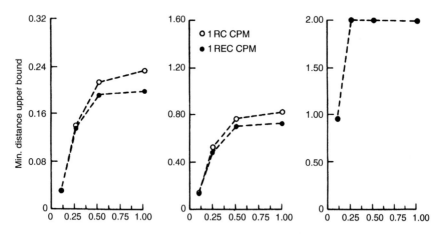

Figure 6.16 Normalized minimum distance against one-sided brickwall filter bandwidth WT, for filtered CPFSK and 1RC binary CPM codes. Modulation indices 0.125, 0.25, 0.5, left to right. RF bandwidth $= 2WT$. (From Seshadri and Anderson [20], copyright IEEE 1988; used with permission.)

behavior. Seshadri [19,20] calculated many cases of the filtered CPM minimum distance. Using ideal lowpass filters of differing widths, he computed the distance via Eq. (6.5-5) and normalized by the post-filtering average energy in Eq. (6.5-6). In order to find $d_{\min}$ he tried a large number of difference sequences, out to a length much greater than the one that led to the apparent $d_{\min}$.[13]

Some typical outcomes of this calculation, taken from [20], are shown in Fig. 6.16. The binary CPM schemes 1REC (i.e. CPFSK) and 1RC are shown for the three modulation indices $h = 0.125, 0.25, 0.5$. In all cases the pattern is the same: Brickwall filtered schemes have the unfiltered-scheme minimum distance until W in the RF bandpass $[f_0 - W, f_0 + W]$ falls below a critical value; thereafter the distance drops rapidly. This is the classic FTN phenomenon. For example, $d_{\min}$ for 1REC, $h = 0.5$, which is equivalent to MSK with filtering, remains at the full value 2 until $W \approx 0.25/T$ Hz. This RF bandwidth, about $0.5/T$, is far below that associated with MSK,[14] and is in fact half that of binary sinc( ) pulse bandpass modulation.

The FTN phenomenon continues to occur in the $h = 0.25$ and 0.125 cases, but there are two interesting differences. The drop from the wideband distance value actually occurs at *wider* bandwidths, despite the smaller indices. Second, the $d_{\min}^2$ values achieved by the index 0.125 schemes at narrow bandwidths are generally lower, not higher, than those of index 0.25 schemes constrained to the

---

[13] This is a standard technique. Strictly speaking, it gives only an upper bound to $d_{\min}$, but it has proven to be a strongly reliable one. The technique is much improved by certain distance properties, especially those in Section 6.5.3.

[14] The RF 99% bandwidth of MSK is $1.18/T$.

PRS Coded Modulation

same bandwidth; similarly, index 0.25 schemes are poorer than MSK. Lowering the index in CPM, which concentrates the PSD around f_0 while maintaining the constant envelope property, does not necessarily concentrate the spectral location of the minimum distance. Combining a low index with a brickwall bandwidth limitation is apparently not always a good way to synthesize a coded modulation.

The FTN phenomenon occurs because the wideband $d_{\min}$-achieving error difference sequence, which is $\Delta a = +2, -2, 0, 0, \ldots$ in all the cases above, continues to drive $d_{\min}$ even as the bandwidth limitation drops, until a critical limit is reached. Furthermore, the drop in $E_s^{(g)}$ caused by filtering just balances the drop in distance that occurs with Δa.

A roughly similar phenomenon may be observed for brickwall filtering of other CPM schemes. For example, filtered binary 3RC loses no distance over the wideband case until W falls at least to $0.5/T$ Hz, for $h = 0.5, 0.25, 0.125$. In fact binary 3RC with $h = 0.5$ and $W = 0.25/T$ actually has a slight *gain* over the wideband $d_{\min}$ [20]; this apparently occurs because $E_s^{(g)}$ drops faster than the critical difference signal energy. For all binary 3RC and 3REC codes in this h range, the difference sequence leading to $d_{\min}$ at wide bandwidth is $+2, -2, 0, 0, \ldots$. A striking example of distance gain occurs with CPFSK: at index $h = 0.715$ and brickwall filtering to $W = 0.25/T$, $d_{\min}^2$ is apparently 4.2, compared with the wideband value of 2.43.

An explanation of these various gains and losses is as follows. In the CPM schemes with larger h, the critical spectral content that sets $d_{\min}$ seems to be located near the center frequency. Removing outer regions of spectrum does not obscure the differences among closely lying signals, and by removing unimportant spectrum, the energy efficiency of the coded modulation is kept high and may even be improved. Just the opposite happens with small-index CPM: the critical regions for $d_{\min}$ lie far out and are easily lost. The inner regions work mostly to maintain a constant envelope.

It is worth reiterating that PRS and other coded modulations that work by linear filtering of a pulse stream cannot increase the wideband $d_{\min}$ in this way. CPM, however, is not linear.

CPM with Practical Filters

When the brickwall filter in these cases is replaced by a Butterworth filter with the same cut-off frequency, the behavior of $d_{\min}$ is very similar. Figure 6.17, taken from [20], shows 4-pole filtering of the binary 1RC cases in Fig. 6.16. The figure also shows the effect of a depth limit (denoted "N") on the paths that are searched for $d_{\min}$. The minimum distance so obtained increases with the depth limit until it reaches the value for an unrestricted search; this indicates the decision depth L_{dec} of the code, and it appears to be about 10 in the figure's cases.

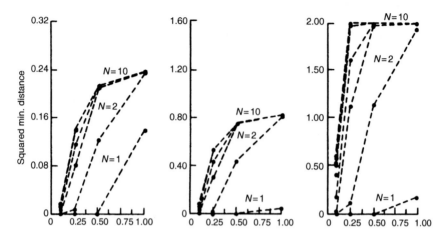

Figure 6.17 Normalized minimum distance against filter cut-off frequency WT, for 1RC binary CPM codes filtered by 4-pole Butterworth filters. Modulation indices 0.125, 0.25, 0.5, left to right. N indicates decoder tree depth limit. (From Seshadri and Anderson [20], copyright IEEE 1988; used with permission.)

With a full depth, Fig. 6.17 is very similar to Fig. 6.16. So also are the same tests with 1REC CPM and with 2-pole filters. A strong FTN phenomenon occurs, with no loss of $d_{\min}$ until the filter cut-off falls to a certain bandwidth.

6.5.3. Critical Difference Sequences at Narrow Bandwidth

What signals are most likely to cause error? When there is no restriction on bandwidth, we have seen several general properties about the signal pair that lies closest and leads therefore to the code minimum distance. We can call such close lying pairs critical difference sequences. Here is a review of the properties encountered so far. In the simpler CPM codes, with moderate or small index, signal pairs with the difference sequence $\Delta a = 2, -2$ lead to $d_{\min}$; this continues to hold under moderate filtering. In PRS coding, or in any other code synthesized via filtering of linear modulation, $d_{\min}$ cannot exceed the antipodal signal bound d_{MF}. When filter bandwidth is moderate or large, $d_{\min}$ is indeed d_{MF} as a rule, and $\Delta a = 2$. As bandwidth drops, other critical sequences lead to $d_{\min}$ and $d_{\min}$ drops.

At very narrow bandwidths, a new property comes to the fore: the critical difference sequences are those whose baseband spectrum have a *DC null*. For a baseband linear coded modulation, this is to say that Δa satisfies $\Delta S(f) = \mathcal{F}\{\sum \Delta a[n] h(t - nT)\} = 0$ at $f = 0$. The null condition is in turn equivalent to

$$\int \sum \Delta a[n] h(t - nT) \, dt = 0. \qquad (6.5\text{-}7)$$

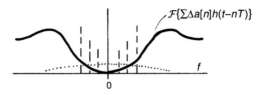

Figure 6.18 Illustration of successively narrower bandlimitation acting on difference spectra of certain error differences. Solid curve has a DC null and must eventually have less distance than the other curve.

A sufficient condition for this is that the difference sequence satisfies

$$\sum_n \Delta a[n] = 0. \tag{6.5-8}$$

Examples of this *zero-sum rule* occur throughout this chapter. An early one is the free distance of the narrowband code $\{b_\ell\} = \{1, 0.9, 0.9\}$ in Example 6.2-1, which occurred with $\Delta a = 2, -2$; the next-nearest neighbor in this example has $\Delta a = 2, -2, 0, 2, -2$. In fact, most error difference sequences that lead to d_f in this chapter – and virtually all when $WT < 0.25$ – satisfy the zero-sum rule.

Why are zero-sum differences the most dangerous ones at narrow bandwidth? The reason is illustrated in Fig. 6.18. Successively narrowed ideal bandlimitation is shown there, acting on a difference sequence spectrum that has a DC null. The dotted curve is a difference spectrum that is low but lacks a null. As the bandwidth tends to zero it is clear that $d(\Delta a)$ becomes small faster than d for any difference sequence that lacks a null. It is obvious that the principle here applies with practical as well as ideal band limitation. A major significance of the zero-sum rule is that it eliminates most candidate sequences during a search for the narrowband code minimum distance.

Figure 6.19 shows a typical[15] narrowband PRS case, the 15-tap optimal-d_f code with 99.9% bandwidth $WT = 0.12$ Hz-s/data bit (the code taps are given in Table 6A.2). The difference power spectrum $|\Delta S(f)|^2$ (defined in Eq. (6.4-4)) is shown for four error difference sequences, two that meet the zero-sum rule and two that do not. It is clear that the last two cannot possibly lead to d_f. d_f is in fact caused by the zero-sum sequence with least energy inside $(-0.12, 0.12)$, and this turns out to be $2, -2, 0, -2, 2, 2, -2$ (the solid curve).

The DC null argument applies equally well to error differences in filtered CPM, except that now both the I and Q difference spectra need to have a null. This is clear from expression (6.5-5) for $\int |\Delta s(t)|^2 dt$ in terms of $\Delta I(f)$ and $\Delta Q(f)$. Let $\psi(t, a^{(1)})$ and $\psi(t, a^{(2)})$ denote the phase relative to carrier for the

---

[15] Especially with 99% spectral concentration, it can happen that the code spectrum itself, Eq. (6.3-26), has a DC null; in that case, the zero-sum rule leads to a higher order DC null, and a more detailed analysis is needed.

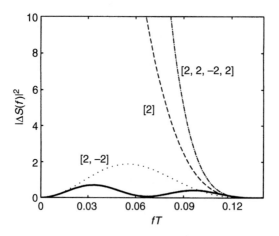

Figure 6.19 Difference power spectra for memory 14 optimal PRS code at 99.9% bandwidth $WT = 0.12$, from indicated error difference sequences with and without zero sum. Heavy curve is sequence $\{2, -2, 0, -2, 2, 2, -2\}$ which attains $d_f^2 = 0.053$. (Orthogonal pulse $v(t)$ assumed to have flat spectrum in bandwidth shown.)

CPM signals created by symbols $a^{(1)}$ and $a^{(2)}$. The I and Q signals, to a constant, are $\cos\psi(t, a^{(1)})$ and $\sin\psi(t, a^{(1)})$ and similarly for $a^{(2)}$. Similar to Eq. (6.5-7), the spectral null condition leads to

$$\int_{-\infty}^{\infty} [\cos\psi(t, a^{(1)}) - \cos\psi(t, a^{(2)})] \, dt = 0$$
$$\int_{-\infty}^{\infty} [\sin\psi(t, a^{(1)}) - \sin\psi(t, a^{(2)})] \, dt = 0. \tag{6.5-9}$$

Seshadri [19,20] derived a number of simple conditions on ψ and the difference $a^{(1)} - a^{(2)}$ that are equivalent to Eq. (6.5-9):

1. For any CPM scheme and error event of length $2K$, a DC null occurs if (i), the sum phase $\psi(t, a^{(1)}) + \psi(t, a^{(2)})$ is symmetric about the half-way point of the event, and (ii), the difference phase $\psi(t, a^{(1)}) - \psi(t, a^{(2)})$ is antisymmetric.

2. For full response CPM, this occurs if and only if symbol stream $a^{(1)}$ is obtained by reading $a^{(2)}$ backwards and reversing the sign.

3. For partial response CPM, it occurs if and only if for some $L > 0$ and $K > 0$

$$a^{(1)}[i] = -a^{(2)}[K - 1 - i], \quad -L + 1 \le i \le K - 1 \tag{6.5-10}$$

with $a^{(1)}[i] = a^{(2)}[i]$ otherwise.

PRS Coded Modulation

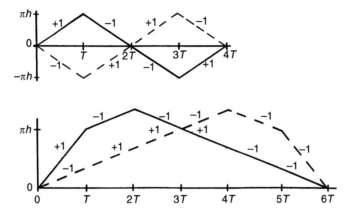

Figure 6.20 Shortest error events with a DC null in CPM (transmitted and neighbor phase trajectories); (*top*) binary CPFSK, (*bottom*) binary 3REC. Events of this type generally lead to $d_{\min}$ with narrow band filtering.

Examples of these rules are shown in Fig. 6.20. The top example shows the shortest binary CPFSK error event with a DC null; the bottom shows the shortest such binary 3REC error event. The actual distance after filtering for these events comes from Eq. (6.5-5), normalized as in Eq. (6.5-6). The simple and clear symmetries here greatly simplify finding $d_{\min}$. Some further analysis shows that among all the events that create a difference spectrum null, the longer ones will have the larger distance, and so the search for narrowband $d_{\min}$ terminates relatively quickly.

By means of a Taylor series analysis, quite accurate estimates of $d_{\min}^2$ at narrow bandwidth may be derived, under the assumption of a certain worst-case error event. For example Seshadri [19] shows that the minimum distance before normalizing for brickwall filtered binary CPFSK is

$$D_{\min}^2 = (32\pi^3 h)^2 (WT)^5 E, \qquad (6.5\text{-}11)$$

for the CPFSK signal $\sqrt{2E/T} \cos(\omega_0 t + \psi(t))$.

6.5.4. Simple Modulation Plus Severe Filtering

The earlier subsections have discussed distances in narrowband PRS codes and filtered CPM. It remains to discuss the synthesis of codes via simple modulation followed by ordinary but severe filtering. This is the idea of the earlier parts of the section, carried to the logical extreme: now the modulation will be QPSK (Case (*ii*) in Section 6.2) or even a simple impulse train (this is Case (*iii*)) and the filter will be an ordinary type from a filter design handbook.

In Chapter 1 these were called modulation + filter (M + F) codes. It is of great interest that such straightforward schemes, when measured in a bandwidth–energy sense, perform as well as traditional, apparently sophisticated coded modulations.

We will demonstrate the power of M + F coding by measurements with 4- and 8-pole Butterworth lowpass filters. The 4-pole filters were featured in Example 2.8-1 (discrete-time models) and Example 6.2-2 (realization as a PRS code). We will set up the encoder as follows. The Butterworth response is $g(t)$. The encoded signal is created by convolution of g with the pulse train $\sum a_n v(t - nT)$, where $\{a_n\}$ are binary data symbols taken from $\mathcal{A} = \{1, -1\}$ and $v(t)$ is one of three unit energy pulses: an impulse, a 30% root-RC pulse defined by Eq. (2.2-7), or the square pulse

$$v(t) = \begin{cases} 1/\sqrt{T}, & -T/2 < t \leq T/2, \\ 0, & \text{otherwise.} \end{cases} \qquad (6.5\text{-}12)$$

By changing the filter cut-off frequency, codes with different bandwidths are created. The square v pulse effectively represents at baseband square-pulse BPSK or QPSK which is filtered by a bandpass Butterworth filter, and similarly for the root-RC pulse. For BPSK, the NBW is twice the baseband positive-frequency bandwidth per bit, and for QPSK it is equal the baseband bandwidth.[16]

With a data difference sequence $\Delta \mathbf{a}$, the Euclidean distance between two signals is calculated from first principles as in Section 6.2.2, or via the autocorrelation of $h = g * v$ as in Section 6.3.2. One obtains the mimimum distance by trying sequences $\Delta \mathbf{a}$ out to some length; the distance of these in general grows with their length and eventually it becomes clear that $d_{\min}$ is found. Technically, the outcome is an upper bound to $d_{\min}$, but experience has shown that it is an exceedingly reliable estimate.

Figure 6.21 shows $d_{\min}^2$ vs 99% bandwidth for 8- (solid line) and 4-pole (dashed) filters as the cut-off frequency changes,[17] with the impulse train as input. As the cut-off drops, $d_{\min}^2$ remains at 2 until a critical cut-off is reached; it is about $0.352/T$ Hz at 8 poles and $0.306/T$ Hz at 4 poles, leading to 99% bandwidths $0.398/T$ and $0.444/T$ Hz. These cut-offs are the point where the antipodal difference signal with $\Delta \mathbf{a} = 2$ ceases to be the minimum distance one, and another $\Delta \mathbf{a}$ leads to $d_{\min}$. For the 4-pole case the new $\Delta \mathbf{a}$ is 2, −2, 2; for the 8-pole case it is 2, −2, 2, −2, 2. Thereafter, $d_{\min}^2$ steadily falls as the cut-off drops, with various $\Delta \mathbf{a}$ leading to $d_{\min}$. For cut-offs below $\approx 0.19/T$ (4 poles) and $\approx 0.23/T$ (8 poles), $\Delta \mathbf{a}$ satisfies the zero-sum condition. The 4-pole distance falls a little short of the 8-pole for the same 99% bandwidth, and below about $0.15/T$ Hz the gap

---

[16] The positive frequency, physical RF bandwidth of BPSK per bit is equal to its NBW.

[17] The 99% bandwidth of a 4-pole Butterworth filter impulse response is 1.45 times its cut-off frequency; for an 8-pole filter it is 1.13 times cut off.

PRS Coded Modulation

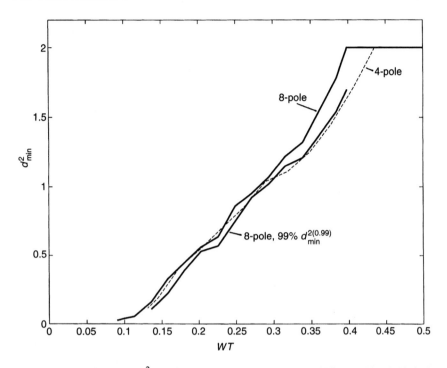

Figure 6.21 Square distance d_{min}^2 against 99% bandwidth for 8-pole (solid line) and 4-pole (dashed) Butterworth modulation + filter codes, with impulse train excitation. Lower solid curve is distance inside 99% bandwidth, $d_{min}^{2(0.99)}$, for the 8-pole codes.

grows more rapidly (this is hard to see on a linear scale). Apparently, the more complex filters are needed to generate a good code below $0.15/T$.

In Section 6.4 we discussed the phenomenon of escaped free distance, which means in this case that a significant part of d_{min} may lie outside the 99% bandwidth. If another signal lies there it will destroy our signal there and the detection will have to make do with signal parts that lie strictly inside the 99% bandwidth. By means of the spectral distance calculation Eq. (6.4-3), the distance $d_{min}^{(0.99)}$ inside the bandwidth can be precisely measured. Some care is required: it can happen that a different Δa leads to $d_{min}^{(0.99)}$. This reduced distance vs 99% bandwidth is also given in Fig. 6.21, as the lower solid line. At very narrow bandwidths it can happen that a considerable fraction of the minimum distance is lost. Error difference events exist that lose 60–80% of their distance to the band outside $W_{0.99}$, and it is often one of those that leads to $d_{min}^{(0.99)}$.

The plot in Fig. 6.21 can be repeated for root-RC and square pulse excitation trains, but the outcome is little different, especially at narrower bandwidths.

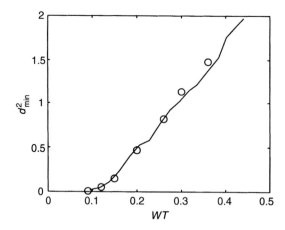

Figure 6.22 Square distance $d_{min}^{2(0.99)}$ inside the 99% bandwidth against 99% bandwidth, for 8-pole Butterworth modulation + filter codes and a selection of PRS codes with best $d_{min}^{(0.99)}$ (circles). All code structures have 30% root-RC pulse excitation.

The reason why can be seen by viewing the total signal $V(f)G(f)$ in the frequency domain: for a narrowband filter $G(f)$, $V(f)$ in the passband for the impulse and root-RC pulses will be identical after energy normalization, and $V(f)$ for the square pulse will be nearly so. Only the filter $G(f)$ really matters in these coded modulations.

It remains to compare modulation + filter coding to the optimal PRS codes of Section 6.4. This is done in Fig. 6.22, which plots the 8-pole Butterworth code square distance inside the bit-normalized 99% bandwidth against that bandwidth. A selection of best PRS codes from Tables 6A.1 and 6A.2, measured in the same way, are indicated by small circles. Some of these are optimal for 99% concentration, some for 99.9%; those shown are the ones that turn out to have best $d_{min}^{(0.99)}$. Both encoding mechanisms are excited by 30% root-RC pulses. There is little difference in these two groups of codes. It is true that the PRS codes are optimized, but they are restricted to short, finite response; the Butterworth codes are not optimized but they are allowed to have a very long effective response, which can grow to 50 symbol intervals. A bandwidth–energy plot for either group is similar, and the plot for the PRS codes, with comparison to TCM and CPM coding, has appeared in Chapter 1. Either kind of code is a way to synthesize excellent narrow bandwidth–high energy coding schemes.

At this writing there is no research to report on modulation + filter coding with quaternary or higher symbol alphabets. If PRS coding is any prediction, such codes should have attractive properties.

6.6. PRS Receivers

The study of receivers for signals of the PRS code/intersymbol interference/filtering type breaks conveniently into three parts: equalizers, trellis-type decoders, and receivers for infinite-response codes. An equalizer can be thought of as a trellis decoder whose path search is reduced to a single path. If the receiver eye is not open after the equalizer filter, reliable detection is not possible, at least for some transmitted sequences (eyes are discussed in Section 2.2.3). If the eye is open but not fully so, more signal energy is needed in general to achieve a given error rate. This loss varies from small to very large, depending on the response of the channel. We need to review equalizers if only to complete the PRS picture, but their types and properties also organize the receiver question in a useful way. This is done in Section 6.6.1. A full search of the PRS trellis is the Viterbi decoder, just as it is for any other trellis code. This implements MLSE detection. With a few exceptions, a great reduction in the trellis search is possible without significant loss in asymptotic error performance. We will explore this by looking at the behavior of the M-algorithm decoder. Trellis decoders, both reduced and not, are the subject of Section 6.6.2.

Infinite-response PRS codes, and particularly the modulation + filter codes, have infinite breadth trellises, and these need something more than the conventional trellis decoder idea. Fortunately, a view based on the filter state variables rather than the data symbols as state variables – plus a greatly reduced search of these new state paths – yields a practical decoder. This receiver is Section 6.6.3.

6.6.1. Review of Equalizers

Equalization is a technique customarily applied to linear modulations of the PAM or QAM type, detected by the PAM-type receiver in Fig. 2.29. The signals have encountered linear intersymbol interference. The continuous-time signal type is thus Case (ii) in Section 6.2, $s(t) = \sqrt{E_s} \sum a_n h(t - nT)$ plus noise, with $h = v * g$, $v(t)$ an orthogonal pulse and $g(t)$ an intersymbol interference. The discrete-time baseband channel model is the PAM receiver model $r_n = \sum_i a_n \gamma_{n-i} + \eta_n$ of Section 2.8 and Eqs (2.8-4) and (2.8-5). We assume white Gaussian noise with PSD $N_0/2$, so that $\{\eta_n\}$ become independent zero-mean Gaussian variates with variance $N_0/2$.

For now, assume that only the receiver knows the PAM model $H(z) = \sum \gamma_k z^{-k}$; this eliminates precoding techniques, because the transmitter cannot know how to precode and sends only $\sum a_n v(t - nT)$. What results from these assumptions is a standard PRS-type signal in which the PRS taps $\{b_\ell\}$ have been exchanged for the PAM model values $\{\gamma_k\}$.

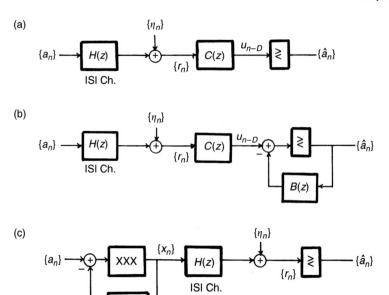

Figure 6.23 Discrete-time system models for equalizers. (a) Linear equalizers, (b) feedback equalizers, (c) Tomlinson–Harashima precoding.

Equalizers may be linear or nonlinear. Within the linear class are two main categories, the zero-forcing equalizer (ZFE) and the least-squares equalizer (LSE). Both are detection via discrete-time linear filtering. A discrete-time system model is Fig. 6.23(a). A major class of nonlinear equalizers is those that feed back decisions, the decision feedback equalizer (DFE) class. All these classes can be seen as single-path reduced trellis receivers. The ultimate nonlinear equalizer is MLSE, which is a full search of the ISI or PRS trellis. Exactly as developed in Sections 6.2.2 and 6.3.2, the tap sequence $\{\gamma_k\}$ implies a minimum distance $d_{\min}$, and the asymptotic MLSE equalizer error probability is $\sim Q(\sqrt{d_{\min}^2 E_b/N_0})$. Since $d_{\min} \leq d_{\mathrm{MF}}$, no equalizer has lower asymptotic probability than $\sim Q(\sqrt{d_{\mathrm{MF}}^2 E_b/N_0})$. We assume in these calculations that the total response h has unit energy. This is sometimes not convenient, for example, in diversity problems; then the effective E_b is scaled by the appropriate factor.

We now review the ZFE, LSE, and DFEs. Others exist, and extensions can also be made to circuits that adapt to a varying channel, and circuits that achieve their own synchronization, or to the case of a channel description known at the transmitter. An elementary treatment of equalization appears in [1] and advanced treatments appear in Proakis [22] and Qureshi [23].

Zero-forcing Equalizers

A ZFE works by forcing the ISI effect on a symbol to zero. A linear recursion always exists that does this [1, pp. 319], but a traditional, stable filter may not. When the model $H(z)$ has an inverse, the ZFE is the filter $C(z) = 1/H(z)$. The filter exists when $H(z)$ has no zeros outside the unit circle in the z-plane. For example, the tap model $1/\sqrt{2.62}\{1, 0.9, 0.9\}$ in Example 6.2-1 has zeros $-0.45 \pm j.84$, which are close to the circle but still inside. $C(z) = 1/H(z)$ has more than 60 significant terms, but the product $C(z)H(z)$ is essentially unity and the output of the model plus equalizer when $\{a_n\}$ is the input is again virtually $\{a_n\}$. Some $H(z)$, such as $1/\sqrt{1.3125}[1 + z^{-1}/2 + z^{-2}/4]$, have a short $C(z)$, but the Butterworth Example 6.2-2 has a zero at -26 and hence no stable zero-forcing $C(z)$.

The ZFE derivation ignores the effect of $C(z)$ on the channel noise and this is often its downfall. The AWGN sequence $\{\eta_n\}$ passes through $C(z)$ as well as the signal. It can be shown that $\{\eta_n\}$ filtered by $C(z)$ grows in variance from $N_0/2$ to $F_{NE}N_0/2$, where the *noise enhancement factor* F_{NE} is

$$F_{NE} = \sum |c_k|^2. \tag{6.6-1}$$

It is easy to show that filter $C(z)$ followed by single threshold interval detection as in Fig. 6.23(a) has asymptotic error $\sim Q(\sqrt{d_{MF}^2 E_b/F_{NE}N_0})$; that is, it has the error probability of the no-ISI case but with noise density $F_{NE}N_0/2$. For channel $1/\sqrt{1.3125}\{1, 1/2, 1/4\}$, F_{NE} is 1.67 (2.2 dB), which would not cause difficulty in a reasonable channel. For $1/\sqrt{2.62}\{1, 0.9, 0.9\}$, F_{NE} is 17.8 (12.5 dB), which is a serious noise enhancement; the problem here is the nearness of the zeros of $H(z)$ to the unit circle. A zero on or outside the unit circle will cause an infinite noise enhancement.

Least-squares Equalizers

A more workable equalizer can be designed if the effect of ISI and noise on $\{\hat{a}_n\}$ are *jointly* minimized. Some ISI can be allowed if F_{NE} is significantly reduced. For a finite response equalizer $C(z)$, the optimization may be set up as follows. Let $C(z)$ produce an estimate $\hat{a}_{n-D}$ for a_{n-D} that lies D before the present symbol. Minimize the expectation of $|\hat{a}_{n-D} - a_{n-D}|^2$ over the choice of $C(z) = \sum_0^L c_k z^{-k}$; that is, find $\{c_k\}$ that achieves

$$\min_{\{c_k\}} \mathcal{E}\left[\left|a_{n-D} - \sum_0^L c_k r_{n-k}\right|^2\right], \tag{6.6-2}$$

in which $r_n = \sum_{i=0}^{m} \gamma_i a_{n-i}$. The optimizing $\{c_k\}$ in Eq. (6.6-2) are the solutions to the set of L equations

$$\mathcal{E}[a_{n-D}r_{n-k}] = \sum_{j=0}^{L-1} c_j \mathcal{E}[r_{n-j}r_{n-k}], \quad k = 0, \ldots, L-1. \quad (6.6\text{-}3)$$

The expectation is over both the data symbols and the noise. The equalizer length L and the estimation delay D are degrees of freedom in the design and so also is the noise PSD, $N_0/2$. Varying these produces a range of equalizers that are useful in different practical situations.

Equation (6.6-3) is a straightforward matrix equation in which the autocorrelation of $\{\gamma_k\}$ plays the prominent role; details appear in [1,22,23]. Further calculations show that the asymptotic detection error probability is $\sim Q(\sqrt{d_{MF}^2 \kappa^2 E_b / N_0 F_{NE}})$, in which F_{NE} is the noise enhancement of $C(z)$, $\kappa^2 \leq 1$ is a constant computed from $H(z)C(z)$, which measures the residual ISI, and the binary d_{MF}^2 is 2. The error performance of the LSE is thus degraded in energy from antipodal signaling by the factor κ^2/F_{NE}. It must be of course that $\kappa^2 d_{MF}^2 / F_{NE} \leq d_{min}^2$. By trying out designs with different L, D, and N_0 parameters, it is very often possible to improve the performance of linear equalization over that of the ZFE, or to find an acceptable $C(z)$ where no stable ZFE exists. The common exception occurs when the zeros of $H(z)$ approach the unit circle.

Feedback Equalizers

When a channel model has zeros near the unit circle or the linear equalizer energy performance lies too far from that predicted by the MLSE benchmark $Q(\sqrt{d_{min}^2 E_b/N_0})$, then a feedback equalizer should be considered. A general diagram of one appears in Fig. 6.23(b). As before, a finite response filter $C(z)$ provides a tentative transmission symbol estimate u_{n-D} after a delay D, which will be converted in a threshold comparison block to a hard estimate $\hat{a}_{n-D}$. But first the effect of some previous decisions $\hat{a}_{n-D-1}, \hat{a}_{n-D-2}, \ldots$ is subtracted; the ISI from these is computed via the order-M FIR filter $B(z)$ from knowledge of the ISI model $H(z)$.

One design for $B(z)$ is zero forcing, which sets $B(z) = 1 - H(z)$. Once started up by a few known symbols, this equalizer makes perfect decisions in the absence of noise. But the design is inordinately sensitive to noise in general, and before long incorrect symbols feed back and propagate more errors. A much better scheme is to design the estimator $C(z)$ from least-square principles, and then design $B(z)$ to cancel the combined effect of $C(z)$ and the ISI $H(z)$. Simple routines exist [1] to carry this out. The result is often a better equalizer. But error propagation can still occur and it is this mechanism that leads to the asymptotic error rate. Theoretical analysis of this propagation has proven to be a challenging problem; all that can be said with certainty is that the asymptotic error rate is not better than $Q(\sqrt{d_{\min}^2 E_b/N_0})$.

A classic example of DFE performance is the model $H(z) = .408 + .817z^{-1} + .408z^{-2}$, originally given by Proakis [22]. Its zeros lie at -1.05 and -0.95, and they render the linear equalizer useless. But the binary-transmission MLSE benchmark $d_{\min}^2$ is 1.33, which is just 1.8 dB below the antipodal signaling value of 2, and we are encouraged to search for a DFE circuit. One is given in [1]; without consideration of error propagation that circuit is shown to have an additional 3 or 4 dB loss beyond the benchmark. Proakis gives evidence that the total extra loss including error propagation is around 5 dB. With such a short ISI model, a full VA search of the 4-state ISI trellis is likely the best detector design, and it should achieve $\sim Q(\sqrt{1.33 E_b/N_0})$.

The art of equalizer design assumes that some relatively mild ISI has occurred and it then tries to find a simple circuit that removes the ISI. Often a simple filter – in our context, a single path trellis search – will do. This scenario might apply to, for example, a fixed telephone channel. In mobile communication, the channel continuously changes, and almost any $H(z)$ may appear. We might play the odds, adopt a simple equalizer structure, and hope that the more challenging $H(z)$ seldom occur. In bandwidth-efficient coding this is a forlorn hope. The very phenomenon that makes equalization hard, zeros of $H(z)$ near the unit circle, is in fact the key to narrowband communication. This is easily observed in the optimal PRS codes of Section 6.4. Consequently, detection of bandwidth-efficient coding must turn to trellis decoders and to reduced-search decoders that have essentially the same performance.

Tomlinson–Harashima precoding. For completeness, and to illustrate precoding, we describe briefly the method proposed independently by Tomlinson and by Harashima and Miyakawa around 1970. As sketched in Fig. 6.23(c), a feedback circuit and a special box XXX come *before* the ISI channel $H(z)$. The box may perform various functions; the original system proposed a mod-2A operation that keeps x_n, the input to the channel, in the range $[-A, A]$. The mod operation is undone in the receiver threshold block. For A reasonably large, the sequence $\{x_n\}$ is close to IID and the combination of the precoder and the channel $H(z)$ approximately performs zero forcing. This time the zero forcing is in advance rather than at the receiver.

When it can be applied, T–H precoding is attractive because it leads to an ISI-free detection. But it is often not applicable. It assumes a Filter-in-the-Medium channel model (see Section 3.6.3) of the sort that is "owned", not shared with others; the feedback circuit approximately multiplies $H(z)$ by $H^{-1}(z)$, which spreads the bandwidth consumption in the physical channel. Furthermore, the power level at the entrance to $H(z)$ may be unacceptable in, for example, a twisted pair channel. The encoding by $H^{-1}(z)$ may not be close to that suggested by the waterfilling argument (3.6-21) and (3.6-22) that computes the Shannon capacity. Finally, the sending end must know $H(z)$.

6.6.2. Reduced-search Trellis Decoders

A finite response PRS code or ISI model with memory m and rate R has a 2^{Rm}-state code trellis. Code trellises were introduced in Section 3.3.2 and Fig. 3.7 is a drawing of a simple 4-state PRS trellis with rate 1, memory 2 and generator tap set $\{1, 0, a\}$. For this section we will assume that the channel is thus modeled in discrete time with z-transform $H(z) = b_0 + b_1 z^{-1} + \cdots + b_m z^{-m}$. The state of the PRS (or equivalent ISI) encoding at time $n - 1$ is composed of the last m transmission symbols $a_{n-m}, \ldots, a_{n-1}$, and the outcome of the PRS (or ISI) encoding process for trellis stage n (the "branch label") is the discrete-time signal value $s_n = \sum_{\ell=0}^{m} b_\ell a_{n-\ell}$. Especially with the ISI problem, it can happen that 2^{Rm} is a small number, and then the full VA MLSE receiver is attractive.

When 2^{Rm} is larger, it is time to think about reduced-search decoding, that is, a decoder that moves stage by stage through the trellis but visits only a fraction of the nodes. An equalizer develops only one path through the trellis. In ISI decoding, equalizer error performance often shows little loss over a full trellis search, despite the drastic reduction. In bandwidth-efficient PRS coding, a single path search is not enough, but it is still true that a dramatic trellis search reduction leads still to essentially MLSE error performance. Reduced-search decoding improves the efficiency of all trellis decoding, but none so much as it does PRS decoding. The topic thus has a natural home in this chapter.

Useful Breadth-first Reduced Decoders

As introduced in Section 3.4.1, breadth-first decoders move only forward in a code trellis. The VA is a breadth-first decoder without a reduced search. Trellis (and tree) searches can also backtrack, and this kind has theoretically a lower average computation, but since PRS decoding is relatively simple, we will discuss here only the more straightforward breadth-first searches. A reduced-search decoder can be as simple as ignoring some early state variables, say the first P of $a_{n-m}, \ldots, a_{n-1}$. The search then visits effectively $2^{R(m-P)}$ states per stage,[18] rather than 2^{Rm}. This idea often works well, because minimum phase bandwidth-efficient tap set values $b_0, b_1, \ldots$ tend to diminish toward zero and the value of the branch label $s_n = \sum_0^m b_\ell a_{n-\ell}$ is relatively unaffected by the P last terms.

A more sophisticated variation of this is to break the sum for s_n into two parts

$$s_n = \sum_{\ell=0}^{m-P} b_\ell a_{n-\ell} + \sum_{\ell=m-P+1}^{m} b_\ell a_{n-\ell}, \qquad (6.6\text{-}4)$$

---

[18] To be precise, this is a full search of a reduced trellis, rather than a reduced search of a full trellis; see also Sections 5.5.1 and 5.5.2 for CPM reduced-trellis receivers.

PRS Coded Modulation

symbols $a_{n-m+P}, \ldots, a_{n-1}$ become the state variables, symbols $a_{n-m}, \ldots, a_{n-m+P-1}$ are frozen and may be given as output, and the second sum value is retained along with the state variables as a residual part of s_n. Now the tail sum is not required to be small, and the reduced scheme will work well so long as freezing the early P state variables has no serious effect. During the extension from a present node to a new one in the trellis, a new residual for s_{n+1} is computed from the old residual part and the term $b_{m-P}a_{n-m+P}$ in Eq. (6.6-4). The value for a_{n-m+P} is taken as the one on the trellis path in storage having minimum distance to the channel output; this value becomes the output for stage $n - m + P$, and one such output occurs for each stage advance in the trellis. A scheme of this sort was first suggested by Duel-Hallen and Heegard [24], who called it delayed decision-feedback sequence estimation. It can be viewed as combining a DFE with a small trellis search. The idea of saving a residual part of s_n plus some "living" transmission symbols will play an important role in decoding infinite state codes in Section 6.6.3.

A better breadth-first decoder results if the strict breakdown into delayed decision and short-term trellis is replaced by a more opportunistic search based on distance. The M-algorithm retains at each trellis stage the M paths that are closest in distance to the received signal. It has a long history as a PRS and ISI decoder that traces back at least to [25]. Other methods have been proposed that retain all paths closer than a certain distance [26] or all paths within a distance band [27]. These schemes are probably more efficient, but the M-algorithm is simple and direct and it is by far the most researched scheme, so it will be featured in what follows. Its path storage requirement may be taken as indicative of the other breadth-first schemes.

The storage of a breadth-first decoder consists of a number of paths – M in the M-algorithm – times the receiver observation width, ideally the decision depth L_{dec} that was defined in Section 3.4.1. If the search has reached stage n, a data bit decision is thus forced at stage $n - L_{\text{dec}}$. The L_{dec} needed by a PRS decoder in order that it attain error probability $\sim Q(\sqrt{d_f^2 E_b / N_0})$ has been studied by Wong [11] for codes up to four taps. These L_{dec} fall generally into the range 5–15, with the exception of a few tap sets with very long L_{dec} (a phenomenon similar to catastrophicity in convolutional coding or weak modulation indices in CPM). For codes with 6–15 taps, L_{dec} can fall in the range 30–100. In most breadth-first decoders that have been tested, the observation width is simply set to a convenient number like 32.

M-algorithm Error Performance with PRS Codes

The path storage, M in the M-algorithm, depends primarily on the bandwidth of the PRS code and secondly on the phase of the tap set. Recall from Section 6.3 that the bandwidth of a code depends only on its tap autocorrelation, and that many tap sets have the same autocorrelation but a different phase characteristic. The dependence of M on the tap phase is simple and dramatic, and so we take this first. The results in the remainder of this section are for the AWGN channel.

Given an autocorrelation, a minimum phase PRS tap set has the least decoder storage requirement and the maximum phase set has the largest. Experiments have shown [9,28] that code column distance profile, among other things, has a strong effect on M, but M is most precisely related to phase through the partial energy function $E(n)$ that was defined in Eq. (6.3-15). $E(n)$ has a clearly defined behavior. It is demonstrated by Fig. 6.24. The figure shows $E(n)$ for all the phase versions of the 10-tap optimal binary PRS codes for concentration 99% and normalized bandwidths 0.12, 0.16, 0.28, 0.36 (the 0.12 and 0.36 minimum phase tap sets appear in Table 6A.1). As an example, the tap set z-transform $\sum b_\ell z^{-\ell}$ for NBW 0.36 has one real zero and four conjugate zero pairs, and since these five objects may be reflected inside or outside the unit circle, there are 2^5 tap sets that lead as a group to the same autocorrelation. (See the spectral factorization discussion in Section 6.3.1.) The minimum phase set maximizes $E(n)$. The 32 partial energy functions fall into four distinct groups. The reason for this is that three of the conjugate zero pairs – namely $z = 0.993$, $-0.741 \pm j.656$, $-0.907 \pm j.304$ – lie very close to the unit circle, and reflecting them causes almost no change in $E(n)$. The partial energies thus form four tight groups of 2^3.

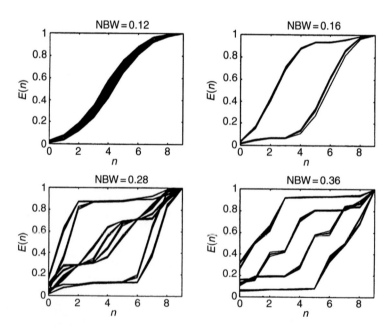

Figure 6.24 Partial energy curves of all phase versions of the optimal tap set for NBW 0.12, 0.16, 0.28, and 0.36 Hz-s/data bit. Curves fall in distinct groups. (Reproduced from Balachandran and Anderson [28], copyright 1997 IEEE; used with permission.)

PRS Coded Modulation 341

Zero placements close to the unit circle are in fact what constrains the bandwidth of these optimal PRS codes. As the bandwidth becomes more constrained, a number of factors conspire to force the partial energies into fewer groups. When the bandwidth reaches the least value possible for the tap number, the energies apparently coalesce more or less into one group [9]. From Eq. (6.3-16), the uppermost $E(n)$ is $E_{\min}(n)$, the partial energy of the minimum phase tap set; the lowermost is the maximum phase $E_{\max}(n)$, and from the figure it appears that $E_{\min}$ and $E_{\max}$ draw together at narrow bandwidth. It can be shown that the max and min phase tap sets are time reversals of each other, and so the fact that $E_{\min}(n)$ and $E_{\max}(n)$ are close is a statement that the most narrowband optimal tap sets are approximately symmetric about their center.

Studies show [28] that the path storage required for the M-algorithm depends closely on the tap partial energy: a higher $E(n)$ means a smaller M. *The minimum phase tap set minimizes storage.* To understand this better, consider a family of tap sets with the same autocorrelation. Suppose that larger and larger M are tested with a given tap set until the error event rate has reached that of the VA and no longer improves. There is a threshold M that first "achieves" the code free distance in this sense. If now the experiment is repeated with another tap set having similar partial energy, the set will have approximately the same threshold M. A tap set with a clearly superior $E(n)$ will have a lower threshold M. Since partial energies occur more or less in groups, there is a set of threshold M, one for each group.

Figure 6.25 shows some typical plots of decoding error event rate vs E_b/N_0 for a PRS code with M-algorithm decoding. The code is the optimal 99% code for bandwidth 0.36 Hz-s/data bit from Table 6A.1; its MDL is 0.19 dB and the figure shows the corresponding estimate $Q(\sqrt{1.91 E_b/N_0})$ as the line labeled "MLSE." The points labeled "min ph" are the M-algorithm event error rate[19] with the minimum phase tap set (the one listed in Table 6A.1). $M = 4$ essentially achieves the MLSE estimate; smaller M perform worse and larger M perform no better. The points labeled "mid ph" are the event error rate with a medium phase tap set, one of the ones in the third group of partial energies, counting from the top in Fig. 6.24; this is the third-worst group and now the threshold M is 32. With a max phase tap set, the threshold M is 128. The VA requires 512 states, and has essentially the "mid ph" event error rate.

A BER curve is shown for reference for the minimum phase tap set. In general, M-algorithm decoding quickly recovers from loss of the correct path from storage, and the ongoing BER lies moderately above the event error rate. Recovery becomes more difficult at higher phase and narrower bandwidth. This will tend to raise the BER curve in the low E_b/N_0 region.

---

[19] Recall that this is the rate that error events begin at a trellis stage, given that no event is in progress already.

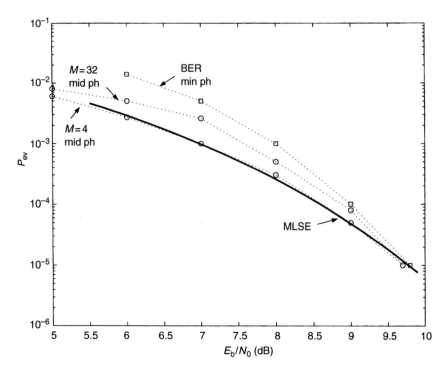

Figure 6.25 Error event rate vs E_b/N_0 for two phase versions of the optimal 10-tap binary 99% PRS code at NBW 0.36, for M-algorithm decoder with threshold M. Minimum phase (taps listed in Table 6A.1) requires $M = 4$; middle phase set requires $M = 32$. A BER curve is shown for minimum phase case. Data taken from [9].

There have been various other reports of M-algorithm decoding. Balachandran [9] reports that only $M = 2$ is required to achieve the VA error rate with a binary minimum phase 10-tap code at bandwidth 0.16 Hz-s/data bit and 99% concentration. Said [5] reports that 15-tap binary minimum phase decoding with 99.9% concentration requires $M = 8$ paths at bandwidth 0.3 (MDL 2.3 dB) and 12 paths at bandwidth 0.2 (MDL 6.4 dB); furthermore, 4QAM decoding at bandwidth 0.3 and 99.9% requires 16 paths.

It is clear that for coding at a given bandwidth and d_f we want to employ the minimum phase tap set that has the required autocorrelation, since this one has the simplest decoder, a decoder that is dramatically simpler than Viterbi decoding. Conversely, if we cannot control the channel model, then the most dangerous model for a given autocorrelation is the one with maximum phase. Its decoder is a sizable fraction of the VA size. Reduced decoding works best when the channel response occurs earliest.

Backwards decoding. A stable allpass filter $A(z)$ always exists that converts $H(z)$ to a maximum phase model $H(z)A(z)$ in the same autocorrelation group. It does this by exchanging zeros inside the unit circle with zeros outside. Furthermore, $A(z)$ leaves the noise variates white. The process does not work in reverse; that is, a stable $A(z)$ does not exist that reduces phase. These facts motivate the following low-complexity decoding strategy: the channel outputs are filtered by the appropriate $A(z)$ to create a maximum phase total model, and the decoder proceeds backwards down blocks. Reversing a transmission converts the effective model from max to min phase. The disadvantages of the scheme are its extra delay and the fact that it cannot be sliding block.

A related problem occurs in the construction of whitened matched filter (WMF) receivers. These work by adding a whitening filter to a certain receive filter; the discrete-time model of the whole chain that results is maximum phase. Very often, there is no compelling reason to use a WMF design, but when it is used, blocking up the received stream and applying backward decoding will greatly reduce receiver complexity. Of course, if the VA complexity is acceptable, then max phase is no particular disadvantage.

Dependence of Storage on Bandwidth and Energy

The above discussion shows that there is a small threshold M that attains the bit error rate (BER) of MLSE, and that it depends on the tap phase. There are also larger dependences of this threshold on the minimum distance attained by the coding system, and most of all, on the code bandwidth.

To obtain a clear picture of these, we will reverse the previous evaluation setup. Instead of increasing the storage M until the BER improves no further, we will fix the storage at a relatively small value and see what the effective code minimum distance seems to be. Asymptotically, the BER will take on some $Q(\sqrt{d^2 E_b/N_0})$. With an effort, d^2 can be discerned from the BER at high E_b/N_0, but much clearer results may be obtained by changing the receiver paradigm to bounded distance decoding. In this view, one sets a fixed decoder storage, complexity, etc. and asks what magnitude of noise is guaranteed always to be corrected. For example, over a BSC, what storage size first guarantees that all combinations of t or fewer errors within the decoder observation window are corrected? Over an AWGN channel, normalized energies of noise outcomes $\eta(t)$ are compared to a threshold δ^2; for what threshold is correct decoding guaranteed for all noises that satisfy $\int |\eta|^2 \, dt/2E_b < \delta^2$? In what follows the decoder is breadth-first but it is not necessarily the M-algorithm. Its constraint is its storage size and when it must drop paths it drops the most distant ones first; it is otherwise unconstrained. M denotes the storage size.

Figure 6.26 plots the log storage size required to guarantee the threshold δ^2 in the bounded distance sense, for a number of PRS codes having 99% bandwidths 0.20 and 0.12 Hz-s/data bit. The codes are various phase versions of those in Table 6A.1, plus certain other codes; codes 20c, 20d and 12c are *bandpass* codes with the 0.20 and 0.12 passband. The data are obtained with a method described

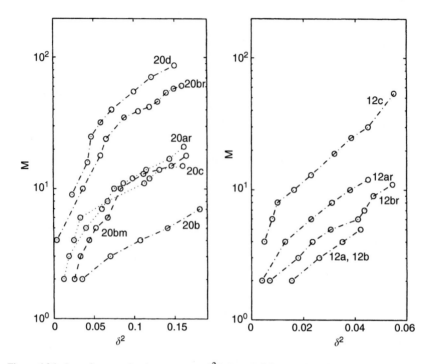

Figure 6.26 Log of storage size that guarantees δ^2 in bounded distance decoding of binary PRS codes, as a function of δ^2. 99% normalized bandwidths 0.20 and 0.12 Hz-s/data bit. Data taken from [30].

Code descriptions, 0.20 bandwidth: 20ar, max phase optimal 8-tap code; 20b, min phase optimal 10-tap code in Table 6A.1; 20 m, a middle phase version of 20b; 20br, max phase version of 20b; 20c, a 10-tap bandpass code; 20d, a 16-tap bandpass code.

Code descriptions, 0.12 bandwidth: 12a, min phase optimal 10-tap code in Table 6A.1; 12ar, max phase version of 12a; 12b, min phase optimal 15-tap code in Table 6A.1; 12br, max phase version of 12b; 12c, a 15-tap bandpass code.

in [30] and come from that work. Note that no $\delta \geq d_f/2$ can be guaranteed (i.e. $\delta^2 \geq d_f^2/4$). The plots show that an approximately exponential law

$$M \approx K' 10^{\zeta \delta^2} \qquad (6.6\text{-}5)$$

applies, in which ζ depends on the bandwidth: for a given bandwidth, storage grows exponentially with the guaranteed square noise correction. Phase causes a vertical offset (a change in K'), but does not alter the slope. The law applies equally to lowpass and bandpass codes.

An exponential relation between storage and error corrrection in breadth-first decoding has been observed rather widely [29–31], and so it is not a surprise to see it here for PRS codes. Equation (6.6-5) is very sensitive to the bandwidth parameter ζ, which is about 3.4 for bandwidth 0.20 and 14 for 0.12. Despite the

PRS Coded Modulation 345

very rapid storage growth at narrow bandwidth, it is still true that δ^2 up to $d_f^2/4$ can be guaranteed with a small M in the range 2–5, providing the encoder is minimum phase. A random coding analysis exists that predicts law (6.6-5) and ζ from the tap set autocorrelation [30]; no analysis yet exists that connects ζ directly to bandwidth.

6.6.3. Breadth-first Decoding with Infinite Response Codes

The usual description of a PRS coded signal is a convolution of the data symbols $\{a_n\}$ and the encoder model $\{b_n\}$, in which the data symbols are the state variables. When the response $\{b_n\}$ is infinite, this becomes problematic. Infinite response arises first and foremost when the model as a z-transform $H(z)$ has poles.[20] It was clear especially in Section 6.5.4 that an $H(z)$ with just a few poles provides as effective an energy–bandwidth coding scheme as an $H(z)$ with many zeros. Of course, the response of a stable $H(z)$ with poles converges to zero, but practical bandwidth-efficient schemes can have a working response of 20 or more symbol intervals.

The fact is that it is easy to dispense with this clumsy symbol-based state description. If we take the ordinary circuit theory state variables of $H(z)$ as the encoder state variables, there will be P encoder state variables, with P the number of poles, in place of a large number of μ-ary variables. The P variables evolve in response to the data and "contain" the data, and the channel output exclusive of noise in $[nT, (n+1)T)$ is determined by them and the present symbol a_n. For an unlimited data stream, there are P variables that take infinitely many values; in contrast, the usual encoder state description would have infinitely many variables taking μ-ary values. To distinguish the older data-based state description from the new one, we will call the new one a *circuit state* description.

To apply the circuit state idea, we return to the continuous-time signaling view and take the case where the signal $s(t)$ is $\sum_{k \leq n} a_k h(t - kT)$, in which $h(t)$ is the impulse response of a filter with P simple poles. This will illustrate a decoder for the Case (iii) modulation + filter codes of Section 6.5.4. $h(t)$ includes the encoder plus channel response, but there is no loss of generality to think of h as residing wholly in the encoder, so that the model state variables are those of the encoder. We need to:

- express the state variables of the system with response $h(t)$,
- express $s(t)$ during the nth interval $[nT, (n+1)T)$ in terms of a state term and a new symbol term, and
- write down an update formula for the state variables.

All of this should be expressed as simply as possible.

---

[20] Example 2.8-1 discusses the modeling of infinite response filters such as the Butterworth. Fig. 2.31 shows one such Butterworth discrete-time model.

A convenient form for the state variables arises if we express $H(s)$, the Laplace transform of $h(t)$, as a partial fraction expansion

$$H(s) = \sum_{i=1}^{P} \frac{K_i}{s + p_i}, \qquad (6.6\text{-}6)$$

with the impulse response

$$h(t) = \sum_{i=1}^{P} K_i e^{-p_i t}, \quad t \geq 0, \qquad (6.6\text{-}7)$$

where $K_1, \ldots, K_P$ are residues. $s(t)$ during $[nT, (n+1)T]$ is

$$s(t) = \sum_{k<n} a_k h(t - kT) + a_n h(t - nT),$$

$$= \sum_{k<n} a_k \sum_{i=1}^{P} K_i e^{-p_i(t-kT)} + a_n \sum_{i=1}^{P} K_i e^{-p_i(t-nT)}.$$

Some rearranging gives

$$s(t) = \sum_{i=1}^{P} K_i \left[\sum_{k<n} a_k e^{-p_i(nT-kT)} \right] e^{-p_i(t-nT)} + \sum_{i=1}^{P} K_i a_n e^{-p_i(t-nT)},$$

and a change of variables gives finally

$$s(t) = \sum_{i=1}^{P} K_i \left[\sum_{\ell \geq 1} a_{n-\ell} e^{-p_i \ell T} \right] e^{-p_i(t-nT)} + \sum_{i=1}^{P} K_i a_n e^{-p_i(t-nT)},$$

$$nT \leq t < (n+1)T. \qquad (6.6\text{-}8)$$

Here the second term is the response to the present symbol a_n, which appears at nT, and the first term is the response to all previous ones. We can define the state vector $\boldsymbol{\sigma}_n$ at a time n as follows. The state variables $\sigma_n(i)$ are the brackets in Eq. (6.6-8); that is,

$$\sigma_n(i) = \sum_{\ell \geq 1} a_{n-\ell} e^{-p_i \ell T}, \quad i = 1, \ldots, P. \qquad (6.6\text{-}9)$$

Equation (6.6-8) becomes

$$s(t) = \sum_{i=1}^{P} K_i [\sigma_n(i) + a_n] e^{-p_i(t-nT)}, \quad nT \leq t < (n+1)T, \qquad (6.6\text{-}10)$$

PRS Coded Modulation

with the simple update formula

$$\sigma_n(i) = [\sigma_{n-1}(i) + a_{n-1}]e^{-p_i T},$$
$$\sigma_0(i) = \sum_{\ell \geq 1} a_{-\ell} e^{-p_i \ell T}. \qquad (6.6\text{-}11)$$

Note that when $h(t)$ and $a_0, a_1, \ldots$ are real, both the residues and the state variables occur in conjugate pairs, thus reducing the complexity of the computation.

As an example we can take $H(s)$ as a 4-pole Butterworth lowpass filter with cut-off $0.5/T$ as in Example 2.8-1 (see also Example 6.2-2 and Fig. 6.21, which feature other cut-offs). Take $T = 1$. The pole positions are $\{p_i\} = \{-1.202 \pm 2.902j, -2.902 \pm 1.202j\}$; the respective residues are $\{K_i\} = \{-1.451 \pm 0.601j, 1.451 \mp 3.503j\}$. The impulse response is then[21]

$$h(t) = [2(-1.451)\cos(-2.903)t + 2(601)\sin(-2.903)t]e^{-1.202t}$$
$$+ [2(1.451)\cos(-1.202)t + 2(-3.503)\sin(-1.202)t]e^{-2.903t}, \quad t \geq 0.$$

Let symbols before time 0 be zero. Let $a_0 = 1$, $a_1 = a_2 = \ldots 0$; then $s(t)$ is the impulse response $h(t)$. The initial state is $\sigma_0 = \mathbf{0}$. During $[0, 1)$,

$$s(t) = \sum_{i=1}^{4} K_i[0+1]e^{-p_i t};$$

after one symbol interval the state vector is σ_1 with components $\{[0+1]e^{-p_i}\}_{i=1}^{4} = \{-0.292 \pm 0.0712j, 0.0198 \pm 0.0512j\}$. During $[1, 2)$,

$$s(t) = \sum_{i=1}^{4} K_i[\sigma_1(i) + 0]e^{-p_i(t-1)};$$

the state vector is σ_2 with components $\{[\sigma_1(i) + 0]e^{-p_i}\}_{i=1}^{4} = \{0.0802 \mp 0.0416j, -0.0022 \pm 0.002j\}$.

The method here extends directly to discrete-time channels, for which $H(z)$ has the expansion $\sum K_i/(p_i + z^{-1})$. It works as well when the input to $H(s)$ is a more general continuous-time function of $a_0, a_1, \ldots$, instead of simply $\sum a_n \delta(t - nT)$. It works at lowpass or bandpass. Seshadri [20,21] develops the equations when the input is a CPM signal; the method is now effectively decoding a concatenated code. Some of his results are collected in Fig. 6.27, which shows BER for an M-algorithm circuit state decoder. The CPM scheme is binary CPFSK with $h = 1/2$ and the filter is 4-pole Butterworth with cut-offs $0.25/T$

---

[21] Each conjugate pole pair contributes a term $[2\Re\{K_i\}\cos y_i t + 2\Im\{K_i\}\sin y_i t]e^{-x_i t}$, where K_i is the residue of the first p_i of the pair and $x_i + jy_i$ is the negative of position p_i.

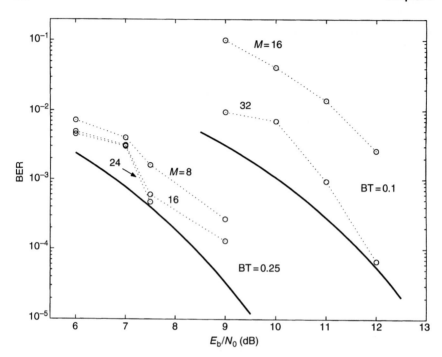

Figure 6.27 BERs for a circuit-state M-algorithm decoding binary CPFSK with 4-pole Butterworth filtering. Filter cut-offs at $BT = 0.25$ and 0.1 Hz-s/data bit. An M is reached that achieves the MLSE bound at moderate E_b/N_0. Data adapted from [21].

and $0.1/T$ Hz-s/data bit. The solid lines show $Q(\sqrt{d_f^2 E_b/N_0})$; d_f^2 is 2 and 0.94, respectively, for the two cut-offs. At bandwidth $0.25/T$, it is clear that M need not exceed 16, and less is perhaps enough; at bandwidth $0.1/T$, around $M = 32$ seems to be enough. Seshadri ran numerous 2-pole tests [21] with binary 2RC and CPFSK and found that $M = 4$–8 was generally enough, with $M = 16$ needed in a few cases.

6.7. Problems

1. When $a_1, a_2, \ldots$ are uncorrelated in Eq. (6.2-1), show that $\int |h(t)|^2 dt = 1$ implies that $\mathcal{E}\{\int |s(t)|^2 dt\} = N E_s$; that is, show that E_s in Eq. (6.2-1) is the average symbol energy.

2. Signals of the form $s(t) = \sqrt{E_s} \sum_1^N c[n] v(t - nT)$ are transmitted. The pulse $v(t)$ is orthogonal to T-shifts. The received signal is $r(t) = s(t) + \eta(t)$, where $\eta(t)$ is AWGN, and $y[n]$ is the outcome of $\int r(t) v^*(t - nT) dt$. Show that $\int |r(t) - s_i(t)|^2 dt = \sum_1^N |y[n] - \sqrt{E_s} c_i[n]|^2$.

PRS Coded Modulation   349

3. Let $h(t)$ be the standard PRS symbol response $\sum_{\ell=0}^{m} b_\ell v(t - \ell T)$, with v orthogonal to T-shifts. Show that the power spectrum $|H(f)|^2$ satisfies the autocorrelation relationship in Eq. (6.3-27).

4. Prove that
$$\sum_{k=-m}^{m} \rho[k] \int_{-\infty}^{\infty} |V(f)|^2 e^{-j2\pi k f T} df = 1,$$
with ρ the autocorrelation sequence of a PRS signal, thus showing that infinite bandwidth always yields unity for the spectral concentration $\mathcal{C}$ defined in Eq. (6.3-28).

5. Consider ordinary PAM transmission $\sum a_n v(t - nT)$ with T-orthogonal v and modulation symbols a_n chosen from an M-ary equispaced alphabet. Derive the matched filter bound parameter d_{MF} as a function of M. What is the asymptotic behavior of d_{MF} as M grows?

6. Example 6.2-2 derives a PRS code by considering the PAM model of a 4-pole Butterworth lowpass filter with cut-off $0.4/T$. Repeat the example and find the taps for cut-offs $0.3/T$, $0.2/T$, and $0.1/T$, and summarize the outcome. Be sure that the PRS code so found is minimum phase. Estimate $d_{\min}$; this can be done approximately by using Fig. 6.21 and the fact that the 99% bandwidth of the 4-pole filter is 1.45 times its cut-off frequency.

7. Using Table 6A.1, investigate the z-plane zero locations of minimum phase binary optimal codes with 99% concentration. Specifically:

    (a) at bandwidth 0.20, compare zero locations for the 6, 10, and 15 tap codes.

    (b) for 10 tap codes, compare zero locations for bandwidths 0.12, 0.15, 0.20, 0.26.

8. Using Table 6A.2, compute and plot the partial energy curves for all phase versions of the optimal 99.9% concentration binary code with 15 taps and bandwidth $WT = 0.12$. Compare the outcome with the 99% codes in Fig. 6.24.

9. Suppose that an FTN linear modulation system is constructed with parameter θ, $0 < \theta \le 1$; that is, the modulation is $s(t) = \sqrt{E_s} \sum a_n v(t - nT)$, with $v(t) = \sqrt{\theta/T} \text{sinc}(\theta t/T)$. Derive a discrete-time model for this system. Let the model assume a sequence $\{a_n\}$ as input, a system z-transform $H(z)$, and AWGN.

10. Let $H(z) = h_0 + h_1 z^{-1}$ be a model for ISI with two taps.

    (a) Show that the binary MLSE benchmark d_f^2 is 2 for any such model.

    (b) When does $H(z)$ have a stable inverse (so that a ZFE can be used)?

    (c) Give the noise enhancement factor for these cases.

11. Let $H(z) = h_0 + h_1 z^{-1} + h_2 z^{-2}$ be a model for ISI with three taps. Show that the worst-case binary MLSE benchmark d_f^2 lies 2.3 dB below $d_{MF}^2 = 2$.

12. Consider the two ISI models $1/\sqrt{1.3125}\{1, 0.5, 0.25\}$ and $1/\sqrt{2.62}\{1, 0.9, 0.9\}$ that are featured in Section 6.6.1.

    (a) Confirm that they are stable and find the ZFE for both.

    (b) Find the noise enhancements of the equalizers.

(c) Assuming single interval detection, give the expressions for the asymptotic error probability.

13. Set up a circuit state decoder for the 1-pole encoder response $e^{-at}u(t)$, where $u(t)$ is the unit step. Find d_{min}^2 for a few values of a above and below $1/2$.

Bibliography

References marked with an asterix are recommended as supplementary reading.

[1] *J. B. Anderson, *Digital Transmission Engineering*. IEEE Press, New York, 1999.
[2] A. Lender, "The duobinary technique for high speed data transmission," *IEEE Trans. Commun. Tech.*, **82**, 214–218, 1963.
[3] E. R. Kretzmer, "Generalization of a technique for binary data communication," *IEEE Trans. Commun. Tech.*, **COM-14**, 67–68, 1966.
[4] S. Pasupathy, "Correlative coding: a bandwidth-efficient signaling scheme," *IEEE Commun. Mag.*, **15**, 4–11, 1977.
[5] A. Said, "Design of optimal signals for bandwidth-efficient linear coded modulation," PhD Thesis, Department of Electrical, Computer and Systems Engineering, Rensselaer Poly. Inst., Troy, NY, Feb. 1994; *see also* "Tables of optimal partial-response trellis modulation codes," *Commun., Inf. Voice Processing Report Series*, Report TR93-3, *ibid.*, Sept. 1993.
[6] A. Said and J. B. Anderson, "Bandwidth-efficient coded modulation with optimized linear partial-response signals," *IEEE Trans. Inf. Theory*, **IT-44**, 701–713, Mar. 1998.
[7] D. G. Luemberger, *Linear and Nonlinear Programming*, 2nd edn. Addison-Wesley, Reading, Mass., 1984.
[8] K. Balachandran, "Mismatched receivers for linear coded modulation," PhD Thesis, Deptartment of Electrical, Computer and Systems Engineering, Rensselaer Poly. Inst., Troy, USA, Aug. 1996.
[9] K. Balachandran, "Effect of encoder phase on sequential decoding of partial response coded modulation," MSc Thesis, Department of Electrical, Computer and Systems Engineering, Rensselaer Poly. Inst., Troy, USA, Feb. 1994.
[10] A. V. Oppenheim and R. W. Schafer, *Discrete Time Signal Processing*. Prentice-Hall, Englewood Cliffs, NJ, 1989.
[11] C. W.-C. Wong and J. B. Anderson, "Optimal short impulse response channels for an MLSE receiver," *Conf. Rec. Int. Conf. Commun.*, Boston, Mass., 25.3.1–25.3.5, June 1979.
[12] S. A. Fredericsson, "Optimum transmitting filter in digital PAM systems with a Viterbi detector," *IEEE Trans. Inf. Theory*, **IT-20**, 479–489, July 1974.
[13] F. R. Magee, Jr. and J. G. Proakis, "An estimate of the upper bound on error probability for maximum-likelihood sequence estimation on channels having a finite-duration pulse," *IEEE Trans. Inf. Theory*, **IT-19**, 699–702, Sept. 1973.
[14] R. R. Anderson and G. J. Foschini, "The minimum distance of MLSE digital data systems of limited complexity," *IEEE Trans. Inf. Theory*, **IT-21**, 544–551, Sept. 1985.
[15] P. E. Gill, W. Murray, and M. H. Wright, *Practical Optimization*. Academic, New York, 1981.
[16] J. Mazo, "Faster than Nyquist signaling," *Bell Sys. Tech. J.*, **54**, 1451–1462, Oct. 1975.
[17] J. E. Mazo and H. J. Landau, "On the minimum distance problem for faster-than-Nyquist signaling," *IEEE Trans. Inf. Theory*, **IT-36**, 289–295, Mar. 1990.
[18] D. Hajela, "On computing the minimum distance for faster than Nyquist signaling," *IEEE Trans. Inf. Theory*, **IT-34**, 1420–1427, Nov. 1988.
[19] N. Seshadri, "Error performance of trellis modulation codes on channels with severe intersymbol interference," PhD Thesis, Department of Electrical, Computer and Systems Engineering, Rensselaer Poly. Inst., Troy, USA, Sept. 1986.

[20] N. Seshadri and J. B. Anderson, "Asymptotic error performance of modulation codes in the presence of severe intersymbol interference," *IEEE Trans. Inf. Theory*, **IT-34**, 1203–1216, Sept. 1988.
[21] N. Seshadri and J. B. Anderson, "Decoding of severely filtered modulation codes using the (M,L) algorithm," *IEEE J. Sel. Areas Communs.*, **SAC-7**, 989–995, Aug. 1989.
[22] J. G. Proakis, *Digital Communications*, 3rd edn. McGraw-Hill, New York, 1995.
[23] S. Qureshi, "Adaptive equalization," *Proc. IEEE*, **73**, 1349–1387, Sept. 1985.
[24] A. Duel-Hallen and C. Heegard, "Delayed decision-feedback sequence estimation," *Proc. Allerton Conf. Communs., Control and Computing*, Monticello, Ill., Oct. 1985; also under same title, *IEEE. Trans. Communs.*, **COM-37**, 428–436, May 1989.
[25] F. L. Vermeulen and M. E. Hellman, "Reduced state Viterbi decoding for channels with intersymbol interference," *Conf. Rec. Int. Conf. Commun.*, Minneapolis, 37B-1–37B-9, June 1974.
[26] G. J. Foschini, "A reduced state variant of maximum likelihood sequence detection attaining optimum performance for high signal-to-noise ratios," *IEEE Trans. Inf. Theory*, **IT-23**, 605–609, Sept. 1977.
[27] S. J. Simmons, "Breadth-first trellis decoding with adaptive effort," *IEEE Trans. Communs.*, **COM-38**, 3–12, Jan. 1990.
[28] K. Balachandran and J. B. Anderson, "Reduced complexity sequence detection for nonminimum phase intersymbol interference channels," *IEEE Trans. Inf. Theory*, **IT-43**, 275–280, Jan. 1997.
[29] J. B. Anderson, "Limited search trellis decoding of convolutional codes," *IEEE Trans. Inf. Theory*, **IT-35**, 944–955, Sept. 1989.
[30] J. B. Anderson, "On the complexity of bounded distance decoding for the AWGN channel," *IEEE Trans. Inf. Theory*, **IT-48**, 1046–1060, May 2002.
[31] R. Johannesson and K. Sh. Zigangirov, *Fundamentals of Convolutional Coding*. IEEE Press, New York, 1999.

Appendix 6A: Tables of Optimal PRS Codes

This appendix is a selection of optimal PRS codes of the 2PAM, 4PAM, and 4QAM types, for in-band spectral concentrations $\mathcal{C}$ equal to 99% and 99.9%. These are organized as Tables 6A.1–6A.6 that follow. The data are selected from compilations by Said [5]. Each table is organized by the NBW range $(-WT, WT)$ in which $\mathcal{C}$ lies (T here is the data bit time in each of the I and/or Q channels; see the definition of NBW in Eq. (6.4-2)). Within each bandwidth are listed tap sets at a representative set of memories m. The MDL is given for each tap set, in dB below binary antipodal signaling (MDL is defined in Eq. (6.4-1)). For 4QAM codes, $(-WT, WT)$ is not centered on zero, and an NBW-normalized offset φ is given for each code.

Table 6A.1 A selection of optimal 2PAM PRS codes with 99% power concentration and ≤15 taps. MDL is normalized square free distance loss in dB, compared to antipodal signaling or BPSK; WT is normalized baseband positive frequency bandwidth (denoted NBW in text) in Hz-s/data bit; m is encoder memory. At each bandwidth, the first code shown is the one achieving WT with smallest odd m (Except when 15-tap is sole code shown); entries are those that first show significant MDL improvement as m increases

MDL(dB)	$m+1$ Taps	c_1	c_2	c_3	c_4	c_5	c_6	c_7	c_8	c_9	c_{10}	c_{11}	c_{12}	c_{13}	c_{14}	c_{15}
At $WT = 0.36$																
1.16	4	0.586	0.766	0.047	−0.262											
0.89	6	0.518	0.398	−0.471	−0.582	0.057	0.103									
0.54	8	0.527	0.415	−0.469	−0.562	0.015	0.053	−0.050	0.095							
0.19	10	0.570	0.442	−0.388	−0.502	0.086	0.051	−0.025	0.093	−0.075	−0.229					
At $WT = 0.30$																
2.80	4	0.420	0.741	0.518	0.083											
2.21	6	0.447	0.745	0.485	0.026	−0.078	0.061									
1.77	10	0.453	0.690	0.445	−0.005	−0.028	0.078	−0.118	−0.264	−0.160	0.071					
1.56	15	0.410	0.580	0.366	−0.082	−0.040	−0.049	−0.272	−0.403	−0.040	0.168	−0.054	−0.185	0.009	0.200	0.080
At $WT = 0.26$																
3.82	4	0.325	0.656	0.625	0.271											
3.25	6	0.351	0.677	0.564	0.169	−0.205	−0.173									
2.63	10	0.363	0.676	0.547	0.117	−0.112	0.005	0.185	0.086	−0.140	−0.157					
2.43	15	0.352	0.631	0.505	0.132	−0.124	−0.018	0.258	0.145	−0.057	−0.067	−0.012	−0.035	−0.154	−0.243	−0.096
At $WT = 0.20$																
6.01	6	0.228	0.477	0.607	0.518	0.285	0.051									
4.95	10	0.257	0.433	0.607	0.478	0.245	−0.082	−0.185	−0.213	−0.057	−0.017					
4.50	15	0.270	0.487	0.575	0.444	0.180	−0.138	−0.215	−0.140	−0.011	−0.023	−0.051	−0.129	−0.138	−0.065	0.044
At $WT = 0.15$																
8.01	10	0.175	0.361	0.468	0.499	0.461	0.341	0.178	0.019	−0.075	−0.062					
7.25	15	0.208	0.331	0.414	0.393	0.266	0.080	−0.182	−0.353	−0.387	−0.309	−0.139	−0.041	0.077	0.106	0.058
At $WT = 0.12$																
10.13	10	0.176	0.226	0.331	0.402	0.480	0.425	0.347	0.287	0.163	0.083					
8.95	15	0.157	0.251	0.299	0.339	0.348	0.276	0.147	−0.023	−0.185	−0.299	−0.349	−0.323	−0.276	−0.221	−0.119
At $WT = 0.09$																
11.83	15	0.142	0.105	0.194	0.284	0.290	0.378	0.372	0.372	0.327	0.346	0.242	0.188	0.131	0.085	0.044

Table 6A.2 A selection of optimal 2PAM PRS codes with 99.9% power concentration and ≤15 taps. Notation and conditions as in previous table

MDL(dB)	$m+1$	Taps...														
At $WT = 0.41$																
1.12	4	0.562	0.796	0.088	−0.208											
0.68	6	0.494	0.478	−0.461	−0.558	−0.005	0.064									
At $WT = 0.36$																
2.91	4	0.379	0.744	0.538	0.115											
2.11	6	0.406	0.782	0.467	−0.022	−0.043	0.047									
1.37	10	0.376	0.545	−0.157	−0.629	−0.124	−0.214	0.026	0.093	0.244	0.103					
0.97	15	0.390	0.538	−0.211	−0.596	0.015	0.225	−0.074	0.030	0.019	−0.247	−0.065	0.150	−0.035	−0.110	−0.008
At $WT = 0.30$																
3.89	6	0.270	0.630	0.658	0.304	−0.023	−0.064									
2.58	10	0.277	0.661	0.636	0.166	−0.134	0.003	0.150	0.031	−0.100	−0.062					
2.32	15	0.294	0.663	0.605	0.136	−0.118	−0.007	0.014	−0.154	−0.142	0.048	0.051	−0.109	−0.113	0.009	0.045
At $WT = 0.26$																
5.60	6	0.197	0.485	0.637	0.512	0.237	0.043									
4.51	10	0.200	0.511	0.619	0.317	−0.152	−0.367	−0.221	0.009	0.081	0.033					
3.77	15	0.218	0.541	0.630	0.303	−0.150	−0.295	−0.109	0.044	−0.034	−0.159	−0.119	0.019	0.073	0.025	−0.007
At $WT = 0.20$																
7.15	10	0.142	0.337	0.519	0.571	0.443	0.207	−0.016	−0.123	−0.113	−0.051					
6.44	15	0.143	0.345	0.527	0.548	0.364	0.070	−0.175	−0.255	−0.186	−0.073	−0.014	−0.029	−0.060	−0.060	−0.032
At $WT = 0.15$																
14.07	12	0.092	0.198	0.311	0.414	0.477	0.463	0.377	0.267	0.158	0.065	0.009	−0.004			
12.28	15	0.090	0.197	0.312	0.385	0.377	0.269	0.084	−0.124	−0.293	−0.375	−0.362	−0.280	−0.170	−0.072	−0.016
At $WT = 0.12$																
15.82	15	0.079	0.142	0.219	0.308	0.375	0.417	0.421	0.384	0.320	0.238	0.154	0.085	0.035	0.001	−0.007

Table 6A.3 A selection of optimal 4PAM PRS codes with 99% power concentration and ≤15 taps. Notation and conditions as in previous tables. Note that all 4PAM systems have MDL in excess of 4.0 dB

MDL(dB)	m+1	Taps...														
At WT = 0.15																
7.54	4	0.440	0.663	0.588	0.146											
6.96	6	0.421	0.664	0.556	−0.032	−0.222	−0.151									
6.19	8	0.473	0.720	0.395	−0.126	−0.112	0.169	0.214	−0.021							
5.60	15	0.429	0.610	0.368	−0.034	−0.040	0.006	−0.189	−0.241	−0.010	0.025	−0.242	−0.177	0.125	0.319	0.085
At WT = 0.12																
9.13	6	0.353	0.525	0.630	0.425	0.145	−0.006									
8.67	8	0.352	0.518	0.621	0.367	−0.002	−0.216	−0.179	−0.091							
8.03	15	0.315	0.446	0.448	0.041	−0.263	−0.316	−0.093	0.012	−0.057	−0.125	0.035	0.289	0.371	0.277	0.055
At WT = 0.09																
12.35	6	0.240	0.414	0.548	0.534	0.381	0.199									
11.58	8	0.246	0.341	0.533	0.538	0.422	0.246	0.104	−0.028							
10.45	10	0.260	0.384	0.499	0.497	0.268	−0.011	−0.205	−0.312	−0.239	−0.146					
10.15	15	0.262	0.406	0.509	0.421	0.140	−0.167	−0.331	−0.342	−0.157	0.003	0.093	0.033	−0.032	−0.105	−0.097
At WT = 0.065																
13.68	10	0.196	0.223	0.349	0.423	0.487	0.413	0.345	0.243	0.155	0.044					
12.90	15	0.179	0.209	0.303	0.354	0.387	0.239	0.121	−0.042	−0.199	−0.337	−0.352	−0.323	−0.260	−0.160	−0.095
At WT = 0.045																
15.81	15	0.142	0.106	0.195	0.284	0.290	0.380	0.373	0.371	0.347	0.326	0.240	0.187	0.133	0.084	0.041

Table 6A.4 A selection of optimal 4PAM PRS codes with 99.9% power concentration and ≤15 taps. Notation and conditions as in previous tables

MDL(dB)	$m+1$	Taps...														
At $WT = 0.20$																
5.57	4	0.514	0.810	0.266	−0.097											
5.08	6	0.515	0.730	−0.084	−0.435	−0.028	0.071									
At $WT = 0.15$																
11.97	6	0.274	0.630	0.659	0.298	−0.028	−0.070									
9.58	10	0.277	0.585	0.465	−0.105	−0.453	−0.295	−0.053	−0.099	−0.188	−0.121					
8.97	15	0.269	0.543	0.405	−0.118	−0.336	−0.095	0.022	−0.254	−0.418	−0.182	0.112	0.155	0.109	0.082	0.066
At $WT = 0.12$																
14.14	8	0.188	0.445	0.623	0.543	0.266	−0.001	−0.095	−0.061							
13.15	10	0.182	0.407	0.535	0.351	−0.039	−0.375	−0.423	−0.256	−0.068	0.018					
12.65	15	0.181	0.413	0.547	0.387	0.036	−0.267	−0.334	−0.242	−0.167	−0.149	−0.094	0.018	0.137	0.142	0.080
At $WT = 0.09$																
17.51	10	0.115	0.238	0.389	0.494	0.511	0.420	0.277	0.136	0.044	−0.014					
16.60	15	0.122	0.277	0.439	0.529	0.489	0.299	0.078	−0.101	−0.142	−0.068	0.061	0.140	0.163	0.117	0.052
At $WT = 0.065$																
20.68	15	0.081	0.149	0.246	0.346	0.423	0.449	0.430	0.359	0.258	0.153	0.056	−0.005	−0.032	−0.035	−0.031

Table 6A.5 A selection of optimal 4QAM PRS codes with 99% power concentration and ≤15 taps. Optimal offset for center frequency of code spectrum is φ/T. Positive-frequency RF bandwidth is $2WT$; NBW is $2WT$ Hz-s/QAM symbol ÷ 2 data bits/QAM symbol = WT Hz-s/data bit. Taps are complex except for the first, and do not occur in complex pairs. Notation otherwise as in previous tables.

MDL(dB)	$m+1$	φ	Taps ...							
At $WT = 0.35$										
0.09	4	0.075	0.569	$0.697+$ $j.345$	$0.134+$ $j.131$	$-0.001-$ $j.191$				
At $WT = 0.30$										
2.19	4	0.050	0.418	$0.707+$ $j.222$	$0.418+$ $j.305$	$0.032+$ $j.084$				
1.22	6	0.075	0.452	$0.626+$ $j.306$	$0.196+$ $j.239$	$-0.051-$ $j.270$	$0.081+$ $j.357$	$0.004-$ $j.075$		
0.81	8	0.085	0.448	$0.554+$ $j.342$	$0.109+$ $j.156$	$0.012-$ $j.383$	$0.199-$ $j.322$	$-0.030+$ $j.062$	$-0.177-$ $j.064$	$-0.054+$ $j.079$
0.23	15	0.095	0.370	$0.492+$ $j.297$	$0.141+$ $j.142$	$0.107-$ $j.202$	$-0.143-$ $j.098$	$-0.187+$ $j.113$	$-0.163+$ $j.121$	$-0.043+$ $j.228$
At $WT = 0.26$										
2.16	6	0.055	0.366	$0.625+$ $j.205$	$0.456+$ $j.303$	$0.125+$ $j.026$	$0.043-$ $j.260$	$0.080-$ $j.203$		
1.76	10	0.060	0.352	$0.558+$ $j.221$	$0.343+$ $j.230$	$0.022-$ $j.103$	$0.032-$ $j.413$	$0.042-$ $j.228$	$-0.154-$ $j.068$	$-0.164-$ $j.100$
1.44	15	0.070	0.333	$0.455+$ $j.185$	$0.249+$ $j.202$	$-0.010-$ $j.161$	$0.103+$ $j.422$	$0.025-$ $j.149$	$-0.177+$ $j.047$	$-0.245-$ $j.005$
At $WT = 0.20$										
5.13	6	0.040	0.238	$0.460+$ $j.128$	$0.529+$ $j.294$	$0.373+$ $j.352$	$0.150+$ $j.245$	$-0.004+$ $j.054$		
4.23	8	0.045	0.261	$0.473+$ $j.129$	$0.473+$ $j.325$	$0.218+$ $j.384$	$-0.053-$ $j.213$	$-0.161-$ $j.081$	$-0.095-$ $j.227$	$-0.010-$ $j.162$
3.14	15	0.050	0.282	$0.423+$ $j.135$	$0.379+$ $j.333$	$0.171+$ $j.296$	$-0.046-$ $j.043$	$-0.059-$ $j.251$	$0.013-$ $j.251$	$-0.024-$ $j.037$
At $WT = 0.15$										
7.59	8	0.090	0.203	$0.248+$ $j.164$	$0.191+$ $j.359$	$-0.082-$ $j.511$	$-0.283-$ $j.369$	$-0.353+$ $j.123$	$-0.245-$ $j.072$	$-0.099-$ $j.081$
6.98	10	0.025	0.191	$0.341+$ $j.068$	$0.443+$ $j.148$	$0.460+$ $j.248$	$0.362+$ $j.287$	$0.204+$ $j.243$	$0.054-$ $j.133$	$-0.026+$ $j.014$
5.65	15	0.035	0.192	$0.326+$ $j.074$	$0.386+$ $j.162$	$0.339+$ $j.287$	$0.206+$ $j.299$	$0.046+$ $j.188$	$-0.080-$ $j.013$	$-0.122+$ $j.170$
At $WT = 0.12$										
8.67	10	0.090	0.172	$0.202+$ $j.141$	$0.135+$ $j.276$	$-0.070+$ $j.402$	$-0.291+$ $j.351$	$-0.419-$ $j.122$	$-0.333-$ $j.116$	$-0.190-$ $j.207$
7.82	15	0.095	0.170	$0.188+$ $j.121$	$0.133+$ $j.268$	$-0.049+$ $j.380$	$-0.221+$ $j.292$	$-0.313+$ $j.138$	$-0.234-$ $j.001$	$-0.146+$ $j.015$

Taps (cont.)						
$-0.025-$ $j.104$	$-0.089-$ $j.086$	$-0.237-$ $j.034$	$-0.141-$ $j.158$	$0.094+$ $j.110$	$-0.042+$ $j.323$	$-0.078+$ $j.078$
$0.041-$ $j.154$	$0.136-$ $j.030$					
$-0.040+$ $j.002$	$-0.011+$ $j.064$	$-0.017-$ $j.006$	$-0.013+$ $j.080$	$-0.106+$ $j.232$	$-0.254-$ $j.195$	$-0.174-$ $j.045$
$-0.197-$ $j.093$	$-0.242-$ $j.074$	$-0.130+$ $j.051$	$0.027+$ $j.111$	$0.084+$ $j.176$	$0.022+$ $j.174$	$-0.045+$ $j.038$
$-0.063-$ $j.064$	$-0.034-$ $j.081$					
$-0.105-$ $j.214$	$-0.102-$ $j.158$	$-0.136-$ $j.080$	$-0.197-$ $j.013$	$-0.218-$ $j.016$	$-0.198-$ $j.014$	$-0.122-$ $j.039$
$-0.035-$ $j.194$	$0.046-$ $j.080$					
$-0.143+$ $j.069$	$-0.219+$ $j.090$	$-0.292-$ $j.049$	$-0.238-$ $j.161$	$-0.093-$ $j.220$	$0.035-$ $j.185$	$0.071-$ $j.091$

Table 6A.6 A selection of optimal 4QAM PRS codes with 99.9% power concentration and ≤15 taps. Notation and conditions as in previous table

MDL(dB)	$m+1$	φ	Taps ...
At $WT = 0.36$			
2.27	4	0.050	0.367, 0.711+j.227, 0.439+j.320, 0.061+j.098
0.70	6	0.080	0.405, 0.657+j.355, 0.158+j.238, −0.043−j.334, 0.081−j.274, −0.002−j.030
At $WT = 0.30$			
2.81	6	0.035	0.268, 0.613+j.143, 0.588+j.300, 0.205+j.205, −0.060−j.020, −0.053−j.073
2.04	8	0.060	0.279, 0.598+j.225, 0.458+j.354, 0.119+j.003, 0.078−j.327, −0.057−j.016, −0.085+j.014, 0.085−j.187
1.41	15	0.075	0.261, 0.512+j.247, 0.275+j.271, 0.119−j.223, 0.329−j.299, −0.212−j.188, −0.076+j.035, 0.134−j.148, 0.038+j.045, −0.019−j.095, 0.061−j.127, −0.008−j.057, −0.142+j.016, −0.035−j.097, −0.047−j.037
At $WT = 0.26$			
4.75	6	0.035	0.180, 0.458+j.113, 0.566+j.297, 0.388+j.351, 0.128+j.209, 0.003−j.046
3.51	8	0.045	0.180, 0.478+j.122, 0.559+j.308, 0.304+j.268, 0.022−j.048, −0.031−j.293, 0.024−j.069, 0.026−j.239
2.45	15	0.050	0.195, 0.485+j.120, 0.527+j.251, 0.295+j.107, 0.122−j.191, 0.088−j.202, −0.031+j.069, −0.187−j.186, −0.128−j.059, 0.041−j.048, 0.053−j.083, −0.035−j.151, −0.029−j.175, 0.028−j.083, 0.025+j.003
At $WT = 0.20$			
7.25	8	0.020	0.112, 0.296+j.042, 0.465+j.130, 0.512+j.225, 0.415−j.259, 0.234−j.204, 0.077+j.103, 0.006+j.026
6.21	10	0.025	0.130, 0.324+j.052, 0.489+j.164, 0.497+j.225, 0.328+j.287, 0.086+j.168, −0.083−j.006, −0.121−j.121, −0.074−j.122, −0.022+j.057
5.14	15	0.030	0.126, 0.334+j.049, 0.501+j.166, 0.495+j.268, 0.307+j.234, 0.069+j.049, −0.067−j.152, −0.071−j.207, −0.036−j.095, −0.037+j.056, −0.055−j.079, −0.009+j.011, 0.020+j.014
At $WT = 0.15$			
12.30	12	0.020	0.081, 0.187+j.037, 0.291+j.104, 0.370+j.186, 0.395+j.272, 0.342+j.320, 0.239+j.298, 0.135+j.226, 0.056+j.138, 0.006+j.053, −0.006+j.003, −0.000−j.012
10.76	15	0.070	0.076, 0.177+j.071, 0.227+j.230, 0.147+j.402, −0.034−j.463, −0.183−j.360, −0.181−j.176, −0.038+j.059, 0.113+j.087, 0.149+j.194, 0.069+j.259, −0.032+j.219, −0.071+j.117, −0.046+j.033, −0.012+j.001
At $WT = 0.12$			
14.71	15	0.025	0.066, 0.134+j.020, 0.206+j.068, 0.261+j.139, 0.296+j.222, 0.294+j.300, 0.242+j.354, 0.165+j.355, 0.083+j.310, 0.026−j.235, −0.014−j.156, −0.026−j.083, −0.016−j.024, −0.002−j.006, 0.004−j.011

Appendix 6B: Said's Solution for Optimal Codes

A. Said [5,6] in 1994 published an efficient linear programming method that solves for PRS codes of a given type that have minimal energy and bandwidth. He made the key observation that energy and bandwidth can be viewed as linear constraints, so that their optimization may be performed by powerful linear programming methods. Earlier methods depended on exhaustion, nonlinear programming, or were not fully carried out (see respectively [11,12,14]). Our purpose here is to hit the high points of Said's approach and summarize its range of application. Representative solutions for the basic FIR encoder cases are given in Appendix 6A, and their bandwidth–energy performance was summarized in Section 6.4. The details of Said's method appear in his PhD thesis [5].

First, some basics of optimization by linear programming. A standard formulation is as follows:

$$\text{Minimize } b'v$$
$$\text{subject to } Av = c \quad v \geq 0. \tag{6B-1}$$

This is called the *primal problem*. v is a vector of variables to be optimized, b and c are constant vectors, and A is a matrix. The formula $Av = c$ forms the set of linear constraints on v, and $b'v$ a weighting of v that is to be optimized, called the objective function. The facts are that the range of allowed v in the problem (6B-1), called the constraint region, forms a mathematical simplex, and the optimal v must lie on one of the vertices around its periphery. In our particular problem the number of variables far exceeds the number of constraints, and for cases like this a scheme called the revised simplex method is employed to find the optimal vertex of the constraint region.

Within the basic problem just presented, there are many extensions and special techniques attuned to them. From the solution to Eq. (6B-1) can be directly obtained the solution to the *dual problem*:

$$\text{Maximize } w'c$$
$$\text{subject to } w'A \leq b. \tag{6B-2}$$

This is essentially Eq. (6B-1) with c and b reversed. Problem (6B-2) may be simpler. Furthermore, the PRS optimization is in fact a "semi-infinite" linear program with finitely many variables and an infinite number of constraints defined over a compact set. It has dual form

$$\text{Maximize } w'c$$
$$\text{subject to } w'A \leq b, \quad w'g(u) \leq 0, \quad \text{over } u \in [0, 1], \tag{6B-3}$$

where $g(u)$ is a vector of functions of u. Such a problem can be solved by the so-called cutting planes method.

The PRS Code Linear Program

Formulas developed in Section 6.3 provide the constraints for the particular problem at hand, which is: What is the distance-maximizing PRS encoder $\{b_0, \ldots, b_m\}$ of memory m at a given bandwidth $(-W, W)$? We will formulate the linear program in successive refinements. A first formal statement is as follows. ρ is the autocorrelation of a tap set and Δa is an error difference sequence. We seek:

$$d_{f,\text{opt}}^2 = \max_{\rho} \min_{\Delta a} \; d^2(\rho, \Delta a), \tag{6B-4}$$

in which

$$d^2(\rho, \Delta a) = \frac{1}{2} \log_2 M \sum_k \sum_j \Delta a[j]\rho[k-j]\Delta a[k]^* \quad \text{(from Eq. (6.3-18))}$$

subject to the linear constraints

$$\text{Bandwidth:} \quad \sum_{k=-m}^{m} \rho[k]\chi[k]^* = \mathcal{C} \quad \text{(from Eq. (6.3-30))}$$

Total energy: $\rho[0] = 1$

Admissibility: $R_D(f) \geq 0, \quad f \in [0, 1)$ (from Eq. (6.3-5))

From the solution for ρ, the minimum phase tap set $\{b_\ell\}$ is found by locating the zeros of the polynomial $\sum \rho[k]z^{-k} = 0$, as discussed in Section 6.3.1. The signal bandwidth $(-W, W)$ is defined by the sequence χ.

The first constraint imposes on the taps the spectral power concentration $\mathcal{C}$. The second insists that the transmitted energy is always unity; this is needed because non-normalized distance can be scaled to any value by changing the energy. The third requires that the signal power spectrum be real and positive. This is necessary because even a small negative region for $R_D(f)$ means that $\{b_\ell\}$ cannot be recovered from ρ. Since there will always be small computation errors, it is safest to replace the constraint with $R_D(f) \geq \epsilon$ for some very small $\epsilon > 0$; ordinarily this has no effect on the solution.

Several details need more attention. To put the optimization in the standard dual form (6B-2), it is necessary to express the first two constraints as inequalities, namely $\mathcal{C} \geq \sum \rho[k]\chi[k]^*$ and $\rho[0] \leq 1$. On physical grounds, an autocorrelation with less compaction than $\mathcal{C}$ in $(-W, W)$ or less energy than 1 will not in any case be the optimizing ρ. A more serious detail is how to incorporate the min over the difference sequence Δa. Ordinarily, just a few error sequences in a critical set, call it $\mathcal{S}$, actually constrain the minimum distance, with the remainder having no interplay at all. If this is the case, we can eliminate the min over Δa in Eq. (6B-4) by the standard method (see [15], Chapter 4) of adding a special distance variable d_0. For each $\Delta a \in \mathcal{S}$, the distance constraint $d^2(\rho, \Delta a) \geq d_0$ applies. We wish to maximize d_0 itself over the choice of ρ; the objective function consists of just d_0. Rewriting the previous problem statement, we have the following dual-form problem:

$$d^2_{f,opt} = \max_{\rho} \min_{d_0} d_0 \qquad (6\text{B-5})$$

subject to

$$\text{Distance: } d^2(\rho, \Delta a) \geq d_0, \quad \text{all } \Delta a \in \mathcal{S}, \qquad (6\text{B-6})$$

$$\text{Concentration in}(-W, W): \sum_{k=-m}^{m} \rho[k]\chi[k]^* \geq \mathcal{C}, \qquad (6\text{B-7})$$

$$\text{Energy: } \rho[0] \leq 1, \qquad (6\text{B-8})$$

$$\text{Admissibility: } R_D(f) \geq \epsilon, \quad f \in [0, 1) \qquad (6\text{B-9})$$

One difficulty remains: the optimizing Δa may not be in the set $\mathcal{S}$; the true d_f^2 for the solution ρ may be less that the maximal d_0 value, and/or a better ρ with higher d_f^2 may exist for a larger set $\mathcal{S}$. To deal with this, Said suggested the following overall scheme, based on the cutting plane method, which keeps adding to $\mathcal{S}$ until it finds a self-consistent solution.

Final PRS code optimization algorithm. Select a symbol alphabet $\mathcal{A}$, a memory m, a spectral concentration $\mathcal{C}$, and a baseband bandwidth $(-W, W)$. Form the first difference set $\mathcal{S}$, denoted $\mathcal{S}^{(1)}$, with a small list of error difference sequences. Then repeat the following for $j = 1, \ldots$.

(i) Solve linear program (6B-5)–(6B-9) to find an optimal jth stage $d_0^{(j)}$ and $\rho^{(j)}$ for the present set $\mathcal{S}^{(j)}$.

(ii) For this $\rho^{(j)}$, find d_f^2 and its corresponding error sequence Δa, using a method like those in Section 6.3.2 or otherwise.

(iii) If this d_f^2 is d_0, stop; $\rho^{(j)}$ is the desired optimal autocorrelation for bandwidth $(-W, W)$ and d_f is its free distance. Otherwise, add Δa to $\mathcal{S}^{(j)}$ to form $\mathcal{S}^{(j+1)}$ and return to (i).

(iv) By spectral factorization, find the minimum phase (or other) tap set with autocorrelation $\rho^{(j)}$.

In applications of this procedure, a good initial difference set $\mathcal{S}^{(1)}$ is often well known from a previous optimization, for example, one at a closely related $(-W, W)$. The Δa that achieves the free distance for an optimal ρ generally remains the same over a range of bandwidths, and the optimal ρ and its d_f evolve slowly, in a well-behaved manner; some of this is evident in Section 6.4. There are of course many details that we have omitted, and linear programming is in fact an art. A great many such points are discussed by Said [5]. Said also proposed new and much faster algorithms to solve for the PRS free distances needed in (ii).

Generalizations

The foregoing solution may be generalized in many ways. One that does not require any modification to the mathematics is the extension to simultaneous coding of I and Q signals, what we have called QAM coding. The optimization may be performed simply by allowing a complex transmission symbol alphabet. As discussed in Section 6.4, I and Q coding is more powerful – it allows a higher d_f in the same RF bandwidth, or a narrower bandwidth at the same d_f – and its spectrum is not symmetric about the carrier frequency. Many other generalizations are possible by adding constraints or small changes to the solution algorithm.

1. Spectral nulls in the signal may be forced by adding linear constraints. Other spectral properties may be inserted such as a Chebyshev stopband criterion.
2. The presence of oversampling in the receiver processing, for example, fractional spaced equalizing, may be inserted through changes to constraint (6B-6) and the distance calculation.
3. Certain extensions to IIR code generation filters are possible. By means of a spectral mapping technique, strictly bandlimited pulses $h(t)$ with concentration C may be studied. This is particularly important at very narrow bandwidth, since, as discussed in Section 6.4, free distance tends otherwise to escape outside $(-WT, WT)$.

7

Introduction to Fading Channels

7.1. Introduction

In this chapter, some important characteristics of fading channels will be discussed and their influence on the performance of a coded modulation system will be given. Some counter measures that can be used to improve the performance of fading channels will also be introduced. Fading appears due to attentuation of the radio signal when it passes objects on its way from the transmitter to the receiver, and due to reflection and scattering of the transmitted signal. Furthermore, the received signal power is reduced in proportion to the distance between the transmitter and the receiver. In Section 7.2, the average power loss between the antennas are discussed and several important models for it are introduced. This power loss can be relatively easily compensated for by increasing the transmitted power by the same amount, and has no dramatic influence on the receiver performance. On the contrary, shadowing, scattering, and reflections give rise to large power variations in the received signal. These variations may cause extreme degradations in receiver performance. Compared to the non-fading case, the error probability decreases much more slowly with increasing power when such variations are experienced. This is all covered in Section 7.3.

In Section 7.4, the fading is characterized in more detail. When designing coded modulation systems, it is important to know the correlation properties over time and frequency of the fading. Section 7.5 is devoted to fading simulators. For more complex channels and coded modulation systems, it is sometimes not possible to find analytical expressions for the error probability. Then, one way to obtain performance estimates is through simulation. The basic methods for simulation of fading channels are given here. The error probability of coded modulation systems over fading channels is discussed in Section 7.6. Here we clearly show the large degradations in performance that appear with fading. Measures to improve the performance are given in Section 7.7.

7.2. Propagation Path Loss

A basic measure that sets the performance of a coded modulation system on a radio channel is the *path loss*. The path loss is a measure of the average power that is lost between the transmit antenna and the receive antenna. Formally, path loss is defined as

$$l \triangleq \overline{P_r}/P_t, \qquad (7.2\text{-}1)$$

where P_t denotes the transmitted power in watts and $\overline{P_r}$ denotes the average received power also in watts. An overbar will be used throughout this chapter to denote an average value. The reason for using an average value of the received power will become clear later when we discuss the variations due to fading and shadowing in the received power. In practice, it is much more common to specify path loss as a positive quantity in decibels, that is,

$$L = -10 \log l = 10 \log P_t - 10 \log \overline{P_r} = P_t^{[dB]} - \overline{P}_r^{[dB]}, \qquad (7.2\text{-}2)$$

the difference between the transmitted power in dB and the average received power in dB.

The exact value of the path loss depends on the transmission medium and the propagation characteristics between the transmitter and receiver. Simple models for the path loss only exists in a few special cases, of which some will now be discussed.

7.2.1. Free Space Path Loss

The notion of free space is used for a medium where the transmitted wave propagates without obstruction between the transmitter and receiver. This means that one single plane wave arrives at the receiver antenna, in contrast to some cases discussed later. Typical examples of systems where the free space path loss model is a good approximation are satellite communication systems, when the satellite is at a rather high elevation, and microwave line-of-sight radio links, when the transmitter and receiver antennas are high above the ground. The free space path loss is given by Friis free space formula [1–4], as

$$l = \left(\frac{\lambda}{4\pi d}\right)^2, \qquad (7.2\text{-}3)$$

where λ denotes the wavelength of the carrier and d the distance between the transmitter and the receiver. In most situations it is more practical to evaluate the path loss in decibels[1] from

$$L = 20 \log f + 20 \log d + 32.44, \qquad (7.2\text{-}4)$$

---

[1] In some literature, the path loss in dB is denoted L_p.

Introduction to Fading Channels

where the distance d is given in km and carrier frequency f in MHz. Here we have used the fact that $\lambda = c/f$, where c is the speed of light in m/s. From this formula we clearly see that a doubling of the distance or carrier frequency increases the path loss by 6 dB.

7.2.2. Plane Earth Path Loss

In many practical situations, the free space model gives a very bad estimate of the actual path loss. Especially in situations when the received signal is a superposition of many reflected waves, the average received power is in most cases much lower than predicted by the free space model. This is the case in many mobile radio applications, where the receiver antenna is typically very close to the ground. To give some insight, let us consider a very simplified situation, with one direct line-of-sight transmission path and one indirect path that is reflected by the ground, as shown in Fig. 7.1. The received signal is now equal to the sum of two signals, the direct signal and the indirect signal. Both are subject to a path loss given by the free space model, but the indirect signal has traversed a slightly longer distance. The indirect signal has a phase difference relative to the direct signal, corresponding to its somewhat longer distance and to the phase shift at the ground reflection.

To simplify the path loss derivation, it is common practice to assume that the ground reflection is lossless and causes a phase change of 180 degrees. Simple geometry then shows that the phase difference between the direct and indirect signal is

$$\Delta \varphi = \frac{2\pi d}{\lambda} \left[\sqrt{\left(\frac{h_t + h_r}{d}\right)^2 + 1} - \sqrt{\left(\frac{h_t - h_r}{d}\right)^2 + 1} \right], \quad (7.2\text{-}5)$$

where d is the distance between the antennas as before, and h_t and h_r are the transmit and receive antenna heights, respectively. When the distance between the

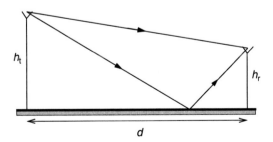

Figure 7.1 Two receiver paths over a flat-reflecting surface.

two antennas is assumed much larger that the antenna heights, which is true in most practical situations, this phase difference is approximately

$$\Delta\varphi \approx 4\pi \frac{h_t h_r}{\lambda d}. \qquad (7.2\text{-}6)$$

The path loss may now be approximated as

$$l = 4\left(\frac{\lambda}{4\pi d}\right)^2 \sin^2\left(\frac{2\pi h_t h_r}{\lambda d}\right). \qquad (7.2\text{-}7)$$

This result can be further simplified by the approximation $\sin(x) \approx x$, when d is much larger than the antenna heights, as

$$l = \left(\frac{h_t h_r}{d^2}\right)^2. \qquad (7.2\text{-}8)$$

This is an inverse fourth-power loss formula in distance, as compared to an inverse second-order loss formula for the free space case. A more detailed derivation of this result may be found in [3]. In decibel scale it becomes

$$L = -20\log h_t - 20\log h_r + 40\log d \qquad (7.2\text{-}9)$$

and here it is clearly seen that a doubling of the distance now leads to a 12 dB loss in average received power instead of 6 dB as in free space. Moreover, we see that increasing one of the antenna heights by a factor of 2 leads to a 6 dB power gain. With the approximations made, there is no carrier frequency effect in this formula. The reason for the increased path loss is that the two signals add destructively.

7.2.3. General Path Loss Model

The two models given above are for very specific cases and they do not apply directly in many practical situations. Still they give important insight in modeling of path loss. Several other models are available in the literature, such as the Okumura model which is based on measurements done in Tokyo, the Hata model which is an empirical set of formulas obtained from Okumura's model, and the PCS extension to the Hata model, just to mention a few [1,3]. Based on these it is possible to define a more general path loss model as

$$L = -10\log K - 20\log h_t - 10x\log h_r + 10\alpha\log d + 10y\log f, \qquad (7.2\text{-}10)$$

where K is a constant that depends on the environment and the units of the other variables, h_t and h_r are antenna heights, d the distance, and f the carrier frequency. In most cases K is not important since we are only interested in the relative path

Introduction to Fading Channels

loss over different distances, frequencies, or antenna heights. The units for the heights, the distance, and the frequency can be chosen in the most convenient way, since all scale factors are included in K. From this relation we see that in most practical situations, the path loss decreases with 6 dB per transmitter antenna height doubling. The variation when changing the receiver antenna height is somewhat different; a scale factor $10x$ is used, where it is common practice to assume that

$$x = \begin{cases} 1, & \text{for } h_\text{r} < 3; \\ 2, & \text{for } 3 \leq h_\text{r} \leq 10. \end{cases} \qquad (7.2\text{-}11)$$

This means that an antenna height doubling decreases the path loss by 3 dB for antennas close to the ground and 6 dB for antennas above 3 m. Even better models can be obtained by estimating the value of x from measurements, but this is in most cases not needed. The distance dependence is given by α which lies typically in the interval $2 \leq \alpha \leq 4$, where the bounding values correspond to free space and ground reflection, respectively, as discussed before. In extreme cases it may be even larger and in some indoor situations it may be as small as 1.6 [1]. The frequency dependence is given by y which is typically in the interval $2 \leq y \leq 3$. Whenever more precision is needed in the factors x, α, and y, one can estimate them from measurements taken in the environment of interest.

An example of how this path loss model can be used is shown in Fig. 7.2, where we show the path loss in dB vs distance in km. At a distance smaller than 15 km, the path loss exponent α is 4, then it is 2 between 15 and 50 km and finally it is 3 for distances longer than 50 km. This model makes it very easy to model an exponent that varies over distance. The model can take account of varying antenna heights and carrier frequency, if needed.

For more details on path loss models, refer to [1–3].

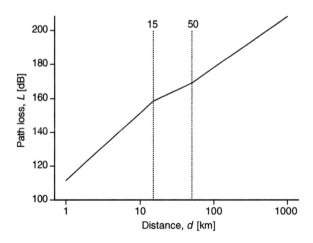

Figure 7.2 An example of path loss vs distance including three different path loss exponents.

7.3. Fading Distributions

From the path loss models above, it seems that the received power is always the same at a constant distance from the transmitter antenna, but this is not the case in most practical systems for radio communication. Instead one typically sees quite a large variation in the power, as shown in Fig. 7.3. In the figure it is clear that the path loss is larger at 20 km as compared to 10 km, but there is a significant variation among measurements of the received power at each distance. This variation would of course translate to different bit error probabilities in the corresponding receivers, and some receivers may not fulfill the requirement on performance given by the dotted line in Fig. 7.3. The path loss in Section 7.2 is a model only for the mean value of all the measurements taken. Received power varies for many reasons. Two important ones are:

- shadowing from objects located between the two antennas, and
- multipath propagation from reflections in the vicinity of the receiving antenna.

These two situations will be discussed now.

In order to obtain knowledge on how these power variations affect the receiver performance, we need to describe them by appropriate statistical models, in the form of probability density functions (pdfs). When the pdf of the (instantaneous) received power is known, we can calculate the probability that the received power is below a given threshold (compare Fig. 7.3). This measure is referred to as the outage probability. It is often used as a design criterion for radio communication

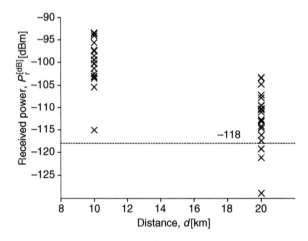

Figure 7.3 Variation of the received power at two different distances from the transmitter antenna. Twenty measurements are taken at each position. The dashed line indicates the received power level to give acceptable performance in a receiver.

Introduction to Fading Channels

systems. In this book we will not go into the details of outage probability, but the interested reader may find more in [1–3]. Later in this chapter we will also explain how the average bit error probability of a given modulation system can be evaluated once the pdf of the received power is known.

7.3.1. Shadow Fading Distribution

Shadowing is due to objects located between the transmit and receive antennas. When the electromagnetic wave propagates through an object, it is attenuated by $\exp(-\alpha_i \Delta r_i)$ where α_i is the attenuation constant for the material of object number i and Δr_i is its thickness. Typically, the electromagnetic wave traverses through many different objects of different size on its way to the receive antenna. This means that the received power P_r is equal to the transmitted power P_t multiplied by a product of a number of negative exponentials, each having a unique exponent, which can be expressed mathematically as

$$P_r = P_t \prod_{i=1}^{\nu} \exp(-\alpha_i \Delta r_i), \tag{7.3-1}$$

where ν is the number of objects. Here it is common practice to assume that the exponents $-\alpha_i \Delta r_i$ can be regarded as random variables. The received power in decibel scale $P_r^{[dB]}$, is now given by

$$P_r^{[dB]} = P_t^{[dB]} - 10 \log e \sum_{i=1}^{\nu} \alpha_i \Delta r_i. \tag{7.3-2}$$

If the number of objects ν is large enough, we can use the central limit theorem to conclude that the pdf of $\sum_{i=1}^{\nu} \alpha_i \Delta r_i$ approaches a normal distribution. The pdf of the received power in dB is therefore normally distributed as

$$p(P_r^{[dB]}) = \frac{1}{\sqrt{2\pi} \sigma_{dB}} \exp \left[-\frac{1}{2} \left(\frac{P_r^{[dB]} - \overline{P}_r^{[dB]}}{\sigma_{dB}} \right)^2 \right], \tag{7.3-3}$$

where $\overline{P}_r^{[dB]}$ is the average received power in dB as given by the path loss models, and σ_{dB} is the standard deviation of the received power in dB. The pdf of P_r (in watts) is referred to as the lognormal distribution and this fading is therefore referred to as lognormal fading. An expression for the pdf of the lognormal distribution can also be given, but is rarely used, and we therefore omit it. The interested reader is referred to [1–3] for more details.

The standard deviation of the received power in dB depends on the type of objects for each particular location, and lies between 0 and 8 dB in most practical situations. It is rather straight forward to estimate the standard deviation from power measurements, such as shown in Fig. 7.3. For any value of the standard deviation and a given average received power level, it is easy to calculate the outage probability as the area under the tail of the normal distribution. It is a bit more complicated to calculate the outage probability over an area,[2] for example, the coverage area of a transmitter, since the average received power is a function of the distance. It can be calculated using the area under the tail of a normal distribution in each small area segment and then integrating over the whole coverage area. Details appears in [1,3]. In some situations also the standard deviation of the received power may vary over the coverage area.

7.3.2. Multipath Fading Distribution

In many situations, there is no line-of-sight signal between the transmit and receive antennas. Instead all the received power is from reflections at local scatterers in the vicinity of the receiver, as shown in Fig. 7.4. The received signal is a sum of many plane waves, each from a single scatterer. Each of these plane waves will have a different amplitude due to the reflection from the scatterer, and a different carrier phase and delay due to the different distance that the wave has traversed. This is commonly referred to as multipath fading. A commonly made assumption for this situation is that the carrier phase of each scatterer is uniformly distributed between 0 and 2π, since a small difference in distance corresponds to a large phase difference at high carrier frequency. Moreover, the amplitudes and the carrier phases are assumed statistically independent from each other. Finally, since the scatterers are located close to the receiver, the relative delays between the different waves are small compared to the symbol period of the modulation method, and therefore no significant intersymbol interference is introduced.[3] This

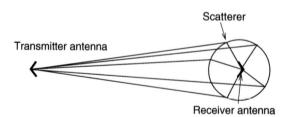

Figure 7.4 An example of local scattering around the receiver antenna.

---

[2] This is sometimes also referred to as area coverage.

[3] It is more correct to say that the relative delays are very short compared to the inverse signaling bandwidth, since this makes the statement correct also for spread spectrum systems, but for the time being we prefer to relate to the symbol period of the modulation method since spread spectrum and code-division multiple-access (CDMA) do not appear until later in this book.

Introduction to Fading Channels

kind of channel is referred to as a *flat fading channel* or a *frequency non-selective fading channel*, since the spectrum of the channel is constant within the bandwidth of the signal. The opposite case, with intersymbol interference, will be discussed later.

Now let us discuss what happens when a single carrier is transmitted from the transmitter. The received signal at time t will then be a sum of carriers, given by

$$s_\mathrm{r}(t) = \sum_{j=1}^{v} a_j \cos(2\pi f_\mathrm{c} t + \varphi_j), \qquad (7.3\text{-}4)$$

where f_c is the frequency of the transmitted carrier, a_j is the amplitude of the jth scattered signal, φ_j the phase of the same signal, and v is the number of scatterers. The inphase and quadrature phase components of this received signal at time t are

$$I_\mathrm{r} = \sum_{j=1}^{v} a_j \cos(\varphi_j),$$
$$Q_\mathrm{r} = \sum_{j=1}^{v} a_j \sin(\varphi_j). \qquad (7.3\text{-}5)$$

With the assumptions given, the inphase and quadrature phase components are therefore a sum of many random contributions and from the central limit theorem we may conclude that I_r and Q_r will be normally distributed. It is straightforward to show that they have zero mean, have identical standard deviation, and are uncorrelated. The amplitude

$$r = \sqrt{I_\mathrm{r}^2 + Q_\mathrm{r}^2} \qquad (7.3\text{-}6)$$

of the received signal at time t may be shown to be Rayleigh distributed, with a pdf given by

$$p(r) = \begin{cases} \dfrac{r}{\sigma_r^2} \exp\left(-\dfrac{r^2}{2\sigma_r^2}\right), & r \geq 0; \\ 0, & r < 0, \end{cases} \qquad (7.3\text{-}7)$$

where $2\sigma_r^2 = \overline{P}_\mathrm{r}$ is the average power of the received signal. The corresponding cumulative distribution function (cdf) is given by

$$P(r) = \int_0^r p(z)\,\mathrm{d}z = \begin{cases} 1 - \exp\left(-\dfrac{r^2}{2\sigma_r^2}\right), & r \geq 0; \\ 0, & r < 0. \end{cases} \qquad (7.3\text{-}8)$$

The phase of the received signal is uniformly distributed between 0 and 2π. This condition is referred to as Rayleigh fading. The pdf and the cdf of the Rayleigh distribution are shown in Fig. 7.5.

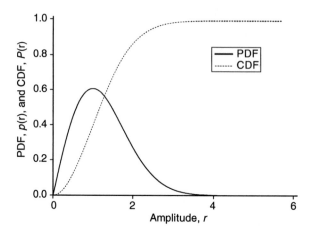

Figure 7.5 The pdf (solid line) and cdf (dashed line) of the Rayleigh distribution, when $\sigma_r = 1$.

Later on, we will be interested in the corresponding distribution of the received power. Since power is proportional to squared amplitude, it is easy to show that the received power is exponentially distributed with pdf

$$p(P_r) = \begin{cases} \dfrac{1}{\overline{P_r}} \exp\left(-\dfrac{P_r}{\overline{P_r}}\right), & P_r \geq 0; \\ 0, & P_r < 0. \end{cases} \quad (7.3\text{-}9)$$

The corresponding cdf is given by

$$P(P_r) = \begin{cases} 1 - \exp\left(-\dfrac{P_r}{\overline{P_r}}\right), & P_r \geq 0; \\ 0, & P_r < 0. \end{cases} \quad (7.3\text{-}10)$$

These are shown in Fig. 7.6. The main problem with this kind of fading is that the probability that the received power lies below a certain value (such as the vertical line in Fig. 7.6) is in many situations quite large. The cdf increases too slowly with power, and many receivers will experience a bad performance. This will be clear in Section 7.6 where we derive the bit error probability for the Rayleigh fading channel.

7.3.3. Other Fading Distributions

The lognormal and the Rayleigh fading models are the most commonly used. Other distributions may be more suitable in some situations. The assumption behind the Rayleigh fading model is that all power is received from scatterers

Introduction to Fading Channels

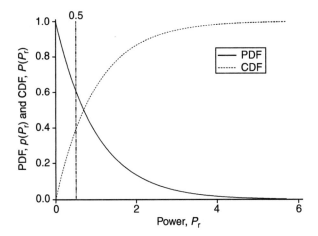

Figure 7.6 The pdf (solid line) and cdf (dashed line) of the exponential distribution, when $\overline{P}_r = 1$.

and that there is no line-of-sight component, but in many situations one exists in addition to the scattered components. Now the inphase signal component in Eq. (7.3-5) will have a constant term in addition to the sum of the scattering contributions. It will still be normally distributed under the assumption of many scatterers, but the mean value of the inphase component will be nonzero. In this case the amplitude may be shown to follow a Ricean probability density given by [2]

$$P(r) = \begin{cases} \dfrac{2r(K+1)}{\overline{P}_r} \exp\left[-K - \dfrac{r^2(K+1)}{\overline{P}_r}\right] I_0\left(2r\sqrt{\dfrac{K(K+1)}{\overline{P}_r}}\right), & r \geq 0; \\ 0, & r < 0, \end{cases}$$

(7.3-11)

where K is the so-called Rice factor that expresses the ratio between the power in the line-of-sight component and the scattered component. The function $I_0()$ is the modified zeroth-order Bessel function. The limiting cases of $K = 0$ and $K \to \infty$ correspond to the Rayleigh fading and the no fading cases, respectively. The pdf of the received power in Ricean fading is given by

$$P(P_r) = \begin{cases} \dfrac{K+1}{\overline{P}_r} \exp\left[-K - \dfrac{P_r(K+1)}{\overline{P}_r}\right] I_0\left(2\sqrt{\dfrac{P_r K(K+1)}{\overline{P}_r}}\right), & P_r \geq 0; \\ 0, & P_r < 0 \end{cases}$$

(7.3-12)

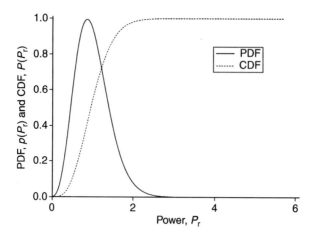

Figure 7.7 The pdf (solid line) and cdf (dashed line) of the power on a Ricean fading channel, when $\overline{P}_r = 1$ and $K = 10$ (10 dB).

and is shown together with its cdf in Fig. 7.7, when $K = 10$ (10 dB). Here it is clearly seen that the area under the left tail of the pdf is reduced as compared to the exponential distribution in Fig. 7.6 and it will reduce further with larger K. Ricean fading is not so detrimental to performance as Rayleigh fading.

In practice, the channel experiences both shadowing and multipath fading.[4] Then the mean value of the exponentially distributed received power (due to Rayleigh fading) can be regarded as a local mean value, which is lognormally distributed. The exponential distribution becomes a conditional distribution and the distribution of the received power is obtained by averaging the exponential distribution over the lognormally distributed local mean. The pdf of the received signal amplitude can be obtained by a similar reasoning and is often referred to as the Suzuki distribution. An integral expression can be found in [3].

Finally we will mention the Nakagami distribution. It was introduced by Nakagami in the early 1940s and is selected to fit empirical data for HF channels. One advantage of this distribution is that it provides a closer approximation to some empirical data than the Rayleigh, Ricean, lognormal, or Suzuki fading models. The Nakagami model can also model cases that are both less and more severe than the Rayleigh fading model. Another advantage is that the Nakagami distribution often leads to closed form expressions for outage probability and bit error probability, which is not the case for Suzuki and lognormal fading models [5].

---

[4] Even if the channel in practice experiences both shadowing and multipath fading, in many cases one of these effects is dominating. Then, the Rayleigh and lognormal models may be preferable to use, since they are simpler.

7.4. Frequency Selective Fading

The variations in received power, modeled by the fading distributions discussed above,[5] are typically the result of multipath fading due to scatterers in the vicinity of one of the transceivers.[6] This is often the situation in narrowband transmission systems where the inverse signaling bandwidth is much larger than the time spread of the multipath propagation delays. In wideband transmission systems, the inverse signaling bandwidth may be of the same order or even smaller than the time spread of the propagation delays. In this situation, the channel introduces amplitude and phase distortion to the message waveform, and the spectrum of the channel varies over the signaling bandwidth. This causes interference between different transmitted symbols and degrades the receiver performance unless countermeasures are taken. Equalization as discussed in Chapter 6, is one countermeasure. In this section we will briefly discuss frequency selective fading channels and some important characteristics of such channels.

7.4.1. Doppler Frequency

In many situations, the transmitter, the receiver, and/or one or several of the scatterers are moving. When the receiver is moving with speed v m/s in a given direction, the experienced frequency at the receiver of a transmitted carrier will be slightly different than the actual transmitted frequency. The change of frequency is

$$f_D = \frac{v}{\lambda} \cos\theta, \qquad (7.4\text{-}1)$$

where f_D is referred to as the *Doppler frequency*, v is the speed of the receiver and θ is the angle in 3-dimensional space between the received plane wave and the direction along which the receiver moves, as illustrated in Fig. 7.8. The actual frequency in the receiver is $f_c - f_D$, where f_c is the frequency of the transmitted carrier. The maximum frequency appears when the receiver is moving towards the transmitter in the horizontal plane of the antenna and the minimum frequency appears when the receiver is moving away. The largest frequency change is referred to as the maximum Doppler frequency $f_m = v/\lambda$. In practice, the Doppler frequency is changing with time, since the angle θ is typically time-varying. The same effect also appears when the transmitter, or one or several of the scatterers move.[7]

---

[5] Except the lognormal distribution which is used to model shadowing.

[6] Normally the mobile terminal in a mobile radio system, since the mobile antenna is closer to the ground than the base station antenna.

[7] The discussion and definition of Doppler frequency presented here is simplistic. The exact solution is given by relativity theory and is discussed in [3].

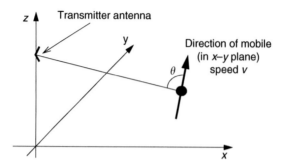

Figure 7.8 Illustration of Doppler frequency. The receiver is moving at speed v in the x–y plane, and the transmitter antenna is located along the z-axis.

7.4.2. Delay Spread

When we discussed Rayleigh fading in Section 7.3.2, we motivated it by the multipath fading due to scatterers in the vicinity of one of the terminals. Sometimes, multipath fading originates from other reflectors which are located further from the transmitter and/or receiver. These could be mountains or large buildings far away from the transmitter and receiver. The signal that is reflected by these objects will finally reach the receiver antenna after also being reflected by the local scatterers in the vicinity of the receiver antenna. Therefore, the received signal due to one specific reflector will be a sum of many scattered paths, just as in the flat fading case in Section 7.3.2. An example with one reflection is shown in Fig. 7.9. The reflected waves traverse a much longer distance than the non-reflected waves, and will typically arrive in the receiver with a delay which is in the order of the inverse signaling bandwidth or larger. This means that the received signal is the sum of many signals, but these are typically clustered such that signals within a cluster have almost the same delay (due to local scattering), while signals between different clusters have significantly different delays (due to far away reflections). In most applications, it is then necessary to model these effects in a different way, since the fading is no longer constant over the signaling bandwidth.

The most general way of describing the channel is by a time-varying impulse response $h(t, \tau)$, which is the response in the receiver at time t to a transmitted impulse at time $t - \tau$. The complex envelope of the received signal[8] is then given by the convolution

$$s_\mathrm{r}(t) = \int_{-\infty}^{\infty} u(t - \tau) h(t, \tau) \, d\tau, \qquad (7.4\text{-}2)$$

---

[8] Since the channel also introduces a phase shift, it is easier to consider the complex envelope of the signal here. The complex envelope of the signal $I(t)\cos(2\pi f_c t) - Q(t)\sin(2\pi f_c t)$ is defined as $I(t) + jQ(t)$.

Introduction to Fading Channels

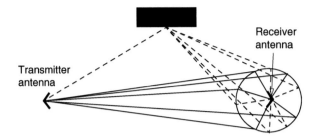

Figure 7.9 Example of a channel with delay spread. The transmitter antenna is at the left and the receiver antenna is at the right. A building is reflecting the signal at the top. The received signal consists of two clusters of signals, each being scattered locally in the vicinity of the receiver.

where $u(t)$ and $s_r(t)$ denote the complex envelopes of the transmitted and received signals, respectively. The impulse response is in general complex-valued. Since a receiver typically cannot resolve received paths which appear in the receiver with a relative delay of less than the inverse signaling bandwidth, it is common to model the time-varying channel by a discrete-time tapped delay line model in which the delay between the closest taps is equal to the inverse signaling bandwidth and the number of taps are selected such that the major part of the received energy is included in the model. The tap coefficients are time-varying and are modeled as random processes. The pdfs discussed in Section 7.3 can be used to model the distribution of each sample of the individual coefficients. With this model, the complex envelope of the received signal can be expressed as

$$s_r(t) = \sum_{j=1}^{n} \chi_j(t) u(t - (j-1)T_i - T_0), \qquad (7.4\text{-}3)$$

where T_0 is the delay of the first received path,[9] T_i is the inverse signaling bandwidth or the time-resolution of the receiver, $\chi_j(t) = a_j(t)\exp[j\varphi_j(t)]$ are the time-varying complex-valued channel coefficients, and n is the number of taps. This expression assumes that almost all of the received energy is contained within the interval $[T_0, (n-1)T_i + T_0]$.

Since the channel impulse response is random, a channel is more commonly specified by its power delay profile which is given by

$$\bar{h}(t) = \sum_{j=1}^{n} \mathcal{E}\left\{|\chi_j(t)|^2\right\} \delta(t - (j-1)T_i - T_0), \qquad (7.4\text{-}4)$$

where $\mathcal{E}\{\}$ is the expected value over the statistics of the coefficients χ_j and $\delta(t)$ is the impulse function. These expected values express the power in each resolvable

---

[9] The delay T_0 depends on the distance between the transmitter and receiver.

Table 7.1 Typical macrocellular urban and bad urban power delay profiles. The tap index refers to j in Eqs (7.4-3) and (7.4-4) when $T_0 = 0$ and $T_i = 0.1\,\mu\text{s}$. This data is adapted from [6] and other sources

	Typical urban			Bad urban	
Delay [μs]	Tap index $T_0 = 0$ $T_i = 0.1$ [μs]	Av. power [dBm]	Delay [μs]	Tap index $T_0 = 0$ $T_i = 0.1$ [μs]	Av. power [dBm]
0.0	1	−7.235	0.0	0	−7.852
0.2	3	−4.214	0.3	4	−5.331
0.5	6	−6.216	1.0	11	−8.327
1.6	17	−10.223	1.6	17	−10.269
2.3	24	−12.147	5.0	51	−7.328
5.0	51	−14.318	6.6	67	−9.318

path. In order to fully specify a frequency selective channel, the set of powers, the corresponding delays, the distribution of the individual path-samples, and the power spectrum of the path processes must be given. Two examples of power delay profiles are given in Table 7.1. The table shows the delay relative to the delay of the first received path and the relative average received power in each path (total power is normalized to one), for two International Telecommunications Union (ITU) channel models [6]. These would be implemented in Eq. (7.4-3) by using $T_0 = 0$ and $T_i = 0.1\,\mu\text{s}$, and average powers $\mathcal{E}\{|x_j(t)|^2\} = \mathcal{E}\{[a_j(t)]^2\}$ according to the table. The corresponding tap indices j are also given in the table. These channels are however not fully specified until the statistics of the tap coefficients (amplitude and phase) and the power spectrum of the tap coefficient processes are given. One possibility for the statistics is to assume that all coefficients are experiencing Rayleigh fading. We will return to the power spectrum shortly.

Other common measures of frequency selective fading channels are the mean time delay given by

$$\overline{T} = \frac{\sum_{j=1}^{n} \mathcal{E}\{a_j^2\}[(j-1)T_i + T_0]}{\sum_{j=1}^{n} \mathcal{E}\{a_j^2\}} \qquad (7.4\text{-}5)$$

and the delay spread

$$\sigma_T = \sqrt{\overline{T^2} - \overline{T}^2}, \qquad (7.4\text{-}6)$$

where

$$\overline{T^2} = \frac{\sum_{j=1}^{n} \mathcal{E}\{a_j^2\}[(j-1)T_i + T_0]^2}{\sum_{j=1}^{n} \mathcal{E}\{a_j^2\}} \qquad (7.4\text{-}7)$$

Introduction to Fading Channels

Table 7.2 Mean delay, delay spread, and coherence bandwidth of the typical macrocellular urban and bad urban channels [6]

	Typical urban	Bad urban
Mean delay [μs]	0.082	2.083
Delay spread [μs]	0.184	2.408
Coherence bandwidth [kHz]	865	66

is the second moment of the time delays. The mean time delay and the delay spread of the two channels in Table 7.1 are given in Table 7.2. The coherence bandwidth is also given, and it will be defined in the next subsection.

On practical channels, the delay spread can be anything from fractions of microseconds to many microseconds. The delay spread is usually longer in urban areas, while it is shorter in suburban areas, open areas, and indoors. Delay spread will gain importance in the next subsection, where we will introduce the coherence bandwidth and coherence time.

7.4.3. Coherence Bandwidth and Coherence Time

The coherence bandwidth and the coherence time[10] of a frequency selective channel are important to know when designing coded modulation systems, since the bit error probability will be heavily influenced by them. The coherence bandwidth is a measure of the bandwidth over which the fading can be considered highly correlated. One can equivalently say that the fading is almost the same on all the frequency components over the coherence bandwidth. The coherence time is in a similar manner a measure of the time over which the fading is highly correlated or almost constant. Unfortunately, it is quite involved to exactly calculate these measures for a particular channel, since it involves 4-dimensional pdfs. Therefore, it is common practice to calculate them from formulas that are only valid for one specific channel, which makes the mathematical derivations easier but still lengthy. We will omit here the mathematical details, and the reader is referred to [3] for details.

In order to calculate the coherence bandwidth and time, one has to consider the following scenario. Two carriers on two different frequencies f_c and $f_c + \Delta f$, respectively, are transmitted. Each of them experiences frequency selective fading according to the model given in Section 7.4.2, which means that the received signal consists of a number of multipath components where the jth

---

[10] These are also referred to as correlation bandwidth and correlation time, respectively, and we will use these names as synonyms.

component is delayed by $(j-1)T_i$ compared to the first component with delay T_0. Each component also has its own unique Doppler frequency. The received signal at frequency f_c is observed at time t and the received signal at frequency $f_c + \Delta f$ is observed at time $t + \tau$. Given this scenario, we want to calculate the normalized cross-covariance between the amplitudes of the two observed signals. The inphase and quadrature phase components of the signal at frequency f_c are given by

$$I_r(t) = \sum_{j=1}^{n}\sum_{k=1}^{v_j} a_{j,k} \cos\left(2\pi \frac{v}{\lambda} \cos(\theta_{j,k})t - 2\pi f_c([j-1]T_i + T_0)\right),$$

$$Q_r(t) = \sum_{j=1}^{n}\sum_{k=1}^{v_j} a_{j,k} \sin\left(2\pi \frac{v}{\lambda} \cos(\theta_{j,k})t - 2\pi f_c([j-1]T_i + T_0)\right),$$

(7.4-8)

where j is an index running over the paths in the power delay profile, and k is an index running over the local scatterers of each path in the power delay profile. The total number of scatterers in path j is denoted v_j. The amplitude of each received waveform is $a_{j,k}$ and its angle of arrival is $\theta_{j,k}$. The delays of the paths in the power delay profile are as before $[j-1]T_i + T_0$. In a similar way, the inphase and quadrature phase component of the signal at frequency $f_c + \Delta f$ are given by

$$\tilde{I}_r(t) = \sum_{j=1}^{n}\sum_{k=1}^{v_j} a_{j,k} \cos\left(2\pi \frac{v}{\lambda} \cos(\theta_{j,k})t - 2\pi (f_c + \Delta f)([j-1]T_i + T_0)\right),$$

$$\tilde{Q}_r(t) = \sum_{j=1}^{n}\sum_{k=1}^{v_j} a_{j,k} \sin\left(2\pi \frac{v}{\lambda} \cos(\theta_{j,k})t - 2\pi (f_c + \Delta f)([j-1]T_i + T_0)\right).$$

(7.4-9)

Now assume:

1. that there are no line-of-sight signals (the amplitude of each coefficient in the impulse response is Rayleigh distributed and its phase is uniformly distributed),
2. that the time arrival of the power delay profile is exponentially distributed with mean and variance $\sigma_T^2 (T_0 = 0)$,
3. that the angle of incident power $\theta_{j,k}$ is uniformly distributed, and
4. that the transmit and receive antennas are in the same horizontal plane.

Then the normalized cross-covariance ρ_r between the amplitudes[11]

$$\sqrt{[I_r(t)]^2 + [Q_r(t)]^2} \qquad (7.4\text{-}10)$$

---

[11] The normalized amplitude (or envelope) cross-covariance is also known as the amplitude (or envelope) correlation coefficient.

Introduction to Fading Channels

and

$$\sqrt{[\tilde{I}_r(t+\tau)]^2 + [\tilde{Q}_r(t+\tau)]^2} \quad (7.4\text{-}11)$$

can be closely approximated by

$$\rho_r \approx \frac{J_0^2(2\pi f_m \tau)}{1 + (2\pi \Delta f \sigma_T)^2}, \quad (7.4\text{-}12)$$

where $J_0(\ )$ is the Bessel function of the first kind and order zero [3]. f_m is the maximum Doppler frequency and σ_T is the delay spread of the channel as defined before. Here we see that the normalized amplitude cross-covariance only depends on the products $f_m \tau$ and $\Delta f \sigma_T$. This means that the cross-covariance obtained at τ for a given maximum Doppler frequency f_m is identical to the cross-covariance obtained at half the lag ($\tau/2$) for double the frequency ($2f_m$), when the frequency separation and delay spread are kept constant. The same relation is true for the frequency separation Δf and the delay spread σ_T. Therefore we can conclude that the bandwidth over which a channel is highly correlated depends inversely on the delay spread, such that a large delay spread corresponds to a narrow correlation bandwidth. Similar arguments can be applied to the correlation time, such that a large Doppler frequency corresponds to a small correlation time. The amplitude cross-covariance over time[12] ($\Delta f \sigma_T = 0$) and frequency ($f_m \tau = 0$), respectively, are shown in Fig. 7.10 and Fig. 7.11. Here we clearly see that the amplitude cross-covariance over frequency decreases monotonically

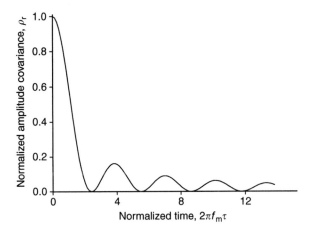

Figure 7.10 Normalized amplitude covariance ρ_r vs normalized time $2\pi f_m \tau$ of the received carrier.

---

[12] This is the covariance over time for the received amplitude when one single carrier is transmitted.

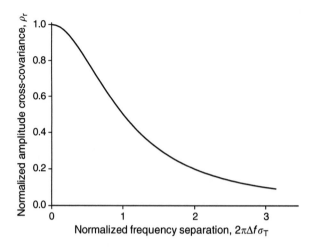

Figure 7.11 Normalized amplitude cross-covariance ρ_r vs normalized frequency separation $2\pi \Delta f \sigma_T$ between two different carriers.

with increasing frequency separation. The amplitude covariance over time is also decreasing with increasing lag τ but there is some oscillation.

One way of defining the correlation or coherence time is to set it equal to the time interval over which the normalized amplitude covariance ρ_r is at least 0.5, when the frequency separation Δf is zero (one received carrier only). From Eq. (7.4-12) it is straightforward to find that this is the solution to

$$J_0^2(2\pi f_m T_c) = 0.5, \qquad (7.4\text{-}13)$$

where T_c is the coherence time. This can be shown to be approximately

$$T_c \approx \frac{9}{16\pi f_m}. \qquad (7.4\text{-}14)$$

Other definitions of coherence time exist, and the main difference is the value of the covariance for which they are defined. In the same way the coherence bandwidth may be defined as the frequency interval over which the normalized amplitude cross-covariance ρ_r is at least 0.5 when the time difference τ is zero (two carriers at different frequencies but observed at the same time). This can again be obtained from Eq. (7.4-12) and is given by

$$B_c = \frac{1}{2\pi \sigma_T}, \qquad (7.4\text{-}15)$$

where B_c denotes the coherence bandwidth.

To exemplify these measures, we take a look at a typical mobile radio channel. The coherence bandwidths for the ITU channels discussed earlier are

Introduction to Fading Channels

given in Table 7.2. In a bad urban area, typical values of speed and delay spread may be 10 km/h and 2.4 µs. A carrier frequency of 900 MHz leads to a maximum Doppler frequency of 8.3 Hz. Then we find that the coherence time is around 173 ms, and the coherence bandwidth is approximately 66 kHz. With the 270 kbps channel data rate and 200 kHz bandwidth of GSM [7], we find that the coherence time corresponds to almost 47000 bit periods and the coherence bandwidth is approximately 1/3 of the signaling bandwidth. The conclusion is that the fading over time changes very slowly compared to the bit rate, which is a disadvantage since it is difficult to exploit time-diversity (see Section 7.7). But, the fading is changing over the bandwidth of the GSM signal, which means that we can exploit frequency diversity. The signal at the receiver will suffer from intersymbol interference since the power delay profile given in Eq. (7.4-4) will typically have several taps. The corresponding taps in the impulse response of the channel will however experience fading with low correlation, making it possible for an equalizer to utilize frequency diversity. We will discuss diversity in more detail in Section 7.7.

The normalized phase cross-covariance can be obtained in a way similar to the normalized amplitude cross-covariance. Correlation time and correlation bandwidth can then be derived from the normalized phase cross-covariance. This will not be done here, but it is worth mentioning that the phase cross-covariance normally decreases faster with increasing time lag τ and increasing frequency difference Δf, and therefore the corresponding correlation bandwidth and correlation time become smaller and do not normally limit the performance of the modulation system. Further details appear in [3].

7.4.4. Fading Spectrum

In order to fully characterize the multipath fading process, we also need to know its spectrum. Here we will only consider the Rayleigh fading process. From the discussion in Section 7.3.2, we know that the inphase and quadrature phase components in the receiver are uncorrelated and normally distributed, when a single carrier is transmitted. The mean values of these components are zero when the channel is Rayleigh fading. We have also discussed the correlation of the amplitude fading process in Section 7.4.3, and it is well known that the spectrum of a process is equal to the Fourier transform of the autocorrelation function. The most important spectrum is however the spectrum of the received signal and not the spectrum of the amplitude process. One reason for this is that the signal spectrum is needed when one wants to simulate such a fading process, as we will discuss in Section 7.5.

During the derivation of the normalized amplitude cross-covariance, as given in Eq. (7.4-12), the autocorrelation of the received signal when a single unmodulated carrier is transmitted is obtained. We omitted this derivation, but the details may be found in [3]. The normalized autocorrelation function of $I_r(t) + jQ_r(t)$ is given by

$$\rho = J_0(2\pi f_m \tau). \qquad (7.4\text{-}16)$$

Thus, the spectrum of the received bandpass signal is

$$S(f) = \begin{cases} K \dfrac{1}{\sqrt{1 - ((f - f_c)/f_m)^2}}, & |f - f_c| < f_m; \\ 0, & |f - f_c| \geq f_m. \end{cases} \quad (7.4\text{-}17)$$

where K is a constant which is related to the total power in the process and f_c is the carrier frequency. This spectrum is shown in Fig. 7.12.[13]

The fading spectrum can also be derived in another way. Since the Doppler frequency is directly related to the angle θ via the cosine function, it is clear that all received power within a small angle interval $d\theta$ must correspond to the power within a small frequency interval df of the power spectrum. Since the Doppler spectrum is an even function of θ, the spectrum must be an even function of f, and based on this the following equality can be specified:

$$S(f)|df| = C[G(\theta)p(\theta) + G(-\theta)p(-\theta)]|d\theta|, \quad (7.4\text{-}18)$$

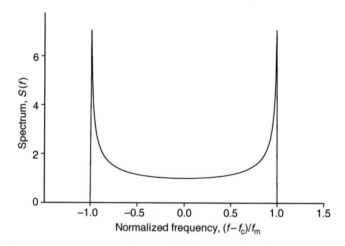

Figure 7.12 Spectrum of a receiver signal when a single unmodulated carrier is transmitted over a Rayleigh fading channel. The frequency axis is normalized and shifted such that the carrier appears at zero frequency, which means that the value on the horizontal axis is $(f - f_c)/f_m$.

---

[13] The reason that the spectrum goes to infinity at $|f - f_c| = f_m$, is the somewhat unrealistic assumption that all scatterers are in the horizontal plane of the receiver antenna (the angle θ in the definition of the Doppler frequency is assumed uniformly distributed in the horizontal plane). In practice there would also be a vertical component and then the spectrum would not go to infinity. More details on this can be found in [8,9].

Introduction to Fading Channels

where C is a constant, $p(\theta)$ is the pdf of the angle and $G(\theta)$ is the antenna pattern. An expression for df can be obtained from $f(\theta) = f_c + f_m \cos\theta$ as

$$df = -f_m \sin\theta \, d\theta = f_m \sqrt{1 - \left(\frac{f - f_c}{f_m}\right)^2} \, d\theta. \qquad (7.4\text{-}19)$$

Assuming a uniform phase θ results in

$$S(f) = \begin{cases} \dfrac{C}{f_m} \dfrac{G(\theta)+G(-\theta)}{\sqrt{1-((f-f_c)/f_m)^2}}, & |f - f_c| < f_m; \\ 0, & |f - f_c| \geq f_m. \end{cases} \qquad (7.4\text{-}20)$$

For an omnidirectional antenna this simplifies to the spectrum given in Eq. (7.4-17).

Here we have not at all considered the spectrum of the magnetic field components. Details about these can be found in [3]. We have also not considered the spectrum of the amplitude fading process, which is discussed in some detail in [2].

7.4.5. Types of Multipath Fading

We have seen that coherence time and coherence bandwidth are important measures of a fading channel. Figure 7.13 shows how these can be used to classify fading channels. In other words, fading can be classified in terms of the relation between coherence bandwidth and signaling bandwidth as *flat* when the signaling bandwidth is smaller than the coherence bandwidth, and as *frequency selective* when the signaling bandwidth is larger than the coherence bandwidth. Fading can also be classified in terms of the relation between coherence time and symbol period as *fast* when the coherence time is smaller than the symbol period, and as *slow* when the coherence time is larger than the symbol period.

It is not possible here to give a complete picture of how this classification affects modulation systems, but it can be summarized as follows:

1. *Slow and flat fading channels* (upper right corner in Fig. 7.13) lead to relatively simple channel estimation and detection, but the error performance is poor since there is no frequency diversity available and time diversity can only be obtained if very long delays are acceptable.
2. *Fast and flat fading channels* (upper left corner in Fig. 7.13) make channel estimation challenging, which degrades performance, and have no frequency diversity, which means simpler detectors but no diversity gains. But time diversity can be obtained without long delays.
3. *Slow and frequency selective fading channels* (lower right corner in Fig. 7.13) lead to more complex channel estimation since several taps must be estimated, but the slow variation makes channel estimation feasible.

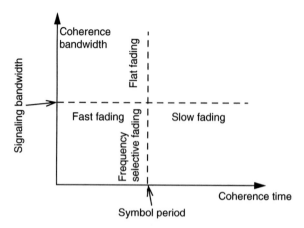

Figure 7.13 Classification of fading channels. As an example, a channel in the upper right corner is flat and slow fading.

Furthermore such channels have frequency diversity, which improves performance at the expense of complex detection, but achieving time diversity is difficult unless long delays are acceptable.

4. *Fast and frequency selective fading channels* (lower left corner in Fig. 7.13) are very difficult since both detection and channel estimation are complex and challenging, but they have a large diversity order, which has the prospect of significantly improving performance.

The signaling bandwidth and the symbol rate are in general proportional to the data rate, which means that high data rate systems tend to make channels more slow and more frequency selective.

7.5. Fading Simulators

When designing coded modulation systems, we always want to evaluate their error probability.[14] In many situations it is difficult to obtain analytical expressions of the error probability. Therefore, one often has to perform simulations in order to obtain estimates of the receiver performance. In this section we will discuss how to devise fading channel simulators for the channels discussed in this chapter. Throughout, we will assume that the simulation is to be performed at baseband and that the modulation scheme has both an inphase and a quadrature

---

[14] Here we do not specify what kind of error probability since it may sometimes be block error probability and sometimes bit error probability.

Introduction to Fading Channels

phase component. If the modulation has only an inphase component, the quadrature component is set to zero. Since the fading channel has a random phase, the received signal will nevertheless contain both an inphase and a quadrature component.

7.5.1. Flat Rayleigh Fading by the Filtering Method

A flat Rayleigh fading channel has only a single coefficient in the power delay profile and the corresponding fading coefficient is a sample from a 2-dimensional Gaussian random process. The inphase and quadrature phase components are uncorrelated and have zero-mean, and the autocorrelation function of the complex fading process is $J_0(2\pi f_m \tau)$. Such a process can be generated by filtering a white 2-dimensional Gaussian random process in a properly selected filter. The filter must have a transfer function (at baseband) equal to

$$|H(f)| = \begin{cases} K_f \dfrac{1}{\sqrt[4]{1-(f/f_m)^2}}, & |f| < f_m; \\ 0, & |f| \geq f_m, \end{cases} \quad (7.5\text{-}1)$$

where K_f is a constant. The filter should have linear phase. An arbitrarily scaled impulse response of such a filter is given by [10]

$$h(t) = \begin{cases} J_{1/4}(2\pi f_m |t|)/\sqrt[4]{|t|}, & t \neq 0; \\ \sqrt[4]{f_m \pi}/\Gamma(5/4), & t = 0, \end{cases} \quad (7.5\text{-}2)$$

where $J_{1/4}(\ )$ is the Bessel function of the first kind and order 1/4, and $\Gamma(\ )$ is the Gamma function. This impulse response is shown in Fig. 7.14. The filter has infinite support, and therefore the response has to be truncated symmetrically around zero and delayed such that it becomes causal. The truncation must be done carefully or the spectrum of the fading process will be significantly changed. It is also possible to use a window function for the truncation to keep the sidelobes of the filter spectrum low. It is difficult to give rules for the truncation, but generally the longer the support the better the approximation. A long filter leads to a more time consuming filtering process, when done in the time domain. According to our own experience, it is safest to truncate the filter such that it has support in $-3.938 \leq f_m t \leq 3.938$; the decimal number is approximately equal to a zero crossing in the impulse response and this is advantageous for reducing spectral sidelobes. An example of a Rayleigh fading process generated by such a filter is shown in Fig. 7.15.

There is also the problem of selecting a proper sampling frequency, since the simulation model will be implemented in discrete time. The sampling frequency is normally defined by the symbol rate of the modulation method, since decision variables have to be generated at this rate. When the receiver consists of a single

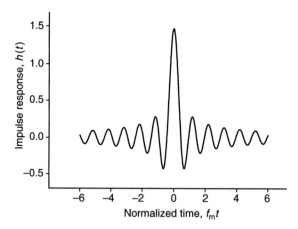

Figure 7.14 Impulse response of filter for generating flat Rayleigh fading. The horizontal axis is normalized to maximum Doppler frequency.

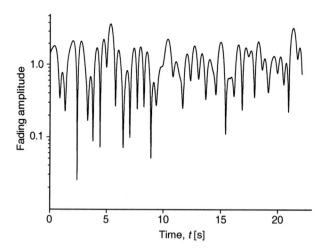

Figure 7.15 Amplitude (in logarithmic scale) of a Rayleigh fading process generated by the filtering method, when the maximum Doppler frequency is one. The filter is truncated to 7.876 s and the sampling frequency is 200 Hz (the number of taps in the FIR filter is 1575).

matched filter as in a simple linear modulation system, the statistics of the matched filter outputs are usually known. These outputs can then be directly generated in the simulator, which means that the sampling rate of the fading simulator can be equal to the symbol rate of the modulator. With more complex modulation, the simulator may have to run at a faster rate to accurately calculate the decision variables. A general rule of thumb in this case is to use a sampling rate that is 5–10

Introduction to Fading Channels

times the symbol rate. With low Doppler frequency, sampling at symbol rate or higher, may lead to a very long finite impulse response (FIR) filter in the Rayleigh simulator. With the truncation length given above and a sampling rate of 10 k symbols/s,[15] which is rather low, the number of FIR filter taps becomes 7876 at a maximum Doppler frequency of 10 Hz. At this sampling rate and with this filter length, the convolution becomes time consuming to do in the time domain. One way of circumventing this problem is to generate fading samples at a lower speed[16] followed by interpolation. Another possibility is to generate fading samples by the method described in Section 7.5.2.

The filtering may also be done in the frequency domain by the fast fourier transform (FFT). Generally the FFT method is to be preferred when the fading is relatively slow (small Doppler frequency) to avoid the very long impulse response. In fast fading, both methods can be used. The length of the FFT must be chosen such that a good resolution of the filter transfer function is obtained. With a sampling rate of 10 k symbols/s and a 10 Hz maximum Doppler frequency, the number of nonzero samples in the FFT of the filter will be approximately 16 with an FFT length of 8192.

It is appropriate to estimate the autocorrelation function or spectrum of the generated fading process, in order to check the accuracy of the simulator. It is most important to have a good match at high correlations, but there are situations when the whole correlation function is important.

A block diagram of a flat Rayleigh fading simulator is shown in Fig. 7.16 and a block diagram of the complete simulator is shown in Fig. 7.17. In the simplest case, the modulator in fact generates mean values of the output of the receiver matched filter[17] at the symbol rate, while in more involved cases it generates samples of the transmitted waveform at a higher rate. When the modulator runs at the symbol rate, the block at the receiving side is typically only a detector, while in other cases it is a complete receiver with filtering.

What remains now to construct a complete simulator is to select the power of the modulator, the power of the fading process, the path loss, and the noise power to obtain a given signal-to-noise ratio (SNR) in the receiver. We will assume here that we want to perform a simulation for a given value of the average received E_b/N_0, and we will denote this value by Γ_{sim}. White Gaussian noise is generated in the same way as the random Gaussian samples for the fading process. Now assume that the individual samples of the inphase and quadrature phase components of the Gaussian noise have standard deviation σ_{noise} (total noise

---

[15] The maximum allowed symbol rate is now 10 k symbols/s.

[16] The rate at which fading samples are generated must be at least twice the maximum Doppler frequency, which is normally far smaller than the symbol rate.

[17] The name of the blocks in Fig. 7.17 may be somewhat misleading in this case, since the receiver filter is actually included in the modulator block and the filter effect is precalculated.

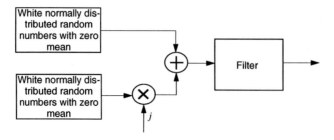

Figure 7.16 Block diagram for generating flat Rayleigh fading.

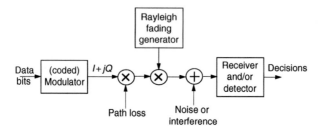

Figure 7.17 Block diagram for simulation with flat Rayleigh fading.

power is $2\sigma_{noise}^2$). Then it is straightforward to show that

$$\sigma_{noise} = \sqrt{\frac{\Xi}{2}} \frac{A}{\sqrt{R\Gamma_{sim}}}, \qquad (7.5\text{-}3)$$

where Ξ is the number of samples per symbol interval in the simulator, and R is the code rate in information bits per transmitted symbol. The variable A relates to the average received energy per symbol as

$$A = \sqrt{E_s l_{tot} 2\sigma_{fading}^2}, \qquad (7.5\text{-}4)$$

where E_s is the average energy per symbol in the transmitter, l_{tot} is the total loss between transmitter and receiver (including antenna gains, path loss, and other losses), and $2\sigma_{fading}^2$ is the power of the fading samples. The power of the fading is equal to

$$2\sigma_{fading}^2 = 2\sigma_{norm}^2 \sum_{i=0}^{L-1} |h_i|^2, \qquad (7.5\text{-}5)$$

where σ_{norm} is the standard deviation of the Gaussian samples used to generate the fading process, $\{h_i\}$ is the set of FIR filter taps, and L is the number of FIR filter

Introduction to Fading Channels

taps. In conclusion, this means that

$$A = \sqrt{E_s l_{tot} 2\sigma_{norm}^2 \sum_{i=0}^{L-1} |h_i|^2}, \qquad (7.5\text{-}6)$$

where we have complete freedom to choose the variables on the right-hand side. A simple way is to set A to one; this is obtained if the FIR filter is normalized to power one ($\sum_{i=0}^{L-1} |h_i|^2 = 1$), the Gaussian samples for the fading process is generated with standard deviation $1/\sqrt{2}$, the path loss is set to one, and the modulation constellation is normalized to unit average energy per symbol.

7.5.2. Other Methods for Generating a Rayleigh Fading Process

Jakes method

The methods described in Section 7.5.1 have the disadvantage that they may be very time consuming. Jakes proposed another method to obtain a correlated Rayleigh fading waveform [4]. Rayleigh fading appears when a large number of scattered waveforms with uniformly distributed angles of arrival are summed in the receiver antenna. The Jakes fading model is based on this assumption but uses a limited number of scattered waveforms to calculate the fading. If we assume N scattered waveforms with equal strength, arriving at angles $2\pi n/N, n = 1, \ldots, N$, the inphase and quadrature phase components of the received signal can be expressed as

$$\begin{aligned}
I_r(t) = K &\left\{ \frac{1}{\sqrt{2}} \cos(\alpha) \cos(2\pi f_m t + \theta_0) \right. \\
&\left. + \sum_{n=1}^{M} \cos(\beta_n) \cos[2\pi f_m \cos(2\pi n/N)t + \theta_n] \right\}, \\
Q_r(t) = K &\left\{ \frac{1}{\sqrt{2}} \sin(\alpha) \cos(2\pi f_m t + \theta_0) \right. \\
&\left. + \sum_{n=1}^{M} \sin(\beta_n) \cos[2\pi f_m \cos(2\pi n/N)t + \theta_n] \right\},
\end{aligned} \qquad (7.5\text{-}7)$$

where K is a constant, α and β_n are phases, θ_n are initial phases usually set to zero, and $M = (N/2 - 1)/2$. N is an even number. It is desirable that the phase of the fading is uniformly distributed, that the power of the inphase and quadrature phase components are equal, and that the inphase and quadrature phase components are uncorrelated. This can almost be accomplished when $\alpha = 0$ and $\beta_n = \pi n/M$.

The powers now become $(M+1)K^2/4$ and $MK^2/4$ for the inphase and quadrature phase components, and the correlation between them becomes zero.

Sometimes several uncorrelated fading waveforms are needed. This is easily obtained with the filtering method described in Section 7.5.1 by using uncorrelated Gaussian samples for the different waveforms. Jakes also proposed a method to obtain multiple waveforms by using the expressions in Eq. (7.5-7). This is done by selecting $\theta_n = \beta_n + 2\pi(k-1)/(M+1)$, where $k = 1, \ldots, M$ is the waveform index. A disadvantage is that only some pairs of waveforms are almost uncorrelated, while other pairs can be significantly correlated [11].

In [11] a remedy to this problem is proposed. The idea is to weight the different frequency components with orthogonal Walsh–Hadamard codewords before summing, but this only works if the frequency terms in Eq. (7.5-7) have equal power. This can be obtained by a rotation of the arrival angles by $\pi n/N$, such that there are four scattered waveforms that contribute to all frequency components; the angles become $\pi(2n-1)/N$. Equation (7.5-7) becomes

$$I_r(t) = \sqrt{\frac{2}{M}} \sum_{n=1}^{M} \cos(\beta_n) \cos[2\pi f_m \cos(2\pi n/N)t + \theta_n],$$

$$Q_r(t) = \sqrt{\frac{2}{M}} \sum_{n=1}^{M} \sin(\beta_n) \cos[2\pi f_m \cos(2\pi n/N)t + \theta_n],$$

(7.5-8)

where $M = N/4$ and N must be a multiple of four. The scaling is here chosen such that the power in the fading process becomes one, that is, $\sqrt{[I_r(t)]^2 + [Q_r(t)]^2} = 1$. By using $\beta_n = \pi n/M$, the inphase and quadrature phase components have equal power and are uncorrelated. A random selection of θ_n gives different waveforms. Since the cross-correlation between waveforms is determined by the sum of the products of the coefficients for the different frequency terms, zero cross-correlation can be obtained by multiplying the frequency components with coefficients from orthogonal sequences as shown in

$$I_{r,k}(t) = \sqrt{\frac{2}{M}} \sum_{n=1}^{M} A_k(n) \cos(\beta_n) \cos[2\pi f_m \cos(2\pi n/N)t + \theta_n],$$

$$Q_{r,k}(t) = \sqrt{\frac{2}{M}} \sum_{n=1}^{M} A_k(n) \sin(\beta_n) \cos[2\pi f_m \cos(2\pi n/N)t + \theta_n],$$

(7.5-9)

where M is a power of two, k is the waveform index, and $A_k(n)$ is the nth entry in Walsh–Hadamard sequence number k. This gives M uncorrelated waveforms and they can efficiently be obtained using a fast Walsh transform [11]. The authors of [11] also explain how a smaller set of waveforms can be obtained. By using $M = 8$, a realization similar to the one in Fig. 7.15 can be obtained.

Introduction to Fading Channels

The autocorrelation function of a fading process generated by Jakes model typically agrees well with the theoretical autocorrelation function for small delays, while it deviates for larger delays. By increasing the number of frequency components used, the deviation point can be made to appear at larger delays. Some examples of autocorrelation functions are shown in [2,11]. One advantage of the Jakes model is its low complexity compared to the filtering method, especially for low Doppler frequency. Another advantage is that it is deterministic and thus always gives the same fading sample for a given value of t, which is useful when a given simulation has to be repeated for the same channel realization.

Other filters

One of the problems with the filtering method is the long FIR needed when a good approximation of the ideal autocorrelation function is to be obtained. One alternative is to use infinite impulse response (IIR) filters. The simplest is a first-order low-pass IIR filter. The output of such a filter is a Markov process, and its autocorrelation, when the input is white Gaussian noise, is given by

$$\rho = K\zeta^{|n|}, \tag{7.5-10}$$

where K is a contant, n is the lag, and ζ the pole. Equation (7.5-10) approximates rather roughly the ideal autocorrelation given by the Bessel function $J_0(2\pi f_m n T_s)$, where T_s is the sampling interval. The ideal baseband spectrum is confined to the interval $-f_m \leq f \leq f_m$, while the spectrum of a first-order IIR filter is infinitely wide. In [2] it is proposed to set the corner frequency of the IIR filter equal to $f_m/4$. In this way it is possible to approximate the ideal autocorrelation for low delays, but the approximation becomes poor for large delays. This might be acceptable in simulations where only the high correlation values at small lags matter.

It is of course possible to use higher order IIR filters, but there is no simple way to relate the filter parameters to the maximum Doppler frequency.

7.5.3. Fading with Other Distributions

Ricean fading

Ricean fading is obtained when a Rayleigh faded component is added to a non-faded component. A block diagram for a simulator with Ricean fading is shown in Fig. 7.18. Here, the Rice factor should be set to the ratio between the power of the unfaded and Rayleigh faded signals.

Lognormal shadowing

Lognormally distributed shadowing can be simulated as in Fig. 7.19. Now a lognormally distributed variate is used to scale the amplitude of the received

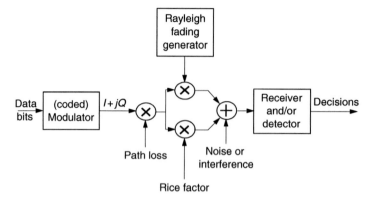

Figure 7.18 Block diagram for simulation with flat Ricean fading. The power of the Rayleigh fading is set here to unity.

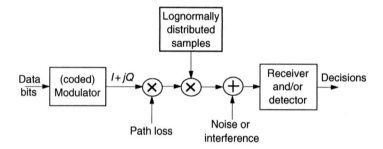

Figure 7.19 Block diagram for simulation with lognormally distributed shadowing.

signal. Since the logarithm of this sample is normally distributed, it can easily be generated. The challenge is to account for the spatial correlation of the shadows. One simple and realistic proposal for spatial correlation is described in [12] (see also [2]), where lognormal shadowing is modeled as Gaussian white noise that is filtered in a first-order IIR low-pass filter. The spatial autocorrelation between two samples of the signal power can now be expressed as

$$\mathcal{E}\{P_r^{[\text{dB}]}(k) P_r^{[\text{dB}]}(k+n)\} = \sigma_{\text{dB}}^2 \xi_D^{(vT_s/D)|n|}, \qquad (7.5\text{-}11)$$

where σ_{dB}^2 is the variance of the shadow fading, ξ_D is the correlation between two points located at distance D meters, v the speed of the receiver (we assume that it is the receiver which is moving) in m/s, $1/T_s$ is the sampling rate, and n is the lag. Here ξ_D can be chosen to adjust the correlation of the shadows. For typical suburban propagation at 900 MHz, $\sigma_{\text{dB}} = 7.5$, and $\xi_D = 0.82$ when $D = 100$ m according to experiments done in [12]. Results reported for a microcellular environment at 1700 MHz are $\sigma_{\text{dB}} = 4.3$, and $\xi_D = 0.3$ when $D = 10$ m.

Introduction to Fading Channels

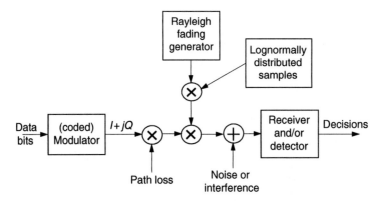

Figure 7.20 Block diagram for simulation with lognormally distributed shadowing and flat Rayleigh fading.

Lognormal shadowing and Rayleigh fading

Finally, in Fig. 7.20, we show a block diagram for a simulator when we include both Rayleigh fading and lognormal shadow fading. As mentioned before, this is also called Suzuki fading. Now the mean value of the Rayleigh fading power is lognormally distributed, and is obtained by multiplying the Rayleigh fading output by lognormally distributed shadowing. (Rayleigh fading is generated as in Section 7.5.1 and lognormal shadowing is generated as in Section 7.5.2.)

7.5.4. Frequency Selective Fading

With frequency selective fading, the received signal is a linear combination of delayed versions of the transmitted signal, each having its own time-varying scaling coefficient, as expressed in Eq. (7.4-3). The average powers of the different delayed paths are given by the power delay profile as defined in Eq. (7.4-4). In the Rayleigh fading case, each component is obtained as a separate uncorrelated Rayleigh fading waveform. Then all these components are summed together with proper delays and scaled according to the power delay profile. There may be situations when the fading of the different components needs to be correlated to obtain an accurate channel model, but there is no simple way to generate several Rayleigh fading waveforms with given correlations, so simulators mostly assume uncorrelated fading between the components.

Other fading distributions than Rayleigh fading can be handled in a similar way. In all cases the power of the interference or receiver noise has to be selected to give the desired average SNR. Simulation of communication systems in general are covered in detail in [13].

7.6. Behavior of Modulation Under Fading

In Chapter 2, we discussed the bit error probability of some basic modulation methods, when received in additive white Gaussian noise (AWGN). On a fading channel, the received signal power will vary locally around an average. The average bit error probability over the fading distribution is the appropriate performance measure with such channels. Here we will show how it can be obtained.

Assume that the bit error probability in AWGN is known. On a fading channel, it represents the conditional bit error probability, and we will refer to it as $P_b(\text{error}|\gamma_b)$, where γ_b is the instantaneous received E_b/N_0. Since γ_b is just a scaled version of the received power, it is straightforward to find the pdf of γ_b, once the pdf of the received power is known. The expected value of the SNR over the fading distribution $\mathcal{E}\{\gamma_b\}$ will be denoted Γ_b, and will be referred to as the average E_b/N_0. The average bit error probability is

$$\overline{P}_b = \mathcal{E}\{P_b(\text{error} \mid \gamma_b)\} = \int_0^\infty p(\gamma_b) P_b(\text{error} \mid \gamma_b) \, d\gamma_b. \tag{7.6-1}$$

$\overline{P}_b$ is implicitly a function of Γ_b.

For BPSK and QPSK with coherent detection, the conditional bit error probability is $P_b(\text{error} \mid \gamma_b) = Q(\sqrt{2\gamma_b})$, and the average bit error probability on a Rayleigh fading channel becomes

$$\overline{P}_b = \frac{1}{2}\left(1 - \sqrt{\frac{\Gamma_b}{1+\Gamma_b}}\right) \tag{7.6-2}$$

with the limiting value

$$\overline{P}_b \to \frac{1}{4\Gamma_b} \quad \text{as } \Gamma_b \to \infty. \tag{7.6-3}$$

We see that Γ_b must be increased by about a factor of 10 (10 dB) to reduce the average bit error probability by 10. From Chapter 2, the non-fading bit error probability decreases exponentially with E_b/N_0 and a factor of 10 decrease costs little in E_b/N_0. This is clear from Fig. 7.21, where the bit error probability of coherent BPSK is shown for non-fading and Rayleigh fading channels. At probability 10^{-3}, the degradation from Rayleigh fading is more than 17 dB. It is more than 43 dB at 10^{-6}.

The formula in Eq. (7.6-1) can be used to obtain an expression for the average bit error probability for any modulation/detection scheme and any fading distribution, as long as the probability in AWGN and the pdf of the received power are known. Examples for some simple modulation/demodulation schemes are given in Table 7.3.

Introduction to Fading Channels

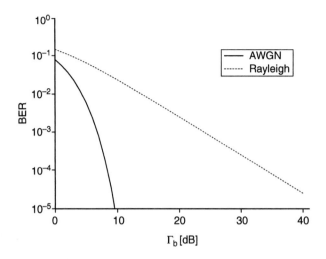

Figure 7.21 Average bit error probability vs average E_b/N_0 for BPSK.

Table 7.3 Average bit error probability expressions for some simple modulation/demodulation schemes on Rayleigh fading channels. The differential schemes are differentially encoded and differentially detected

Scheme	Average bit error probability $\overline{P}_b$
Coherent binary FSK	$\frac{1}{2}\left(1 - \sqrt{\frac{\Gamma_b}{2+\Gamma_b}}\right)$
Non-coherent binary FSK	$\frac{1}{2+\Gamma_b}$
Differential binary PSK (DBPSK)	$\frac{1}{2(1+\Gamma_b)}$
Differential QPSK (DQPSK)	$\frac{1}{2}\left(1 - \frac{\sqrt{2}\Gamma_b}{\sqrt{2(1+\Gamma_b)^2 - 1}}\right)$

Other probabilities such as symbol error probability can be obtained by simply replacing the conditional bit error probability with the desired conditional probability. Many such expressions are known and we refer to [5] for an extensive collection. The common element in all such expressions is that the error probability decreases much more slowly compared to the non-fading case. How slowly depends on the type of fading and its parameters. Very much higher transmitted power must be used on fading channels. In many situations this is not acceptable and we need countermeasures against fading. We will return to such methods in the next section.

In Chapter 3, it has been shown that channel coding can be used to reduce the bit error probability of a given modulation/detection scheme over AWGN channels. Is this also true over a fading channel and what are the coding gains? To see what happens, we consider a binary block code for simplicity. The word error probability for a block code with hard decisions is given in Eq. (3.2-14), and this expression is valid when errors of the symbols in a codeword are uncorrelated. When there is no fading, p in Eq. (3.2-14) should be replaced by the bit error probability of the modulation/detection scheme. On a fading channel the upper bound on the word error probability becomes

$$\overline{P}_\text{w} \leq \mathcal{E}\left\{ \sum_{n=t+1}^{N} \binom{N}{n} \prod_{i=1}^{n} P_\text{b}(\text{error}|\gamma_{si}) \prod_{i=n+1}^{N} (1 - P_\text{b}(\text{error}|\gamma_{si})) \right\}, \quad (7.6\text{-}4)$$

where γ_{si} for $i = 1, \ldots, n$ are the instantaneous received E_s/N_0 for the coded bits in error and γ_{si} with $i = n+1, \ldots, N$ for the correct coded bits. The expectation is here over the N-dimensional distribution of $\gamma_{s1}, \gamma_{s2}, \ldots, \gamma_{sN}$. It is assumed that errors due to noise are independent. In one extreme when the SNRs are independent between the different bits in a codeword, Eq. (7.6-4) evaluates to

$$\overline{P}_\text{w} \leq \sum_{n=t+1}^{N} \binom{N}{n} (\overline{P}_\text{b})^n (1 - \overline{P}_\text{b})^{N-n}, \quad (7.6\text{-}5)$$

where $\overline{P}_\text{b}$ is given in Eq. (7.6-1). In the other extreme when the fading is very slow, $\gamma_{si} = \gamma_\text{s}$ for all i and Eq. (7.6-4) becomes

$$\overline{P}_\text{w} \leq \mathcal{E}\left\{ \sum_{n=t+1}^{N} \binom{N}{n} (P_\text{b}(\text{error}|\gamma_\text{s}))^n (1 - P_\text{b}(\text{error}|\gamma_\text{s}))^{N-n} \right\}. \quad (7.6\text{-}6)$$

The expected value in this equation mostly have to be evaluated numerically.

As a demonstration, consider a Golay (23, 12) code that corrects up to 3 errors in a codeword, and BPSK modulation with coherent detection. Since this code is perfect, the upper bounds given above become equalities. As a reference we will take uncoded BPSK modulation with coherent detection. In both cases we evaluate the word error probability over 12 bit information words. Figure 7.22 shows the word error probability of the coded system on a non-fading channel (left curve), and the coded and uncoded systems on a slow Rayleigh fading channel where the whole codeword experience the same SNR (the two curves to the right). Coding does not help very much against slow fading. The coded word error

Introduction to Fading Channels

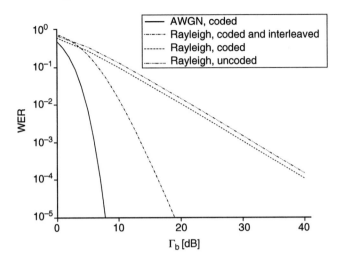

Figure 7.22 Average word error probability vs average E_b/N_0 for a Golay (23,12) code with BPSK.

probability on the Rayleigh fading channel is proportional to $1/\Gamma_b$ just as the bit error probability is. The coding gain is approximately the same as on the non-fading channel (1.4 dB at a word error probability of 0.001), but this is dwarfed by the coded system loss due to fading, which is more than 24 dB at a word error probability of 0.001.

There is, however, a fourth curve in Fig. 7.22 which is significantly improved compared to the two rightmost curves: it is the coded performance of the same Golay code in Rayleigh fading when the bits in the codeword experience uncorrelated fading as obtained from Eq. (7.6-5). The gain of this system at a word error probability of 0.001 is almost 17 dB compared to the uncoded system on a Rayleigh channel, and its loss compared to coding over a non-fading channel is reduced to 7.5 dB. This happens when perfect interleaving (see Section 7.7) is used.

In uncorrelated fading the coding gain becomes significantly larger than in slow (highly correlated) fading. The difference grows at smaller word error probabilities. From the asymptotic form of the word error probability given in Eq. (3.2-15), we can easily obtain an asymptotical expression for a Rayleigh fading channel. Since the bit error probability on a fading channel is approximately equal to K/Γ_s, where Γ_s denotes the average E_s/N_0, and K is a constant, the word error probability satisfies

$$\overline{P}_w \to \left(\frac{K}{\Gamma_s}\right)^{t+1} = \left(\frac{K}{R\Gamma_b}\right)^{t+1} = \frac{K_1}{(\Gamma_b)^{t+1}}, \quad \text{as } \Gamma_b \to \infty; \qquad (7.6\text{-}7)$$

K_1 is another constant, and R is the code rate. We clearly see that the channel code is able to change the slope of the word error probability curve; it is also clear in Fig. 7.22, where $t + 1 = 4$. Now the probability decreases by $10(t + 1)$ when the average received power increases 10 dB. This effect is an example of *diversity*, and will be discussed in more detail in the next section.

To end this section, we look briefly at soft decision decoding. One can show that the average word error probability satisfies

$$\overline{P}_w \to \frac{K}{(\Gamma_b)^d}, \quad \text{as } \Gamma_b \to \infty \tag{7.6-8}$$

on uncorrelated Rayleigh fading channels, where d is the minimum Hamming distance of the block code [14]. Here the average word error probability decreases even faster with increasing Γ_b, since the exponent d is about twice t. In summary, we can conclude that the fading rate significantly influences the word error probability on fading channels. A similar conclusion applies to the bit error probability of coded systems, which we will return to in Section 8.5.

7.7. Interleaving and Diversity

From Section 7.6 it is clear that the error probability deteriorates dramatically with fading. There is great need for methods that improve error performance. Channel coding is one of them and we have already seen in Section 7.6 that it can improve performance significantly in fast fading channels. In this section we will discuss some other methods.

7.7.1. Diversity Combining

The major cause of the poor performance on fading channels is the occasional deep fades. Even at high average SNRs, there is a significant probability that the instantaneous error probability becomes large, due to the left tail of the pdf of the instantaneous received power. We would like to change the pdf so that the tail is smaller.

In this section, we assume uncoded modulation with a matched filter receiver and symbol by symbol detection. This will significantly simplify the treatment of diversity and the methods and results can quite easily be extended to other cases.

One way to improve the error probability is to avoid making decisions for symbols which are transmitted when the fading amplitude is small. This can be done by making more than one replica of each transmitted symbol available at the receiver, and making sure that the fading on these replicas has a low correlation. Assuming that a symbol is available at the receiver from two uncorrelated but otherwise identical fading channels, we can easily conclude that the probability that both have a signal power below a certain level P_r is reduced to $[P(P_r)]^2$,

Introduction to Fading Channels

where $P(P_r)$ is defined in Eq. (7.3-10) for Rayleigh fading. When $P(P_r)$ is small, the reduction is significant. There is a similar improvement even when the fading samples of the two replicas are mildly correlated.

Selection diversity

In the simplest diversity scheme, the best out of the available replicas is selected and used for the decision. This is *selection diversity*. Formally this means that the selected decision variable λ is given by

$$\lambda = \lambda_j, \quad \text{where } \gamma_j = \max(\gamma_1, \gamma_2, \ldots, \gamma_M). \tag{7.7-1}$$

Here λ_i is the decision variable of the ith replica and M is the number of replicas available. The instantaneous E_b/N_0 of replica i is γ_i. The probability that the selected replica has an instantaneous SNR smaller than a given value γ is given by

$$P_{\text{SEL}}(\gamma) = \Pr[\gamma_1, \gamma_2, \ldots, \gamma_M \leq \gamma]. \tag{7.7-2}$$

In general this probability function has to be evaluated by an integral of the M-dimensional pdf. With Rayleigh fading, independent and identically distributed replicas, this expression evaluates to

$$P_{\text{SEL}}(\gamma) = \begin{cases} \left[1 - \exp\left(-\dfrac{\gamma}{\Gamma_c}\right)\right]^M, & \gamma \geq 0; \\ 0, & \gamma < 0, \end{cases} \tag{7.7-3}$$

where $\Gamma_c = \mathcal{E}\{\gamma_1\} = \mathcal{E}\{\gamma_2\} = \cdots = \mathcal{E}\{\gamma_M\}$. The corresponding pdf is

$$p_{\text{SEL}}(\gamma) = \begin{cases} \dfrac{M}{\Gamma_c}\left[1 - \exp\left(-\dfrac{\gamma}{\Gamma_c}\right)\right]^{M-1} \exp\left(-\dfrac{\gamma}{\Gamma_c}\right), & \gamma \geq 0; \\ 0, & \gamma < 0, \end{cases} \tag{7.7-4}$$

It is also useful to know the average E_b/N_0 *after selection* (at the detector), which is the average value of the pdf in Eq. (7.7-4). This is

$$\Gamma_b = \Gamma_c \sum_{i=1}^{M} \frac{1}{i}. \tag{7.7-5}$$

From these equations, we can see that the performance gain has two contributions. The first and in most cases the biggest contribution is the shape of the pdf in Eq. (7.7-4). This gain is illustrated in Fig. 7.23, where the pdf in Eq. (7.7-4) and cdf in Eq. (7.7-3) are shown for selection diversity with two replicas. As a reference, the corresponding pdf in Eq. (7.3-9) and cdf in Eq. (7.3-10) for the no diversity case are also shown. All the distributions are scaled to obtain $\Gamma_b = 1(0 \text{ dB})$. This means that $\Gamma_c = 2/3$ with selection diversity. The pdf with selection diversity

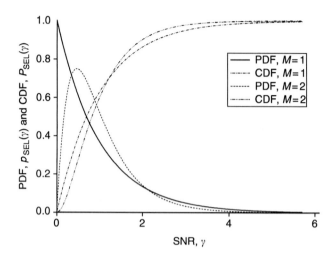

Figure 7.23 The pdf and cdf without selection diversity and with selection diversity between 2 replicas, when $\Gamma_b = 1$ (0 dB).

starts at zero when γ is zero, unlike the pdf without fading. The cdf increases more slowly at small SNRs. A second performance gain is due to the increase Γ_b/Γ_c in average E_b/N_0 that is obtained by the selection of one replica out of several replicas, as shown in Eq. (7.7-5). This is a pure SNR gain and would scale the horizontal axis of the pdf and cdf for selection diversity in Fig. 7.23 if the curves were instead plotted for average E_b/N_0 *per replica before combining* $\Gamma_c = 1$. With two replicas, this gain is 1.76 dB.

At the end, for digital modulations we are most interested in the average error probability after selection diversity. From Eq. (7.6-1), the bit error probability can now be evaluated, by using the pdf $p_{\text{SEL}}(\gamma)$ instead of $p(\gamma_b)$. After some tedious integration steps, the bit error probability of coherent BPSK[18] can be expressed as

$$\overline{P}_b = \frac{1}{2}\sum_{i=0}^{M}(-1)^i \binom{M}{i} \Big/ \sqrt{1+i/\Gamma_c}. \qquad (7.7\text{-}6)$$

The average bit error probability is here expressed for a given average E_b/N_0 per replica. It is straightforward to express it for a given average energy after selection, by replacing Γ_c in Eq. (7.7-6) with $\Gamma_b / \sum_{i=1}^{M} \frac{1}{i}$. The asymptotic (large SNR)

---

[18] Coherent BPSK may not be the most appropriate modulation/detection method used with selection diversity, since it requires a carrier phase estimate, which is not necessary with selection diversity. Selection diversity is mostly used with noncoherent detection. The example still serves the purpose of demonstrating the SNR gains by selection diversity.

Introduction to Fading Channels

expression for the average bit error probability can be shown to be

$$\overline{P}_b \propto \left(\frac{1}{\Gamma_c}\right)^M \propto \left(\frac{1}{\Gamma_b}\right)^M, \qquad (7.7\text{-}7)$$

where $\propto$ means proportional to. The exponent M is called the diversity order. Selection diversity thus makes the bit error probability decrease much faster with SNR. When the number of replicas M approaches infinity, the bit error probability approaches that of a non-faded channel, since the SNR variation due to fading is completely removed.

In Fig. 7.24, we illustrate the improvement in average bit error probability for coherent BPSK. The SNR on the horizontal axis is Γ_b and all detectors are working at the same average SNR per bit. The gain with 2 replicas compared to only one replica is 9.63 dB at an average bit error probability of 0.001, and the additional gain with 3 replicas is 2.63 dB more. These are the gains in Γ_b. Sometimes the gain in Γ_c is more interesting, and then we have to add 1.76 dB for the 2 replica case and 2.63 dB for the three replica case, which means that 2 replicas gain 11.39 dB compared to one replica, and the additional gain with 3 replicas is 3.5 dB more.

Weight combining

With more than one replica of each transmitted symbol available in the receiver, one may think that it is a waste of information just to select one of them

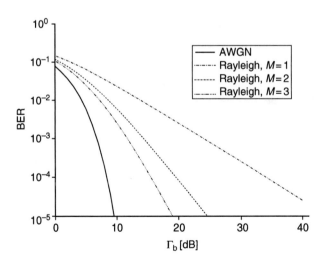

Figure 7.24 Average bit error probability vs average received E_b/N_0 for BPSK with selection diversity.

for the decision. Weight combination is a method where the replicas are summed together before the decision is made. The decision variable is given by

$$\lambda = \sum_{i=1}^{M} a_i e^{J\varphi_i} \lambda_i, \qquad (7.7\text{-}8)$$

where a_i is a scale factor and φ_i is a phase rotation for replica i. The optimum weight selection is the one that optimizes the instantaneous SNR at the detector.[19] The instantaneous SNR of the decision variable is equal to the power of the noise-free signal divided by the noise power,

$$\gamma' = \frac{\overline{\lambda}}{2\sigma_\lambda^2} = \frac{\left|\sum_{i=1}^{M} a_i e^{J\varphi_i} \overline{\lambda}_i\right|^2}{\sum_{i=1}^{M} |a_i e^{J\varphi_i}|^2 2\sigma_{\lambda_i}^2}, \qquad (7.7\text{-}9)$$

where $\overline{\lambda} = \mathcal{E}\{\lambda\}$ is the desired (noise-free) part of the decision variable and $2\sigma_\lambda^2 = \mathcal{E}\{|\lambda - \overline{\lambda}|^2\}$ is the noise power. The expected value $\mathcal{E}\{\ \}$ is over the noise distribution. Note here that $\overline{\lambda}_i = \mathcal{E}\{\lambda_i\}$ is the output of the matched filter in the receiver of replica i when there is no noise, and it includes a phase shift due to the fading channel; it is equal to the transmitted symbol, scaled by the fading amplitude and phase shifted by the fading phase. The phase φ_i can now be selected such that the numerator is maximized, since the denominator does not depend on φ_i. Since

$$\left|\sum_{i=1}^{M} z_i\right|^2 \leq \sum_{i=1}^{M} |z_i|^2, \qquad (7.7\text{-}10)$$

where z_i is complex-valued, with equality when

$$\angle z_i = \angle z_j, \quad \forall i, j, \qquad (7.7\text{-}11)$$

we should select φ_i such that all the terms in the numerator line up in the same direction in the complex plane. There are several ways of doing this, and since any direction in the complex plane is valid, one can select

$$\varphi_i = -\angle \overline{\lambda}_i. \qquad (7.7\text{-}12)$$

In this way we remove the phase due to the fading and modulation on each replica and the remaining desired signal is located along the real axis in the complex plane. With this choice of φ_i, the SNR can be rewritten as

$$\gamma' = \frac{\left(\sum_{i=1}^{M} a_i |\overline{\lambda}_i|\right)^2}{\sum_{i=1}^{M} a_i^2 2\sigma_{\lambda_i}^2}. \qquad (7.7\text{-}13)$$

---

[19] Here we use γ' and γ_i', respectively, to denote signal power divided by noise power after combining and of each individual replica. γ and γ_i will as before be used to denote the corresponding instantaneous E_b/N_0.

Introduction to Fading Channels

We can now use the Schwartz inequality to write

$$\gamma' = \frac{\left(\sum_{i=1}^{M} a_i \sqrt{N_i}(|\bar{\lambda}_i|/\sqrt{N_i})\right)^2}{\sum_{i=1}^{M} a_i^2 N_i} \leq \frac{\sum_{i=1}^{M} a_i^2 N_i \sum_{i=1}^{M}(|\bar{\lambda}_i|^2/N_i)}{\sum_{i=1}^{M} a_i^2 N_i} = \sum_{i=1}^{M} \frac{|\bar{\lambda}_i|^2}{N_i} = \sum_{i=1}^{M} \gamma_i', \quad (7.7\text{-}14)$$

where we have used N_i as a short-hand notation for $2\sigma_{\lambda_i}^2$. Equality is obtained when

$$a_i \propto \frac{|\bar{\lambda}_i|}{2\sigma_{\lambda_i}^2} \quad (7.7\text{-}15)$$

and the maximum instantaneous SNR is equal to the sum of the instantaneous SNRs of the individual replicas. When all the replicas have the same average noise power, that is, $2\sigma_{\lambda_i}^2 = 2\sigma_{\lambda}^2$ for all i, the absolute values of the weights a_i are simply proportional to the fading amplitudes of the individual replicas, since the fading amplitude is all that differs in $\bar{\lambda}_i$ for the different i-values. This diversity method is referred to as *maximal ratio combining* (MRC). It is more complex to implement than selection diversity, since complete channel estimates are needed for all replicas.

The pdf of the received power with MRC is given by the Chi-square distribution with $2M$ degrees of freedom,

$$p_{\text{MRC}}(\gamma) = \begin{cases} \dfrac{\gamma^{M-1} \exp(-\gamma/\Gamma_c)}{\Gamma_c^M (M-1)!}, & \gamma \geq 0; \\ 0, & \gamma < 0, \end{cases} \quad (7.7\text{-}16)$$

when the replicas experience uncorrelated Rayleigh fading and have identical average SNRs Γ_c [3]. This can be derived by using the fact that the desired signal power is equal to the sum of powers of the inphase signal and the quadrature signal. The corresponding cdf is given by

$$P_{\text{MRC}}(\gamma) = \begin{cases} 1 - \exp\left(-\dfrac{\gamma}{\Gamma_c}\right) \sum_{i=1}^{M} \dfrac{(\gamma/\Gamma_c)^{i-1}}{(i-1)!}, & \gamma \geq 0; \\ 0, & \gamma < 0. \end{cases} \quad (7.7\text{-}17)$$

The average E_b/N_0 after combining is given by

$$\Gamma_b = M\Gamma_c. \quad (7.7\text{-}18)$$

The error probability can now again be obtained from Eq. (7.6-1). For BPSK with coherent detection, the bit error probability evaluates to [14]

$$\bar{P}_b = q^M \sum_{i=0}^{M-1} \binom{M-1+i}{i} (1-q)^i, \quad (7.7\text{-}19)$$

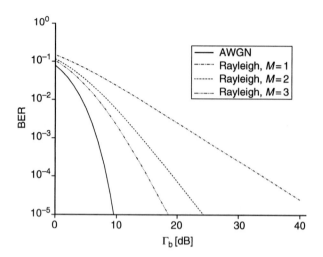

Figure 7.25 Average bit error probability vs average received E_b/N_0 for BPSK with MRC.

where

$$q = \frac{1}{2}\left(1 - \sqrt{\frac{\Gamma_c}{1+\Gamma_c}}\right). \tag{7.7-20}$$

This can be shown to be

$$\overline{P}_b \propto \left(\frac{1}{\Gamma_b}\right)^M \tag{7.7-21}$$

for large SNRs. Therefore, the bit error probability of MRC has the same asymptotic behavior as the bit error probability of selection diversity. In Fig. 7.25, we show the bit error probability with two and three replicas. As a comparison, the bit error probability without fading and with Rayleigh fading but no diversity are also shown. The SNR on the horizontal axis is Γ_b, the average E_b/N_0 *after combining*. The SNRs needed to obtain two different bit error probabilities are given in Table 7.4 for both MRC and selection diversity.

We conclude that MRC improves Γ_b over selection diversity by 0.24 dB with 2 replicas and 0.39 dB with 3 replicas. If we instead make the same comparison for Γ_c, the gains are 1.48 dB with 2 replicas and 2.53 dB with 3 replicas. Here we clearly see that the biggest performance improvement of MRC compared to selection diversity is due to the larger ratio Γ_b/Γ_c while the improvement due to the shape of the pdf is minor.

A simpler combining scheme than MRC is *equal gain combining* (EGC). Here the scaling weights a_i in Eq. (7.7-8) are all equal, while the phase shifts φ_i remain as with MRC in Eq. (7.7-12). This scheme is somewhat simpler to implement, since estimates of the fading amplitudes are not needed. EGC is more difficult to analyze and the pdf and cdf with uncorrelated Rayleigh fading are only

Table 7.4 The required Γ_b in dB, to obtain the given bit error probabilities for coherent BPSK. In the last row, the ratio Γ_b/Γ_c in dB is given

		No diversity	Rayleigh fading					
				MRC		Selection diversity		
BER	No fading		$M=2$	$M=3$	$M=4$	$M=2$	$M=3$	$M=4$
10^{-3}	6.79	23.97	14.10	11.32	10.06	14.34	11.71	10.53
10^{-6}	10.50	53.93	29.37	22.02	18.65	29.63	22.46	19.25
Γ_b/Γ_c	0	0	3	4.77	6.02	1.76	2.63	3.19

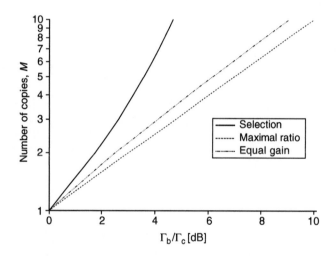

Figure 7.26 Number of replicas M vs average SNR gain Γ_b/Γ_c.

known for 2 replicas [3]. The average E_b/N_0 is easier to find [3] and is

$$\Gamma_b = \Gamma_c \left[1 - (M-1)\frac{\pi}{4}\right]. \qquad (7.7\text{-}22)$$

The bit error probability of coherent BPSK with EGC is almost the same as with MRC, for a given Γ_b. Figure 7.26 shows the diversity order vs average SNR gain Γ_b/Γ_c for the discussed diversity methods. Here we clearly see that EGC performs much closer to MRC than to selection diversity.

7.7.2. Ways to Obtain Diversity

Large performance gains are obtainable when more than one replica of each transmitted symbol is available at the receiver. The replicas should preferably experience independent fading. The question now arises as to how to obtain these replicas.

Antenna diversity

One way of obtaining diversity is to use more than one antenna at the transmitter and/or receiver. This is called antenna diversity. In this case the fading correlation depends on the distance between the antennas and whether the antennas are separated in the horizontal or vertical plane. The situation is different at the base station antenna and at the mobile antenna, since all the local scatterers are in the vicinity of the mobile antenna. The normalized covariance over time for the amplitude of the fading has previously been approximated as $\rho_r \approx J_0^2(2\pi f_m \tau)$, see Eq. (7.4-12) with $\Delta f = 0$. This expression is valid at the mobile receiver, when the mobile and the antennas are all in the horizontal plane. Since

$$f_m \tau = \frac{v\tau}{\lambda} = \frac{d}{\lambda}, \qquad (7.7\text{-}23)$$

where $d = v\tau$ is the separation distance at speed v between the two signals, the normalized covariance can also be expressed in terms of distance as

$$\rho_r \approx J_0^2\left(2\pi \frac{d}{\lambda}\right). \qquad (7.7\text{-}24)$$

Uncorrelated signals are obtained when this covariance is zero and since the first null of the Bessel function appears at argument 2.404, the value $d \approx 0.38\lambda$ gives uncorrelated signals. In practice, the performance loss due to correlated diversity replicas is very small as long as the correlation is below 0.5, which means that a horizontal spacing between the antennas of

$$d \geq \frac{9}{16\pi}\lambda \approx 0.18\lambda \qquad (7.7\text{-}25)$$

is required in the mobile terminal. With a carrier frequency of 900 MHz ($\lambda = 1/3$ m), this is 6 cm, which is still a large distance for a small pocket-sized terminal, but there are arrangements that give significant antenna diversity gains at a mobile terminal even at smaller antenna separations. Distances have to be larger when the antennas are separated in the vertical plane; typically a distance of two to three wavelengths is needed [8].

The antenna spacing at the base station needs to be much larger. The reason is that the scatterers tend not to be uniformly distributed around the base

Introduction to Fading Channels

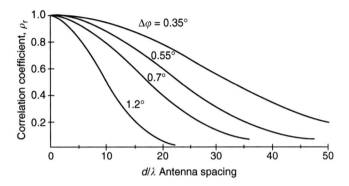

Figure 7.27 Normalized amplitude covariance between horizontally separated base station antennas for different 3 dB beamwidths $\Delta\varphi$ (from [15], copyright Wiley 1993; used with permission).

station antenna, as is the case with the mobile antenna. In Fig. 7.27 we show one example of how the normalized amplitude covariance of the fading depends on the antenna spacing for horizontally separated base station antennas. An antenna spacing of typically 10–30 wavelengths is needed with horizontal separation, while even bigger separations are needed with vertical separation. For further details, refer to [9,15].

Frequency diversity

Several replicas of the same symbols can be made available at the receiver by transmitting them simultaneously on different frequencies. This is referred to as frequency diversity. The main drawback of this method is its low bandwidth efficiency since the transmitted bandwidth is multiplied by the diversity order. The frequency spacing must be at least equal to the coherence bandwidth defined in Eq. (7.4-15). The spacing depends on the channel and to guarantee a given diversity order for all channel conditions, it has to be large. Frequency diversity is sometimes not possible. With smaller coherence bandwidths, it is easier to implement, but then other means of diversity may also be available, especially if the signaling bandwidth is larger than the coherence bandwidth. In this case an equalizer or a RAKE receiver may provide additional diversity.

Time diversity

The dual of frequency diversity is time diversity: each symbol is repeated in time instead of in frequency. Now the delay between the repetitions must be at least equal to the coherence time, which is defined in Eq. (7.4-14). This time depends on the maximum Doppler frequency and very long intervals between repetitions must be used for slow fading channels with small maximum Doppler frequency. The delay will be at least equal to

$$T_{\text{delay}} \geq 2(M-1)T_c, \qquad (7.7\text{-}26)$$

where M is the diversity order to be obtained. The factor of 2 appears because the same delay is experienced in both the transmitter and receiver. With slow fading it soon becomes unacceptable for real time services. Time diversity is also bandwidth inefficient, since each symbol is repeated. Note that time diversity is a simple form of channel coding diversity.

Interleaving and diversity through channel coding

It is well known that repetition codes are not the most powerful codes from a minimum distance point of view at a given rate. If there is to be time diversity, optimum codes should be used instead of simple repetition as with time diversity described above. We have already demonstrated in Section 7.6 that the diversity order obtained with channel codes on uncorrelated fading channels is $t+1$ for hard decision decoding and d for soft decision decoding. The problem is slow fading channels, for which the time diversity order approaches one. In order to obtain low correlation between the code symbols and maximum diversity order, the delay becomes $T_{\text{delay}} \geq 2(M-1)T_c$ with M equal to the code block length N. This is valid for a block code. Another relation holds for convolutional codes.

In practice, the time separation between code symbols is often obtained by an interleaver. The simplest form of interleaving is block interleaving, and ordinarily each row in the block can be a codeword. The stored symbols are read columnwise to the waveform generator and sent to the channel. The number of rows in the block should satisfy

$$r \geq \frac{T_c}{T_s}, \qquad (7.7\text{-}27)$$

where $1/T_s$ is the symbol rate on the channel, so that the code symbols have a low correlation. In the receiver, the inverse operation must be performed before decoding. The decoding delay becomes

$$T_{\text{delay}} = 2rNT_s \geq 2NT_c \qquad (7.7\text{-}28)$$

to give a diversity order of maximum d (with soft decoding). Since d is less than N (equal to N for a repetition code), this means a longer delay compared to a repetition code for the same diversity order, but a factor of K better spectral efficiency, where K is the number of data bits mapped to a codeword. In practice, spectral efficiency has to be traded against delay. Delay can also be traded against power efficiency, by using a small interleaver (smaller r); the code symbols experience a higher correlation, which reduces the power efficiency.

The selection of the interleaver dimensions is not as easy with convolutional or trellis codes, since coding is done sequentially. Here it is important to make fading uncorrelated over the most likely error events. Their length is related to the constraint length of the code and therefore typically the number of columns is chosen to be a number of constraint lengths. Since it is unusual that an interleaver can be long enough to break up the fading correlation completely, an interleaver often must be designed by trial and error.

Interleaving can also be done by a convolutional interleaver. These are better matched to convolutional codes and lead generally to smaller delays for a given performance. More details can be found in [16,17].

Finally, interleaving sometimes may be done over the frequency domain. This is the case when multicarrier modulation or orthogonal frequency division multiplexing (OFDM) [14] is used on a channel whose signaling bandwidth greatly exceeds the coherence bandwidth. Different code symbols may be interleaved to different subcarriers, and if the subcarriers are further apart than the coherence bandwidth, they will experience uncorrelated fading and a maximum diversity order may be obtained. In the more general case, code symbols may be spread over both time and frequency to utilize both time and frequency diversity.

Diversity provided by a channel

In some situations, (frequency) diversity is already available on the channel, and signal processing in the receiver is all that is needed in order to get diversity improvements on the receiver performance. This type of diversity is sometimes also referred to as path diversity. It is the case when the signaling bandwidth is wider than the coherence bandwidth.[20] It can appear when large data rates are transmitted or when spread spectrum or CDMA is used by the transmitter. Especially when large data rates or CDMA are combined with channels with significant delay spread, there are big opportunities for diversity gains simply by using proper signal processing in the receiver. When the signaling bandwidth is larger than the coherence bandwidth, a proper channel model is Eq. (7.4-4). Now the received signal consists of several replicas of the transmitted signal, and therefore each transmitted symbol is replicated several times in the receiver. In most situations, the received symbol replicas experience fading with low correlation and by combining them before the detector, performance gains due to diversity will occur. In non-spread systems, an equalizer may be used in the receiver to combine the power of the different symbol replicas. In spread spectrum system, a RAKE receiver may be used for the same purpose (See Section 8.6.1 for details). The principle of a RAKE receiver is to use a matched filter for each received signal component. At the

---

[20] The OFDM example in the previous paragraph is actually a special case of this type of diversity. With OFDM the channel is used in a very special way to avoid frequency selectivity on individual symbols. In this section we discuss the more general case.

output of the matched filter, the different replicas can be resolved if the time delay between them is larger than the inverse signaling bandwidth, which is mostly the case. The replicas can be combined according to MRC, if individual channel estimates are available for the different replicas, or the best may be selected according to selection diversity, if individual SNR estimates are available for the different replicas. Channel estimates or SNR estimates can be obtained from either pilot channels/symbols or by decision feedback methods.

A RAKE receiver in spread spectrum or CDMA is identical to diversity combining and the performance improvements will be similar. The difference in performance is that in the other diversity methods, the individual replicas normally have the same average SNR, while different paths on a frequency selective channel quite often have different average SNRs (proportional to the values in the power delay profile). The analysis of MRC and selection diversity can be done also for this case but is more involved. An error probability analysis is given in [14]. The main result is that the diversity order will be the same as in the case of equal SNR per replica, and it is equal to the number of resolvable paths on the channel. Similar results are obtained with an equalizer in a non-spread system. One problem with this kind of diversity is that the worst performance then appears when the channel is flat within the signaling bandwidth, since there is no frequency or path diversity available. This means that a system normally has to include some other means of diversity to guarantee a performance criterion. The most commonly used methods are diversity by channel coding with interleaving and/or antenna diversity.

Bibliography

[1] T. S. Rappaport, *Wireless Communications: Principles and Practice*, 2nd edn. Prentice-Hall, Englewood Cliffs, NJ, 2002.
[2] G. L. Stüber, *Principles of Mobile Communication*, 2nd edn. Kluwer, Boston, 2001.
[3] M. D. Yacoub, *Foundations of Mobile Radio Engineering*. CRC Press, Boca Raton, Fla., 1993.
[4] W. C. Jakes (ed.), *Microwave Mobile Communications*. IEEE Press, New York, 1993.
[5] M. K. Simon and M.-S. Alouini, *Digital Communication over Fading Channels: A Unified Approach to Performance Analysis*. Wiley, New York, 2000.
[6] COST 207 TD(86)51-REV 3 (WG1), "Proposal on channel transfer functions to be used in GSM tests late 1986," Sept. 1996.
[7] A. Mehrotra, *GSM System Engineering*. Artech House, Boston, 1997.
[8] J. D. Parsons and A. M. D. Turkmani, "Characterisation of mobile radio signals: model description," *IEE Proc.-I*, **138**(6), 549–556, Dec. 1991.
[9] A. M. D. Turkmani and J. D. Parsons, "Characterisation of mobile radio signals: base station crosscorrelation," *IEE Proc.-I*, **138**(6), 557–565, Dec. 1991.
[10] D. Verdin and T. C. Tozer, "Generating a fading process for the simulation of land-mobile radio communications," *Electronics Lett.*, **29**(23), 2011–2012, Nov. 1993.
[11] P. Dent, G. E. Bottomley, and T. Croft, "Jakes fading model revisited," *Electronics Lett.*, **29**(13), 1162–1163, June 1993.
[12] M. Gudmundson, "Correlation model for shadow fading in mobile radio systems," *Electronics Lett.*, **27**(23), 2145–2146, Nov. 1991.

[13] M. C. Jeruchim, P. Balaban, and K. S. Shanmugan, *Simulation of Communication Systems: Modeling, Methodology, and Techniques*, 2nd edn. Kluwer/Plenum, New York, 2000.
[14] J. G. Proakis, *Digital Communications*, 4th. edn McGraw-Hill, New York, 2000.
[15] W. C. Y. Lee, *Mobile Communications Design Fundamentals*, 2nd edn. Wiley, New York, 1993.
[16] J. L. Ramsey, "Realization of optimum interleavers," *IEEE Trans. Inf. Theory*, **IT-16**(3), 338–345, May 1970.
[17] G. D. Forney Jr., "Burst correcting codes for the classic bursty channel," *IEEE Trans. Communics. Tech.*, **COM-19**(5), 772–781, Oct. 1971.

8

Trellis Coding on Fading Channels

8.1. Introduction

The major part of this chapter is devoted to new convolutional codes and rate compatible convolutional (RCC) codes and their applications for rate matching, in direct-sequence spread spectrum (DSSS) and code-division multiple-access (CDMA), and in systems with a feedback channel. The theme throughout is how convolutional codes can respond to changing channel conditions and how they can be used to obtain low error probabilities on fading channels. This chapter also includes a section on the design and performance of set-partition codes (trellis coded modulation, TCM) on fading channels. We have already seen from Chapter 7, that the diversity order becomes much more important than Euclidean distance on fading channels, and this has a great impact on how to design good coded modulation schemes. There is a clear disadvantage of narrowband schemes since these most likely will have no inherent frequency diversity, and instead must rely completely on time diversity through coding and interleaving, which may cause long delays. Set-partition codes are such codes and we will here show that it is difficult to design these codes with high diversity order. Set-partition codes are in general spectrum efficient which is a big advantage as opposed to convolutional codes with low rate which require a large bandwidth. A large bandwidth is however not necessarily spectrum inefficient in a multiuser system, if it can be efficiently shared between many users. The major problem with many users sharing a frequency band is the inter-user interference that cannot in general be avoided and the system has to be designed to cope with this interference. For some situations, low rate convolutional codes represent a good solution to this problem, while in other situations orthogonal codes are better, as will be demonstrated in this chapter.

Convolutional codes that maximize the free distance have been known for a long time and are tabulated in many textbooks. Over time varying channels, these codes do not always give the best performance. In Section 8.2, optimum distance spectrum (ODS) convolutional codes are introduced and it is demonstrated that these codes give a better performance on fading channels.

In many applications it is useful to have not only one single code but a family of codes that can be decoded by the same decoder. RCC codes represent one family of such codes, and various kinds of such codes are introduced in Section 8.3. Section 8.4 is devoted to rate matching, which is the process of "mapping by coding" a given source data rate into an available channel data rate. This can be done by simple repetition and puncturing, but improved performance is obtained by using more sophisticated techniques.

In Section 8.5, the error event probability of TCM is derived and it is shown that good TCM schemes for additive white Gaussian noise (AWGN) channels may not be good on fading channels. Instead of optimizing the minimum Euclidean distance, the design should concentrate on optimizing the Hamming distance between codewords. Set-partition principles for such designs are briefly discussed.

DSSS and CDMA are introduced in Section 8.6 and described from a channel coding perspective. This clearly demonstrates the similarities between these systems and channel coding. New systems based on improved codes are also discussed. It is also clearly demonstrated that extremely large diversity orders can be obtained with such coded CDMA systems. Finally, in Section 8.7 channel coding in systems with a feedback channel from the receiver to the transmitter is discussed. The feedback channel opens many new possibilities for error protection but the discussion here is restricted to Automatic Repeat Request (ARQ) type of systems and especially to those using convolutional codes. Systems with feedback of this kind, provides an interesting possibility to increase spectrum efficiency by using coding only when necessary without having channel information available in the transmitter.

8.2. Optimum Distance Spectrum Feed-forward Convolutional Codes

Tables of maximal free distance (MFD) convolutional codes can be found in many textbooks and journal papers [1–5]. These have been found by extensive computer search. Since the number of codes increases exponentially with the memory of the code and the number of generator polynomials, the search was typically terminated once a code with free distance equal to the Heller bound was found [1]. From Chapter 3, it is clear that MFD codes minimize the error probability when E_b/N_0 approaches infinity, but not necessarily for other values of E_b/N_0. The reason is that the number of error events at higher distances also influences the error probability.

The error event probability is overbounded by $\sum A_d P_d$, where A_d is the number of alternative decodings at distance d and P_d is the error event probability of an event at distance d, as given in Chapter 3. In a similar way, the bit error probability is overbounded by

$$P_b < \sum_{d=d_f}^{\infty} c_d P_d, \qquad (8.2\text{-}1)$$

where c_d is the sum of the bit errors (the information error weight) for error events at distance d and d_f is the free distance of the code. For an AWGN channel, P_d decreases exponentially with both d and E_b/N_0, and therefore $c_{d_f} P_{d_f}$ dominates the sum for high enough E_b/N_0. Thus asymptotically, (E_b/N_0 goes to infinity) the bit error probability only depends on the free distance. For Rayleigh fading channels however, the error event probability P_d is obtained similarly to the average bit error probability in Section 7.7.1, but with the number of replicas M replaced by d (for soft decision decoding). With, for example, BPSK, coherent soft decision decoding, and perfect interleaving, the error event probability is obtained from Eqs (7.7-19) and (7.7-20) as

$$P_d = q^d \sum_{k=0}^{d-1} \binom{d-1+k}{k} (1-q)^k, \qquad (8.2\text{-}2)$$

with

$$q = \frac{1}{2}\left(1 - \sqrt{\frac{\Gamma_c}{1+\Gamma_c}}\right), \qquad (8.2\text{-}3)$$

where $\Gamma_c = \Gamma_b R$ and Γ_b is the average E_b/N_0 [6]. Thus, P_d decrease as $(E_b/N_0)^{-d}$, and many terms in the upper bound contribute significantly for all values of the average E_b/N_0. For a given channel, the upper bound can be reduced by reducing $\{c_d\}$. This is true for any channel, but becomes more or less important depending on the behavior of P_d vs average E_b/N_0. In any case, codes should have maximum free distance and low coefficient values to perform well on all types of channels. One such class of codes is ODS codes [7–11]. These are defined through the following definitions.

Definition 8.2-1. A feed-forward encoder with information weights c_d, giving a code with free distance d_f has a *superior distance spectrum* to a feed-forward encoder with information weights $\tilde{c}_d$ giving a code with free distance $\tilde{d}_f$, if one of the following conditions is fulfilled:

1. $d_f > \tilde{d}_f$ or
2. $d_f = \tilde{d}_f$ and there exists an integer $l \geq 0$ such that $c_d = \tilde{c}_d$ for $d = d_f, d_f+1, \ldots, d_f+l-1$, and $c_d < \tilde{c}_d$ for $d = d_f + l$.

Definition 8.2-2. An *optimum distance spectrum* (ODS) convolutional code is generated by a feed-forward encoder and has an equal or superior distance spectrum to all other convolutional codes generated by a feed-forward encoder with the same rate and memory.

An ODS code has the same free distance as a feed-forward MFD code for the same rate and memory, but equal or lower information error weights for the most significant error events. ODS codes for rates 1/2, 1/3, and 1/4 have been presented in [7–11]. In Table 8.1 we give the currently known rate 1/2 ODS codes. The information error weight spectrum can be found from [9–11]. The generator polynomials are given in octal notation as defined in Chapter 3.

In Fig. 8.1 the upper bound given by Eq. (8.2-1) with Eqs (8.2-2) and (8.2-3) is shown for a few selected ODS and MFD codes. The sum is truncated to 15 terms. The channel is a perfectly interleaved flat Rayleigh fading channel. The ODS codes perform better than the MFD codes and the largest gain is approximately 0.6 dB. There is no difference at all in implementation complexity between the ODS and MFD code, so there is no reason not to use the ODS code instead of the MFD code, although the gain may be relatively small.

The performance of convolutional codes is not directly given by the distance spectrum. A code design based on minimum error probability would

Table 8.1 Rate 1/2 ODS codes. The generator polynomials are given in octal notation as defined in Chapter 3 (64 = 1101, 74 = 1111). Note that the octal notation is defined differently in [9–11]. The memory 6 code is a well-known code mentioned in many books and publications. With the octal notation used in [9–11] and other sources, this code has generator polynomials 133, 171. The codes marked with * are previously known as MFD codes [1,12]. All codes in this table are collected from [9,11]

Memory	Generators	Free distance
2	5, 7*	5
3	64, 74*	6
4	46, 72*	7
5	53, 75*	8
6	554, 744*	10
7	516, 762*	10
8	561, 753*	12
9	4644, 7654	12
10	6712, 7426	14
11	5261, 7173	15
12	53734, 72304	16
13	56522, 76746	16
14	63057, 44735	18
15	533514, 653444	19

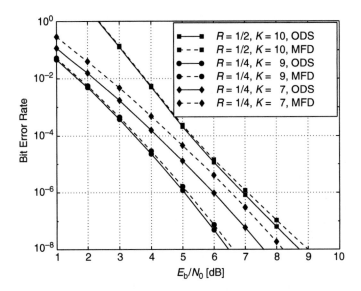

Figure 8.1 Upper bound on the bit error probability on a perfectly interleaved flat Rayleigh fading channel for ODS and MFD codes. K denotes the constraint length of the code (memory plus one). (From [9], copyright IEEE 1999; used with permission.)

therefore be advantageous. In [13,14] this approach has been taken and gains of 0.1–0.4 dB are reported. A code designed according to this criterion is optimum for one channel with a given E_b/N_0 but not necessarily good for other channels and/or other values on E_b/N_0. ODS codes on the other hand are channel independent and will result in low error probability for most channels and most E_b/N_0.[1]

8.3. Rate Compatible Convolutional (RCC) Codes

Rate compatible codes are codes where the coded bits of a higher rate code are also used by a lower rate code. This means that the higher rate code is embedded into the lower rate code. This is a restriction of the code, but it has been shown that good codes can still be found within the family of rate compatible codes. Such codes allow transmission of incremental redundancy in ARQ/Forward Error Correction (FEC) schemes (see Section 8.7) and continuous rate variation to match the source rate to the channel rate and/or change the level of error protection within a data frame (see Section 8.4). They have also proven useful for varying the spreading factor in spread spectrum (SS) and CDMA systems (see Section 8.6).

---

[1] There may exist channels where P_d for $d = d_f, d_f + 1, \ldots$ behave in such a way that ODS codes do not give a small error probability, but such channels are not common.

In this section we will discuss three different methods for obtaining RCC codes based on feed-forward encoders. The same techniques can also be used for block codes and other convolutional codes, but the criterion for finding them may have to be modified.

8.3.1. Rate Compatible Punctured Convolutional Codes

Punctured codes are normally used to obtain high rate codes. Viterbi decoders for (non-punctured) rate b/c (with $b > 1$ and $c > b$) convolutional codes are more complex than a decoder for a rate $1/c$ convolutional code. However, an alternative way to obtain such a code with a simpler decoder is to puncture a rate $1/c$ code as described in Chapter 3 [15–18]. The code to be punctured is referred to as the mother or parent code. Puncturing means that some of the coded bits are removed and never transmitted over the channel. The bits to be punctured are specified by a puncturing matrix P with c rows and p columns. $P_{i,j} = 0$ defines that the coded bits $x_i[j+np]$ as defined in (3.3-4), n an integer, are not transmitted; bits to be transmitted are identified by $P_{i,j} = 1$. The range of code rates that can be obtained by puncturing a mother code of rate $1/c$ is

$$R = \frac{p}{p+l}, \quad l = 1, \ldots, (c-1)p, \tag{8.3-1}$$

that is, between $p/(p+1)$ and $1/c$. Puncturing is a restriction but it has been shown that good punctured codes exist. Rate compatible punctured coding is a further restriction in that the coded bits of a high rate code are also included among the coded bits of a lower rate code [19–21].

Definition 8.3-1. A punctured convolutional code with puncturing matrix P and rate $R = p/(p+l)$ is *rate compatible* with a punctured convolutional code with puncturing matrix P' and rate $R' = p/(p+l')$ if they have the same mother code and $P'_{i,j} \geq P_{i,j}$ for all $i = 1, \ldots, c$ and $j = 1, \ldots, p$, with $(c-1)p \geq l' \geq l \geq 1$.

It is possible to generate a whole family of rate compatible punctured convolutional (RCPC) codes with rates spanning between $p/(p+1)$ and $1/c$. With large enough values of c and p, the family spans many rates.

Computer search based on the ODS criterion[2] has been used to find RCPC code families for many different mother codes [11,19,20]. As an example, we give a family of codes in Table 8.2 with rates between 1/4 and 8/9 and memory 8. The mother code has rate 1/4 and the code polynomials are given (in octal) as

---

[2] The computer search started with an ODS mother code. Further, at each rate, the ODS code among all codes fulfilling the rate compatibility criterion was found. The found code is however not necessarily the best RCPC code in the ODS sense at that rate, since starting the search from another mother code may lead to a better code. It is however believed that the whole code family is a good code family.

Table 8.2 A family of RCPC codes based on a mother code with polynomials (473, 513, 671, 765). These numbers are collected from [11,20]

Rate	Punctured position	Free distance
8/32	–	24
8/31	(2, 8)	22
8/30	(2, 4)	22
8/29	(1, 5)	21
8/28	(2, 1)	20
8/27	(1, 7)	20
8/26	(1, 3)	19
8/25	(2, 6)	18
8/24	(1, 2)	17
8/23	(4, 8)	16
8/22	(4, 1)	15
8/21	(3, 5)	15
8/20	(3, 3)	14
8/19	(4, 4)	13
8/18	(3, 2)	12
8/17	(3, 6)	11
8/16	(3, 8)	10
8/15	(4, 7)	9
8/14	(1, 6)	8
8/13	(4, 3)	8
8/12	(4, 2)	7
8/11	(4, 5)	6
8/10	(1, 4)	5
8/9	(3, 1)	4

(473, 513, 671, 765). The weight distribution is also given in the references. In the column denoted "punctured position" we give the indices (row, column) of the puncturing matrix where the *additional* 0 is to be placed to obtain the rate of that particular row[3]. As an example, the puncturing matrix for rate 8/30 has $P_{2,8} = 0$ and $P_{2,4} = 0$ and all other entries equal to 1. The best known ODS code with rate 1/2 and 1/3, respectively, and memory 8 has free distance 12 and 18, while the codes with corresponding rate in the family of RCPC codes in Table 8.2 have free distance 10 and 17. This shows typical losses due to the rate compatibility

---

[3] Note that rows are numbered from 0 to $c - 1$ and columns from 0 to $p - 1$ in [11,20].

constraint. The actual difference in bit error rate (BER) performance is small in most cases [11,20]. For the codes compared above, the degradation in E_b/N_0 at a bit error probability of 10^{-6} is less than 0.1 dB on an AWGN channel and less than 0.3 dB on a Rayleigh fading channel. The loss for a given rate typically increases with bigger code family and therefore a too large code family (too low rate of mother code and too large puncturing period) should be avoided. As a rule of thumb, a mother code rate of 1/4 with puncturing period $p = 8$ seems to be a reasonable choice for a larger code family [11,20].

8.3.2. Rate Compatible Repetition Convolutional Codes

To avoid too large performance degradations for RCPC codes compared to non-punctured convolutional codes, a too low mother code rate should be avoided. The lowest rates may instead be obtained by repetition coding [22–24]. A family of rate compatible repetition convolutional (RCRC) codes can be used together with a family of RCPC codes to increase the span of code rates if both families are based on the same mother code. A repetition code is defined in a similar manner as a punctured code, but instead of the puncturing matrix, a repetition matrix Q with c rows and p columns is used. The entries in this matrix are integers from one to infinity and define the number of times the coded bit is to be repeated. As an example, $Q_{i,j} = 3$ implies that $x_i[j + np]$ is transmitted three times over the channel. The rates of a repetition code family are

$$R = \frac{p}{pc + l}; \quad l = 0, \ldots, \infty, \qquad (8.3\text{-}2)$$

for a mother code rate of $1/c$.

Definition 8.3-2. A repetition convolutional code with repetition matrix Q and rate $R = p/(pc + l)$ is a rate compatible with a repetition code with repetititon matrix Q' and rate $R' = p/(pc + l')$ if the two codes have the same mother code and $Q'_{i,j} \geq Q_{i,j}$ for all $i = 1, \ldots, c$ and $j = 1, \ldots, p$ with $\infty > l' \geq l \geq 0$.

The currently best known RCRC codes are given in [24,25]. These are obtained in a similar way as the RCPC codes in Section 8.3.1, and the ODS criterion has been used at each rate in the search for the best code. As an example, in Table 8.3 we show part of the RCRC code family based on the rate 1/4 mother code with memory 8 and generator polynomials (473, 513, 671, 765) (the lowest rate in [24,25] is 1/20). The number of columns p of the repetition matrix is 2. The weight distribution is also given in the references. In the column denoted "repetition position" we give the indices of the repetition matrix where the 1 is to be added to obtain the rate of that particular row. As an example, the repetition matrix for rate 2/10 has $Q_{3,1} = 2$ and $Q_{3,2} = 2$ and all other entries are 1. These codes are in general not MFD codes except if the mother code has low enough rate.

Trellis Coding on Fading Channels

Table 8.3 A family of RCRC codes based on a mother code with polynomials (473, 513, 671, 765). Numbers are from [24,25]

Rate	Repetition position	Free distance
2/9	(3, 1)	27
2/10	(3, 2)	30
2/11	(1, 1)	33
2/12	(1, 2)	36
2/13	(3, 1)	39
2/14	(3, 2)	42
2/15	(1, 1)	45
2/16	(1, 2)	48
2/17	(3, 1)	51
2/18	(3, 2)	54
2/19	(1, 1)	57
2/20	(1, 2)	60

The codes given in Table 8.3 are in most cases one unit below the Heller bound for the higher rates and two units below the Heller bound for the lower rates. The degradation in bit error probability from this is usually small.

8.3.3. Rate Compatible Nested Convolutional Codes

One drawback with RCRC codes is that they are in general not MFD codes. Very low rate MFD codes are very difficult to find by a full search since the number of search parameters is very large. A simpler way to obtain low rate codes is by nesting [11,13,20,26–28]. These codes are obtained by extending a rate $1/c$ code to a rate $1/(c+1)$ code by finding an addition generator polynomial $g_{c+1}(D)$ by some proper performance criterion. This code is rate compatible with all higher rate codes in the same family and all can be found with limited search complexity. In [11,20,28] nested codes for very low rates are presented (the lowest rate is 1/512). The criteria used here for finding the additional code generator polynomial is the ODS criterion. The nested codes themselves are not necessarily ODS or MFD since a full search for the given rate is not done. However, all the found codes based on the ODS mother code with rate 1/4 turn out to be MFD codes. In Table 8.4, an example of a family of nested codes is given. The mother code has rate 1/4 and memory 8 with generator polynomials (473, 513, 671, 765).

Table 8.4 A family of nested codes based on a mother code with polynomials (473, 513, 671, 765). Numbers are from [11,20]

Rate	Additional polynomial	Free distance
1/5	657	31
1/6	745	37
1/7	753	44
1/8	517	50
1/9	473	56
1/10	657	62

The weight distribution is also given in the references. In the column denoted "additional polynomial" we give the last generator polynomial, which together with all the previous polynomials specify the complete code.

Nested codes at very low rates are unequal repetition codes in the sense that some generator polynomials are repeated many times while others are only used once [28]. Interestingly enough, it often turns out that some of the generator polynomials of the mother code are not repeted at all. This repetition structure may be useful in some applications and simplifies the metric calculations in the decoder.

Another criterion for finding good nested codes is used in [26,27]. This leads to a simplified code search but the codes are not MFD codes. The free distance is, however, reported to be quite close to the Heller bound for rates down to 1/128.

As a final remark, it can be added that nested codes are always to be preferred to repetition codes, since nesting allows new polynomials for lower rates while repetition only reuses previous polynomials. However, it can be somewhat less complex to find a repetition code as compared to a nested code. The decoding complexity is approximately the same.

8.4. Rate Matching

Modern communications systems must convey information from many different services. These can have quite different requirements on quality, data rate, delay, etc. The physical layer (excluding channel coding) on the other hand carries one or a few data rates at a set performance. In GSM as an example, each time slot of length 576.92 μs carries 114 coded bits [29]. If one user is allocated one time slot out of the eight in total, this gives a coded data rate of 24.7 kbps. By allocating

Trellis Coding on Fading Channels

more than one time slot to each user, GSM is able to provide each user a multiple of 24.7 kbps, but the largest rate is 197.6 kbps. In practice, not all time slots can be used simultaneously in one direction, which means that the maximum coded rate is actually lower. One could think of higher coded rates in GSM if each user was allocated more than one carrier, so in general a system like GSM can provide any multiple of 24.7 kbps to a user. Third generation systems like Wideband CDMA are designed in a similar way to provide a limited number of coded data rates on the physical layer, although these provide a larger set of coded rates [30,31].

Rate matching is used here as a notation for the operation of matching the service data rate to one of the available channel data rates in such a way that the performance requirements are fulfilled. Since a service may deliver almost any data rate, a flexible data rate conversion is needed. RCC codes represent one good solution to this problem, since they also provide good performance. The major advantage is that the channel decoder is not service dependent but the same Viterbi decoder can be used for all rates.

Example 8.4-1. To illustrate the idea, GSM is used as an example. As stated before, each time slot provides a coded data rate of 24.7 kbps to the services. Here it will be assumed that each service can be allocated between 1 and 5 time slots. The RCC codes discussed in Section 8.3 will be used for rate matching. Specifically, RCPC codes with puncturing period 8 are used above rate 1/4 and nested codes are used below rate 1/4. The mother code has rate 1/4, code memory 8, and generator polynomials (473, 513, 671, 765). In Table 8.5 we illustrate the information data rate and performance that may be obtained. Performance is measured by free distance. By using Fig. 8.2, the required E_b/N_0 in dB for a bit error probability of 10^{-6} can be obtained for two different channels. These results are for BPSK but the GMSK as used in GSM is only a few tenths more. From the table and the figure it is concluded that this system is able to provide data rates between 2.47 and 109.8 kbps with a required E_b/N_0 on a flat Rayleigh fading channel spanning between 4 and 18 dB. By increasing the puncturing period, an even higher rate resolution can be obtained above rate 1/4, at the expense of a small penalty in performance. For rates lower than 1/4, RCRC codes can be used instead of nested codes to increase the rate resolution but this will also be at a small performance degradation.

With the currently available RCC codes, it is clear that very flexible rate matching can be obtained. But even with a large family of RCC codes, an exact match between the source rate and the channel rate will not always be possible. This gap can easily be filled by repeating or puncturing one or a few bits over the time slot. Since the amount of additional rate matching is so small, it will have little influence on the performance also when it is not done in an optimal way.

Table 8.5 An example of rate matching with RCC codes in GSM

Code rate	Free distance memory 8	Data rate in kbps with				
		1 slot	2 slots	3 slots	4 slots	5 slots
1/10	62	2.47	4.94	7.41	9.88	12.35
1/9	56	2.744	5.489	8.233	10.98	13.72
1/8	50	3.087	6.175	9.262	12.35	15.44
1/7	44	3.529	7.057	10.59	14.11	17.64
1/6	37	4.117	8.233	12.35	16.47	20.58
1/5	31	4.94	9.88	14.82	19.76	24.7
1/4	24	6.175	12.35	18.52	24.7	30.88
8/31	22	6.374	12.75	19.12	25.5	31.87
4/15	22	6.587	13.17	19.76	26.35	32.93
8/29	21	6.814	13.63	20.44	27.26	34.07
4/14	20	7.057	14.11	21.17	28.23	35.29
8/27	20	7.319	14.64	21.96	29.27	36.59
4/13	19	7.6	15.2	22.8	30.4	38
8/25	18	7.904	15.81	23.71	31.62	39.52
1/3	17	8.233	16.47	24.7	32.93	41.17
8/23	16	8.591	17.18	25.77	34.37	42.96
4/11	15	8.982	17.96	26.95	35.93	44.91
8/21	15	9.41	18.82	28.23	37.64	47.05
2/5	14	9.88	19.76	29.64	39.52	49.4
8/19	13	10.4	20.8	31.2	41.6	52
4/9	12	10.98	21.96	32.93	43.91	54.89
8/17	11	11.62	23.25	34.87	46.49	58.12
1/2	10	12.35	24.7	37.05	49.4	61.75
8/15	9	13.17	26.35	39.52	52.69	65.87
4/7	8	14.11	28.23	42.34	56.46	70.57
8/13	8	15.2	30.4	45.6	60.8	76
2/3	7	16.47	32.93	49.4	65.87	82.33
8/11	6	17.96	35.93	53.89	71.85	89.82
4/5	5	19.76	39.52	59.28	79.04	98.8
8/9	4	21.96	43.91	65.87	87.82	109.8

8.5. TCM on Fading Channels

This section is devoted to the performance and design of TCM on fading channels [32–36] and [37, Chapter 5]. The definition of TCM is given in Chapter 4 and will not be repeated here. A common design principle for TCM on AWGN channels is a set-partition principle that maximizes the minimum Euclidean distance of trellises based on convolutional codes as described in Sections 4.2 and

Trellis Coding on Fading Channels

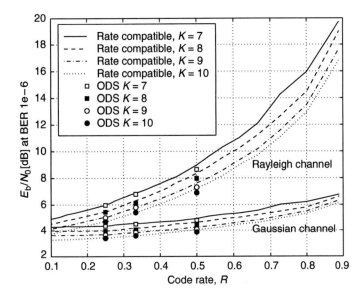

Figure 8.2 Required E_b/N_0 in dB to obtain a bit error probability of 10^{-6} with BPSK on flat Rayleigh fading and Gaussian channels vs the code rate R. K denotes the constraint length of the code (memory plus one). (From [20], copyright IEEE 1999; used with permission.)

4.3. Throughout this section the same trellis structure is assumed, but as we will see, the set partitioning used for AWGN channels is not always a good design principle in fading channels.

8.5.1. Performance of TCM on Fading Channels

On an AWGN channel, the bit error and error event probabilities of TCM are dominated by the free Euclidean distance d_f for moderate to large E_b/N_0 as seen in Eqs (4.3-5) and (4.3-6). d_f will influence the performance to some extent also on fading channels, but the minimum Hamming distance will be more important. This will now be shown, by deriving an upper bound on the error event probability of TCM on a flat fading channel. The sample at the output of the matched filter in the receiver at $t = nT$ is

$$y[n] = \Psi[n]x[n] + \eta[n], \qquad (8.5\text{-}1)$$

where $\Psi[n]$ is the deinterleaved complex channel gain, $x[n]$ the transmitted TCM symbol, and $\eta[n]$ the normally distributed complex white noise. For simplicity, it is assumed that the transmitted symbols are normalized to unit power, such that $\mathcal{E}\{|x[n]|^2\} = 1$. In an ideal case, the complex channel gains are completely known to the detector. This situation is referred to as *perfect* or *ideal side information*. The task of the maximum likelihood detector is then to find the transmitted sequence of TCM symbols (codeword) $\tilde{\mathbf{x}} = \{\tilde{x}[0], \tilde{x}[1], \ldots, \tilde{x}[N]\}$ that is closest to the received sequence $\mathbf{y} = \{y[0], y[1], \ldots, y[N]\}$ in Euclidean distance; that is,

$$\|\mathbf{y} - \mathbf{\Psi}\tilde{\mathbf{x}}\| \leq \|\mathbf{y} - \mathbf{\Psi}\mathbf{x}\|, \qquad (8.5\text{-}2)$$

for any allowed code sequence $\mathbf{x}$, and where

$$d_E^2(\mathbf{z}) = \|\mathbf{z}\|^2 = \sum_{n=0}^{N} |z[n]|^2, \qquad (8.5\text{-}3)$$

is the squared Euclidean distance and $\mathbf{\Psi} = \{\Psi[0], \Psi[1], \ldots, \Psi[N]\}$. Here $N+1$ is the length of the sequence.

The error event probability[4] $P(\mathbf{x} \to \hat{\mathbf{x}})$ is the probability that an incorrect code sequence $\hat{\mathbf{x}}$ is chosen by the detector instead of the correct sequence $\mathbf{x}$. The conditional error event probability for a given channel is obtained as

$$P(\mathbf{x} \to \hat{\mathbf{x}} | \mathbf{\Psi}) = \Pr\left[d_E^2(\mathbf{y} - \mathbf{\Psi}\hat{\mathbf{x}}) \leq d_E^2(\mathbf{y} - \mathbf{\Psi}\mathbf{x})\right]$$

$$= \Pr\left[\sum_{n=0}^{N}\left\{|y[n] - \Psi[n]x[n]|^2 - |y[n] - \Psi[n]\hat{x}[n]|^2\right\} \geq 0\right]. \qquad (8.5\text{-}4)$$

This error probability can be upper bounded by using the Chernoff bound, which states that

$$\Pr\left[\sum_{n=0}^{N} z[n] \geq 0\right] \leq \prod_{n=0}^{N} \mathcal{E}\{\exp(\lambda z[n])\}, \qquad (8.5\text{-}5)$$

where $z[n]$ is independent of $z[m]$ for $n \neq m$. $\lambda \geq 0$ is a parameter to be optimized and $\mathcal{E}\{\}$ denotes the expected value over the statistics of $z[n]$. Using this expression in Eq. (8.5-4) results in

$$P(\mathbf{x} \to \hat{\mathbf{x}} | \mathbf{\Psi}) \leq \prod_{n=0}^{N} \exp\left(-\lambda |\Psi[n]|^2 |x[n] - \hat{x}[n]|^2\right) \qquad (8.5\text{-}6)$$
$$\times \mathcal{E}\left\{\exp\left(2\lambda \mathrm{Re}\left\{\Psi[n](\hat{x}[n] - x[n])(\eta[n])^*\right\}\right)\right\}.$$

---

[4] This is also referred to as the pairwise error probability. In other parts of this book, the error event probability is denoted P_d since it only depends on the distance d, but since this is not the case here, a more general notation will be used.

After performing the expectation, one gets

$$P\left(\mathbf{x} \to \hat{\mathbf{x}} | \boldsymbol{\Psi}\right) \leq \prod_{n=0}^{N} \exp\left(-\lambda |\Psi[n]|^2 |x[n] - \hat{x}[n]|^2 \left(1 - \lambda 2\sigma_\eta^2\right)\right), \quad (8.5\text{-}7)$$

where $2\sigma_\eta^2$ is the noise power. The optimum Chernoff parameter that minimizes the upper bound is given by

$$\lambda_{\text{opt}} = \frac{1}{4\sigma_\eta^2}, \quad (8.5\text{-}8)$$

which leads to

$$P\left(\mathbf{x} \to \hat{\mathbf{x}} | \boldsymbol{\Psi}\right) \leq \exp\left(-\frac{1}{8\sigma_\eta^2} d_E^2 \left(\boldsymbol{\Psi}\mathbf{x} - \boldsymbol{\Psi}\hat{\mathbf{x}}\right)\right). \quad (8.5\text{-}9)$$

For a Rician fading channel with perfect interleaving, such that $\Psi[n]$ is independent of $\Psi[m]$ when $m \neq n$, the unconditional error event probability is obtained by averaging Eq. (8.5-9) over all the individual pdfs of $|\Psi[n]|$ (given by Eq. (7.3-11)), resulting in

$$P\left(\mathbf{x} \to \hat{\mathbf{x}}\right) \leq \exp\left(-\frac{E_s}{4N_0} d^2\left(\mathbf{x}, \hat{\mathbf{x}}\right)\right), \quad (8.5\text{-}10)$$

with

$$d^2\left(\mathbf{x}, \hat{\mathbf{x}}\right) = \sum_{n=0}^{N} \left\{d_1^2\left(x[n], \hat{x}[n]\right) + d_2^2\left(x[n], \hat{x}[n]\right)\right\}, \quad (8.5\text{-}11)$$

where

$$d_1^2\left(x[n], \hat{x}[n]\right) = \frac{K |x[n] - \hat{x}[n]|^2}{1 + K + E_s/(4N_0) |x[n] - \hat{x}[n]|^2}, \quad (8.5\text{-}12)$$

and

$$d_2^2\left(x[n], \hat{x}[n]\right) = \left(\frac{E_s}{4N_0}\right)^{-1} \ln\left(\frac{1 + K + E_s/(4N_0) |x[n] - \hat{x}[n]|^2}{1 + K}\right).$$

$$(8.5\text{-}13)$$

The average received signal-to-noise ratio per coded symbol is denoted $E_s/N_0 = RE_b/N_0$.

It is easy to verify that the bound is in agreement with the results obtained in Chapter 4 on a nonfading channel, by including $K \to \infty$ in

Eqs (8.5-10)–(8.5-13). On a Rayleigh fading channel, on the other hand, $K = 0$ and $d_1^2(x[n], \hat{x}[n]) = 0$ while

$$d_2^2(x[n], \hat{x}[n]) = \left(\frac{E_s}{4N_0}\right)^{-1} \ln\left(1 + \frac{E_s}{4N_0}|x[n] - \hat{x}[n]|^2\right). \quad (8.5\text{-}14)$$

For large E_s/N_0, 1 in the expression above can be neglected when $x[n] \neq \hat{x}[n]$ and thus the error event probability becomes

$$P(\mathbf{x} \to \hat{\mathbf{x}}) \leq \frac{1}{d_P^2(\mathbf{x}, \hat{\mathbf{x}})} \left(\frac{E_s}{4N_0}\right)^{-d_H(\mathbf{x}, \hat{\mathbf{x}})}, \quad (8.5\text{-}15)$$

where $d_H(\mathbf{x}, \hat{\mathbf{x}})$ is the Hamming distance between the codewords[5] and

$$d_P^2(\mathbf{x}, \hat{\mathbf{x}}) = \prod_{n=0; x[n] \neq \hat{x}[n]}^{N} |x[n] - \hat{x}[n]|^2 \quad (8.5\text{-}16)$$

is the product of the nonzero squared Euclidean symbol distances along the error event. The last will be referred to as the *product distance*. This type of expression is recognized from Section 7.7 as the asymptotic error probability with $d_H(\mathbf{x}, \hat{\mathbf{x}})$-fold diversity on a perfectly interleaved Rayleigh fading channel. To make the upper bound small, it is most important to have a large Hamming distance between codewords, and second most important to have a large product distance. The set partitioning for AWGN channels from Chapter 4 gives large squared Euclidean distance but does not necessarily give large Hamming and product distances, as seen later.

The Chernoff bound is in general not tight, although it has a correct exponent. Thus the results have to be used with some care. The difference compared to the true error probability may be large. For example, with BPSK, the Chernoff bound, is 6 dB pessimistic. An exact expression for the error event probability in Rayleigh fading is given in [39] and it can be expressed as the upper bound in (8.5-15) multiplied by a correction factor. The value of the correction factor depends on the poles of the two-sided Laplace transform of the pdf of the decision variable. For an eight-state 8-PSK TCM scheme with 2 bits per symbol, the exact result is 3.6 dB better than the Chernoff bound.

An upper bound on the error event probability of the full TCM scheme can be obtained by a union bound as in Section 8.2. N in the error event probability calculation then becomes the length of the error event and the distance measures are for a given error event. It is clear that the union bound is dominated by the minimum Hamming distance at large E_s/N_0, and the minimum Hamming distance specifies the diversity order of the TCM scheme. The diversity order can be further increased

---

[5] The Hamming distance between $\mathbf{x}$ and $\hat{\mathbf{x}}$ is equal to the number of positions in which $|x[n] - \hat{x}[n]| \neq 0$ [38, Definition 4-3].

by combining TCM with other diversity methods as described in Section 7.7. It is also noted that the squared Euclidean distance over the error events does not appear at all in the union bound. Since the design criteria for TCM on AWGN channels is to maximize the minimum squared Euclidean distance, this means that a good TCM scheme on an AWGN channel may not be good on a Rayleigh fading channel, and we will soon see that it is quite often poor. In this respect TCM is quite different from convolutional codes, since a convolutional code which maximizes the minimum Hamming distance also maximizes the minimum squared Euclidean distance. Consequently, a convolutional code which is good on an AWGN channel is also good on a Rayleigh fading channel. For Rician fading channels the importance of the different distances depend on K. The squared Euclidean distance becomes more important the bigger K becomes, while the Hamming and product distances become less important. Finally it should be noted that independence between channel gains is essential on fading channels, since otherwise the diversity order is reduced. This independence is normally obtained by interleaving.

The situation when channel state information is not available can be solved in a similar way. The decision rule is to find a decision $\tilde{\mathbf{x}}$ such that

$$\|\mathbf{y} - \tilde{\mathbf{x}}\| \leq \|\mathbf{y} - \mathbf{x}\|, \tag{8.5-17}$$

for any allowed codeword $\mathbf{x}$. The error event probability can be upper bounded in a similar way for this detector, but unfortunately the optimum Chernoff parameter cannot be found before the averaging over the fading distribution is performed. Therefore it becomes more difficult to analyze this detector. Since this detector has less information available, its performance will be worse than the performance of the detector with perfect channel state information. Details on the analysis with no side information and for MDPSK can be found in [32–34,37,39–44].

8.5.2. Design of TCM on Fading Channels

In this subsection we will briefly discuss some design principles of TCM on Rayleigh fading channels [32–37,45–49]. Many publications are available on this topic and space does not permit us to go into details on all the design principles.

From Section 8.5.1 it is clear that TCM on Rayleigh fading channels must maximize the minimum Hamming distance rather than the minimum squared Euclidean distance. Many of the TCM schemes for AWGN channels are not at all good. For example, the simple TCM scheme shown in Fig. 4.1 has minimum Hamming distance one since each trellis transition is labeled with a subset consisting of two points. This is a so-called parallel transition. An uncoded bit is used to select one of the two points in the subset chosen by the convolutional subset selection. If the decoder makes an error on this symbol, an error event of length one symbol may be the result. The TCM scheme therefore has minimum Hamming distance one and no diversity gain on a Rayleigh fading channel. This is true for all set-partition codes of the types shown in Figs 4.4 and 4.5 with $R - b \geq 1$. The commonality of all these configurations is that some uncoded bits are used in selecting the modulation symbol for a trellis transition, leading to error events of length one trellis section and thus minimum Hamming distance of one. A code that fully encodes all R bits[6] is necessary to obtain a diversity order larger than one on a Rayleigh fading channel. However, even for such a code, the optimum labeling is likely to be different on a fading channel as compared to an AWGN channel.

Multiple Trellis Coded Modulation

Multiple trellis coded modulation (MTCM) is a generalization of TCM in which $k > 1$ QAM or PSK symbols are transmitted on every branch in the trellis. In Fig. 8.3 a simple MTCM example is shown. The trellis has two states and there are eight transitions in parallel between each pair of states. Each transition is labeled with two successive 8PSK symbols, denoted by the integer pair i, j. The symbol sets used are defined as

$$E = \begin{Bmatrix} 0,0 \\ 1,5 \\ 2,2 \\ 3,7 \\ 4,4 \\ 5,1 \\ 6,6 \\ 7,3 \end{Bmatrix} \quad F = \begin{Bmatrix} 0,4 \\ 1,1 \\ 2,6 \\ 3,3 \\ 4,0 \\ 5,5 \\ 6,2 \\ 7,7 \end{Bmatrix} \quad G = \begin{Bmatrix} 0,2 \\ 1,7 \\ 2,4 \\ 3,1 \\ 4,6 \\ 5,3 \\ 6,0 \\ 7,5 \end{Bmatrix} \quad H = \begin{Bmatrix} 2,0 \\ 3,5 \\ 4,2 \\ 5,7 \\ 6,4 \\ 7,1 \\ 0,6 \\ 1,3 \end{Bmatrix} \quad (8.5\text{-}18)$$

where 8PSK symbol i has phase $i\pi/4$. The minimum Hamming distance within each of the sets is two and this results in a diversity order of two for this MTCM code. The minimum squared Euclidean distance of the code is 2.343 as compared to 4 for the optimum two-state TCM scheme with two bits per symbol. This leads to an asymptotic loss of 2.3 dB on an AWGN channel, but a big gain on a Rayleigh fading channel. The Ricean-channel bit error probability for large E_b/N_0 is given by

$$P_b \approx \frac{1+K}{4E_b/N_0} \exp(-K), \quad (8.5\text{-}19)$$

for the optimum AWGN scheme with $d_f^2 = 4$, while it is

$$P_b \approx \frac{(1+K)^2}{2(E_b N_0)^2} \exp(-2K) \quad (8.5\text{-}20)$$

---

[6] The trellis for a code that fully encodes all R bits has at least 2^R states and the number of transitions per trellis section is 2^R. Such a code does not have parallel transitions.

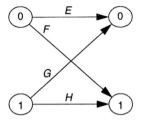

Figure 8.3 A trellis section of a simple MTCM scheme with 2 8PSK symbols per trellis branch. Each transition corresponds to a subset of 8 parallel transitions and the labels are given in Eq. (8.5-18).

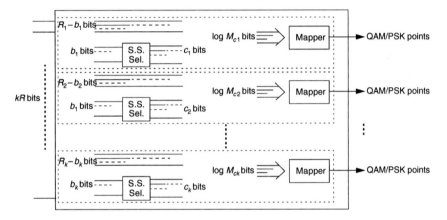

Figure 8.4 A generic MTCM encoder with kR bits at the input and k symbols at the output.

for the optimum Rayleigh fading MTCM scheme [32]. The MTCM scheme in Fig. 8.3 is also not the best such MTCM scheme of its kind on an AWGN channel, which has a minimum squared Euclidean distance of 3.172 [32]. This example clearly illustates the differences in TCM design principles on fading and nonfading channels.

A generic MTCM encoder is shown in Fig. 8.4, compared to a generic TCM encoder in Fig. 4.4. It basically consists of k generic TCM encoders in parallel, where k is referred to as the multiplicity of the code. At the input, $kR = \sum_{i=1}^{k} R_i$ bits are divided into k groups each entering a generic TCM encoder, where R is the average rate of the code in bits/QAM symbol. The generic encoder produces $\log_2 M_{ci} = R_i - b_i + c_i$ bits that are mapped into a PSK/QAM master constellation of size M_{ci}. These master constellations may be of different size. An important special case is when $b_i = b$, $R_i = R$, and $M_{ci} = M_c$ for all i, and the example shown in Fig. 8.3 belongs to this class. It should be noted that MTCM schemes may have some advantanges also on AWGN channels, as described in [32, Chapter 7] and [46].

The design of MTCM on a Rayleigh fading channel consists of labeling a trellis in such a way that the minimum Hamming distance is maximized. Moreover, each error event should have a large product distance. On a Rician channel, the minimum squared Euclidean distance should also be considered. Whenever the trellis has parallel transitions, the diversity order is limited by k. Two set-partitioning approaches for MPSK have been proposed for this design in [32, Chapter 10] and [45]. The simplest of them performs the set partition from root to leaf as for TCM on AWGN channels. With multiplicity $k = 2$ it starts with a twofold Cartesian product of the two master constellations. In the first step, the Cartesian product set is divided into M subsets such that two purposes are accomplished. One is that a minimum Hamming distance of two between the sets is guaranteed. The other is that the product distance is maximized. The four sets in Eq. (8.5-18) have been obtained in this way, and correspond to four of the eight sets obtained when 8PSK is partitioned. In the lower levels of the partitioning, each subset is split into two subsets. This is done such that the product distance is maximized at each level.

The other set partitioning method proceeds from leaf to root. The details of it can be found in [32, Chapter 10]. Yet another code contruction for fading channels is given in [34], while [45] presents an MTCM code construction for AWGN channels.

Other Trellis Codes for Fading Channels

One problem with TCM, as illustrated in the previous section, is that different codes perform well on AWGN and fading channels. Therefore it is a difficult trade-off to find a code that performs reasonably well on all channels. In [36] *bit-interleaved coded modulation* is suggested and studied as an alternative to TCM. This scheme is studied in more detail in [48]. The main difference is that the symbol interleaving done in TCM is replaced by bit-interleaving between the trellis encoder and the mapper. This is illustrated in Fig. 8.5. The main disadvantage with this scheme is that the metric computation in the receiver becomes more complicated. The advantage is that the performance is improved for a given complexity compared to ordinary TCM.

Another technique is *multilevel trellis coded modulation* [47,49]. The inputs to the symbol mapper are in this case obtained from $c = \log_2 M_c$ different binary convolutional codes, as illustrated in Fig. 8.6. The rate of the scheme

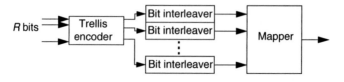

Figure 8.5 The principle of trellis coded bit-interleaved modulation.

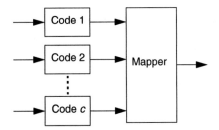

Figure 8.6 The principle of multilevel trellis coded modulation.

becomes equal to the sum of the rates of the individual convolutional codes. Multilevel codes can be used with both bit-interleaving and symbol-interleaving. Again, the main disadvantage compared to TCM is that the decoder becomes more complex. The number of states in the optimum decoder with symbol-interleaving is equal to the product of the number of states in the individual convolutional code trellises, which often becomes too large. Therefore good suboptimum decoders are needed. Multilevel codes are normally iteratively decoded, one code after the other, with information fed back between the iterations.

8.6. DSSS and CDMA

In SS systems, the channel bandwidth is expanded to enable operation at lower signal-to-noise ratios E_s/N_0 [50–53]. From the capacity formula given in Eq. (3.6-10), it is clear that E_s/N_0 decreases with increasing bandwidth at a fixed capacity. Although the capacity formula does not say anything about how to design such a system, it is clear that expanding the channel bandwidth makes it possible to transmit a given data rate at a lower E_s/N_0. The larger bandwidth also facilitate a large diversity order since the channel becomes frequency selective, and this diversity order is quite easily obtained in SS system. This makes SS systems very robust in applications where interference and frequency selective fading are present. The main applications are in military antijam communications and multiuser mobile radio communications.

8.6.1. DSSS

Conventional channel coding leads to bandwidth expansion, and can be used in SS systems. In DSSS, bandwidth expansion is typically obtained by low rate repetition (block) coding, followed by pseudo-random scrambling to make the transmitted spectrum white. This is not a very common way of describing DSSS, but it is quite practical especially in a book on coded modulation systems. We will here limit ourselves to SS systems using BPSK modulation, in order not to complicate the description.[7] Details on other kinds of DSSS systems can be found in [50–53].

A repetition code is a block code where the transmitted symbol is repeated N times before transmission. The coded symbols $x[i]$, which takes values from $\{\pm 1\}$ after mapping to antipodal symbols, are called *chips* in DSSS. The code rate of the repetition code is $R = 1/N$. The chip rate $1/T_c$, where T_c is the chip time, is the rate at which these chips are generated. The nth codeword of length N will here be denoted

$$\mathbf{x}_{nN} = \{x[nN], x[nN+1], \ldots, x[(n+1)N-1]\}, \tag{8.6-1}$$

where $n = 0, 1, \ldots$, and the complete sequence of codewords is given as $\mathbf{x} = \{\mathbf{x}_0, \mathbf{x}_N, \mathbf{x}_{2N}, \ldots\}$.

Before modulation the codeword may be interleaved. This is normally not done in DSSS, but we include it for completeness. After interleaving, the codeword is multiplied by a binary antipodal scrambling sequence $\mathbf{C} = \{C[0], C[1], \ldots\}$, such that $\mathbf{X} \odot \mathbf{C} = \{X[0]C[0], X[1]C[1], \ldots\}$ is transmitted.[8] Here $\mathbf{X} = \pi(\mathbf{x})$ denotes the interleaved codeword sequence, where π is the notation for interleaving.[9] The scrambling sequence is often referred to as the spreading code in DSSS. Most families of scrambling sequences are periodic. When the period is equal to the spreading factor they are often referred to as short codes, while a much longer period is referred to as a long code. A block diagram of the transmitter is shown in Fig. 8.7. An outer encoder is also included in the block diagram for completeness.

A block diagram of the DSSS receiver is displayed in Fig. 8.8. Here, the signal is passed through a matched filter[10] after downconversion. The matched filter output $V(t)$ is then sampled at chip rate to obtain $V[i] = V(iT_c + T_0)$, where $i = 0, 1, \ldots$, and T_0 denotes the delay between a transmitted symbol and the corresponding sample at the matched filter output. The sequence of deinterleaved sampled matched filter outputs $\pi^{-1}(\mathbf{V}) = \mathbf{v}$, where π^{-1} denotes deinterleaving, is referred to as $\mathbf{v} = \{\mathbf{v}_0, \mathbf{v}_N, \mathbf{v}_{2N}, \ldots\} = \{v[0], v[1], \ldots\}$ in the same way as

---

[7] Other DSSS systems can also with some simple modifications be described in a channel coding context.

[8] In the more general case, the repetition code can be defined in any finite field, and the scrambling code may be real-valued or complex-valued with components from any finite alphabet. The coded symbols are mapped to real-valued or complex-valued symbols before scrambling. There also exist several ways to map coded or scrambled symbols to inphase and quadrature phase components of the modulation. They can all be described in the context of repetition coding when DSSS is used.

[9] We use uppercase letters for the variables between the interleaver/deinterleaver pair closest to the channel and lowercase for the others. The only exception is the notation for the channel gain, where we use Ψ between the interleaver and deinterleaver pair and χ outside. The reason is that χ is used in Chapter 7 and uppercase χ cannot be distinguished from X.

[10] This is commonly referred to as chip-matched filtering, since the corresponding waveform carries information about the chips.

Trellis Coding on Fading Channels

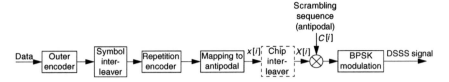

Figure 8.7 A block diagram of a DSSS transmitter. One of the interleaving blocks are dashed since it is not that often used in DSSS.

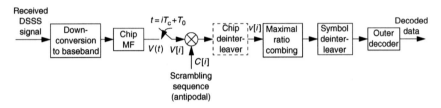

Figure 8.8 A block diagram of a DSSS receiver. Without outer encoding, the two right-most blocks becomes a threshold detector. One of the interleaving blocks are dashed since it is not that often used in DSSS. For a frequency selective channel, the maximal ratio combing block also includes RAKE filtering.

the codeword sequence. The sampled matched filter output is descrambled before decoding. The optimum decoder is equivalent to maximal ratio combining of the soft information for the repeated bits, which means that the decisions on an AWGN channel[11] are given by

$$\mathrm{sgn}\left(\mathbf{v}_{nN} \mathbf{c}_{nN}^{T}\right) = \mathrm{sgn}\left(\sum_{i=0}^{N-1} v\left[i + nN\right] c\left[i + nN\right]\right); \quad n = 0, 1, \ldots, \quad (8.6\text{-}2)$$

where $(\ )^T$ denotes vector transpose, and $\mathbf{c} = \pi^{-1}(\mathbf{C})$ is the deinterleaved scrambling sequence. DSSS is therefore not different from any other coded modulation system, except that the code rate typically is much lower.

Since $Rd_{\min} = 1$ for a repetition code, it is clear from Section 3.2.3 that no coding gain is obtained, but E_s/N_0 is reduced by a factor equal to the inverse code rate as compared to E_b/N_0, that is, $E_b/N_0 = R^{-1} E_s/N_0$. The factor $R^{-1} = N$ is referred to as the *spreading factor* or the *spreading gain*. In practice and as shown in Fig. 8.7, most DSSS systems use an outer channel code with the purpose of obtaining also a coding gain. Such a system is an example of a concatenated coding scheme, where the inner code is the repetition code and the outer code may be any block or convolutional code. The total spreading gain is then equal to $\left(R^o R^i\right)^{-1}$, where R^o and R^i denote the code rates of the outer and inner codes.

---

[11] Maximal ratio combining is equal to equal gain combining on an AWGN channel.

The maximum coding gain is equal to $R^o d^o_{min}$ where d^o_{min} is the minimum distance (or free distance) of the outer code. This is obtained when soft information in the form of $\mathbf{v}_{nN} \mathbf{c}^T_{nN}$ for $n = 0, 1, \ldots$ is fed from the inner decoder to the outer decoder.

On a flat fading channel the decision variable for the repetition code becomes

$$\text{sgn}\left(\mathbf{v}_{nN} \left(\boldsymbol{\Psi}_{nN} \odot \mathbf{c}_{nN}\right)^\dagger\right) = \text{sgn}\left(\sum_{i=0}^{N-1} v[i+nN] \Psi^*[i+nN] c[i+nN]\right),$$

$$n = 0, 1, \ldots, \tag{8.6-3}$$

where $(\ )^\dagger$ denotes hermitian transpose,

$$\boldsymbol{\Psi} = \{\boldsymbol{\Psi}_0, \boldsymbol{\Psi}_N, \ldots\} = \{\Psi[0], \Psi[1], \ldots\} \tag{8.6-4}$$

is the deinterleaved sequence of sampled complex channel gains $\chi[i] = \chi(iT_c + T_0)$, and $(\ )^*$ denotes complex conjugate. In practice the complex channel gains have to be estimated and the estimates are used in the decision variable. From Section 7.7.2 it is known that a diversity gain may be obtained from channel coding but this requires that the complex channel gains corresponding to different chips in a codeword are not completely dependent. This is the case when chip interleaving[12] is used, and the maximum diversity order becomes $R^{-1} = N$. With concatenated coding, the soft decision fed from the inner decoder to the outer decoder becomes $\mathbf{v}_{nN} (\boldsymbol{\Psi}_{nN} \odot \mathbf{c}_{nN})^\dagger$ for $n = 0, 1, \ldots$. Now the maximum diversity order becomes $d^o_{min} N$ with chip interleaving, while it is limited to d^o_{min} with symbol interleaving. In addition to the diversity gain, a spreading gain is obtained as on an AWGN channel.

Traditionally in DSSS, only symbol interleaving but not chip interleaving, is employed. It should be noted that some properties of DSSS used for aquisition and tracking are destroyed by chip interleaving. In systems with a special synchronization channel like WCDMA [30,31], there is no such reason not to use chip interleaving. One drawback however is that the interleaver memory becomes much larger with chip interleaving and this may be hard to implement.

Since low rate channel coding (and DSSS) leads to a wideband signal, the channel is typically frequency selective. In this case, an optimum receiver also needs to employ a filter matched to the channel. This filtering is typically done after the chip-matched filtering and is referred to as a *RAKE receiver* (or *RAKE filter*). As described in Chapter 7, a frequency selective channel is often modelled as a tapped-delay line and then the channel-matched filter combines the contributions from the different taps (multipath components) in a maximal ratio combining sense. Given the channel model in Eq. (7.4-3) with L channel taps,

---

[12] Chip interleaving is used here to denote interleaving between inner coding (repetition coding in this section) and modulation, as seen in Figs 8.7 and 8.8. The interleaving between outer and inner coding will be referred to as symbol interleaving.

Trellis Coding on Fading Channels

the matched filter output has to be sampled at the chip rate, with relative delays corresponding to the relative delays of the multipath components. The samples obtained are denoted

$$V^{(j)}[i] = V\left(iT_c + (j-1)T' + T_0\right), \quad i = 0, 1, \ldots; \, j = 1, \ldots L \quad (8.6\text{-}5)$$

and the corresponding sequences are given by $\mathbf{V}^{(j)}$ for $j = 1, \ldots, L$. The deinterleaved sequences are referred to $\mathbf{v}^{(j)}$ for $j = 1, \ldots, L$. T' is the time resolution of the system and in most cases it is equal to T_c. Note that with $T' = T_c$, $\mathbf{V}_i^{(j)} = \mathbf{V}_{i+j-1}^{(1)}$, that is, one sequence is a shift of another sequence, the complexity is significantly reduced since no extra sampling is required. The soft decision variable now becomes

$$\sum_{j=1}^{L} \mathbf{v}_{nN}^{(j)} \left(\mathbf{\Psi}_{nN}^{(j)} \odot \mathbf{c}_{nN}\right)^{\dagger}, \quad (8.6\text{-}6)$$

where $\mathbf{\Psi}_{nN}^{(j)} = \{\Psi^{(j)}[nN], \Psi^{(j)}[nN+1], \ldots, \Psi^{(j)}[(n+1)N]\}$ denotes the deinterleaved complex channel gains of path j.

In this case the maximum diversity order increases to $d_{\min}^o NL'$, where $L' \leq L$ is the number of nonzero values in the power delay profile (see Eq. (7.4-4)). Zero taps in the power delay profile will not contribute to the decision variable above since the complex channel gain is zero. With estimated complex channel gains, these taps should not be estimated but removed from the sum. To fully utilize the channel diversity order L' and avoid intersymbol interference (ISI), long spreading codes are required (see Section 8.6.3). It is clear that a large diversity gain can relatively easily be obtained.

8.6.2. Direct-Sequence CDMA

Direct-Sequence CDMA (DS-CDMA) is a multiple access technique where individual users employ DSSS as described in the previous section. In this case each user employs the same repetition code but has its own unique scrambling code such that the composite transmitted codeword when all users are synchronized and transmitted from the same transmitter becomes

$$\sum_{k=1}^{K} A_k \mathbf{X}^{(k)} \odot \mathbf{C}^{(k)}, \quad (8.6\text{-}7)$$

where A_k, $\mathbf{X}^{(k)}$, and $\mathbf{C}^{(k)}$ are the transmitted amplitude, the interleaved codeword sequence, and the scrambling sequence of user k, respectively, and K is the total number of users. This situation is typical when the transmission is from a common transmitter (base station) to a single user receiver (mobile terminal), and it is called the *downlink*. The composite codeword is transmitted on a common channel to the receiver. After matched filtering, descrambling with $\mathbf{C}^{(l)}$, possibly deinterleaving, RAKE combining, and maximal ratio combining, a decision variable is obtained for user l. This decision variable is similar to Eq. (8.6-6), except that the scrambling sequence corresponds to user l. It consists of a term proportional to the desired symbol $x^{(l)}[nN]$, which is equal to

$$x^{(l)}[nN] A_l \sum_{i=1}^{L} \mathbf{\Psi}_{nN}^{(i)} \left(\mathbf{\Psi}_{nN}^{(i)}\right)^{\dagger}, \qquad (8.6\text{-}8)$$

an ISI term

$$A_l \sum_{i=1}^{L} \sum_{\substack{j=1 \\ j \neq i}}^{L} \left(\mathbf{x}^{(l)} \odot \mathbf{c}^{(l)} \odot \mathbf{\Psi}^{(i)}\right)_{nN+j-i} \left(\mathbf{c}^{(l)} \odot \mathbf{\Psi}^{(j)}\right)_{nN}^{\dagger}, \qquad (8.6\text{-}9)$$

a multiple access interference (MAI) term

$$\sum_{i=1}^{L} \sum_{\substack{j=1 \\ j \neq i}}^{L} \sum_{\substack{k=1 \\ k \neq l}}^{K} A_k \left(\mathbf{x}^{(k)} \odot \mathbf{c}^{(k)} \odot \mathbf{\Psi}^{(i)}\right)_{nN+j-i} \left(\mathbf{c}^{(l)} \odot \mathbf{\Psi}^{(j)}\right)_{nN}^{\dagger}, \qquad (8.6\text{-}10)$$

and a noise term. Here the subscripts are used to denote a subsequence of length N as defined in Eq. (8.6-1), $\mathbf{x}^{(k)} = \pi^{-1}(\mathbf{X}^{(k)})$ the codeword sequence of user k, and $\mathbf{c}^{(k)} = \pi^{-1}(\mathbf{C}^{(k)})$ the deinterleaved scrambling sequence of user k. The ISI and MAI terms can to a certain extent be controlled by properly choosing the scrambling sequences (see Section 8.6.3).

In the opposite direction, with transmission from a single user transmitter to a common receiver, denoted the *uplink*, users typically lie at different distances from the receiver and therefore synchronous transmission becomes difficult, especially when the transmitters are mobile. Moreover the signals from the different users to the common receiver are transmitted on different channels, since the transmitters are at different positions. This will not affect the ISI of the decision variable, since ISI is generated by the desired user and therefore only one channel is involved, but the MAI term now becomes

$$\sum_{i=1}^{L} \sum_{\substack{j=1 \\ j \neq i}}^{L} \sum_{\substack{k=1 \\ k \neq l}}^{K} A_k \left(\mathbf{x}^{(k)} \odot \mathbf{c}^{(k)} \odot \mathbf{\Psi}^{(i,k)}\right)_{nN+j-i+m} \left(\mathbf{c}^{(l)} \odot \mathbf{\Psi}^{(j,l)}\right)_{nN}^{\dagger}, \qquad (8.6\text{-}11)$$

where a second superscript has been added to the complex channel gains $\mathbf{\Psi}$ to identify the different user channels and m takes care of the relative time delays between users.

8.6.3. Code Design

There are two different codes involved in DSSS and DS-CDMA systems. One of them is the error correction code, with the main purpose of reducing the bit error probability. The other is the scrambling code, with the purpose of making the transmitted signal noise-like and reducing the interference. From the previous section it is clear that the interference terms depend on both codes, so a joint design should be advantageous. However, a joint design is difficult, and in practice the codes are designed more or less independently. The design is somewhat simpler in a single user system since there is no MAI to worry about. In this section, the code design will be discussed and some examples of codes are given.

Scrambling Codes

The soft decision variable for DSSS and DS-CDMA is given in various forms in the previous section. The ISI term only exists on frequency selective channels ($L > 1$) and the MAI term only exists in a multiuser system ($K > 1$). Each term contributing to ISI for user l in Eq. (8.6-9) is a random variable and depends on the scrambling code of user l, the complex channel gains of the different multipath components of the lth user's channel, and the transmitted codeword sequence of user l. Since a joint design of error correcting code and scrambling code is difficult, we will here basically consider the error correcting code as given and look at the properties of the scrambling code. It is desirable to minimize every outcome of the ISI, but it is clear that this would require a scrambling code that depends on the codewords, the chip interleaving (if present), and the channel gains, and this is not feasible.

Instead the average power of the ISI can be minimized when parameters of the channel gain distribution are known. These depend on whether chip interleaving or symbol interleaving is performed. With symbol interleaving on relatively slow fading channels, the channel gains for each codeword are essentially constant, and their influence on each term in Eq. (8.6-9) is a complex multiplication. This reduces the complexity in finding a scrambling code that minimizes the ISI power. With perfect chip interleaving, the channel gains for each codeword are uncorrelated random variables, and the power of the ISI depends on the $2NL$-dimensional pdf of all the gains of the channels. With practical chip interleaving, the channel gains are correlated. Finding scrambling codes that minimize the ISI power in general does not seem very feasible and such codes have the disadvantage that they are channel dependent and not very flexible.

In the special case of no chip interleaving and almost constant channels within one codeword, the maximum length or pseudonoise (PN) sequences are quite good. These sequences are generated by a shift register with feedback [6, Section 13.2.5], [52, Chapter 5], and have a normalized periodic autocorrelation function

$$\phi(n) = \begin{cases} 1, & n = 0, 2^m - 1, 2\,(2^m - 1), \ldots, \\ \frac{1}{2^m-1}, & \text{otherwise,} \end{cases} \qquad (8.6\text{-}12)$$

where m is the length of the shift register ($2^m - 1$ is the period of the sequence). This means that each ISI term in the sum in Eq. (8.6-9) is attenuated by the sequence length but this occurs only when a constant codeword sequence is transmitted. With a random codeword sequence the autocorrelation properties are changed, resulting in an increased ISI power.

In a multiuser DS-CDMA system, the scrambling codes also affect the MAI, as seen in Eqs (8.6-10) and (8.6-11). On a flat slowly varying channel (constant over one codeword) with symbol interleaving and synchronized users, the $K - 1$ terms in Eq. (8.6-10) become zero when

$$\mathbf{c}_{nN}^{(k)} \mathbf{c}_{nN}^{(l)} = 0, \qquad k = 1, \ldots K, \; k \neq l. \qquad (8.6\text{-}13)$$

One such family is the Walsh–Hadamard codes [6, Section 8.1.2]. They can be generated as

$$\mathbf{H}_2 = \begin{bmatrix} +1 & +1 \\ +1 & -1 \end{bmatrix},$$

$$\mathbf{H}_{2n} = \begin{bmatrix} \mathbf{H}_{2(n-1)} & \mathbf{H}_{2(n-1)} \\ \mathbf{H}_{2(n-1)} & -\mathbf{H}_{2(n-1)} \end{bmatrix}, \qquad n = 2, 3, \ldots \qquad (8.6\text{-}14)$$

The spreading gain is limited to a power of two and the maximum number of users is equal to the spreading gain. Due to their recursive structure, the Walsh–Hadamard code can also be used in variable spreading gain systems as is done in WCDMA [30,31].

With chip interleaving or faster varying channels (not constant over one codeword), Walsh–Hadamard codes do not any longer make the MAI term equal to zero for all realizations of codewords and channel gains. There are in fact no known scrambling codes that make both the ISI and MAI powers equal to zero on frequency selective channels. It does not matter whether the users are synchronous or asynchronous, or whether chip or symbol interleaving is used. There are also no known codes that make the MAI power equal to zero except for the case given above. What remains is to jointly minimize the powers in the ISI and MAI, but this is even more difficult than minimizing only the ISI power, since it involves all K scrambling codes.

The conclusion from the discussion above is that it is only on flat and slowly varying channels (with coherence time much larger than symbol time), with symbol interleaving and synchronous users, that scrambling codes with zero ISI and MAI powers exist. The codes that make the ISI power small, however, make

Trellis Coding on Fading Channels

the MAI power quite large and the opposite is also true. Quite strong restrictions on the codeword sequence are needed to reduce ISI power. In practice, a scrambling code design is difficult and instead one has to design the system to cope with the nonzero ISI and MAI powers.

The MAI power can be reduced to some extent by *power control* of the different transmitters. Since at least some of the terms in Eq. (8.6-11) in general are nonzero, and the transmitters are at different distance from the receiver, the system suffers from the so-called *near–far effect*. If all transmitters use the same power, a signal from a nearby transmitter will be received at a much higher power than a signal from a distant user. This strong signal will cause a large interference power to the signal of the distant user when the correlation between those two signals are nonzero. This can be partly solved by controlling the powers of the transmitters such that all signals are received at the same power level. This is done in all CDMA systems used in practice and requires a feedback link for control information.

The spectrum of the interference (MAI plus ISI) should also be made as white as possible to avoid error bursts for the outer decoder. From this respect, long scrambling codes with a period much longer than N have an advantage over short scrambling codes with period N, especially in slowly varying flat channels with asynchronous users and symbol interleaving. With short codes, the scrambling code is identical for all codewords. The correlation time[13] of the interference process may thus be significant when the channel gains vary slowly with time and the memory of the symbol interleaver is small. With frequency selective fading and/or chip interleaving, the correlation time becomes smaller even with short codes and error bursts will become less likely. Thus, overall performance is improved. Long codes further reduce the correlation time of the interference process and are therefore preferable. With more advanced detectors that try to estimate the contributions in the MAI (see Section 8.6.4), short codes are in general preferable, since they often reduce the computation complexity of such detectors [54]. Since these in fact remove most of the MAI, errors become uncorrelated anyway, since random noise is the major cause of errors. Power control is also not that important when more advanced detectors are used [54].

So far we have concentrated on the properties of the scrambling codes from the viewpoint of reducing ISI and MAI powers. The scrambling codes will also significantly influence the transmitted spectrum. This is not that important in civilian applications but is very important in military applications. The scambling code should typically be noise-like to generate a white spectrum and the number of zeroes and ones should be almost the same. The PN sequences and most of the Walsh–Hadamard codes fulfil these requirements. Many other scrambling codes

---

[13] The correlation time is equal to the time lag at which the normalized covariance of the interference process is reduced to a certain value, i.e. 0.5, as defined in Chapter 7.

have been proposed in the literature, with Gold and Kasami codes being two of the most commonly used. Details on these and other codes may be found in [6,50–53].

Optimum Free Distance Coding in Spread Spectrum

From the previous section it is clear that in general it is quite difficult to design scrambling codes for DSSS and DS-CDMA such that the ISI and MAI powers become small. MAI and ISI will in effect add to the receiver noise so that the detector will have to operate at a lower signal-to-noise/interference ratio. It is also clear that chip interleaving and long codes make the total interference quite uncorrelated. This leads to the conclusion that powerful outer and inner codes can be used in these systems to reduce the effects of the errors caused by noise and interference. It is important to note that the discussion on scrambling codes in the previous section assumed DSSS (as shown in Figs 8.7 and 8.8) and DS-CDMA, which means an inner repetition code. Once this code is changed, both the PN sequences and the Walsh–Hadamard codes will lose their advantage in the special cases discussed.

Instead of using a repetition code as in DSSS and DS-CDMA, it is possible to use any other channel code with improved performance [26–28,53,55–64]. Viterbi has suggested orthogonal and super-orthogonal convolutional codes as inner codes, and in this way the inner code gives a coding gain on a Gaussian channel, and a larger diversity gain on a fading channel. In [58,59] the performance of these codes has been studied. In [60] a comparison between orthogonal codes and trellis codes is given. A drawback with orthogonal and super-orthogonal codes is that they only exist for code rates $1/2^i$ where i is an integer, so that the code rate and the memory of the code cannot be changed independently. In [62,63] the quasi-optimum nested convolutional codes from [26,27] are analyzed for CDMA. A comparison between trellis codes and convolutional codes in SS and CDMA is done in [64]. Frenger *et al.* have found maximum free distance low rate codes which are suitable in SS and CDMA. In [28] such a system is referred to as Code-Spread CDMA (CS-CDMA). Code-Spread SS (CSSS) will denote the special case with only one user. A block diagram of the transmitter of CSSS is shown in Fig. 8.9 and the corresponding receiver for one user is displayed in Fig. 8.10.

CS-CDMA consists of an outer code, a symbol interleaver, a general inner code (a repetition code can be considered a special case), a chip interleaver, and scrambling followed by modulation. The repetition code case will be called DS-CDMA. From a performance point of view in a single user system, there is no need to use concatenated encoding, but it may be advantageous from a complexity point of view. In a multiuser system the situation is different, since a proper choice of inner code can reduce the interference in the system, while the outer code can combat noise and remaining interference. As we will see, there is no single design strategy that is always the best.

Trellis Coding on Fading Channels

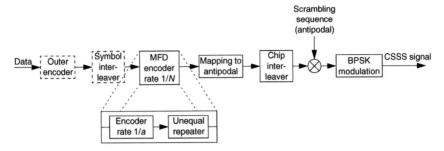

Figure 8.9 A block diagram of a CSSS transmitter. The dashed blocks may be used to reduce complexity but are not necessary. The split of the MFD encoder into two blocks for MFD low rate nested codes is also shown. a is typically much smaller than N when N is large.

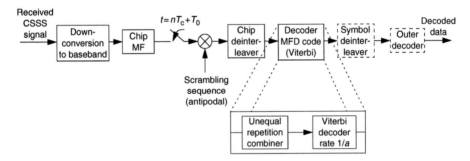

Figure 8.10 A block diagram of a CSSS receiver. The dashed blocks are optional. For a frequency selective channel, the MFD decoder also includes RAKE filtering. A split of the MFD decoder into two blocks for MFD low rate nested codes is also shown.

One example of an inner code is a low rate nested convolutional code based on a feed-forward encoder as described in Section 8.3.3 and this kind will be considered here. They have been found for rates $1/4, 1/5, \ldots, 1/512$ and for memory 10 or less [11,28], and the found codes are MFD codes. The free distances of these codes are shown in Fig. 8.11. Compared to the free distance of repetition codes, they provide a significant improvement. The MFD codes are also better than the super-orthogonal codes since they provide higher free distance at equal code memory[14], and they exist for all rates $1/c$, c an integer less equal 512, and memory less equal 10. It is important to mention that these codes have an unequal repetition structure which simplifies both encoding and decoding as illustrated in Figs 8.9 and 8.10 (see Section 8.3.3). It should be noted that a concatenated coding scheme consisting of an outer MFD convolutional code at rate 1/4 with an inner pure repetition code has a free distance which is only slightly smaller than that of the low rate nested convolutional code itself, when the memory and the total

---

[14] At code memory two, super-orthogonal and MFD codes have the same free distance, but codes with this low memory are not used in practice.

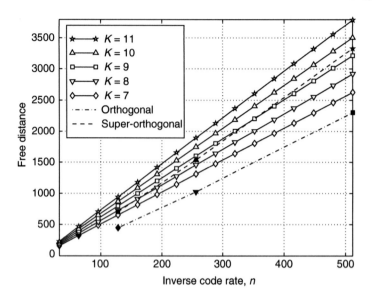

Figure 8.11 Free distance of nested low rate convolutional codes, orthogonal convolutional codes, and super-orthogonal convolutional codes. (From [28], copyright IEEE 2000; used with permission.)

rate are the same [65,66]. For complexity reasons there may be some advantage in using the concatenated scheme.

In Fig. 8.12, the bit error probability vs E_b/N_0 of a CS-CDMA system is shown. The inner code here is a rate 1/32 memory 6 nested convolutional code and the chip interleaver is perfect. No outer encoding and symbol interleaving are done. The channel is a flat Rayleigh fading channel. The system has perfect average power control such that all the users are received at the same average power. The scrambling codes are chosen at random to model an uplink. Both simulated and analytical bit error probability results are shown. The analytical results are calculated from a union bound as given in Eq. (8.2-1) with P_d as given in Eq. (8.2-2). A Gaussian approximation has been used for the MAI, which means that q in Eq. (8.2-2) is given by Eq. (8.2-3) with

$$\Gamma_c = \left((\Gamma_b R)^{-1} + \frac{2(KL-1)}{3\beta}\right)^{-1}, \qquad (8.6\text{-}15)$$

where $\beta = 1$ with chip interleaving and $\beta = 1/R^i$ without chip interleaving. For the results in Fig. 8.12, $\beta = 1$ and $L = 1$. Here we see that CSSS obtains low bit error probabilities at very low E_b/N_0 and CS-CDMA is capable of offering low bit error probability to many users in the system.

Trellis Coding on Fading Channels

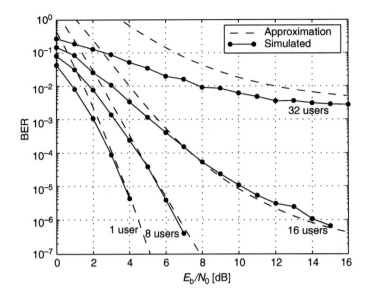

Figure 8.12 Simulated and analytical bit error probability for CS-CDMA with inner code rate 1/32 and memory 6, perfect chip interleaving, and no outer coding. (From [28], copyright IEEE 2000; used with permission.)

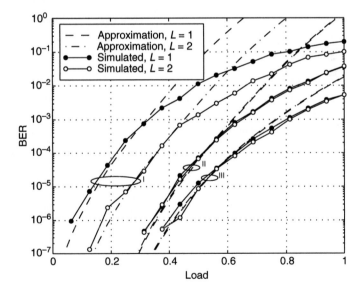

Figure 8.13 Simulated and analytical bit error probability for three different CS-CDMA systems vs the system load when E_b/N_0 is 10 dB. (From [28], copyright IEEE 2000; used with permission.)

In Fig. 8.13 the performances of three different systems are compared when E_b/N_0 is 10 dB. System I and II are both concatenated systems with an outer convolutional ODS code with rate 1/2 and memory 6 and an inner repetition block code with rate 1/16. System I uses perfect symbol interleaving but no chip interleaving and system II uses both perfect symbol and perfect chip interleaving (compare Figs 8.7 and 8.8). System III is the same system as considered in Fig. 8.12 (compare Figs 8.9 and 8.10). The scrambling codes are random for all systems. The channels are a flat Rayleigh fading channel ($L = 1$) and a two-path ($L = 2$) frequency selective channel with a constant power delay profile. The two path gains are uncorrelated. All systems have perfect average power control. Analytical results for system I and II are shown, and here $\beta = 16$ in Eq. (8.6-15). The load on the horizontal axis is defined as $K R^i R^o$, which means that load one corresponds to a number of users equal to the total spreading factor. Here the additional diversity gains of system III compared to system II, and system II compared to system I are clearly seen. For system I, diversity gains are only obtained from the outer coding and the channel when $L = 2$, and the result is that the normalized load of the system is limited to below 0.3 for the considered channels at bit error probability 10^{-5}. With system II, which is system I with chip interleaving, a much higher diversity gain is obtained, and the extra diversity on the channel does not improve performance much. Now the load is around 0.4 at bit error probability 10^{-5}. Replacing the inner repetition code by a MFD rate 1/32 code and removing the outer encoder gives an even better performance, with a load of 0.5 at bit error probability 10^{-5}. It should be noted that the Viterbi decoders (outer decoder for systems I and II) have almost the same complexity for all three systems. The number of states are the same but the incremental metrics are calculated a bit differently.

The main problem with the chip interleaver is its length. In [28], the performance with interleavers of limited size has also been measured. It is clear that some degradation occurs with a limited interleaver size, but there is still a significant gain unless the interleaver becomes too small or the channel varies too slowly.

Other Error Correcting Codes in CS-CDMA

In the previous section mainly convolutional error correcting codes of low rate have been considered. It is of course possible to use other low rate codes in a CS-CDMA system too. Examples of such codes are low rate turbo codes [67,68], ordinary turbo codes [69], super-orthogonal turbo codes [70], low rate block codes [71], low rate trellis/Reed-Muller codes [72], and low density parity check codes (Gallager codes) [73,74] to mention only some. Low rate convolutional codes were also recently considered in [75] where a different design strategy was used. CS-CDMA and similar systems can also be combined with space-time coding and transmit diversity [76–78]. There is much other related work.

General Conclusion on Code Design

The results presented for CSSS and CS-CDMA are for the asynchronous uplink case. Here it is always better to use an MFD inner code as compared to a repetition inner code, since no scrambling codes exist that significantly reduce MAI power. As discussed above, however, it is somewhat easier to limit the MAI power in a synchronous downlink when the channel is relatively flat and slowly varying, by using an inner repetition code (DS-CDMA) with orthogonal scrambling codes and no chip interleaver. The disadvantage with such a design is that the diversity gains (if available) are significantly reduced in trade for reduced MAI power. The general conclusion for a CDMA system, however, seems to be that a concatenated coding scheme with most of the bandwidth expansion in an inner repetition code and no chip interleaving is preferable in most cases [28,79,80]. For more frequency selective channels MFD coding again becomes the better alternative. In the single user scenario, MFD coding is always better than a concatenated code with an inner repetition code.

8.6.4. Multiuser Detection in CS-CDMA

The single user detector and the RAKE receiver considered in the previous section are optimum in AWGN but not when MAI exists. The performance of CDMA can be improved by using more advanced multiuser detection [54,81,82]. The optimum decoder can be implemented by the Viterbi algorithm and makes a joint decision on the information transmitted from all users [83,84]. Its main disadvantage is that the number of states in the Viterbi algorithm increases exponentially with the number of users and the number of multipath components. Therefore many suboptimum multiuser detectors with lower complexity have been proposed in the literature [54,81,82]. In this book, we will not go into details on multiuser detection in general, but we want to show the improvements that are possible for CS-CDMA [61].

One particular family of multiuser detectors is interference cancelation (IC). The basic idea is to estimate the ISI and MAI contributions given in Eqs (8.6-9)–(8.6-11). Each of the contributing terms may be estimated from tentative decisions on the data symbols from each user and the complex channel gains for all involved channels, and knowledge of the scrambling codes of all users. The tentative decision can, for example, come from single user decoders of the inner codes. Once these MAI estimates are available, they can be subtracted from the received signal before the single user decoder is applied a second time, and so on for several iterations. In Figs 8.8 and 8.10 this means that an interference estimate is subtracted immediately after the chip deinterleaver blocks. A successive IC (SIC) typically decodes the strongest received user and then subtracts the interference from this user before proceeding to the second strongest user. Then it proceeds with all users, one at a time, in a similar way [54,85]. A disadvantage with this serial processing is a long decoding delay. The other approach is to estimate the MAI from all users and all multipaths at once and subtract them in parallel before the next decoding step. This is referred to as parallel IC (PIC) [54].

CS-CDMA with IC has been studied in [61]. With PIC, the decisions of the inner decoder from the previous stage for all users except the desired are re-encoded and used together with the most current channel estimates to obtain an estimate of the MAI in the current stage. The MAI estimate is subtracted from the received signal before making a new decision in the inner decoder in the current stage. In the first stage, no MAI estimates are available. With SIC, the MAI in a given stage is obtained from re-encoded decisions made by the inner decoder in the previous stage for those users that have not been processed in the current stage, plus re-encoded decisions from the inner decoder in the current stage for the other users. These re-encoded decisions are used together with the most current channel estimates to form the MAI estimate, which is subtracted before a new decision from the inner decoder of the current user is made.

In Figs 8.14 and 8.15, the bit error probability on a flat Rayleigh fading channel of a CS-CDMA and a coded DS-CDMA system are shown for the case of SIC. The CS-CDMA and coded DS-CDMA systems are the ones referred to as system III and system I, respectively, in Section 8.6.3. The number of users is 16 for the CS-CDMA system, while it is only 6 for the coded DS-CDMA system. The channel estimates are assumed perfect and perfect average power control is

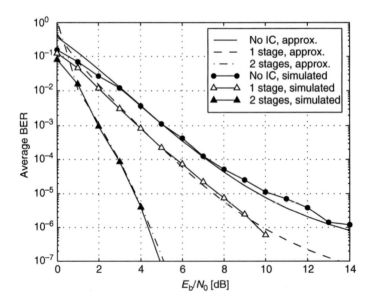

Figure 8.14 Simulated and analytical bit error probability for CS-CDMA vs E_b/N_0 with 16 users in the system. (From [61], copyright IEEE 1999; used with permission.)

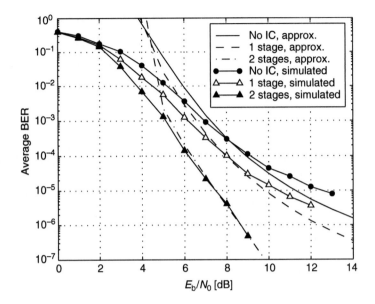

Figure 8.15 Simulated and analytical bit error probability for coded DS-CDMA vs E_b/N_0 with 6 users in the system. (From [61], copyrights IEEE 1999; used with permission.)

assumed. From these results it is clear that the CS-CDMA system can operate at much lower E_b/N_0 with a larger number of users. It should be noted here that the decoding of the outer code in the coded DS-CDMA system is done after SIC and this system has no chip interleaving. The performance would improve by including this in the SIC as proposed in [86,87] and by using chip interleaving, but coded DS-CDMA would still be inferior to CS-CDMA.

The analytical bit error probability in these figures is obtained in a similar way as in the previous section, by using the union bound and assuming that the interference is normally distributed. However, the number of users is modified to take into account both correct and incorrect decisions in previous stages, which affects the MAI estimates [61]. Incorrect decisions of other users' data causes the MAI of the desired user to increase instead of decrease as with correct decisions, and this can be modelled by changing the number of users. The equivalent number of users $\kappa_{s,k}$ seen by user k at stage s can be iteratively calculated from

$$\kappa_{1,1} = K,$$
$$\kappa_{s,k} = \left(1 - P^{(s,k-1)}\right)(\kappa_{s,k-1} - 1)) + P^{(s,k-1)}\left(\kappa_{s,k-1} + 1\right),$$
$$s \geq 1, \quad k = 2, \ldots, K,$$
$$\kappa_{s+1,1} = \left(1 - P^{(s,K)}\right)(\kappa_{s,K} - 1) + P^{(s,K)}\left(\kappa_{s,K} + 1\right), \quad s \geq 2,$$

(8.6-16)

where the factors $\kappa_{s,k-1} - 1$ and $\kappa_{s,K} - 1$ must be at least one. It is assumed that users are processed in increasing order of index at each stage of cancellation, such that user 1 is decoded/re-encoded first, then user 2, etc. $P^{(s,k)}$ is the symbol error probability of user k at stage s, and it can be upper bounded as

$$P^{(s,k)} < \left(R^i\right)^{-1} \sum_{d=d_f}^{\infty} dA_d P_d \leq 1, \qquad (8.6\text{-}17)$$

where it is assumed that the rate R^i of the inner encoder is one divided by an integer. The factor d in this union bound comes from the fact that the error event has d symbol errors. The error event probability P_d for user k at stage s is obtained from Eqs (8.2-2) and (8.2-3) with

$$\Gamma_c = \left((\Gamma_b R)^{-1} + \frac{2(\kappa_{s,k} - 1)}{3\beta}\right)^{-1} \qquad (8.6\text{-}18)$$

with $\beta = 1$ for CS-CDMA and $\beta = (R^i)^{-1}$ for DS-CDMA. The average bit error probability of user k after the last step $s = S$ of SIC can then be calculated from Eqs (8.2-1)–(8.2-3) with

$$\Gamma_c = \left((\Gamma_b R)^{-1} + \frac{2(\kappa_{S,k} - 1)}{3\beta}\right)^{-1}. \qquad (8.6\text{-}19)$$

The average bit error probability shown in the figures is obtained by averaging the bit error probability of an individual user over all users. A similar technique may be used for PIC; see [61] for details. The analysis can also be extended to frequency selective channels. Another DS-CDMA iterative analysis technique for IC with hard decisions is given in [88], and this can also be modified for CS-CDMA.

The capacity improvement with SIC in CS-CDMA system is illustrated in Figs 8.16 and 8.17 for system III defined in Section 8.6.3. First of all it is clear that a much higher load for a given bit error probability can be carried with SIC. Moreover, the bit error probability at low load is defined by the concatenated code and the E_b/N_0. A lower bit error probability can be obtained by increasing the memory in the convolutional code. The E_b/N_0 here takes into account not only receiver noise but also interference from users not considered in the SIC, for example, users from neighboring cells in a cellular system. With powerful codes and a few stages of cancellation, the net result is that the single user bit error probability of the system is kept until the number of users approches a level where the cancellation no longer makes reasonable interference estimates. This level depends on the concatenated code, since the estimates are calculated from tentative decisions. For more users, the bit error probability goes quickly to 0.5.

Trellis Coding on Fading Channels

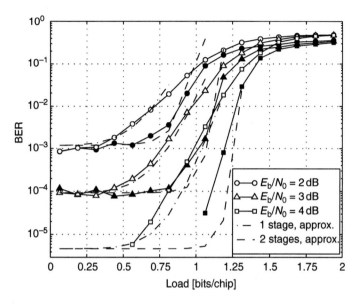

Figure 8.16 Simulated and analytical bit error probability for CS-CDMA vs load with one stage ($S = 1$) (non-filled markers) and two stage ($S = 2$) (filled markers) SIC. (From [61], copyright IEEE 1999; used with permission.)

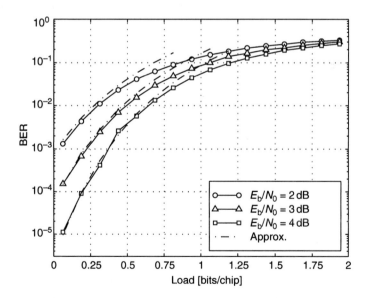

Figure 8.17 Simulated and analytical bit error probability for CS-CDMA without IC vs load. (From [61], copyright IEEE 1999; used with permission.)

8.6.5. Final Remark on SS and CDMA

It is important to point out that an optimum design for spread spectrum and CDMA systems of the kind discussed above is still an open question. What is clear is that in a system without self interference (no ISI and MAI), MFD coding provides a very low error probability. In a system with self interference, the situation is much more complicated. The reason is that error correcting coding, interleaving, scrambling, and sophisticated detection algorithms all influence the bit error probability in such a way that a joint design is necessary. Such a design is complex and is still an open research problem. An individual design of the components may not lead to an overall good design.

8.7. Generalized Hybrid ARQ

On channels with feedback, new designs of error control systems are possible. In this section some of these techniques will be described. Channel conditions will now control the operation of the coding system. In this way, a significant increase of bandwidth efficiency is obtained especially on time varying channels, since the code can be designed for the instantaneous channel instead of the average channel.

Throughout the section, it is assumed that the feedback channel is error free. Moreover, since the intention is not to discuss and analyze complete transmission protocols, but rather code designs and their influence on the performance, all overhead due to protocols is neglected. Thus, the throughputs as given in this section represent an upper bound to the throughput of real protocols which must include overhead. It is also assumed that transmitters and receivers have infinite buffers and that packet numbers are represented by infinite resolution which means that there is no upper limit on the number of retransmissions.

8.7.1. Simple ARQ

The simplest form is ARQ which dates back to the early days of digital computers [38]. In its most basic form a *cyclic redundancy check* (CRC) code, which is a block code designed for pure error detection, is used [4,38,89]. CRC encoders are systematic, and thus they simply append a number of parity bits to the message before transmission. In the decoder, the codeword is accepted and acknowleged (an ACK) if its syndrome is zero, otherwise a retransmission request (an RQ[15]) is sent on a feedback channel to the transmitter. On an RQ, the codeword is retransmitted and this is continued until a codeword is received. The principle is illustrated in Fig. 8.18.

---

[15] This is also referred to as a negative acknowledgment (a NACK).

Trellis Coding on Fading Channels

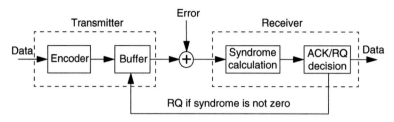

Figure 8.18 The principle of simple ARQ. The error process takes into account modulation, the physical channel and demodulation. A code for pure error detection is used.

One important performance measure is the average accepted codeword error probability, which is defined as the average probability that an accepted codeword is different from the transmitted codeword[16]. This is given by

$$P(E) = \frac{P_e}{1 - P_r} \qquad (8.7\text{-}1)$$

where P_e is the average probability of undetected error (one codeword is changed into another codeword) and P_r is the average probability of a detected error (received word is not a codeword). The second important measure of performance is the throughput.

Definition 8.7-1. The throughput η is defined as the ratio between the average rate of *accepted* received bits divided by the (peak) symbol rate of the transmitter.

For a channel coding scheme without feedback, this definition of throughput is equal to the code rate R in bits/QAM or PSK symbol. With M-ary modulation, the maximum throughput becomes $\log_2 M$.[17] For an RQ protocol, the throughput strongly depends on the average number of times T_r a codeword has to be transmitted before it is accepted in the receiver, which is given by

$$T_r = \frac{1}{1 - P_r}. \qquad (8.7\text{-}2)$$

The throughput depends on how the RQs are handled by the transmitter and receiver [38]. Here we will only consider selective-repeat ARQ (SR-ARQ), which requires infinite buffers in transmitter and receiver but gives the highest throughput. In this scheme, the transmitter sends a continuous stream of codewords and has no idle

---

[16] The averaging is over the fading distribution.

[17] Quite often throughput is defined to have a maximum value of one, when the channel is fully utilized. Here we prefer to have a definition that relates to code rate of channel coding without feedback. Our definition also makes comparison between systems using different modulations feasible. It should however be noted that the symbol rate of the transmitter is the rate used when transmitting (peak rate), and it should not be reduced due to idle time in the transmitter which results for some protocols.

time. When an RQ appears, the transmitter resends the requested codeword, and then resumes transmission of new codewords. Consequently, each RQ results in the retransmission of one codeword only. The throughput now becomes

$$\eta = R\frac{1}{T_r} = R(1 - P_r). \qquad (8.7\text{-}3)$$

Thus the throughput decreases when the channel error probability increases, since this will increase P_r.

To illustrate the performance of ARQ, the following system is considered. Information bits are encoded in a Hamming (31, 26) code which is used for pure error detection. Since Hamming codes have a known weight distribution [38, Eq. (4-12)] both P_r and P_e can be computed (see Chapter 3 and [38, Chapter 10]). The outputs for the Hamming code are BPSK-modulated and transmitted over a channel. Three different channels are compared:

Rayleigh uncorrelated. This is a perfectly interleaved Rayleigh fading channel.

Rayleigh bursty. This has a Rayleigh distributed channel gain within each codeword, and channel gains between codewords are uncorrelated, which means that the channel will give burst errors of length equal to a codeword. This is sometimes referred to as a *block fading channel*.

AWGN. This is an AWGN channel.

The BPSK demodulation is coherent and the phase synchronization is assumed perfect. The channel decoder calculates the syndrome using the last received word and asks for a retransmission whenever the syndrome is nonzero. The feedback channel is assumed error free.

It is straightforward to calculate P_r and P_e for the Rayleigh uncorrelated and AWGN channels by including the BPSK (average) bit error probability for these channels in the expressions for P_r and P_e (which are similar to Eq. (7.6-5)). For the bursty Rayleigh fading channel, the BPSK error probability over an AWGN channel has to be used to evaluate the (conditional) probabilities of undetected error and detected error, respectively, for a given codeword, and these can then be averaged over the exponential SNR distribution to obtain P_r and P_e (compare the word error probability given in Eq. (7.6-6)). The integrals can be evaluated numerically.

In Figs 8.19–8.21 the throughput and accepted codeword error probability are shown in different ways. From Fig. 8.19 it is clearly seen that the throughput is higher on a bursty Rayleigh fading channel as compared to an uncorrelated Rayleigh fading channel. For a given throughput, the gain in average E_s/N_0 (average SNR per coded bit) is about 6 dB. At throughputs below 0.25, the bursty fading channel even has a higher throughput than the AWGN channel. However, the increase in throughput on the bursty channel is at the expense of higher accepted

Trellis Coding on Fading Channels

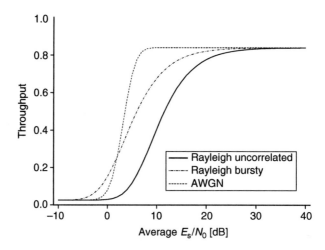

Figure 8.19 An example of throughput vs average E_s/N_0 for ARQ on three different channels.

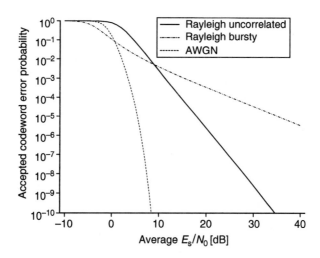

Figure 8.20 An example of accepted codeword error probability vs average E_s/N_0 for the system in Fig. 8.19.

codeword error probability, as is seen in Fig. 8.20. It is clear that there is no diversity gain from the channel coding on the bursty channel since the channel is constant within each codeword. For the uncorrelated Rayleigh fading channel on the contrary, the accepted codeword error probability shows a third order diversity (the minimum Hamming distance is three). In reality an ARQ system is designed for a given accepted codeword error probability and from this point of view it is

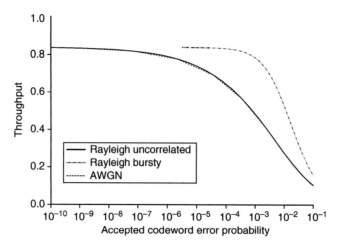

Figure 8.21 An example of throughput vs accepted codeword error probability for the system in Fig. 8.19.

clear from Fig. 8.21 that the bursty channel always gives some advantage but at the expense of having to operate at a higher E_s/N_0. It is quite often concluded that ARQ systems perform better on bursty channels, but as we see from the figures this is only true if the higher E_s/N_0 and/or the higher accepted codeword error probability are acceptable.

8.7.2. Hybrid Type-I ARQ

The main drawback with simple ARQ is that the throughput drops quickly when the error probability on the channel increases (E_s/N_0 is decreased). This is evident in Fig. 8.19. Under these circumstances type-I hybrid ARQ is better [38, Section 15.3]. In this scheme both error correction and error detection are used. One common implementation is a concatenated coding scheme with an outer CRC code and an inner block or convolutional code designed for error correction. Another implementation is to use a single code which is designed for joint error correction and error detection. In this protocol an RQ is only generated if the channel error is such that the received word is not corrected into a valid codeword in the decoder; otherwise an ACK is declared. The principle is illustrated in Fig. 8.22.

The expressions (8.7-1), (8.7-2), and (8.7-3) for accepted codeword error probability, average number of transmissions, and throughput, respectively, are still valid, but P_e and P_r will typically be reduced, since the dominating errors are corrected (see Chapter 3 and [38, Chapter 10]). However, the reduced coding rate

Trellis Coding on Fading Channels

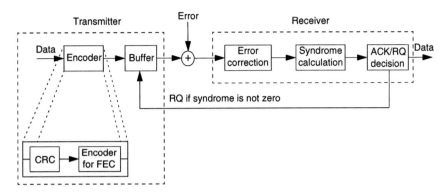

Figure 8.22 The principle of hybrid type-I ARQ. The error process takes into account modulation, the physical channel and demodulation. A typical implementation of the encoder is also shown.

has a threefold implication:

1. throughput is reduced, since it is directly proportional to the coding rate and the maximum throughput becomes significantly lower as compared to simple ARQ;
2. a smaller coding rate makes each codeword longer for a given number of information bits, and a longer codeword leads to a higher probability of detected error P_r and thus lower throughput;
3. the lower coding rate also reduces E_s/N_0 when the information rate and transmit power are constant, and this increases the error probability on the channel and thus the probability of detected error P_r.

The error correcting code therefore has to be powerful enough to counteract these factors. The trade-off is that simple ARQ should be used on good channels, while a well-designed hybrid type-I ARQ scheme gives better throughput below a certain E_s/N_0. In Figs 8.23–8.25 this behavior is illustrated for three different ARQ schemes on an uncorrelated Rayleigh fading channel with BPSK modulation and ideal coherent detection. The simple ARQ system is based on a Hamming (31,26) code which is used for error detection only (same as used in Figs 8.19–8.21). The other two systems use a Golay (23,12) code which corrects $t = 1$ and 2 errors, respectively, while the rest of the ability of the code is used for error detection. Here P_e can be calculated from [38, Eqs (10-11) and (10-12)] with the weight distribution as given in [38, Table 6-2]. In Fig. 8.23 it is demonstrated that the hybrid type-I scheme improves the throughput when the E_s/N_0 becomes smaller, but this is at the expense of a smaller throughput for higher E_s/N_0 due to the reduced coding rate. Figure 8.24 shows that the accepted codeword error

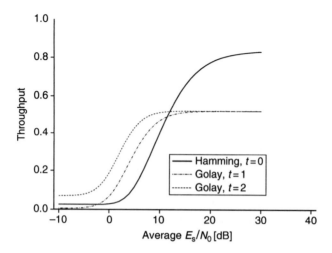

Figure 8.23 An example of throughput vs average E_s/N_0 for ARQ and hybrid type-I ARQ on an uncorrelated Rayleigh fading channel.

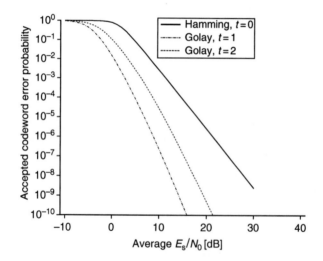

Figure 8.24 An example of accepted codeword error probability vs average E_s/N_0 for the system in Fig. 8.23.

probability is reduced in hybrid type-I ARQ.[18] The throughput and accepted codeword error probability are both shown in Fig. 8.25, and here it is clear that simple

---

[18] From Fig. 8.24 it is seen that the diversity orders are 3, 6, and 5, respectively, for the Hamming code, the Golay code with $t = 1$, and the Golay code with $t = 2$. The diversity order is equal to the number of bit errors in the most significant accepted codeword error.

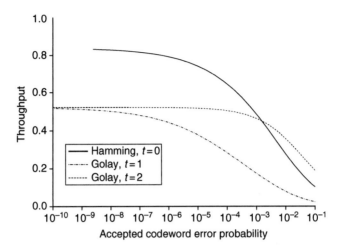

Figure 8.25 An example of throughput vs accepted codeword error probability for the system in Fig. 8.23.

ARQ gives the highest throughput for a given accepted codeword error probability unless the accepted codeword error probability is unacceptably high. This is at the expense of a larger E_s/N_0.

8.7.3. Hybrid Type-II ARQ

Hybrid type-II ARQ is a more powerful protocol since it adapts to changing channel conditions and thereby tries to avoid the reduced throughput of hybrid type-I on good channels, while still offering good throughput on bad channels. Many different type-II schemes have been proposed in the literature [22,23,38,90–93]. A simple form of hybrid type-II ARQ is to make better use of the transmitted codewords in simple ARQ or hybrid type-I ARQ, which can be done by improving the decoder only. Both these protocols concatenate an error correction and/or error detection code with simple repetition coding (through retransmission). The power of the repetition coding is, however, not taken into account in the decoder, since only the last received word is used in the decoder.

One decoding approach is the so called *diversity combining system*. Here, all received words are maximum ratio combined, hard decision detected, and decoded. If soft decision decoding is employed, the hard decision is replaced by metric quantization. Another approach is to jointly decode the concatenated code, which means that all the received words are aggregated into one word for the concatenated code, which is then used in a joint decoder of the concatenated code. Such a scheme is referred to as an *incremental redundancy scheme* since additional coded symbols are transmitted incrementally at each RQ [15]. Another name is *code combining*, since the received symbols are combined into a word in one single code [23]. With unquantized soft decisions in the diversity combining system, these two approaches are identical. The block diagrams in Figs 8.18 and 8.22 are still valid if a buffer with some processing is added immediately after the channel.

Example 8.7-1. In Figs 8.19–8.21, a simple ARQ scheme based on a Hamming (31,26) code was considered. Since this scheme utilizes repetition coding, it can be combined with diversity combining to obtain a hybrid type-II ARQ protocol. The scheme works in the following way. The 26 information bits are encoded by the Hamming code and transmitted over the channel. The syndrome is calculated in the decoder and an RQ is sent back to the transmitter whenever the syndrome is nonzero. When an RQ appears on the feedback channel, the same codeword is retransmitted. The received word is maximum ratio combined with the previous received word before hard decision detection and syndrome calculation. In this step, the code rate is in fact 13/31 and the code is a concatenation of a rate 1/2 repetition code and a rate 26/31 Hamming code. The decoder is a two step decoder which first combines repeated bits, then makes a hard decision, and finally recomputes the syndrome. The scheme then repeats in the same way until an ACK is obtained. The code rate after i transmissions is $26/31i$.

Example 8.7-2. The scheme in Example 8.7-1 can be changed to a code combining or incremental redundancy scheme by modifying the decoder. After one retransmission, the code used is a (62,26) code, where each codeword is equal to one codeword in the (31,26) Hamming code aggregated with itself. By aggregating the two received word in the same way, a hard decision on the the resulting word can be made and used in the syndrome calculation. If syndrome calculation is done based on soft information, hard decisions are replaced by quantized values. After yet another retransmission, each codeword consists of three repetitions of a codeword in the Hamming code and the word used at the syndrome calculation is an aggregation of all three received words. With more retransmission, more codewords are repeated and more received words are aggregated in a similar way.

RCPC and RCRC Codes in Hybrid Type-II ARQ

Hybrid type-II ARQ with repetition coding and diversity combining is a relatively simple scheme. However, since repetition coding is known to have a low minimum distance for the given code rate, MFD codes that allow incremental redundancy should have a superior performance. RCPC codes have these properties as demonstrated in Section 8.3. Given a rate $1/c$ mother code and a puncturing period p, the code rates are as given in Eq. (8.3-1). It is in fact also possible to puncture all but p bits over a puncturing period to obtain code rate 1, but this code cannot correct any errors [94,95]. To be able to detect errors, the RCPC code has to be concatenated with, for example, an (n, k) CRC code. To simplify Viterbi

Trellis Coding on Fading Channels

decoding of the RCPC code, it is common to add m zero tail bits at the end of the CRC codeword.[19] The resulting family of concatenated codes has code rates

$$R = \frac{k}{n+m} \frac{p}{p+l}, \quad l = 0, \ldots, (c-1)p. \tag{8.7-4}$$

One nice property is that the RCPC codes can easily be soft decision decoded for improved performance.

Example 8.7-3. Assume a mother code with rate 1/3 and puncturing period 2. A family of RCPC codes with rates 1, 2/3, 1/2, 2/5, and 1/3 can be constructed and used for error correction. When concatenated with an (112, 100) CRC code, the obtained code rates are 5/6, 5/9, 5/12, 1/3, and 5/18, when the memory of the mother code is $m = 8$. A hybrid type-II ARQ protocol with incremental redundancy based on this code family works in the following way. The incoming data is divided into 100-bit blocks. Each block is CRC encoded and an 8-bit zero tail is appended to the codeword. The resulting codeword is rate 1/3 convolutionally encoded, resulting in 360 bits. The convolutional codeword is then punctured according to the rate 1 pattern. The non-punctured 120 bits are transmitted over the channel. The received word is soft decision decoded in a Viterbi decoder for the rate 1/3 code, where metrics are not calculated for the punctured bits. The zero tail is removed and the syndrome is calculated for the CRC code. If the syndrome is zero an ACK is generated; otherwise an RQ is sent to the transmitter. In the first retransmission the 360 bit codeword of the convolutional code is punctured according to the rate 2/3 pattern. All the non-punctured 180 bits are *not* transmitted because 120 of these where already transmitted in the first transmission (due to the rate compatibility restriction of the code). Therefore only the new 60 bits (incremental redundancy) are transmitted. The corresponding 60 received symbols are aggregated properly with the previous 120 received symbols to form the received word for the rate 2/3 punctured convolutional code. These 180 symbols are Viterbi decoded, the tail is removed, and the syndrome is calculated. This procedure then repeats in a similar way until the syndrome becomes zero.

The maximum number of transmissions with the above protocol is $(c-1)p + 1$ when all rates according to Eq. (8.7-4) are used (which is of course not necessary). To increase the number of transmissions and increase the throughput on bad channels, RCPC codes can be combined with RCRC codes from Section 8.3. The code rates of such a family of codes are given by

$$R = \frac{k}{n+m} \frac{p}{pc+l}, \quad l = -(c-1)p, \ldots, \infty, \tag{8.7-5}$$

---

[19] These m extra tail bits can be avoided by using so called tailbiting codes or tailbiting trellises [96, Section 4.9], [97], but the decoding then becomes a bit more complex.

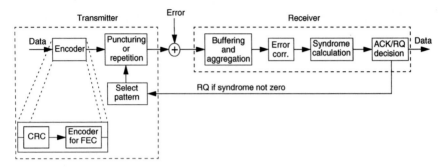

Figure 8.26 The principle of hybrid type-II ARQ based on RCPC and RCRC codes. The error process takes into account modulation, the physical channel, and demodulation.

which allows for an infinite number of transmissions. In practice, a time-out is needed to reduce delay and allow a finite precision in packet numbering. The principle for a hybrid type-II ARQ scheme based on RCPC and RCRC codes is illustrated in Fig. 8.26. Similar schemes can be constructed based on codes other than convolutional codes.

Example 8.7-4. The code family in Example 8.7-3 can be modified by using RCRC codes also to obtain code rates $5/6, 5/9, 5/12, 1/3, 5/18, 5/21, 5/24, 5/27, \ldots$. As before only incremental redundancy is transmitted and the decoding is done as in Example 8.7-3.

The accepted codeword error probability and the throughput for hybrid type-II protocols becomes more complex to calculate. The main reason is that the probability of undetected error P_e and the probability of detected error P_r depend on the number of transmissions, because the code is changed. The accepted codeword error probability can be expressed as

$$P(E) = P_e^{(1)} + P_r^{(1)} P_e^{(2)} + P_r^{(1)} P_r^{(2)} P_e^{(3)} + \cdots = \sum_{i=1}^{\infty} P_e^{(i)} \prod_{j=1}^{i-1} P_r^{(j)}, \quad (8.7\text{-}6)$$

with $\prod_{j=1}^{i-1} P_r^{(j)} = 1$ when $i \leq 1$. The throughput is given as

$$\eta = \frac{k}{\bar{N}}, \quad (8.7\text{-}7)$$

where k is the number of input bits to the error detecting block code and $\bar{N}$ is the average number of transmitted channel symbols given by

$$\bar{N} = N_1\left(1 - P_r^{(1)}\right) + N_2 P_r^{(1)}\left(1 - P_r^{(2)}\right) + N_3 P_r^{(1)} P_r^{(2)}\left(1 - P_r^{(3)}\right) + \cdots, \tag{8.7-8}$$

$$= \sum_{i=1}^{\infty} N_i \left(1 - P_r^{(i)}\right) \prod_{j=1}^{i-1} P_r^{(j)}, \tag{8.7-9}$$

where N_i is the number of channel symbols transmitted in the ith transmission. Here $P_e^{(i)}$ and $P_r^{(i)}$ are the probability of undetected error and the probability of detected error, respectively, after the ith decoding step. These are relatively straightforward to calculate when block codes are used for combined error correction and error detection. They are, however, more difficult to calculate for concatenated CRC coding and convolutional coding. One reason is that the bit error probability can in general only be upper bounded for convolutional codes; another is that bit errors are correlated at the output of the Viterbi decoder, which makes it difficult to analyze the CRC code performance.

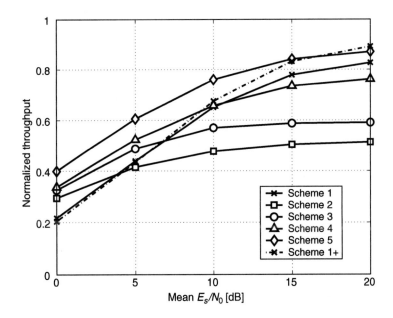

Figure 8.27 Throughput results for hybrid type-II ARQ protocols on Rayleigh fading channels with normalized (with channel data rate) Doppler frequency 0.001. (From [94], copyright Sorour Falahati; used with permission.)

Some throughput simulations for hybrid type-II ARQ schemes are compared in Fig. 8.27. The considered protocols are as follows:

Scheme 1. 68 information bits are encoded in an (80,68) CRC code. The 80 coded bits are transmitted as one packet repeatedly until an ACK is generated. All received words are maximum ratio combined before syndrome calculation. This is equivalent to a simple ARQ except for the maximum ratio combining in the detector.

Scheme 1+. 148 information bits are encoded in a (160,148) CRC code. The 160 coded bits are transmitted as one packet repeatedly until an ACK is generated. The decoding is done as in Scheme 1.

Scheme 2. 62 information bits are encoded in a (74,62) CRC code and then 6 zero tail bits are appended. The resulting 80 bits are encoded in a rate 1/3, memory 6 convolutional encoder before puncturing/repetition. The puncturing/repetition matrix has period 2 and the rates used are 2/3, 1/3, 2/9, 1/6, ... The 120 bits of incremental redundancy in each transmission are transmitted as one packet. All received symbols are aggregated to form a received word for the given rate and applied to a soft input Viterbi decoder. The tail is then removed and the syndrome calculated for the CRC code.

Scheme 3. 142 information bits are encoded in a (154,142) CRC code and then 6 zero tail bits are appended. The same family of RCPC/RCRC codes as in Scheme 2 are used. However, the 240 bits of incremental redundancy in each transmission are divided into three 80-bit packets before transmission on the channel. The decoding is done as in Scheme 2.

Scheme 4. 62 information bits are encoded in a (74,62) CRC code and then 6 zero tail bits are appended. The resulting 80 bits are encoded in a rate 1/3, memory 6 convolutional encoder before puncturing/repetition. The puncturing/repetition matrix has period 2 and the rates used are 1, 1/2, 1/3, 1/4, 1/5, ... The 80 bits of incremental redundancy in each transmission are transmitted as one packet. The decoding is done as in Scheme 2.

Scheme 5. 142 information bits are encoded in a (154,142) CRC code and then 6 zero tail bits are appended. The resulting 160 bits are encoded in a rate 1/3, memory 6 convolutional encoder before puncturing/repetition. The puncturing/repetition matrix has period 2 and the rates used are 1, 2/3, 1/2, 2/5, 1/3, 2/7, 1/4, ... The 160 bits of incremental redundancy in transmission number 1, 6, 11, ... are divided into two packets of length 80 bits which are transmitted. The 80 bits of incremental redundancy in the other transmissions are transmitted as one packet. The decoding is done as in Scheme 2.

Trellis Coding on Fading Channels

The modulation is BPSK in all the schemes. The RCPC codes used above rate 1/3 in Schemes 2–5 are obtained as described in Section 8.3.1 (code details are in [94]). The RCRC codes used below rate 1/3 in Schemes 2–5 are not optimized but pure repetitions of the previous incremental redundancy. The detection is ideal coherent detection. The channel is a flat Rayleigh fading channel with normalized (with coded bit rate) Doppler frequency of 0.001. The fading between transmitted packets are independent. Interleaving is used for Schemes 2–5 and is done over the incremental redundancy is each transmission.

The results in Fig. 8.27 clearly demonstrate that the maximum throughput depends on the code rate used in the first transmission. Scheme 1+ has the highest maximum throughput, followed closely by Scheme 5. The only difference in the first code rate for these two codes is due to the tail bits used in Scheme 5 but not in Scheme 1+. Schemes 1 and 4 have lower maximum throughput due to a smaller number of input bits to the CRC encoder and the tail bits in Scheme 4. Schemes 2 and 3 have the lowest maximum throughput due the initial code rate 2/3 in the RCPC code.

At lower E_s/N_0 it is an advantage not to have too large packets on the channel, as seen by comparing Schemes 1 and 1+. There is also a clear advantage of starting at a high code rate and transmitting a small incremental redundancy in each retransmission (small steps between code rates), as is done in Scheme 5 and to a lesser extent in Scheme 4. In practice, the number of incremental redundancy bits has a lower limit given by the packet overhead needed to transmit the packets. Another advantage is to divide many bits of incremental redundancy in a transmission into several packets, if these can be transmitted on independent channels; this was done in Scheme 3 and some of the transmissions in Scheme 5.

In [94] and its references the above schemes are also modified for a longer bit sequence at the input of the CRC encoder, which means that the packets transmitted on the channel become longer. The advantage is the higher CRC code rate, although 16 parity bits was used. The throughput of these two schemes has also been evaluated for a Rayleigh fading channel with a 10 times higher Doppler frequency and an AWGN channel. The main trends in the results are the same, but there is nothing to gain by longer packets since the increase in P_r counteracts the increase in code rate. With faster fading the error correction becomes more efficient since errors tend to appear more randomly. This means that schemes including error correction (Schemes 2–5) have a higher throughput, especially for intermediate E_s/N_0. Other schemes typically have a reduced throughput. On an AWGN channel the advantage from incremental redundancy generally becomes smaller, but Scheme 5 is still the best from a throughput point of view. The disadvantage with schemes using code rates with small difference is that the average delay tends to be longer.

The RCRC part of the schemes above is not optimized. In [94] and its references evaluations of schemes using optimized RCRC codes have been performed, and the general conclusion is that the gain in throughput is quite small.

One reason is that these codes are not used at rates above the mother code rate and these are the most important transmissions, another is that the simple repetition codes are almost equally good. The influence of the convolutional code memory on throughput has also been evaluated. Even if a larger memory leads to more powerful error correction, the throughput is relatively insensitive to the memory. One reason is that more tail bits are needed with larger memory, which reduces code rate and throughput. The tailbiting decoding techniques proposed in [98], which makes the tail bits superfluous have also been studied with Scheme 5, but no throughput gain was found. The reason seems to be the somewhat higher error probability of the tailbiting decoder.

As a final remark, it should be mentioned that the overhead and idle time of the transmitted (for some protocols) will strongly influence the throughput of a real protocol. In general, the overhead will restrict the number of bits in each retransmission not to be too small, and consequently not too many retransmission should be used. The idle time of the transmitter will seriously degrade the throughput of these protocols [38].

Other Codes in Hybrid Type-II ARQ

Hybrid type-II ARQ can be constructed in a similar way with many other codes. We do not give a complete overview here, but only a few examples. Complementary convolutional codes have a special property that they can be decoded from only the incremental information transmitted in the second transmission, which makes them suitable for hybrid type-II ARQ [99]. Invertible convolutional codes have a similar property [100]. Rate compatible turbo codes can of course also be used in hybrid type-II ARQ [101,102]. A few other examples of codes used in hybrid type-II ARQ are zigzag codes [103], Reed–Solomon codes [104], and concatenated convolutional and Reed–Solomon codes [105]. Hybrid ARQ can also benefit from being combined with sequential decoding, since this makes the error detecting code unnecessary [106–109].

8.7.4. Hybrid Type-II ARQ with Adaptive Modulation

One disadvantage with the hybrid type-II ARQ schemes discussed above is that the throughput is limited to 1 bit per channel symbol. This is due to the binary modulation. By using a signaling constellation of size M, the maximum throughput rises to $\log_2 M$. The throughput of such a scheme, however, deteriorates quickly with decreasing E_s/N_0, since higher constellations are noise sensitive. One solution is to adapt both the code rate and the size of the signaling constellation to the transmission number as proposed in [94,110]. Figure 8.28 shows throughput for Scheme 5, where 16QAM is used for the first transmission and the incremental redundancy in the second transmission, 8PSK is used for the incremental redundancy in the third and fourth transmission, QPSK is used for the incremental

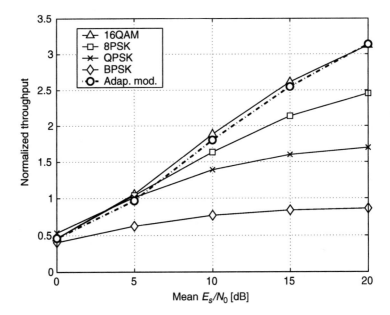

Figure 8.28 Throughput results for hybrid type-II ARQ protocols with different modulations on Rayleigh fading channels with normalized (with channel data rate) Doppler frequency 0.001. (From [94], copyright Sorour Falahati; used with permission.)

redundancy in the fifth, sixth, and seventh transmission, and BPSK is used for all the other incremental redundancy transmission. The decoder is coherent with perfect side information. This is the "Adap. mod." curve. Results are shown for comparison for Scheme 5 with 16QAM, 8PSK, QPSK, and BPSK for all the transmissions. The scheme with adaptive modulation closely follows the upper envelope of the other curves. The throughput is similar for other Doppler frequencies [94]. The signaling constellations were scaled such that E_s was the same for all of them in the adaptive modulation in Fig. 8.28. Some more freedom can be obtained by adapting E_s also and not only the size of each constellation and in this way it is expected that the throughput can be improved somewhat at low average E_s/N_0.

Similar schemes can be designed based on TCM codes. In [111] a rate compatible family of TCM codes are designed which allows code combining and incremental redundancy. As a remark, it should be noted that the scheme presented in this section is very different from adaptive modulation of the kind described in [112] and references therein. These schemes require knowledge of the channel SNR in the transmitter and selects a suitable (coding and) modulation scheme based on this knowledge. Thus, they also require a feedback channel, but instead of sending only ACK/RQ information, predicted SNRs must be made available to the transmitter. These kind of adaptive modulation schemes are outside the scope of this book.

Bibliography

[1] K. J. Larsen, "Short convolutional codes with maximal free distance for rates 1/2, 1/3, and 1/4," *IEEE Trans. Inf. Theory*, **IT-19**(3), 371–372, May 1973.

[2] G. C. Clark, Jr. and J. Bibb Cain, *Error-control Coding for Digital Communications*. Plenum, New York, 1981.

[3] S. Lin and D. J. Costello, Jr., *Error Control Coding*. Prentice-Hall, Englewood Cliffs, NJ, 1983.

[4] A. M. Michelson and A. H. Levesque, *Error-control Techniques for Digital Communication*. Wiley, New York, 1985.

[5] J. B. Anderson and S. Mohan, *Source and Channel Coding*. Kluwer, Boston, 1991.

[6] J. G. Proakis, *Digital Communications*, 4th. edn. McGraw-Hill, New York, 2000.

[7] I. E. Bocharova and B. D. Kudrashov, "Rational rate convolutional codes for soft-decision Viterbi decoding," *IEEE Trans. Inf. Theory*, **43**(4), 1305–1313, July 1997.

[8] J. Chang, D. Hwang, and M. Lin, "Some extended results on the search for good convolutional codes," *IEEE Trans. Inf. Theory*, **43**(5), 1682–1697, Sept. 1997.

[9] P. Frenger, P. Orten, and T. Ottosson, "Convolutional codes with optimum distance spectrum," *IEEE Commun. Lett.*, **3**(11), 317–319, Nov. 1999.

[10] P. K. Frenger, P. Orten, and T. Ottosson, "Comments and additions to recent papers on new convolutional codes," *IEEE Trans. Inf. Theory*, **47**(3), 1199–1201, Mar. 2001.

[11] P. Frenger, P. Orten, T. Ottosson, and A. Svensson, "Multi-rate convolutional codes," Tech. Rep. R021/1998, Department of Signals and Systems, Chalmers University of. Technology, Göteborg, Sweden, Apr. 1998.

[12] J. P. Odenwalder, *Optimal Decoding of Convolutional Codes*, PhD thesis, School Eng. Appl. Sci., Univ. California, Los Angeles, CA, 1970.

[13] P. J. Lee, "New short constraint length rate 1/N convolutional codes which minimize the required SNR for given desired bit error rates," *IEEE Trans. Commun.*, **COM-33**(2), 171–177, Feb. 1985.

[14] P. J. Lee, "Further results on rate 1/N convolutional code constructions with minimum required SNR criterion," *IEEE Trans. Commun.*, **COM-34**(4), 395–399, Apr. 1986.

[15] D. M. Mandelbaum, "An adaptive-feedback coding scheme using incremental redundancy," *IEEE Trans. Inf. Theory*, **IT-20**(3), 388–389, May 1974.

[16] J. B. Cain, G. C. Clark, and J. M. Geist, "Punctured convolutional codes of rate $(n-1)/n$ and simplified maximum likelihood decoding," *IEEE Trans. Inf. Theory*, **IT-25**(1), 97–100, Jan. 1979.

[17] Y. Yasuda, "Development of variable-rate Viterbi decoder and its performance characteristics," in *Proc. 6th Int. Conf. Digit. Satellite Commun.*, Phoenix, AZ, Sept. 1983, XII-24–XII-31.

[18] Y. Yasuda, K. Kashiki, and Y. Hirata, "High rate punctured convolutional codes for soft decision Viterbi decoding," *IEEE Trans. Commun.*, **COM-32**(3), 315–319, Mar. 1984.

[19] J. Hagenauer, "Rate-compatible punctured convolutional codes (RCPC codes) and their performance," *IEEE Trans. Commun.*, **36**(4), 389–400, Apr. 1988.

[20] P. K. Frenger, P. Orten, T. Ottosson, and A. B. Svensson, "Rate-compatible convolutional codes for multirate DS-CDMA systems," *IEEE Trans. Commun.*, **47**(6), 828–836, June 1999.

[21] L. H. C. Lee, "New rate-compatible punctured convolutional codes for Viterbi decoding," *IEEE Trans. Commun.*, **42**(12), 3073–3079, Dec. 1994.

[22] S. Kallel and D. Haccoun, "Generalized type II hybrid ARQ scheme using punctured convolutional coding," *IEEE Trans. Commun.*, **38**(11), 1938–1946, Nov. 1990.

[23] S. Kallel, "Analysis of a type II hybrid ARQ scheme with code combining," *IEEE Trans. Commun.*, **38**(8), 1133–1137, Aug. 1990.

[24] Z. Lin and A. Svensson, "New rate-compatible repetition convolutional codes," *IEEE Trans. Inf. Theory*, **46**(7), 2651–2659, Nov. 2000.

[25] Z. Lin, "Rate compatible convolutional codes and their application to type II hybrid ARQ transmission," MS thesis, Department of Signals and Systems, Chalmers University of Technology, Göteborg, Sweden, 1998.
[26] S. Lefrançois and D. Haccoun, "Search procedures for very low rate quasioptimal convolutional codes," in *Proc. IEEE Int. Symp. Inf. Theory*, Trondheim, Norway, June 1994, p. 278.
[27] S. Lefrançois and D. Haccoun, "Very low rate quasi-optimal convolutional codes for CDMA," in *Proc. Canadian Conf. Electr. Computer Eng.*, Halifax, Canada, Sept. 1994, 210–213.
[28] P. Frenger, P. Orten, and T. Ottosson, "Code-spread CDMA using maximum free distance low-rate convolutional codes," *IEEE Trans. Commun.*, **48**(1), 135–144, Jan. 2000.
[29] A. Mehrotra, *GSM System Engineering*, Artech House, Boston, 1997.
[30] T. Ojanperä and R. Prasad, *Wideband CDMA for Third Generation Mobile Communications*, Artech House, Boston, 1998.
[31] J. P. Castro, *The UMTS Network and Radio Access Technology: Air Interface Techniques for Future Mobile Systems*. Wiley, New York, 2001.
[32] E. Biglieri, D. Divsalar, P. McLane, and M. K. Simon, *Introduction to Trellis-coded Modulation with Applications*. Macmillan, New York, 1991.
[33] S. H. Jamali and T. Le-Ngoc, *Coded-modulation Techniques for Fading Channels*. Kluwer, Boston, 1994.
[34] C. Schlegel and D. J. Costello, Jr., "Bandwidth efficient coding for fading channels: code construction and performance analysis," *IEEE J. Sel. Areas Communs.*, **7**(9), 1356–1368, Dec. 1989.
[35] C.-E. W. Sundberg and J. Hagenauer, "Hybrid trellis-coded 8/4 PSK systems," *IEEE Trans. Commun.*, **38**(5), 602–614, May 1990.
[36] E. Zehavi, "8-PSK trellis codes for a Rayleigh channel," *IEEE Trans. Commun.*, **40**(5), 873–884, May 1992.
[37] E. Biglieri and M. Luise (eds), *Coded Modulation and Bandwidth-efficient Transmission*. Elsevier, Amsterdam, 1992.
[38] S. B. Wicker, *Error Control Systems for Digital Communication and Storage*. Prentice-Hall, Englewood Cliffs, NJ, 1995.
[39] J. K. Cavers and P. Ho, "Analysis of the error performance of trellis-coded modulations in Rayleigh-fading channels," *IEEE Trans. Commun.*, **40**(1), 74–83, Jan. 1992.
[40] J. Huang and L. L. Campbell, "Trellis coded MDPSK in correlated and shadowed Ricean fading channels," *IEEE Trans. Veh. Tech.*, **40**(4), 786–797, Nov. 1991.
[41] D. Divsalar and M. K. Simon, "Performance of trellis coded MDPSK on fast fading channels," in *Proc. Int. Conf. Communs.*, Boston, MA, June 1989, 261–267.
[42] F. Edbauer, "Performance of interleaved trellis-coded differential 8-PSK modulation over fading channels," *IEEE J. Sel. Areas Communs.*, **7**(9), 1340–1346, Dec. 1989.
[43] R. G. McKay, P. J. McLane, and E. Biglier, "Error bounds for trellis-coded MPSK on a fading mobile satellite channel," *IEEE Trans. Commun.*, **39**(12), 1750–1761, Dec. 1991.
[44] K. Chan and A. Bateman, "The performance of reference-based M-ary PSK with trellis coded modulation in Rayleigh fading," *IEEE Trans. Veh. Tech.*, **41**(2), 190–198, May 1992.
[45] D. Divsalar and M. K. Simon, "The design of trellis coded MPSK for fading channels: set partitioning for optimum code design," *IEEE Trans. Commun.*, **36**(9), 1013–1021, Sept. 1988.
[46] D. Divsalar and M. K. Simon, "Multiple trellis coded modulation MTCM," *IEEE Trans. Commun.*, **36**(4), 410–419, Apr. 1988.
[47] N. Seshadri and C.-E. W. Sundberg, "Multilevel trellis coded modulations for the Rayleigh fading channel," *IEEE Trans. Commun.*, **41**(9), 1300–1310, Sept. 1993.
[48] G. Caire, G. Taricco, and E. Biglieri, "Bit-interleaved coded modulation," *IEEE Trans. Inf. Theory*, **44**(3), 927–946, May 1998.

[49] N. Seshadri and C.-E. W. Sundberg, "Multi-level trellis coded modulations with large time diversity for the Rayleigh fading channel," in *Proc. Annual Conf. Inf. Sciences Sys*, Princeton, NJ, Mar. 1990, 853–857.
[50] R. C. Dixon, *Spread Spectrum System with Commercial Applications*, 3rd edn. Wiley, New York, 1994.
[51] S. Glisic and B. Vucetic, *Spread Spectrum CDMA System for Wireless Communications*. Artech House, Boston, 1997.
[52] M. K. Simon, J. K. Omura, R. A. Scholtz, and B. K. Levitt, *Spread Spectrum Communications Handbook*, revised edn. McGraw-Hill, New York, 1994.
[53] A. J. Viterbi, *CDMA Principles of Spread Spectrum Communication*. Addison-Wesley, Reading, 1995.
[54] S. Verdú, *Multiuser Detection*. Cambridge University Press, Cambridge, UK, 1998.
[55] A. J. Viterbi, "Very low rate convolutional codes for maximum theoretical performance of spread-spectrum multiple-access channels," *IEEE J. Sel. Areas Communs.*, **8**(4), 641–649, May 1990.
[56] J. Y. N. Hui, "Throughput analysis for code division multiple accessing of the spread spectrum channel," *IEEE J. Sel. Areas Communs.*, **SAC-2**(4), 482–486, July 1984.
[57] A. J. Viterbi, "Orthogonal tree codes for communication in the presence of white Gaussian noise," *IEEE Trans. Commun.*, **COM-15**(4), 238–242, Apr. 1967.
[58] R. F. Ormondroyd and J. J. Maxey, "Performance of low-rate orthogonal convolutional codes in DS-CDMA applications," *IEEE Trans. Veh. Tech.*, **46**(5), 320–328, May 1997.
[59] K. Rikkinen, "Comparison of very low rate coding methods for CDMA radio communication systems," in *Proc. IEEE Int. Symp. Spread Spectrum Techniques Appl.*, Oulu, Finland, Sept. 1994, 268–272.
[60] Y. M. Kim and B. D. Woerner, "Comparison of trellis coding and low rate convolutional codes for CDMA," in *Proc. IEEE Military Commun. Conf.*, Forth Monmouth, NJ, 1994, 765–769.
[61] P. K. Frenger, P. Orten, and T. Ottosson, "Code-spread CDMA with interference cancellation," *IEEE J. Sel. Areas in Communs.*, **17**(12), 2090–2095, Dec. 1999.
[62] D. Haccoun, S. Lefrançois, and E. Mehn, "An analysis of the CDMA capacity using a combination of low rate convolutional codes and PN sequence," in *Proc. Canadian Conf. Electr. Comp. Eng.*, **1**, 32–35, 1996.
[63] D. Haccoun and Z. Gherbi, "On the application of very low rate error control coding to CDMA," in *Proc. Canadian Conf. on Electr. Comp. Eng.*, **2**, 466–469, 1997.
[64] G. D. Boudreau, D. D. Falconer, and S. A. Mahmoud, "A comparison of trellis coded versus convolutionally coded spread-spectrum multiple-access systems," *IEEE J. Sel. Areas Communs.*, **8**(4), 628–640, May 1990.
[65] A. Persson, "On convolutional codes and multi-carrier techniques for CDMA systems," Tech. Rep. 423L, Techn. licentiate thesis, Department of Signals and Systems, Chalmers University of Technology, Göteborg, Sweden, Jan. 2002.
[66] J. Lassing, "Aspects of coding in communication systems," Tech. Rep. 424L, Techn. licentiate thesis, Department of Signals and Systems, Chalmers University of Technology, Göteborg, Sweden, Jan. 2002.
[67] C. F. Leanderson, O. Edfors, T. Maseng, and T. Ottosson, "On the performance of turbo codes and convolutional codes at low rate," in *Proc. IEEE Veh. Tech. Conf.*, Amsterdam, The Netherlands, Sept. 1999, 1560–1564.
[68] C. F. Leanderson, J. Hokfelt, O. Edfors, and T. Maseng, "On the design of low rate turbo codes," in *Proc. IEEE Veh. Tech. Conf.*, Tokyo, Japan, May 2000, 1200–1204.
[69] D. J. van Wyk and L. P. Linde, "A turbo coded DS/CDMA system with embedded Walsh–Hadamard codewords: coder design and performance evaluation," in *Proc. IEEE Int. Symp. Spread Spectrum Techniques Appl.*, Sun City, South Africa, Sept. 2000, 359–363.
[70] P. Komulainen and K. Pehkonen, "Performance evaluation of superorthogonal turbo codes in AWGN and flat Rayleigh fading channels," *IEEE J. Sel. Areas Communs.*, **16**(2), 196–205, Feb. 1998.

[71] N. Alon, J. Bruck, J. Naor, M. Naor, and R. M. Roth, "Construction of asymptotically good low-rate error-correcting codes through pseudo-random graphs," *IEEE Trans. Inform. Theory*, **38**(2), 509–516, Mar. 1992.

[72] J.-P. Chaib and H. Leib, "Very low rate trellis/Reed–Muller (TRM) codes," *IEEE Trans. Commun.*, **47**(10), 1476–1487, Oct. 1999.

[73] V. Sorokine, F. R. Kschischang, and S. Pasupathy, "Gallager codes for CDMA applications. I. Generalizations, constructions, and performance bounds," *IEEE Trans. Commun.*, **48**(10), 1660–1668, Oct. 2000.

[74] V. Sorokine, F. R. Kschischang, and S. Pasupathy, "Gallager codes for CDMA applications. II. Implementations, complexity, and system capacity," *IEEE Trans. Commun.*, **48**(11), 1818–1828, Nov. 2000.

[75] P. D. Papadimitriou and C. N. Georghiades, "On asymptotically optimum rate 1/n convolutional codes for a given constraint length," *IEEE Commun. Lett.*, **5**(1), 25–27, Jan. 2001.

[76] D. J. van Wyk, I. J. Opperman, and L. P. Linde, "Low rate coding considerations for space–time coded DS/CDMA," in *Proc. IEEE Veh. Tech. Conf.*, Amsterdam, The Netherlands, Sept. 1999, 2520–2524.

[77] D. J. van Wyk, I. J. Opperman, and L. P. Linde, "Performance tradeoff among spreading, coding and multiple-antenna transmit diversity for high capacity space-time coded DS/CDMA," in *Proc. IEEE Military Commun. Conf.*, Atlantic City, NJ, June 1999, 393–397.

[78] D. van Wyk and P. van Rooyen, "On the performance of super-orthogonal turbo-transmit diversity for CDMA cellular communication," in *Proc. IEEE Int. Symp. Spread Spectrum Techn. Appl.*, Newark, NJ, Sept. 2000, 127–131.

[79] A. Persson, J. Lassing, T. Ottosson, E. G. Ström, and A. Svensson, "Combined coding and spreading in downlink CDMA systems," in *Proc. IEEE Veh. Tech. Conf.*, Tokyo, Japan, May 2000, 2403–2407.

[80] A. Persson, J. Lassing, T. Ottosson, and E. G. Ström, "On the differences between uplink and downlink transmission in code-spread CDMA systems," in *Proc. IEEE Veh. Tech. Conf.*, Rhodes, Greece, June 2001, 2421–2425.

[81] S. Moshavi, "Multi-user detection for DS-CDMA communications," *IEEE Commun. Mag.*, **34**(10), 124–136, Oct. 1996.

[82] A. Duel-Hallen, J. Holtzman, and Z. Zvonar, "Multiuser detection for CDMA systems," *IEEE Personal Commun. Mag.*, **2**(2), 46–58, Apr. 1995.

[83] S. Verdú, "Minimum probability of error for asynchronous Gaussian multiple-access channels," *IEEE Trans. Inf. Theory*, **32**(1), 85–96, Jan. 1986.

[84] T. R. Giallorenzi and S. G. Wilson, "Multiuser ML sequence estimator for convolutionally coded asynchronous DS-CDMA systems," *IEEE Trans. Commun.*, **44**(8), 997–1008, Aug. 1996.

[85] A.-L. Johansson, *"Successive interference cancellation in DS-CDMA systems,"* Tech. Rep. 344, PhD thesis, Department of Signals and Systems, Chalmers University of Technology, Göteborg, Sweden, 1998.

[86] T. R. Giallorenzi and S. G. Wilson, "Suboptimum multiuser receivers for convolutionally coded asynchronous DS-CDMA systems," *IEEE Trans. Commun.*, **44**(9), 1183–1196, Sept. 1996.

[87] M. R. Koohrangpour and A. Svensson, "Joint interference cancellation and Viterbi decoding in DS/CDMA," in *Proc. IEEE Personal, Indoor, Mobile Radio Conf.*, Helsinki, Finland, Sept. 1997, 1161–1165.

[88] P. Frenger, P. Orten, and T. Ottosson, "Bit error rate calculation for nonlinear interference cancellation," *Electronics (Lett.)*, **35**(18), 1572–1573, Sept. 1999.

[89] G. Castagnoli, J. Ganz, and P. Graber, "Optimum cyclic redundancy-check codes with 16-bit redundancy," *IEEE Trans. Commun.*, **38**(1), 111–114, Jan. 1990.

[90] R. A. Comroe and D. J. Costello, Jr., "ARQ schemes for data transmission in mobile radio systems," *IEEE J. Select. Areas in Commun.*, **SAC-2**(4), 472–481, July 1984.

[91] L. R. Lugand, D. J. Costello, Jr., and R. H. Deng, "Parity retransmission hybrid ARQ using rate 1/2 convolutional codes on a nonstationary channel," *IEEE Trans. Commun.*, **37**(7), 755–765, July 1989.
[92] B. Vucetic, "An adaptive coding scheme for time-varying channels," *IEEE Trans. Commun.*, **39**(5), 653–663, May 1991.
[93] M. Rice and S. Wicker, "A sequential scheme for adaptive error control over slowly varying channels," *IEEE Trans. Commun.*, **42**(4), 1533–1543, Apr. 1994.
[94] S. Falahati, "Adaptive modulation and coding in wireless communications systems with feedback," Tech. Rep. 434, PhD thesis, Department of Signals and Systems, Chalmers University of Technology, Göteborg, Sweden, Sept. 2002.
[95] S. Falahati, T. Ottosson, A. Svensson, and Z. Lin, "Convolutional coding and decoding in hybrid type-II ARQ schemes in wireless channels," in *Proc. IEEE Veh. Tech. Conf.*, Houston, TX, May 1997, 2219–2224.
[96] R. Johannesson and K. Sh. Zigangirov, *Fundamentals of Convolutional Coding*. IEEE Press, New York, 1999.
[97] M. H. Howard and J. K. Wolf, "On tailbiting convolutional codes," *IEEE Trans. Commun.*, **34**(2), 104–111, Feb. 1986.
[98] R. V. Cox and C. -E. W. Sundberg, "An efficient adaptive circular Viterbi algorithm for decoding generalized tailbiting convolutional codes," *IEEE Trans. Veh. Tech.*, **43**(1), 57–68, Feb. 1994.
[99] S. Kallel, "Complementary punctured convolutional (CPC) codes and their use in hybrid ARQ schemes," in *Proc. IEEE Pacific Rim Conf. Commun., Comp. Sig. Proc.*, 1993, pp. 186–189.
[100] S. Kallel, "A variable-redundancy hybrid ARQ scheme using invertible convolutional codes," in *Proc. IEEE Veh. Tech. Conf.*, Stockholm, Sweden, June 1994, 1417–1420.
[101] D. N. Rowitch and L. B. Milstein, "On the performance of hybrid FEC/ARQ systems using rate compatible punctured turbo (RCPT) codes," *IEEE Trans. Commun.*, **48**(6), 948–959, June 2000.
[102] C. -F. Law, C. H. Lau, and T. -M. Ko, "A modified adaptive hybrid FEC/ARQ protocol using turbo codes with incremental redundancy transmission," in *Proc. IEEE Veh. Tech. Conf.*, Amsterdam, The Netherlands, Sept. 1999, 1670–1674.
[103] K. S. Chan, L. Ping, and S. Chan, "Adaptive type II hybrid ARQ scheme using zigzag code," *Electronics (Lett.)*, **35**(24), 2102–2104, Nov. 1999.
[104] U. H. -G. Kressel and P. A. M. Buné, "Adaptive forward error correction for fast data transmission over the mobile radio channel," in *Proc. EUROCON*, Stockholm, Sweden, June 1988, 170–173.
[105] C. F. Bradshaw and D. Wiggert, "Performance of type II hybrid ARQ systems using concatenated convolutional and Reed–Solomon codes," in *Proc. Tactical Commun. Conf.*, 1990, 499–514.
[106] S. Kallel, "Sequential decoding with an efficient incremental redundancy ARQ scheme," *IEEE Trans. Commun.*, **40**(10), 1588–1593, Oct. 1992.
[107] S. Kallel and D. Haccoun, "Sequential decoding with ARQ and code combining: a robust hybrid FEC/ARQ system," *IEEE Trans. Commun.*, **36**(7), 773–780, July 1988.
[108] S. Kallel and D. Haccoun, "Sequential decoding with an efficient partial retransmission ARQ strategy," *IEEE Trans. Commun.*, **39**(2), 208–213, Feb. 1991.
[109] P. Orten, "Channel coding and multiuser detection for CDMA and wireless communications," Tech. Rep. 372, PhD thesis, Department of Signals and Systems, Chalmers University of Technology, Göteborg, Sweden, 1999.
[110] S. Falahati and A. Svensson, "Hybrid type-II ARQ schemes with adaptive modulation systems for wireless channels," *Proc. IEEE Veh. Tech. Conf.*, Amsterdam, The Netherlands, Sept. 1999, 2691–2695.
[111] H. Stenhoff, C. Lindstrand, M. Johansson, U. Hansson, and S. Sjöberg, "Adaptive coded modulation for wireless packet data systems," *Proc. IEEE Veh. Tech. Conf.*, Amsterdam, The Netherlands, Sept. 1999, 1790–1794.
[112] S. T. Chung and A. J. Goldsmith, "Degrees of freedom in adaptive modulation: a unified view," *IEEE Trans. Commun.*, **49**(9), 1561–1571, Sept. 2001.

Index

~ (Asymptotically equal), 30

ACG. *See* Asymptotic coding gain
Alphabet
  in CPM, 193, 239–244
  in PRS, 290–291, 311
  transmission, 17, 39–40
AMF. *See* Average matched filter
Amplifier, RF, 13, 41–44, 192
Amplitude and phase form, 38
Amplitude variation
  in filtered CPM, 323
  in QPSK, 41–44
  in RC pulses, 20
Anderson, J.B., 192, 266
Antipodal signaling distance. *See* Matched filter bound
Antipodal signals, 31–32
A priori information, 26; *in BCJR* 105
A priori probabilities, in CPM, 229, 233–236
Armstrong, E., 5
ARQ (automatic request repeat).
    See Request repeat
Ascheid, G., 265
Asymptotic coding gain (lattice), 180, 182
Aulin, T., 10, 192, 197, 213, 224, 273
Autocorrelation
  and bandwidth, 303ff
  baseband, 63
  channel, *def.* 295–297
  of difference sequence, *def.* 299–300
  and distance, 299ff
  time average, 62–63, 294
Average matched filter, 272–273
AWGN (additive white Gaussian noise).
    See Channel; Noise, Gaussian

B_c. *See* Coherence bandwidth
Backtracking, 95
Baker's rule, 65, 226
Balachandran, Krishna, 342
Balachandran, Kumar, 224

Bandpass
  filtering, 323
  PRS codes, 343–344
Bandwidth. *See also* Spectrum, frequency
  and capacity, 117–126
  and complexity, 343–345
  in conv. codes + PSK/QAM, 185–186
  in CPM, 194–195, 226, 240–244
  per data bit, 6, 17, 61, 117–121
  and ISI, 118
  of modulators, 18, 58–61
  normalized (NBW), *TCM* 135–136, *CPM* 185, *PRS* 310
  of PRS, 303–305, 311–313, 316–318, 341,
    tables 351–357
  per symbol, 2, 6
  in TCM, 135–136, 149, 156–157
  and zero placement, 341
Baseband
  autocorrelation, 63
  pulses, 19–26
  spectrum, 61, 63
  transmission, 19
Basis vectors
  lattice, 172
  in signal space, 27
Baud rate, 17
BCH codes. *See* Coding, binary
BCJR (Bahl-Cocke-Jelinek-Raviv) algorithm, 101–106
BEC (binary erasure channel). *See* Channel
BER. *See* Bit error rate
Biased trellis, 222–223
Binary erasure channel. *See* Channel
Binary symmetric channel. *See* Channel
Bit error rate
  capacity for, 119–121
  in conv. coding, 417–419
  in CPM, 248–250, 348
  in fading, 396, 402–403, 405–407
  in lattice coding, 182
  in PRS, 341–342
  in spread spectrum, 446–447, 451–453
  in TCM, 149, 151, 156, 166–169
Bit time, *def.* 17, 61, *CPM* 195

Bit interleaved coded modulation, 434
Block coded modulation, 186
Bounded distance decoder. *See* Decoder
Branch and bound, 300–303
Brickwall filter
  in capacity derivation, 125–126
  in distance calculation, 323–325
BSC (binary symmetric channel). *See* Channel
BSEC (binary symmetric erasure channel). *See* Channel
Butterworth filter
  applied to CPM, 325–326
  in M + F coding, 330–331, 347–348
  modeling, 70–72, 292–293
  as PRS code, 292–293

C. *See* Concentration, spectral
C_α (CPM spectral decay factor), 228, 213–232
C_W (capacity in bandwidth W). *See* Capacity
Calderbank, A.R., 10, 172
Capacity. *See also* Cutoff rate
  of AWGN channel, 112, 114–117
  in bandwidth W, 117–121
  of BSC, 82, 111–114
  of DMC, 112
  general, 110–111
  under filtering, 121–126
  of QAM channel, 111–112, 115–117, 128
  of soft output channel, 113–114
  of TCM channel, 115–117, 149
Carrier
  in CPM, 244–245
  frequency, 37
Carrier modulation. *See* Modulation, FSK; Modulation, general; Modulation, PSK
Carrier synchronization. *See* Synchronization, carrier
CCITT modem standards, 158–165
Centroid, 31
Channel. *See also* Capacity; Channel, fading
  AWGN, 2–8, 84, 108–109, 114–117
  with bandwidth (Gaussian), 109–110, 117–119
  BEC, 107–108
  BSC, *def.* 76–77, 82–83, 107
  BSEC, 107–108
  DMC, 107–108
  guided media, 2
  hard decision, 83
  mobile, 2
  radio/microwave, 2, 122–123
  in Shannon theory, 107
  soft output, 108
  space, 9
  wireline, 2, 122
  worst case, 318–319

Channel, fading
  coherence time & bandwidth, 379–383
  distributions, 368–375
  frequency selective, 375–386
  path loss, 364–367
  simulation of, 386–395
  slow/fast, 385–386
Class C. *See* Amplifier, RF
Classes, of coded modulation, 11
CNR (synchronizer carrier-to-noise ratio), 262–263
Code spread CDMA, 444–454, *IC receiver* 450–453
Coded modulation, general. *See also under code type:* Continuous-phase modulation; Lattice; Modulation+filter; Trellis-coded modulation
  classes, 11–13
  Gaussian channel, 4–5
  linear coded modulation, 10
  history, 10
  as patterning, 4
Coding, binary. *See also* Convolutional codes; Modulation + parity-check codes; Parity-check codes
  in ARQ, 456–468
  BCH, 8, 81
  as patterning, 4
  Reed-Muller, 8
  Reed-Solomon, 81–82
Coding gain
  of conv. codes + PSK/QAM, 184
  of lattices, 176–178, 180, 182
  of parity-check codes, 83–84
  and RF amplifiers, 192
  of TCM, 139, 145–146, 149, 154–156
Coding theorem, 4–6, *DMC* 112–113
Coherence bandwidth & time, 379–383, *defs.* 382
Column distance, PRS, 340
Concatenation, code, 101–102
Concentration, spectral, 304–309
Configuration, of subsets, 142–143
Constellation. *See* Cross constellation; Master constellation; Signal constellation
Continuous approximation, 178
Continuous-phase modulation codes (CPM). *See also* Difference sequence; Energy-bandwidth tradeoff; Phase trellis; RC CPM codes; REC CPM codes
  alphabet in, 193, 239–244
  bandwidth, normalized, 195
  basics, 51–52, 193–197
  classes of, 194–196
  with conv. codes, 183
  d_B bound, 200–209, 211–212, 257–261
  distance, 193, 197–221
  error events, 99, 224–225, 328–329

Index

Continuous-phase modulation codes (CPM) (*cont.*)
 filtering of, 323–329, 347–348
 frequency response function, g, 52, *def.* 193–196, 240
 from FSK, 51–52, 191
 full response, 193, 197–221
 GMSK, 195, 277
 HCS, 202
 history, 10, 191–192
 index, modulation, 51, *def.* 193, 209–210, 239, 241–243
 index, optimal, 213, 216
 index, weak, 209–210, 214, 217–218
 modeling, 66, 68
 multi-h, 210–212, 218–221, 232
 partial response, *def.* 193, 203–209
 phase response function, q, 51, *def.* 193–196
 rate in, 195
 receivers for
  AMF, 272–273
  discriminator, 251–253
  M-algorithm, 273–275, 347–348
  MSK-type, 275–277
  noncoherent, 245, 251–257
  Osborne-Luntz, 246–257, 253–261
  partially coherent, 245, 253–261
  simplified pulse, 269–272
  Viterbi, 247–251
 with RF amplifiers, 192
 spectrum, *formula* 63–65, 225–240
 SRC, *def.* 195, 209, 229, 234–236, 239
 standard CPM, 193, 231
 state description, 220–221, 271–275
 synchronization for, 57, 261–266
 TFM, *def.* 195, 233–234, 277
 transmitters for, 266–268
Controller canonical form, 150–151, 153
Converse, to coding theorem, 113
Convolutional codes. *See also* Orthogonal convolutional codes; Punctured codes; Rate-compatible codes
 error bounds, 416–417
 error events, 99
 fading error rate, 416–419, 427
 feedback, *def.* 89, 150–152
 general, 8, 85–92
 with PSK/QAM 183–186
 systematic, *def.* 88–90, 151–152
 in TCM, 137–139, 150–155, 183
 trellis for, 90–92
 termination, 98–99
Correlative state, 221–222, 271
Correlator receiver. *See* Receivers
Cosets
 of lattices, 133, 173–175
 in TCM, 133

CPFSK (continuous-phase frequency-shift keying). *See* Modulation, FSK; REC CPM codes
CPM. *See* Continuous-phase modulation codes
CR (cross). *See* Cross constellation
CRC. *See* Cyclic redundancy check
Critical sequence. *See* Difference sequence
Cross constellation, 140–141, 158–160
CS-CDMA. *See* Code spread CDMA
CSSS (code-spread spread spectrum). *See* Code spread CDMA
Cutoff rate, *BSC* 116, 126–128, *QAM* 127
Cyclic redundancy check, 454

d_B (CPM distance bound). *See* Continuous-phase modulation codes
d_e. *See* Partial coherence distance
d_m. *See* Mismatched Euclidean distance
d_{MF}. *See* Matched filter bound
d_{min} (normalized minimum distance), D_{min} (non-normalized minimum distance). *See* Distance
d_{ss} (same subset minimum distance). *See* Distance; Trellis coded modulation
d_{std} (QAM benchmark distance), 177, 181
d_w (worst case minimum distance), 318
Data aided synchronization. *See* Synchronization, carrier
Data noise, 55, 262
Data symbol, 17, *spectra normalized to* 61
DDP. *See* Double dynamic programming
de Buda, R., 172, 191–192, 198, 263
Decision depth
 in conv. codes, 96
 in CPM, 213, 247, 249, 325–326
 in d_f algorithms, 171
 in PRS, 339
Decision regions, 32–33
Decoder. *See also* BCJR algorithm; Decision depth; Receiver; Trellis; Viterbi algorithm; Window, observation
 backtracking in, 95
 backwards, 342–343
 bounded distance, 95, 343–345
 breadth first, 95, 338–339
 circuit state, 345–348
 complexity of, 96
 delayed decision, 339
 error events in, 99–100
 iterative, 101
 M-algorithm, 273–275, 339–342, 347–348
 reduced search, 10, 95, 268–269, 272–275, 338
 RSSD, *footnote* 274
 sequential, 9, 95
 soft output, 101
 syndrome, 78–79
 turbo, 10

Delay spread/profile, 376–379, 381
Detection
  differential, 45, 159–164, 251–252
  MAP, 27–29
  ML, 26–29
  theory of, 9
DFE (decision feedback equalizer). *See* Equalizer
Difference power spectrum, *def.* 314, *after filter* 322
Difference sequence/symbol
  in CPM, 200, 204–206
  critical, 326–329
  in PRS, 290, 298–303, 314–315
  in QAM, 49
Differential encoder/decoder. *See also* Detection
  in CPM, 251–252
  in TCM, 159–164
Digital FM, 253
Direct sequence CDMA, 439–440, *IC receiver* 450–452
Direct sequence spread spectrum, 435–439
Discrete memoryless channel, 107–108
Discriminator. *See* Receiver
Distance. *See also* Branch and bound; Double dynamic programming; Escaped distance; Free distance; Product distance; Spectrum, distance
  bounds to, 308–309, 329
  bounds based on, 100, 165–166
  BPSK/QPSK, 40
  in CPM, 193, 197–221, 240–243, 255–261
  for filtered signals, 321–326, 328
  FSK, 50–51
  for FTN signals, 320–322
  Hamming, 77
  intersubset, 139, 145
  in lattice coding, 176–178, 181–182
  in M + F coding, 331–332
  minimum, parity-check coding, 78–79, 82
  minimum, signal space, 33–34
  MSK-type receiver, 276
  normalized, 30–31
  partial coherence, 255–261
  in PRS, 289–293, 298–303, 308–309, 311–319
  in QAM, 141
  same subset, 137, 142–150, 181
  in TCM, 134–135, 137–139, 145–147, 153–157, 165–167
Diversity
  antenna, 408–409
  by channel coding, 410–411
  equal gain, 406–407
  frequency, 409
  maximal ratio combining, 405–407
  selection, 401–403
  time, 409
  weight, 403–405

Diversity combining, in ARQ, 461–462
DMC. *See* Discrete memoryless channel
Doppler
  frequency, 375
  spectrum. *See* Spectrum, fading
Double dynamic programming, 169–171, 219
DPSK (differential phase-shift keying). *See* Modulation, PSK
DS-CDMA. *See* Direct sequence CDMA
DSSS. *See* Direct sequence spread spectrum
Duel-Hallen, A. 339
Duobinary partial response, 283

E_b (energy per bit). *See* Energy
E_b/N_0. *See* Signal-to-noise ratio
E_s (energy per transmission symbol); $\bar{E}_s$ (average signal energy). *See* Energy
EGC (equal gain combining). *See* Diversity
Energy
  average, 140–141
  per bit, 17–117
  in CPM, 240–244
  in lattices, 176–178
  in PRS, 310
  per symbol, 30–31, 117, 193, *for filters* 322–323
Energy-bandwidth tradeoff, 5–7, 192
  in CPM, 240–244
  in PRS, 310–312, 316–317
Entropy, 106–107
Envelope correlation, 254
Envelope, RF
  amplifier efficiency, 13
  constant, in CPM, 18
  def., 38
  in QPSK, 41–44, 47
  variation, 9
Equalizer, 4, 9, 69
  DFE, 9, 334, 336–337
  least squares, 335–336
  T-H precoding, 334, 337
  ZFE, 9, 334–335
Equivocation, 107, 111
Erasure channel. *See* Channel
Error event, 99–100
  in CPM, 224–225, 327–329
  in PRS, 327
Error exponent, 128–129
Error probability. *See also* Bit error rate; Error event
  antipodal signal, 32
  bounds, 100, 165–166, 224–225
  BPSK/QPSK, 32–33, 39–40
  in conv. coding, 416–419, 427
  in CPM, 224–225, 248–249

Index

Error probability (*cont.*)
  in fading, 396–400, 402–403, 405–406, TCM 427–431
  from minimum distance, 33–34
  in MLSE, 334
  in MPSK, 46
  multi-signal, 32
  in parity check coding, 77, 83–84
  in PRS, 341–342
  in QAM, 46, 48–49
  in signal space, 29–34
  in TCM, 165–169
  transfer function bound, 100
  2-signal, 30–31
Error propagation, 336–337
Escaped distance, 313–316, 331
η (throughput of ARQ), *def.* 455. *See also* Request repeat systems
Euclidean distance. *See* Distance
Euclidean space. *See* Signal space
Excess bandwidth factor, 20
Excess phase, 193
Eye pattern, 24–26

f_D (Doppler frequency), 375
f_m (maximum Doppler frequency), *def.* 375, 381
Fading. *See* Channel, fading
Fast fading, 385–386
  conv. code error rate, 419, 427
Faster than Nyquist signaling, 286, *def.* 320–322, in CPM 324–326
Filters
  Butterworth, 70–72, 292–293
  channel capacity with, 121–126
  in CPM, 232, 323–326
  distance after, 321–326
  eye patterns, 25
  and ISI, 25
  modeling, 68–72, 286, 292–293
  passband, 44–45, 323
  in PRS, 283–284, 292–293
Filter-in-the-medium model. *See* Model
Filter-in-the-transmitter model. *See* Model
Flat fading, 371, 385–386
Folded power spectrum, 295–296
Forney, G.D., Jr., 9, 10, 172
Free space loss, 364
Free distance. *See also* Distance; Minimum distance loss
  algorithms for, 169–171, 301–303
  basics, 91–92, 96
  in bounds, 100, 165–167
  of conv. coding + PSK/QAM, 184
  of CPM, 213–221, 224, 240–243
  in lattice coding, 182
  of PRS, 300–303, 308–313, *worst case* 318–319

Frenger, P., 444
Frequency non-selective fading, *See* Flat fading
Frequency selective fading, 375–385
  coherence bandwidth & time, 379
  delay spread of, 376–379
Friis formula, 364
FSK (frequency-shift keying). *See* Modulation, FSK
FTN. *See* Faster than Nyquist signaling
Fundamental volume, 172

γ_b (instantaneous received E_b/N_0), *def.* 396
Γ_b (average E_b/N_0 over fading), *def.* 396
γ_{cg}. *See* Lattice coding gain
γ_s. *See* Shape gain
Galko, P., 276–277
Gallager exponent. *See* Error exponent
Gaussian channel. *See* Channel
Generator matrix/polynomial
  of conv. code, 80–81, 85–88
  of lattice, 172–173, 175
Gibby, R.A., 21
Gilbert-Varsharmov bound, 82
GMSK (Gaussian minimum-shift keying). *See* Modulation, FSK
Gram-Schmidt procedure, 27, 40
Group codes. *See* Linear codes
GSM telephone system, 424–425
Guided media. *See* Channel

Hadamard. *See* Walsh-Hadamard
Hamming codes. *See* Coding, binary
Hard decision, 83
HCS (half cycle sinusoid). *See* Continuous-phase modulation codes
Heegard, C., 339
Huber, J., 272

IC (interference cancelation) receiver. *See* Receiver
Incremental redundancy, in ARQ, 461–462
Index. *See* Weak index; Modulation, FSK; Continuous-phase modulation codes
Information error weight, 417
Information rate, *def.* 112, 115–117, 122–126
Information source, 106–107
In-phase and quadrature description, 38
Integrate and dump. *See* Receiver
Interleaving, 410–411, *in DSSS* 436–439
Interpolation, 287
Intersubset distance. *See* Trellis coded modulation codes
Intersymbol interference, 4, 9, 18
  eye pattern, 24–26

Intersymbol interference (*cont.*)
  modeling, 68–69, 286
  at passband, 45
  worst case, 318–319
I/Q plot, 41–43
ISI. *See* Intersymbol interference

Jacobs, I.M., 7, 26
Jakes method, 391–393
Jitter, 262, 264

Kassem, W., 172
Kissing number, 172, 179
Kotelnikov, V.A., 4, 8, 26
Kretzmer, E.R., 283

L_{dec}. *See* Decision depth
Lattice
  checkerboard, 173, 179
  cubic, 172, 176, 179
  partition, 173–175
  rotation, 173
  sphere packing, 171, 175–176, 179
Lattice codes
  bit error rate, 182
  distance, 176–178
  history, 10, 133–134
  lattice theory, 172–174
  modeling as, 67, 133
  multidimensional, 179–182, *table* 180
  set partitions (subsets) in, 172, 179–181
Lattice coding gain, 176–182
Lender, A., 283
Linear codes, 76
Lindell, G., 183
Linear programming, 358–359
Linear modulation. *See* Modulation, linear
Linear receiver. *See* Receiver
Liu, W., 272
Lognormal distribution, *def.* 369,
  simulation 393–395
Lucky, R.W., 9

M (alphabet size). *See* Alphabet
M_c (master constellation size). *See* Master constellation
Macdonald, A.J., 224, 261, 264–266
Magee, F.R., Jr., 318
MAI (multiple access interference), 440,
  442–443, *in IC receiver* 449–451
Main lobe, spectral, 60
M-algorithm. *See* Decoder

MAP (maximum a posteriori) receiver.
  See Receiver
Mapper
  in conv. codes + PSK/QAM, 184–186
  natural, *def.* 145, 158, 185
  in TCM, 142, 145, 151, 158–160
Master constellation
  asymmetric, 157
  cross, 140–141, 158–160
  for lattice code, 179–180
  multi-dimen., 179–181
  1-dimen., 156
  in TCM, 133, 136–137, 140–142, 156–157
Matched filter
  AMF, 272–273
  in fading, 387–388
  MF model, 69–72
  in receivers, 35–37
  whitened, 387–388
Matched filter bound, *def.* 290–291, 300, 309,
  in equalizers 334–336
Matched filter receiver. *See* Receiver
Maximum likelihood receiver. *See* Receiver
Mazo, J.E., 320
Mazur, B.A., 264
McLane, P.J., 265
MDL. *See* Minimum distance loss
Merge. *See* Phase merge; Trellis
Memory (conv. code), 85
MF. *See* Matched filter; Receiver
MF model, 69–72, 286
MFD (maximal free distance) codes, *def.* 416,
  table 418
Minimum distance. *See* Distance
Minimum distance loss, *def.* 310, *tables* 351–357
Minimum phase. *See* Phase, min/max
Minkowski-Hlawka theorem, 172
Mismatched Euclidean distance, 269–271
Miyakawa, H., 192
ML (maximum likelihood). *See* Receiver
MLSE (maximum likelihood sequence
  estimation). *See* Receiver
Model
  for CPM, 66, 68
  discrete-time, 248
  filter-in-the-medium, 121–125
  filter-in-the-transmitter, 121–123, 125–126
  for ISI, 68–69
  MF, 69–72, 286
  for orthogonal pulse modulation, 66–68
  PAM, 69–72
  for path loss, 366–367
  for PRS, 68–69, 284–288, 292–293
  for TCM, 66–68, 134–135
Modulation + filter codes, 329–332, 347
Modulation + parity check codes, 11–12

Index

Modulation, FSK
  CPFSK, *def.* 50, 64–65
  distance in, 50
  with fading, 397
  general, 3, 9, 49–51
  GMSK, *def.* 52, 195, 277
  index in, 49–50
  MSK, *def.* 51, 64, 233, 238–239
  narrowband/wideband, 49
  spectrum, *formula* 64–65, 233, 238–239
Modulation, general
  carrier, 18, 37–38
  coded modulation history, 3, 4
  def., 3
  pulse in, 17–26
  quadrature, 38
Modulation, linear, 3, 18, 134–135
  distance in, 290–291
  filtering of, 330–332
Modulation, phase-shift keying
  BPSK, 38–39
  with conv. codes, 183–185
  in CPM generation, 267–268
  DPSK, 45
  error probability, 32–33, 39–40, 45–46
  with fading, 396–397, 402–403, 405–407
  filtered, 330–332
  general, 3, 9, 38–47
  many phases, 46
  modeling, 66
  offset QPSK, 44, 238
  QPSK, 39–47
  spectrum, 59–62, 238
  synchronization for, 56–57
  TCM, basis for, 94, 137–139
Modulation, pulse amplitude, 22
Modulation, quadrature amplitude
  in ARQ, 468–469
  as benchmark, 177
  capacity of, 115–117
  constellations, 47
  distance, 48–49, 141
  error exponent, 128–129
  error probability, 46, 48–49
  I/Q formulation, 47–48
  as a lattice, 172, 176
  modeling, 67–68
  rectangular, 48
  symbol interval, 135
  synchronization for, 57
  in TCM, 12–13, 134–135, 140–141, 145–146, 158
Mother code, 420
MRC (maximal ratio combining). *See* Diversity
MSC (maximum spectral concentration).
  See Concentration, spectral

MSK (minimum-shift keying). *See* Modulation, FSK
MSK-type receiver. *See* Receiver
MTCM. *See* Multiple TCM
Multi-h codes, *def.* 210–212, 218–221, 232, 264–265
Multi-level TCM, 434–435
Multipath fading, 368, 370–372, 385–386.
  See also Frequency selective fading
Multiple TCM, 432–434
Multiplicities. *See* Spectrum, distance
Multi-T codes, 212
Mutual information, 110

N_{dim} (dimensions per symbol time), *CPM* 195, *PRS* 310, *TCM* 195
N_{FE}. *See* Noise enhancement factor
N_0 (Gaussian noise power spectral density). *See* Noise, Gaussian
N_{win}. *See* Window, observation
Nakagami distribution, 374
NBW (normalized bandwidth). *See* Bandwidth
Near-far effect, 443
Neighbor
  in conv. codes + PSK/QAM, 184–185
  in distance calculation, 166–167, 169–170
  in lattice code, 180
Nested codes. *See* Rate-compatible codes
99% power. *See* Power out of band
Noise enhancement factor, 335
Noise equivalent bandwidth, 55, 262, 264
Noise floor, 267
Noise, Gaussian
  in fading simulation, 389–391
  after matched filter, 37
  in PAM & MF models, 70
  PSD, 117
  in signal space, 27–28
Noncoherent receiver. *See* Receiver
Normalized bandwidth. *See* Bandwidth
NRZ (non return to zero). *See* Pulse
Null, *DC* 284, 326–327, *spectral* 361
Nyquist bandwidth, 19
Nyquist, E., 21–22
Nyquist pulse, *criterion* 19–22. *See also* Pulse

Observation window. *See* Window, observation
Observer canonical form, 150–151, 153
Octal notation, *def.* 153
ODS (optimum distance spectrum).
  See Spectrum, distance
Offset QPSK. *See* Modulation, phase-shift keying
Okumura-Hata model, 366
Optimality principle, in VA, 97–98
Orthogonal basis, 27

Orthogonal convolutional codes, 444–446,
    superorthogonal 444–446
Orthogonal pulse
  criterion, 23, 37
  modeling of, 66–68
  in PRS, 285
Orthogonal signal constellation, 31,
    modeling 66–68
Osborne, W.P. (with Luntz), 192, 246–247, 272
Outage probability, 370

Pair state. *See* Double dynamic programming
Pairwise error-probability, 428–429
Palenius (Andersson), T., 274
PAM (pulse amplitude modulation). *See*
    Modulation, pulse amplitude;
    PAM model
PAM model, *def.* 69–72, 285–286, 288–289, 293
Parity-check codes. *See also* Coding, binary;
    Convolutional codes
  in ARQ, 456–468
  over AWGN channel, 84
  basics, 3–4, 7, 11–12, 76–82, 186
  bounds on, 82–84
  over BSC, 76–77, 82–83
  coding gain, 83–84
  fading effect on, 398–400
  generator matrix/polynomial, 80–81
  MAP receiver, 77
  ML receiver, 76–77
  parity-check matrix, 77–79
  rate, 76
  systematic, 76
Partial coherence distance, 256–261
Partial energy function, *def.* 297–299, 340
Partially coherent receiver. *See* Receiver
Partial response signaling codes
  bandpass, 343–344
  bandwidth, *tables* 310–311, 351–357
  basics, 11–13, 283–286
  capacity for, 121–126
  def., 285–286
  distance, 289–293, 299–303, 311–319
  error events, 99, 327
  as filtering, 283–284
  history, 10, 283–284
  impulse response, 311–312
  infinite response, 321
  as ISI, 284, 286
  as linear coded modulation, 10, 285
  as M + F codes, 332
  MDL, 310, 312, 318, *table* 351–357
  models for, 68–69, 284–289, 292–293
  optimal, 311–313, 316–317, *solution* 358–361
  phase, 297–298
  poles/zeros, 296–298, 340

Partial response signaling codes (*cont.*)
  receivers, 338–345
  spectrum, asymmetric, 317–318
  spectrum, shape, 303–307
Partition, lattice, 173–175, 179–181
Passband. *See* Bandpass
Passband filter. *See* Filter
Path loss, 364–367
Pelchat, M.G., 192
Phase ambiguity, 57, 157
Phase cylinder. *See* Phase trellis
Phase difference tree. *See* Phase tree
Phase-lock loop, 53–58
  in CPM generation, 267–268
  first/second order, 54–55
  jitter in, 55, 245, 262
  noise equiv. bandwidth, 55, 262–264
  response, 55
Phase merge, 200, *first etc.* 205–207
Phase, min/max, 297–298, 340–345
Phase state, 220–222
Phase trellis/tree, 197–199, 208–210
  difference tree, 204
Plane earth loss, 365–366
PLL. *See* Phase-lock loop
POB. *See* Power out of band
Power bandwidth. *See* Power out of band
Power–bandwidth tradeoff. *See* Energy–bandwidth
    tradeoff
Power out of band
  in CPM, 238–244
  def., 61–62
  in PRS, 304, *asymmetric* 317–318
Power spectral density, *def.* 58. *See also* Spectrum
Precoding, 123
Prehistory, in CPM, 204
Premji, A., 264–265
Proakis, J.G., 318
Probability of error. *See* Error probability
Product distance, 430
PRS (partial response signaling).
    See Partial response signaling codes
PSD. *See* Power spectral density
Pulse
  at baseband, 19
  modulation, 17–26
  NRZ, 20
  Nyquist, *def.* 19–22, 25
  orthogonal, *def.* 22–24
  RC, *def.* 19–20
  root RC, *def.* 23–24
  shaping, 9
  sinc, *def.* 19–22
  spectrum, 58–60
Punctured codes, 92, 420–422, *in ARQ* 462–466

Index

QAM (quadrature amplitude modulation). *See* Modulation, quadrature amplitude
Q-function, *def.* 30
   with error events, 100
Quadrature receiver/transmitter. *See* Receiver; Transmitter

$R_D(f)$ (channel autocorrelation transform), 295–296
$R_{fold}(f)$. *See* Folded power spectrum
R_0. *See* Cutoff rate
$R_Z(z)$ (channel autocorrelation z-transform), *def.* 295–296, 298
Raised cosine pulse, 19–20
RAKE receiver, 411–412, 438, 449
Rate
   of CPM, 195
   of conv. code, 85, 87
   information rate, *def.* 112, 115–117
   of lattice code, 171
   of TCM, 135, 137, 142–143, 147
   with termination, 98
Rate compatible codes, 420–426
   in ARQ, 462–468
   nested, 423–424, 445
   rate matching in, 424–427
   repetition, 422–423
Rate matching. *See above*
Rayleigh distribution/fading
   in ARQ, 456–458, 465
   coding, effect on, 398–400
      conv. codes, 418–419, 427
   def., 371–372
   modulation, effect on, 396–397
   simulation, 387–393
   in spread spectrum, 450–451
   in TCM, 429–431
RC (raised cosine). *See* Raised cosine pulse; RC CPM codes
RC CPM codes
   approximation, as REC, 269–271
   distance, 202, 207–210, 217–220, 260–261
   energy–bandwidth, 242–243
   error rate, 248–249
   filtered, 324–326, 348
   partial response, 207–209, 217–218, 222–223, 242–243
   receivers, 248–249, 269–277
   spectrum, 233–239, 242–243
   synchronization, 266
   trellis 93–94
RCC (rate compatible convolutional) codes. *See* Rate compatible codes
RCPC (rate compatible punctured convolutional) codes. *See* Punctured codes

RCRC (rate compatible repetition convolutional) codes. *See* Rate compatible codes
REC (rectangular) CPM codes
   basics, 194–196
   distance, 199–206, 213–218, 258–260
   energy–bandwidth, 241
   error events, 329
   error rate, 249–250
   filtered, 324–326, 348
   full response, 213–215, 222–223, 241
   partial response, 203–206, 215–217
   receivers, 249–250, 272–274
   spectrum, 241
   synchronization, 263–265
   tree/trellis, 197–200
Receiver. *See also* Decoder; Differential encoder/decoder; Equalizer; RAKE receiver; Viterbi algorithm
   AMF, 272–273
   correlation, 34–35
   digital FM, 52
   discriminator, 51–52, 251–253
   IC, 449–453
   integrate and dump, 37
   linear, 22–23, 109
   M-algorithm. *See* Decoder
   MAP, 27–29, 35–36, 102, 246–247
   matched filter, 35–37
   ML, 26–29, 34–37, 76–77, 95, 255, 289
   MSK-type, 275–277
   noncoherent, 245, 251–257
   one-symbol, 246–247
   Osborne-Luntz, 246–247
   partially coherent, 245, 253–261
   passband, 224
   quadrature, 41–43
   sampling, 21
   sequence (MLSE), 4, 10, 334, 341–342
   simplified pulse, 269–272
   WMF, 343
Reduced search decoding. *See* Decoder
Regular code, 166, 171, 224
Repetition code, 77, 436–437
Request repeat systems
   with adaptive modulation, 468–469
   basics, 454–458
   combining in, 461–462
   hybrid-I, 458–461
   hybrid-II, 461–468
   incremental redundancy in, 461–464
   RCPC/RCRC in, 462–468
   with TCM, 469
$\rho(\tau)$ (channel autocorrelation), 295–297
Ricean distribution/fading
   def., 373–374
   simulation, 393–394
   in TCM, 429–431

Rolloff factor. *See* Excess bandwidth factor
Root RC pulse
  def., 23–24
  filtered, 70–72
  in PRS, 287, 305–306
  in QPSK, 41–44
  in TCM, 135
Rotation, lattice. *See* Lattice
Rotational invariance. *See* Trellis coded modulation
RSSD (reduced state sequence detection), 274

$\bar{s}(t)$ (time average autocorrelation). *See* Autocorrelation
Said, A., 290, 305–309, 342, 358, 361
Same subset distance. *See* Distance; Trellis coded modulation
Sampling receiver. *See* Receiver
Sampling theorem, 4, 7, 8, 21–22, 287
Schonhoff, T., 192, 246
Scrambling sequence, *def.* 436, 441–444
Selection diversity. *See* Diversity
Selector subset. *See* Trellis coded modulation
Sequence estimation. *See* Receiver
Sequential decoding. *See* Decoder
Seshadri, N., 273, 324, 328, 347–348
Set partitioning/set partition codes. *See* Lattice codes; Trellis coded modulation
Shadowing, 368–370
Shannon, C.E., 4–6, 8, 22, 26, 110, 133, 171, 182
Shannon limit, 120
Shape gain, 176, 178–180
Shift register, as encoder, 85–86, 88–89
Side lobe, spectral, 60, 65
σ_T. *See* Delay spread
Signal constellation. *See also* Master constellation
  antipodal, 31
  orthogonal, 31
  QAM, 17, 140–141
  QPSK, 32
  TCM, 140–141, 156–157
Signal space
  basics, 26–30
  in CPM, 194, 272
  history, 4–5, 8
  in TCM, 134
Signal-to-noise ratio
  Gaussian channel (E_b/N_o) 34, 117
  after matched filter, 36–37
Simmons, S.J., 265, 272, 274
Simulation, of fading, 386–395
Sinc pulse
  in capacity proof, 109, 120–121
  def., 19–22
Slepian, D., 183

Sloane, N.J.A., 10, 172
Slow fading, 385–386
Smith, J.W., 21
SNR. *See* Signal-to-noise ratio
Soft decision, 84
Spectral factorization, 296–298
Spectrum, fading, 383–385
Spectrum, distance, 166-167, *ODS* 417–422
Spectrum, frequency. *See also* Bandwidth; Difference power spectrum
  asymmetric, in PRS, 317–318
  concentration, 304–309
  of CPM, 63–65, 194–195, 225–244, 266–267
  folded, 295–296
  of linear modulation, 58–60
  line, 231–232, 262–264
  main/side lobes, 60, 266
  of nonlinear modulation, 61–63
  null, 326–327, 361
  of NRZ modulation, 59–62
  POB, 61–62
  of PRS, 303–307, 311–318
  PSD (def.), 58, 62
  of PSK, 60–62
  of TCM, 135
  transmitter, 266–267
Sphere packing, 171. *See also* Lattice
Splits, in TCM, 145–149
Spread spectrum. *See under type*: Code spread CDMA; Direct sequence CDMA; Direct sequence spread spectrum;
Spreading code. *See* Scrambling sequence
Spreading gain, 437
Squaring noise, 262
SRC (spectral raised cosine). *See* Continuous-phase modulation codes
Staircase approximation, 248–251
State/state transition
  in conv. codes, 90–92, 97–98
  in CPM, 93–94, 220–222, 271–275, 345
  circuit state, 345–346
  in lattice codes, 180
  in PRS, 93, 338
  in TCM, 94, 138–139, 153–155, 161–164
Subset, in TCM. *See* Trellis coded modulation
Sundberg, C.E.W., 183, 192, 197
Superorthogonal codes. *See* Orthogonal convolutional codes.
Suzuki distribution, 374
Svensson, A., 274
Symbol interval, 2–3, 17, *CPM* 195, *TCM* 134–135
Symbol rate, 17
Symbol timing, 52, 58, 263–265
Synchronization, carrier
  for CPM, 57, 244–245, 261–266
  data aided, 261, 264–266

Synchronization, carrier *(cont.)*
  fourth power, 56–57, 261–264
  open loop, 261–264
  phase ambiguity in, 57, 157
  remodulation loop, 57
  VA, 265–266
Syndrome, 78–79
Systematic code, 76, *conv.* 88–90, 151–153

T. *See* Symbol time
T_c. *See* Coherence time
Tailbiting, 99
Taylor, D.P., 192, 264
TCM. *See* Trellis coded modulation
Termination, trellis, 98–99
TFM (tamed frequency modulation).
    See Continuous-phase modulation codes
Tikhonov distribution, 245
Time average autocorrelation. *See*
    Autocorrelation
Tomlinson-Harashima precoding. *See* Equalizer
Transfer function bound, 100, 166
Transmission alphabet, 17
Transmitter
  CPM, 266–268
  quadrature, 41–43
Trellis
  basics, 9, 84, 90
  conv. code, 90–92
  CPM, 9, 94, 220–224
  ISI, 9, 338
  merge, 91
  PRS, 92–94, 338–339
  TCM, 94, 161–164
  termination, 98–99
Trellis coded modulation (TCM). *See also*
    Bit interleaved coded modulation;
    Lattice codes; Multi-level TCM; Multiple
    TCM; Trellis; State
  in ARQ, 469
  bandwidth, 135–136, 149
  BER, 149, 151, 156, 166–169
  best codes *(tables)*, 154–155, 184
  capacity in, 115–117
  coding gain, 139, 145–146, 149
  error events, 100
  fading BER, 427–432
  free distance, 153–157, 165
  history, 10, 133–134
  intersubset distance,
      def. 138–139, 142–150, 153
  as linear modulation, 18, 133
  mapper, 142, 185–186
  modeling, 66–68, 134–135

Trellis coded modulation (TCM) *(cont.)*
  partitioning in, 133, 136, 158
  PRS, comparison, 156–157
  PSK, based on, 94, 137–139, 153–157
  QAM, based on, 140–141, 145–146, 153–157, 158–160
  rate, 135, 137, 142–143, 147
  rotation invariance, 157–165, 179
  same subset distance, *def.* 137–139, 142–150, 153
  selecting subsets, 138–139, 142–157
  subsets in, 137–139, 160–162
  Ungerboeck rules, 144–146, 151–152
turbo coding. *See* Decoder
Turin, G.L., 9

Ungerboeck, G., 10, 133, 145
Ungerboeck rules, 144–146, 151–152

van Trees, H.L., 9
VCO (voltage controlled oscillator).
    See Phase-lock loop; Synchronization,
    carrier
Viterbi, A.J., 9, 97
Viterbi algorithm, 9, 94–98
  in CDMA, 449
  in CPM, 247–251
  in TCM, 166–169
Voronoi region, 172

W_b (bandwidth per bit). *See* Bandwidth
Walsh-Hadamard (transform), 392, 442
Waterfilling, 123–125
Weak index, 209–210, 214, 217–218
Wei, L.-F., 159–163, 172
Whitened matched filter, 343
Wiener, N., 8
Window, observation, 96
  in CPM, 200, 213–218, 255, 257–261, 325–326
Window, sliding, 247
Wong, C.C.-W., 339
Word error probability, 83, *fading* 398–400
Wozencraft, J.M., 9, 26

Zero crossing criterion, 19, 23
Zero sum rule, 327
Zetterberg, L., 183
ZFE (zero forcing equalizer). *See* Equalizer
Zhang, W., 184–186